D1587353

STRUCTURAL MASONRY DESIGNERS' MANUAL

W. G. CURTIN M.Eng., F.I.C.E., F.I.Struct.E., M.Cons.E.

G. SHAW C.Eng., F.I.Struct.E., M.Cons.E.

J. K. BECK C.Eng., M.I.Struct.E.

W. A. BRAY B.Eng., C.Eng., M.I.C.E., M.I.Struct.E.

W. G. Curtin & Partners
Consulting Structural & Civil Engineers

GRANADA
London Toronto Sydney New York

Granada Publishing Limited – Technical Books Division
Frogmore, St Albans, Herts AL2 2NF
and
36 Golden Square, London W1R 4AH
866 United Nations Plaza, New York, NY 10017, USA
117 York Street, Sydney, NSW 2000, Australia
100 Skyway Avenue, Rexdale, Ontario, Canada M9W 3A6
61 Beach Road, Auckland, New Zealand

British Library Cataloguing in Publication Data

Curtin, W. G.
    Structural masonry designers' manual
    1. Masonry
    I. Title
    624.18'3   TA670

ISBN 0-246-11208-5

First published in Great Britain 1982 by Granada Publishing Limited

Phototypesetting by Parkway Group, London and Abingdon

Printed in Great Britain by Robert MacLehose & Co. Ltd., Glasgow

CONTENTS

ACKNOWLEDGEMENTS

We appreciate the help given by many friends in the construction industry, design professions and organisations. We learnt much from discussions (and sometimes, arguments) on site, in design team meetings and in the drawing office. To list all who helped would be impossible – to list none would be churlish. Below, in alphabetical order, are some of the organisations and individuals to whom we owe thanks:

Brick Development Association
British Standards Institution   } for
Building Research Establishment   general
Cement and Concrete Association } assistance

Professor Heyman for permission to quote from his book, *Equilibrium of Shell Structures*.
Mr. J. Korff, Deputy Structural Engineer, GLC for advice on accidental damage.
Mr. W. Sharp, County Structural Engineer, Lancashire County Council for particular help on strapping and tying.

Material from BS 5628 is included by permission of The British Standards Institution, 2 Park Street, London, W1A 2BS, from whom complete copies can be obtained.

PREFACE

When we designed our first loadbearing brick structure more than twenty years ago, we did so without much enthusiasm or interest, since we felt that bricks were not a 'proper' material for self-respecting professional engineers to be using. Most engineers at that time used mainly concrete and steel and thought that bricks had been used in the past by Victorian engineers only because they had no alternative, and that their present use was restricted to low-rise housing and some cladding. We have changed our attitude since those early days, but our previous thinking is still common amongst many engineers and graduates.

With experience, we have learnt to appreciate the value of the 'new' engineering materials of brick and block, and probably have derived more interest and excitement from them than we have from concrete or steel.

As we have learnt more and more, and have applied to masonry the advances of research and practice associated with other structural materials, the calculations have become more precise but also more complex. The calculations for the first major job we designed in brickwork covered less than three-quarters of a page of foolscap, whereas the last one has run into many sheets. Not that we wish to imply that there is any correlation between the *quantity* and complexity of the calculations and the *quality* of the design.

We first started this book as an internal design manual for the young graduates in our practice who were recruited from various universities where most of them had been taught practically nothing about structural masonry. It was not difficult to teach them, since designing structural masonry uses the same basic engineering principles which they had learnt. However, word of the manual's production spread to other design engineers, and they began to ask for copies, with the result that we decided to adapt this book for the use of undergraduates, young engineers and established practising engineers alike.

The book is a joint effort of four members of our practice, who are on the board designing and out on the site supervising and, we hope, passing on our experience in structural masonry design. We are very grateful to many past and present members of our enthusiastic staff (particularly Tim Dishman, Dave Fowler, Steve Hunt, Merlyn Saunders and Lena Skovsted) for their suggestions, criticisms and advice. We are indebted to Irene Mussell for her patience, interest and care in typing and retyping the manuscript and to Debbie Banfield for her untiring help in the final typing. We appreciate the help given by Ron Adams who not only carried out the editing with meticulous care but produced many ideas to improve the clarity and flow of the text.

We appreciate that many engineers may disagree with our interpretation of the Codes of Practice, our methods, priorities, etc., just as we disagreed amongst ourselves and with our own staff. But, since engineers, like other people, never wholly agree about everything, this is not surprising. The manual is a reasonable consensus of our opinions, and we know from our experience that the ideas work and the resulting buildings fulfil their purpose well.

We have worked, where possible, mainly from BS 5628, because this is the most up-to-date Code and is based on experience and recent research. We have tried to keep the manual as simple as possible but have not attempted to make it 'idiot proof' – which would have required not a manual but a whole library.

NOTATION

A	cross-sectional area
A_s	area of tensile reinforcement
A_{sc}	area of compressive reinforcement
A_{sv}	area of shear reinforcement
A_t	total area of reinforcement $(A_s + A_{sc})$
a	depth of stress block
a_v	shear span (distance from support to concentrated load)
B	width of bearing under a concentrated load
B_r	centre to centre of cross-ribs in diaphragm wall
b	width of section
b_c	breadth of compression face
b_r	clear dimension between diaphragm cross-ribs
C	compressive force
C_c	total compressive force
C_s	compressive force in reinforcement
C_{pe}	wind, external pressure coefficient
C_{pi}	wind, internal pressure coefficient
D	overall depth of diaphragm wall section or depth of arch
Dia	diameter of reinforcing bar
d	effective depth to tensile reinforcement and depth of cavity (void) in diaphragm wall
d_n	depth to neutral axis
d_2	depth to compression reinforcement
E	Young's modulus
E_b	Young's modulus for masonry
e	eccentricity
e_a	additional eccentricity due to deflection in wall
e_{ef}	effective eccentricity
e_m	the larger of e_x and e_t
e_{max}	maximum eccentricity which can be practically accommodated in section
e_t	total design eccentricity at approx. mid-height of wall
e_x	eccentricity at top of wall
F_k	characteristic load
F_m	average of the maximum loads carried by two test panels
F_t	tie force
f_b	characteristic anchorage bond strength
f_{bs}	characteristic local bond strength
f_c	design axial stress due to minimum vertical load
f_k	characteristic compressive strength of masonry
f_{ki}	characteristic compressive strength of masonry at age when post-tensioning force is applied
f_{kx}	characteristic flexural strength (tensile) of masonry
$f_{kx\ par}$	value of f_{kx} when plane of failure is parallel to bed joints
$f_{kx\ perp}$	value of f_{kx} when plane of failure is perpendicular to bed joints
f_t	theoretical flexural tensile stress or flange thickness

f_{uac}	design axial compressive stress
f_{ubc}	flexural compressive stress at design load
f_{ubt}	flexural tensile stress at design load
f_v	characteristic shear strength of masonry
f_{vb}	characteristic flexural shear strength of masonry
f_w	flange width
f_y	characteristic tensile strength of steel
f_{yv}	characteristic strength of shear reinforcement
G_k	characteristic dead load
g_A	design vertical load per unit area
g_d	design vertical dead load per unit area
H_z	thrust at crown of arch
h	clear height of wall or column between lateral supports
h_a	clear height of wall between concrete surfaces or other construction capable of providing adequate resistance to rotation across the full thickness of the wall
h_{ef}	effective height or length of wall or column
h_L	clear height of wall to point of application of lateral load
I	second moment of area/moment of inertia
I_{na}	second moment of area about neutral axis
K	stiffness coefficient
K_a	constant term relating design strengths of steel and masonry
K_1	shear stress coefficient for diaphragm walls
K_2	trial section stability moment coefficient for diaphragm walls
k	multiplication factor for lateral strength of axially loaded walls
k_1	$\dfrac{1 - \sin \theta}{1 + \sin \theta}$ from Rankines formula for retained materials
L	length
L_a	a span in accidental damage design
L_f	spacing of fins, centre to centre
l_a	lever arm
M_A	applied design bending moment
M_b	design bending moment at base of wall
MR	moment of resistance
MR_s	stability moment of resistance
M_{rb}	moment of resistance of a balanced section
M_{rc}	moment of compressive resistance
M_{rs}	moment of tensile resistance
M_w	design bending moment in height of wall
N	design vertical axial load
N_b	design vertical axial strength at balanced condition
N_d	design vertical axial strength
N_0	design vertical axial strength when loaded on the centroidal axis
N_s	number of storeys in building
n	axial load per unit length of wall, available to resist arch thrust
n_w	design vertical load per unit length of wall
P	design post-tensioning force
P_k	characteristic post-tensioning force
P_{lim}	acceptance limit for compressive strength of units
P_0	specified compressive strength of units
P_u	mean compressive strength of units
p_{ubc}	allowable flexural compressive stress
p_{ubt}	allowable flexural tensile stress
Q_k	characteristic superimposed load
q	dynamic wind pressure
q_{lat}	design lateral strength per unit area
q_1	design horizontal pressure at any depth (from retained material)

R	constant term for design flexural strength of masonry in compression or radius of arch curve
r	ratio of area of reinforcement to area of section or radius of gyration
r_d	projection of rib (or fin) beyond flange (in a T profile)
r_t	rib (or fin) thickness (in a T profile)
S	clear span of arch
SR	slenderness ratio
S_d	section depth
S_v	spacing of link reinforcement
T	total tensile force or thickness of diaphragm leaf or flange
t	thickness of wall (or depth of section)
t_{ef}	effective thickness of wall
t_p	thickness of a pier
t_r	thickness of a cross-rib in a diaphragm wall
UDL	uniformly distributed load
V	shear force
v_h	design shear stress
W	own weight of effective area of fin wall per m height
W_k	characteristic wind load
W_{k1}	design wind pressure, windward wall
W_{k2}	design wind pressure, leeward wall
W_{k3}	design wind pressure uplift (on roof)
w_s	width of stress block
x_n	depth to neutral axis from top of beam
Y_1	fin dimension, neutral axis to end of fin
Y_2	fin dimension, neutral axis to flange face
Y_u	deflection of test wall in mid-height region
Z	section modulus
Z_1	minimum section modulus of fin
Z_2	maximum section modulus of fin
α	bending moment coefficient for laterally loaded panels
β	capacity reduction factor
γ_f	partial safety factor for loads
γ_m	partial safety factor for materials
γ_{mb}	partial safety factor for bond between reinforcement and mortar or grout
γ_{ms}	partial safety factor for steel reinforcement
γ_{mv}	partial safety factor for masonry in shear
δL	short linear measurement
ϵ	strain in reinforcement
μ	orthogonal ratio
Σu	sum of the perimeters of the tensile reinforcement
ψ_m	reduction factor for strength of mortar
ψ_u	unit reduction factor
Ω	trial section coefficient for fin walls

CHAPTER 1

INTRODUCTION

The successful 'Brick is Beautiful' campaign mounted by the Brick Development Association, the public's rejection of the concrete jungle, the increase in research over the last decade, the issue of a revised Code of Practice and the energy crisis – all coinciding fortuitously – could well lead to a renaissance in the use of structural masonry.

It is now generally accepted that brickwork forms an attractive durable cladding with good thermal and acoustic insulation, excellent fire resistance, etc. What is not so widely appreciated is that it is an economical structural material that can often be built faster, cheaper and more easily than its main rivals, steel and concrete. In many respects, the same is true of concrete blocks in that improvements have been made in their manufacture, strength and use.

1.1 PRESENT STRUCTURAL FORMS

The Code of Practice, BS 5628: Part 1: 1978, has increased the ability of the two most common forms of multi-storey structural masonry – crosswall and cellular construction (see Figures 14.13 and 14.37) – to show as much as 10% reduction in overall construction costs and time, compared with other materials. Crosswalls have been extensively used in school classroom blocks where one brick thick walls have been spaced at about 7 m centres. In halls of residence, hotel bedroom blocks and similar applications, half brick thick walls have been used, spaced at about 3 m centres. These walls are not only space dividers and acoustic barriers; they also form the structure and completely eliminate the need for columns and beams.

One of the reasons for the speed of erection, mentioned earlier, is illustrated in Figure 1.1, which demonstrates the essential simplicity of brickwork and blockwork structures. A further reason is to be found in the fact that there is a continuous 'follow on' of other trades. Several contractors have successfully used the 'spiral' method to speed construction, shown in Figure 1.2, and this is described in some detail in Chapter 9 (see Figure 9.29).

Useful and economical though they are for a range of applications, since both crosswall and cellular construction demand repetitive floor plans, they tend not to be suitable for buildings such as office blocks requiring large, flexible, open spaces.

1.2 NEW STRUCTURAL FORMS

The masonry 'spine wall' (see Figure 14.36), could be developed for office blocks where precast prestressed concrete floor units can span up to 8 m onto the corridor walls or spine. There has, in fact, been a move away from the once fashionable open-plan office building – partly due to the high energy costs in lighting, heating and air-conditioning, and partly to people's natural desire to be able to see out of a window, and to use the window as an environmental control. Recent trends show that the depth of space now asked for has decreased to not more than 6 m.

The large spaces required for multi-storey warehouses and department stores can be achieved by using columns of high strength bricks or blocks supporting concrete plate floors. However, it is in single-storey wide-span structures such as factories, garages, sports halls, etc., that structural masonry is likely to make a major breakthrough.

Industrial shed-type buildings have for many years, by tradition, been formed of steel or concrete portal frames, clad with corrugated sheeting. When thermal insulation is required for this type of building, the cladding has to be backed by an insulating lining and this, in turn,

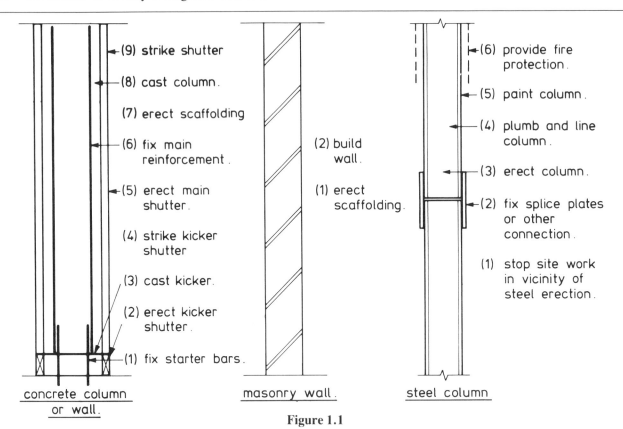

concrete column
or wall.

- (9) strike shutter
- (8) cast column.
- (7) erect scaffolding
- (6) fix main reinforcement.
- (5) erect main shutter.
- (4) strike kicker shutter
- (3) cast kicker.
- (2) erect kicker shutter.
- (1) fix starter bars.

masonry wall.

- (2) build wall.
- (1) erect scaffolding.

steel column

- (6) provide fire protection.
- (5) paint column.
- (4) plumb and line column.
- (3) erect column.
- (2) fix splice plates or other connection.
- (1) stop site work in vicinity of steel erection.

Figure 1.1

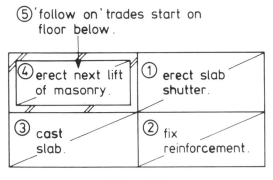

⑤ 'follow on' trades start on floor below.

④ erect next lift of masonry.

① erect slab shutter.

③ cast slab.

② fix reinforcement.

Figure 1.2

needs to be backed by a hard lining to protect it from damage. The sheeting, insulation and lining frequently require a subsidiary steel frame to support them, and to provide wind bracing to the wall. Thus four different materials have to be provided to carry out four different functions, and this involves several suppliers and specialist sub-contractors. This is in complete contrast to the situation that arises with brick (or block) cavity walls stiffened by piers, diaphragm walls and fin walls. Each of these economical and efficient forms of construction can provide the structure, the cladding, the insulation and the lining, in one material erected by the main contractor using only one trade.

Pier-stiffened cavity walls are now economical up to about 5 m in height but, above that, diaphragm and fin walls are more suitable. The diaphragm wall shown in Figure 13.1 has proved very satisfactory in a number of sports halls, gymnasia, swimming pools, factories, a theatre, a church and several mass retaining walls designed by the authors' practice.

A diaphragm wall consists of two half-brick leaves separated by a wide cavity stiffened by brick cross-ribs. The structural action is of a series of connected I or box sections. The cladding function is performed by the outer leaf, the insulation by the cavity, and the lining by the inner leaf. Over twenty such buildings have been constructed in the North-West of England since 1967 and, during the intervening period, have been subjected to the worst gales, the hottest summer and the wettest autumn months on record, plus the very severe winter of 1978/9. They have suffered no distress and little or no deterioration, and have required minimal maintenance. It should be added that their design was chosen on its economic advantages alone, in strict competition with steel and concrete structures. In service, they are almost certainly showing economic advantages in heating and maintenance.

The fin wall (see Figure 13.41), which acts structurally as a series of connected T sections, has been found to be highly efficient for tall single-storey structures, and could well be found useful for multi-storey work – particularly for the column warehouse-type structure. Architects have welcomed the dramatic visual effect of the fin wall. Fins readily lend themselves to post-tensioning. Both post-tensioned brick fins and diaphragm walls have been built up to 10 m in height and, with the results of diaphragm wall research, it is evident that post-tensioned fins and diaphragms could be built to an even greater height.

Engineers will probably be interested in the simplicity of diaphragm and fin wall design, contractors will welcome the elimination of sub-contractors and suppliers, and architects will enjoy the wide choice of architectural treatments. Clients are likely to be pleased with good-looking buildings with lower heating costs, and which are durable and maintenance-free. Some cladding manufacturers proudly guarantee their products a twenty year life. Most brickmakers would be quite happy to guarantee a much longer life than that.

1.3 REINFORCED AND POST-TENSIONED MASONRY

Brickwork and blockwork, like concrete, have high compressive strength but relatively low tensile resistance. So, as with concrete, reinforcing and post-tensioning can be used to carry or relieve the tensile stresses. Reinforced brickwork has been used in India and Japan since the First World War, and in America since the second. In this country, a number of progressive engineers, architects and contractors have used reinforced brickwork for the occasional beam, lintol, etc., but, in the last decade or so, a fair number of reinforced brick retaining walls, a prestressed brick tank and reinforced brick cantilevers have been built.

The authors' practice assisted in the development of a standardised construction system for schools which made extensive use of post-tensioned low-rise walls subject to bending due to lateral wind forces. Over a hundred schools were built using this system, and it was further developed for taller walls in libraries and other buildings. Post-tensioned masonry has been found to be particularly effective for retaining walls. There has also been a recent increase in the use of reinforced and post-tensioned concrete blockwork.

With the new Code of Practice, and the attendant follow-up, applied research structural desig-

ners will undoubtedly develop widespread applications of these techniques.

1.4 ARCHES AND VAULTS

Not only is interest awakening and applied research beginning into new techniques of masonry construction, but there is also a revival of interest in the old structural form of the arch. Already, there is a small but growing demand amongst modern architects for arch construction.

So far, the modern use of arches has been principally for their aesthetic appeal and has been limited in use to churches, link-corridors, colonnades, etc., of small span. But, even so, some brickmakers have found the demand sufficient to make it worthwhile to manufacture segmental bricks. Although these can be three times the cost of ordinary bricks, the increase in the total cost of the job is minimal, but the visual impact is maximal. On the other hand, standard format bricks and blocks could be used for major arches. Research work on the limit state design of masonry arches is continuing. The results of this research should further stimulate architects and engineers into investigating the potential applications of arches and vaults.

Obviously, it is much too early to predict the comparative economics of brick or block arch construction. Nevertheless, it can be confidently stated that they should possess great durability, require little or no maintenance and can have more aesthetic appeal than steel or concrete beams.

1.5 THE ROBUSTNESS OF MASONRY STRUCTURES

Before leaving building forms, it may be interesting to note that had Ronan Point been built in correctly designed structural masonry, there would probably have been no 'progressive collapse'. Indeed, the familiar phrase might never have been coined. Much structural research was carried out after the Ronan Point disaster. This, together with the results of subsequent explosions and accidents, and the experience of bombed buildings during the war, has demonstrated the practical immunity of structural masonry to progressive collapse.

1.6 PREFABRICATION

The argument that precast columns, cladding, etc., could save most of the site operations shown in Figure 1.2 is true, but ignores the off-site manufacturing costs, and transportation to the site – not to mention the capital investment tied up in the production factories and site erection plant. Despite the theoretical advantages of factory mass production, the experience of the 'prefabricated' housing at the end of the war, the 'industrialised' building of the 1950s and the 'system' building which followed, has shown that they have yet to prove satisfactory in performance and economical in construction. The only apparent advantage has been speed of erection.

Whether this speed of construction is due mainly to prefabrication alone, is doubtful. More likely, it is due to the intensive pre-planning and systematic programming that are necessary to make the technique economically viable.

When the same amount of forethought is given to structural masonry construction, the speed results can be equally dramatic. In 1975, the Brick Development Association sponsored a housing scheme carried out by an average-sized contractor. Nine days after the site start on prepared foundations, tenants were moving into completed brick houses.

1.7   INTERACTION WITH OTHER MATERIALS

This is a field that needs attention. There is a real need to study the interaction between brickwork and blockwork and other materials. For example, it would be helpful to designers if they had more information on the strapping of timber floors, and the tying down of lightweight roofs, to brick walls. Further information on cavity walls with brick outer linings and lightweight concrete block inner linings would also be valuable. Impartial bodies such as BRE and DOE could well examine these problems to the benefit of industry and society.

1.8   FUTURE LABOUR

An understandable fear is often expressed that when the current serious recession in the construction industry is over, there will not be enough brick and block layers to go around. There probably will not. Neither may there be enough steel fabricators, concrete shutterers, plasterers or any other building tradesmen. There remains reason for hope and qualified optimism in the fact that the government's Training Services Agency is finding that bricklayers

competent enough to tackle straightforward work can be trained in six months.

Concern has also been expressed that, as construction site labour costs rise, structural masonry will eventually be priced out of existence and that factory-produced buildings will take over. This ignores the obvious fact that factory costs will also rise.

1.9   ENGINEERING EDUCATION

At the beginning of the Victorian era, bricks were the main civil and structural engineering medium. Sir Marc Brunel used reinforced brick rings for the shafts of the Blackwall Tunnel. His son, Isambard Kingdom Brunel, used brick arches of over 100 ft span to bridge the Thames at Maidenhead. Stephenson carried out research into the compressive strength of brickwork when he was designing and building the Conway Bridge. Jesse Hartley made extensive use of structural brickwork in the construction of the superb Albert Dock in Liverpool, and Telford did the same in the elegant and recently modernised St Katharine's Dock in London. The Victorians used bricks to retain canals and railway cutting banks, for aqueducts, tunnel and sewer linings, deep manholes and inspection chambers, road foundations, bridges, warehouses, cotton mills, factories, railway stations, churches, houses – every conceivable type of building and engineering structure.

However, the advent of steel and reinforced concrete, with their superior tensile and bending strength, marked the decline of structural brickwork. Engineers adopted the new materials with great enthusiasm and, since the end of the last century, the decrease in the use of structural brickwork has been so sharp that few, if any, contemporary engineering graduates can design in the medium. This ignorance on the part of the engineers is creating a log-jam in the revival of structural masonry. Architects want to design in it, contractors want to build with it and the public wants to live with it – but many engineers cannot design with it.

There is now a desperate and urgent need in the universities and polytechnics to wean the staff and students away from their traditional preoccupation with structural steel and concrete. A major education drive by the brick and block industries will be necessary to break this log-jam, and to educate and stimulate engineers to investigate the civil engineering possibilities of structural masonry. To take just a minor example – the majority of public health engineers use precast concrete rings for inspection chambers, manholes and sewage works. Mostly, it would seem, because they always have done and because it is so quick and easy to look up the section required in the manufacturer's catalogue. Yet there must be many cases where brickwork or blockwork would be cheaper and more satisfactory. If only engineers were educated in (and had access to) the existing data, then they would probably put it to good and effective use.

The past decade or so has seen much fundamental research into the structural behaviour of masonry. Although, in all engineering, there is always a need for some continuing fundamental research, sufficient knowledge has now been gained to switch the emphasis and resources to applied research in order to give designers and constructors the data and information they require.

There is no mystique or esoteric flummery in structural masonry design – any of the professional design team, whether he be an architect, contractor or surveyor, who understands that stress equals load over area, that $f = M/Z$, and appreciates the concept of slenderness ratio, should be able to prepare a preliminary design and approximate feasibility study of a masonry structure. Detailed design does require the ability of a competent and experienced engineer – as do steel, concrete, timber and any other engineering material. But, in the authors' practice, it has been found that engineers can be more quickly trained in brickwork and blockwork than in any other structural materials.

Added to the natural conservatism of human beings, there is an understandably increased reluctance on the part of the building industry to try out new techniques after the Ronan Point disaster, high alumina cement, calcium chloride, glu-lam, etc. But brickwork is scarcely a new technique – it has the longest continuous history of any building material. All we need to do is to re-learn the techniques of construction; to study the wealth of data poured out by BRE, BDA, CACA and others; to educate our young engineers, architects, surveyors and contractors; and to carry out applied structural research based on modern and proven structural theories. If we can accomplish these tasks, the built environment can be improved with robust durable economical buildings with a human scale and aesthetic appeal. We may well see the eventual demise of the harsh concrete jungle and the eye-wearying steel and glass rent-slabs.

CHAPTER 2

ADVANTAGES AND DISADVANTAGES OF STRUCTURAL MASONRY

The durability of masonry, when used correctly, is excellent. However, as with other materials, the proper use of masonry requires an understanding of its physical characteristics, its strengths and weaknesses, the methods of construction and the availability of various shapes and textures together with relative costs. The advantages which follow are based on the proper use of masonry.

2.1 ADVANTAGES

2.1.1 Cost

It is notoriously difficult to obtain accurate and comprehensive costs for building elements – let alone completed buildings. Too often, costs reflect the current state of the building market, and nearly always provide only the cost of erecting the building, and not the long-term cost of the building over its life.

The argument that masonry structures are labour intensive compared to steel or concrete, and are, therefore, uneconomical in a high wage situation, is not borne out by the facts. The experience of the authors' practice has always been that where a masonry structure is appropriate, it has inevitably been cheaper than the other structural alternatives.

The authors have found as follows:

(1) In steel and concrete frame structures, masonry or other materials are used to form the partition, staircase and corridor walls, etc. In so many instances, if these partition and other walls are designed in calculated loadbearing masonry they can be made to carry the loads and dispense with the need for columns and beams.
(2) The general contractor can usually erect a masonry structure, whereas steel and some other materials normally require specialist sub-contractors. Experience has shown that generally the less the amount of work put out to sub-contractors, the lower are the construction costs – always assuming, of course, that the job is within the overall capability of the main contractor. With masonry structures, not only is the number of sub-contractors reduced, but there is also a reduction in the number of site operations, trades and materials. The possibility of delays whilst awaiting fabrication is also overcome.
(3) Masonry buildings tend to be faster to erect, resulting in lower site overhead costs.
(4) The maintenance costs of masonry are minimal.
(5) A high degree of fire protection, thermal and sound insulation, exposure protection, etc., is automatically provided for within the structural requirements of masonry buildings which are, therefore, much more economical in these respects.

The best way to determine the differences for a particular building is to design and cost compare an appropriate structure using the various structural possibilities, including masonry, ensuring that the most economical scheme has been chosen for each material.

2.1.2 Speed of erection

This, as was noted in the Introduction (1.1), is one of the main advantages of masonry construction. Unfortunately, it tends to be underrated – principally due, it would seem, to the widely held though erroneous assumption that because the prefabricated frame of a building can be erected to a high level in a short time, this must result in an early completion of the whole project. Frequently though, a steel frame suddenly appears on a site, rapidly rises to roof level, and then stands rusting away waiting for the follow-on trades to work their way through the building.

Ignoring the fabrication time, it is true, of course, that a steel frame has a short site erection time. On the other hand, it should be appreciated that no other construction work can take place during the erection period. This is not the case with masonry structures, where other trades can quickly follow on thus achieving a faster overall construction time for the whole building.

A masonry wall can easily be built in one day, and support a floor load soon after. Compare this with a concrete column where the time taken to fix reinforcement, erect shuttering, cast concrete, cure, prop and then strike the shutter is often more than a week.

In conclusion, it is worth pointing out that the speed of masonry construction is achieved without the same planning constraints that limit the application of system building.

2.1.3 Aesthetics

This should be mentioned, even though this aspect of building is not usually considered as being within the the province of engineers. Certainly, many engineering colleges do not bother to teach it.

The aesthetic appeal of a building is a complex amalgam of many factors: form, massing, scale, elevational treatment, colour, texture, etc. Masonry provides the human scale, is available in a vast range of colours and textures, and, due to the small module size of bricks and blocks, is extremely flexible in application in that it can be used to form a great variety of shapes and sizes of walls, piers, arches, domes, chimneys, etc. Masonry structures tend to wear well and mellow

with time. In our climatic and environmental conditions, many other materials perform conspicuously less well.

2.1.4 Durability

The excellent durability of masonry is obviously a great advantage. Many historic buildings and engineering structures provide living proof of this quality. It must be emphasised again, however, that this considerable functional and environmental benefit applies only to properly designed masonry (see Disadvantages 2.2). Provided that masonry structures are designed and built with competence and care, they should last much longer than their required life.

2.1.5 Sound insulation

The majority of noise intrusion is by airborne sound, and the best defence against this is mass – the heavier the partition, the less the noise transmitted through it. It is an added bonus if the mass structure is not too rigid. Brickwork and blockwork provides the mass without too much rigidity. Typical sound insulation values for a range of brick walls are given in Table 2.1.

2.1.6 Thermal insulation

The good thermal properties of cavity walls have long been recognised and, more recently, have become critical in the attempts to conserve energy. Cavity walls and diaphragm walls can easily be insulated within the void to provide further improved thermal values. However, care is required in both the choice of insulation material and the details employed. Thermal insulation values for some typical masonry walls are shown in Table 2.2.

Table 2.1 Typical sound insulation values of masonry walls

Material and construction	Thickness (mm)	Weight (kg/m^2)	Approximate sound reduction index (dB)
Brick wall plastered both sides with a minimum of 12.5 mm thick of plaster	215	415	49.5
Brick wall plastered both sides with a minimum of 12.5 mm thick of plaster	102.5	220	46
Type A concrete block to BS 2028, 1363. Wall plastered both sides with a minimum of 12.5 mm thick of plaster	180	340	47
Cavity wall with outer leaf of brick not less than 100 mm thick and inner leaf of 90 mm thick solid Type A concrete to BS 2028, 1364. Wall plastered both faces with a minimum of 12.5 mm thick of plaster	250 (including 50 mm cavity)	380	53

Table 2.2 Typical thermal insulation values of masonry walls

Construction and materials	Min. thickness (mm)	'U' values external walls (N/m² °C)
Single leaf of bricks of clay, concrete or sand-lime	327.5	1.5
Single leaf of bricks of clay, concrete or sand-lime	215	2.0
Single leaf of Type A concrete blocks to BS 2028, 1364	190	2.6
Single leaf of Type A concrete blocks to BS 2028, 1364	90	3.0
Single leaf of Type B concrete blocks to BS 2028, 1364	100	2.9
Cavity walls with outer leaf of bricks or blocks of clay, concrete or sand-lime not less than 100 mm thick and		
(a) inner leaf of Type A concrete blocks to BS 2028, 1364	100	1.7
(b) inner leaf of bricks of clay, concrete or sand-lime	100	1.5

Notes: (a) All walls are unplastered.
(b) All materials have a 5% moisture content.

2.1.7 Fire resistance and accidental damage

Charles II was no fool to insist on brick and stone buildings after the Great Fire of London in 1666. The Victorians lit fires in their mills and warehouses, yet these were suprisingly free from being burnt down. In the bombing of the Second World War, brick structures suffered less damage than steel or concrete buildings – which fact provides evidence of not only the high fire resistance of masonry structures, but also of their inherent capacity to resist accidental damage (see Chapter 8).

Brickwork and blockwork are incombustible and could not start or spread a fire. Masonry is rarely seriously damaged by fire; it does not buckle like steel, spall like reinforced concrete or burn like timber. The Building Regulations 1976, Schedule 8, give the following fire resistance values for masonry shown in Table 2.3.

2.1.8 Capital and current energy requirements

The staggering increase in oil prices during the 1970s concentrated world attention on the energy crisis. It seems probable that the world will exhaust its fossil fuels by the end of the century, or soon after, by which time we can only hope that man's ingenuity will have learnt how to extract energy from other sources such as the waves, the sun and the wind.

Table 2.3 Typical fire resistance values of masonry walls

Construction and material	Min. thickness (mm)	Max. period of fire resistance (h)
Bricks of clay, concrete or sand-lime	100	2
Solid or hollow concrete blocks of Class 1 aggregate	100	2
Solid concrete blocks of Class 2 aggregate	100	2
Hollow concrete blocks, one cell in wall thickness of Class 1 aggregate	100	2
Cavity wall with outer leaf of bricks or blocks of clay, concrete or sand-lime not less than 100 mm thick and		
(a) inner leaf of bricks or blocks of clay, concrete or sand-lime	100	4
(b) inner leaf of solid or hollow concrete bricks or blocks of Class 1 aggregate	100	4

Notes: (a) All walls are unplastered.
(b) All walls are loadbearing.
(c) In the case of a cavity wall, the load is assumed to be on both leaves.

Over half the energy used in this and other western countries goes into the construction and running of buildings. Cars, by comparison, consume a relatively insignificant quantity of fuel. Of the total consumed by buildings, 10–15% is used in constructing buildings (capital energy) and the remainder in running buildings (current energy). The bulk of the current energy goes in heating, and smaller amounts in lighting, operating lifts, etc. It has been shown in a number of studies that brick structures require the lowest capital and current energy.

2.1.9 Resistance to movement

We live in a substantially brick-built environment, and it may certainly be claimed that load-bearing brickwork structures have been subjected to the most intensive full scale testing, over a longer period, than any other present-day building material. The results are impressive, readily visible but, unfortunately, not very well documented.

Wartime damage, mining settlement, earthquake movement and demolition have taught engineers a great deal about the behaviour of brickwork when subjected to large deformations. Buildings can often be seen undergoing demolition, or in areas severely affected by mining, containing deformations and cantilevered projections which would be extremely difficult to justify by calculation. Observations of such cases can teach us a great deal about the use of masonry where deformation is expected, and the part which mortar strength plays in controlling the cracking of masonry under such severe conditions.

It is essential when using masonry not to use too strong a mortar, relative to the strength of the brick or block used in the wall. The mortar joint should always be the weak link, in order to retain any cracking within the numerous bed and perpendicular joints between the bricks or blocks. A correct relationship between the mortar and the brick or block strengths will result in the total effects of the movement being distributed amongst numerous fine cracks. Such cracks are largely concealed and can be easily pointed over without becoming unsightly (see Appendixes 2 and 3).

Had the Ronan Point tower block been built in loadbearing masonry, it is unlikely that the Fifth Amendment to the Building Regulations would exist. Progressive collapse would not have occurred if brickwork or blockwork had been the major structural material, and damage would have been reduced. As one would expect, compliance with the Amendment presents very few problems for the designer of masonry buildings, and this is dealt with in more detail in Chapter 8: Accidental Damage.

2.1.10 Repair and maintenance

Properly designed masonry requires little or no maintenance and is extremely economical in terms of maintenance costs. With reference to its use in areas of possible high deformation, such as mining areas, a well-designed building will contain the majority of damage within the mortar and movement joints, and repointing of the masonry will repair most of the defects.

2.1.11 Ease of combination with other materials

The main structural quality of masonry is its ability to resist compression forces. However, this does not prevent its use in locations where bending and tension conditions have to be resisted. In most situations, sufficient precompression exists to prevent tension occurring, and in areas where this is not the case, post-tensioning, reinforcing or composite action can be used to provide the combined use of masonry with high tensile resisting materials to overcome the problem of high tensile stresses. This is dealt with in more detail in Chapter 15.

The ability of masonry to act compositely with other materials has long been known, but not fully exploited by engineers. Demolition and building contractors continually take advantage of brickwork's true abilities, based on their experience, and it is unfortunate that many engineers are, if anything, lagging behind the more practical man.

2.1.12 Availability of materials and labour

The normal module size of bricks and blocks and the ready availability of their raw materials means that they can be mass produced in many locations and stocked in standard sizes. Modern transportation and packaging enable speedy delivery of bulk supplies of bricks and blocks, and reduce the number damaged in transit to a minimum. Similarly, the materials used in mortar are available in many locations and are easily transported.

Being a well-established trade, skilled bricklayers are normally available in most areas. Early discussions with the tradesmen on the site regarding the structural requirements will result in a proper understanding of the constructional and engineering needs. The inspection of completed work can be made immediately – an advantage over concrete which only tends to reveal its defects when the shutters are struck.

2.2 DISADVANTAGES

2.2.1 Lack of education in masonry

This was referred to in 1.9. However, there seems justification for mentioning it again, since it must be regarded as a genuine disadvantage at the present time.

The many uses to which masonry can be put, the wide range of materials and material behaviour, and the great strides forward in the structural use of masonry, require the backing of a good sound education to prevent misuse and to ensure the maximum economy and efficiency in design and construction. Unfortunately, education has been lagging behind the developments, and this has left the construction industry in a situation where it cannot fully exploit masonry's capabilities until it is geared up to the new techniques and applications.

This essential gearing up applies as much to an attitude of mind as to anything else. Industry must get rid of the attitude that masonry is an old-fashioned out of date material, and encourage a modern philosophy.

Running parallel to a new philosophy must be a willingness to learn from the past. The advantage of durability, for example, must not be taken for granted, since no material is proof against poor design or bad workmanship. Consider the case of a parapet wall, badly built with absorbent masonry, inadequate mortar and no protective coping. Such a wall will become saturated and suffer from frost damage during the first severe winter. Yet the experienced designer can design a suitable wall which will survive its intended life without trouble. The durability of masonry depends on the quality of design and construction, and these, in turn, depend upon suitably educated and experienced designers and construction operatives.

2.2.2 Increase in obstructed area over steel and reinforced concrete

Although masonry units can be obtained with extremely high crushing strengths, the design compressive strengths of masonry walls are generally lower than for steel or reinforced con-

crete. It follows, therefore, that for a particular loading condition, masonry will require a greater cross-sectional area.

In locations where large unobstructed areas are required, this can create problems which might make masonry unacceptable. It should be noted,

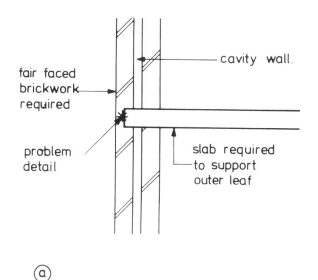

(a)

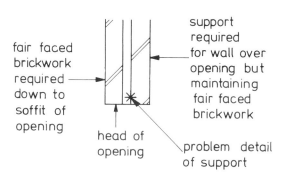

(b)

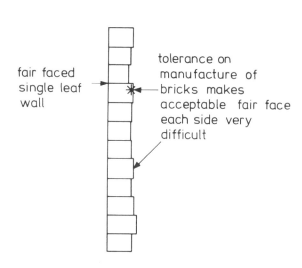

(c)

Figure 2.1 Vertical cross-sections

however, that careful design and detailing can frequently produce an acceptable and economical scheme for many applications, and masonry should not be completely ignored for buildings requiring large unobstructed areas. See Chapter 9: Structural Elements and Forms.

2.2.3 Problems with some isolated details
Like many other materials, masonry can give rise to problems in the achievement of satisfactory isolated details. For example, fair-faced masonry often creates local detailing difficulties (see Figure 2.1).

These apparently minor problems require care and forethought if a satisfactory result is to be achieved. In the case of detail (a), differential movement of the two leaves must be allowed for if the brick face to the slab is to remain (see Appendix 3: Movement Joints). In the case of detail (b), reinforced brickwork or a suitably combined lintol can overcome the problem and, in the case of (c), bricks manufactured to tight tolerances can be obtained, and these should be specified.

2.2.4 Foundations
Since one of masonry's main advantages over concrete is that it does not require expensive shuttering, it follows that in situations where shuttering is reduced to a small percentage of the cost, the competitive use of concrete comes into its own. Foundations come into this category, and masonry will generally be found inferior to concrete in situations where the soffit and sides of the excavations form, in effect, the majority of the shuttered faces (see Figure 2.2).

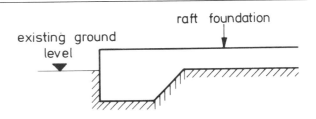

///// indicates shutter formed in excavation or on blinded hardcore surface

Figure 2.2

2.2.5 Large openings
In situations where large openings are to be formed and a level soffit is required, reinforced concrete or steel beams are generally found to be the most economical means of support. They can be combined with the composite action of any masonry above and, unless fair-faced masonry is a particular requirement for the soffit of the support, they will usually provide a more economical solution than the masonry alternative.

It must be pointed out, however, that where the soffit can be in the form of an arch, and where the horizontal reactions from such a form can be accommodated, masonry may prove more economical (see Figure 2.3).

2.2.6 Beams and slabs
The use of masonry in situations where the dead weight of the material is the major portion of the load, and where a level soffit support is required, can often be uneconomical, and most beam and slab situations fall into this category.

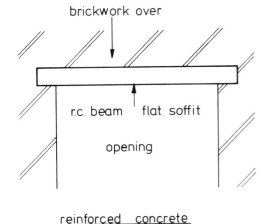

reinforced concrete beam

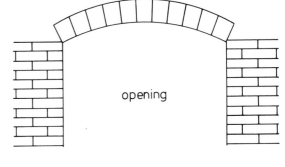

buttressed masonry arch

Figure 2.3

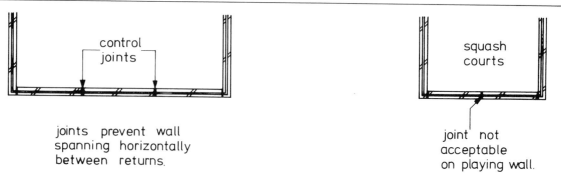

Figure 2.4 Plan on walls showing unacceptable control joints

The merits of masonry should nevertheless be considered for each individual scheme, taking into account recent changes in material costs and construction methods and, from this point of view, situations may well arise where the use of reinforced or post-tensioned masonry can be exploited.

2.2.7 Control joints

In some forms of masonry construction, the need for relatively close spacing of the control joints necessary to prevent cracking from the effects of shrinkage and/or expansion can be difficult to accommodate, due to structural, visual and other constraints (see Figure 2.4).

It should be remembered, however, that masonry is often required for partition walls when an independent structure is employed, and the introduction of a frame in a different material can often cause even greater problems from differential movement.

CHAPTER 3

DESIGN PHILOSOPHY

The main underlying aim should always be to keep the solutions simple, to see that the construction methods and the effects of the design upon them are carefully considered, and to ensure that the design is based upon masonry as a material in its own right, and not simply as a variation on the design of concrete structures.

3.1 STRENGTH OF MATERIAL

To exploit the structural potential of any material, it is essential to understand its strengths and its weaknesses. Masonry is strong in compression and weak in tension and, in order to use the material economically, engineers must exploit the strength and overcome the weakness.

Consider the critical loading conditions for the material, two examples of which are indicated in Figure 3.1.

Condition 'A' is a masonry wall at a cross-section where only a small downward load, W, exists at a time when a large uplift force, U, is applied.

Condition 'B' is a masonry wall at a cross-section where only a small downward load, W, exists at a time when a large bending moment, M, is applied.

In both cases the resulting stress in part or the whole of the cross-sections will be tensile and, therefore, critical to a simple masonry form.

Consider next the further critical loading condition indicated in Figure 3.2.

Condition 'C' is a slender masonry wall subjected to a large axial load.

In this case, the load which the wall can carry is severely restricted by the wall's tendency to buckle, and the condition is therefore critical to the proper exploitation of the material.

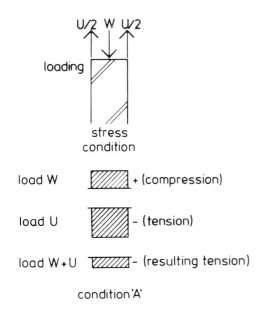

condition 'A'

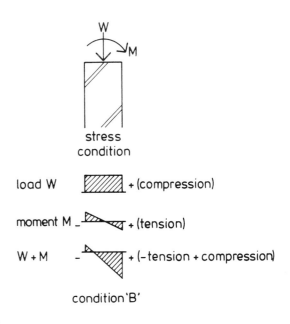

condition 'B'

Figure 3.1

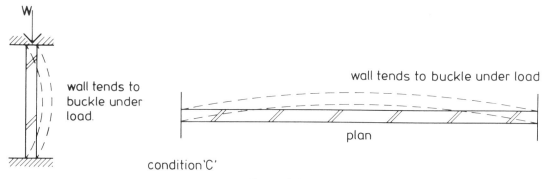

Figure 3.2

3.2 EXPLOITATION OF CROSS-SECTION

In the case of condition 'A', there are a number of methods of solving the problem. One is to increase the dead loading above the level considered. This can often be achieved by changing the construction, for example roof structures can often be changed to a heavier form, but it is important to check that this results in the most economical overall cost.

Alternatively, strapping or tying the elements, to which the uplift is applied, down to a level where sufficient dead weight exists to satisfy the safety requirements can be considered. See Figure 3.3 for typical roof strapping details.

The use of concrete capping beams and/or post-tensioning rods can also be used to 'anchor' down these forces (see Figure 3.4).

In the case of condition 'B', there are a number of engineering approaches which can produce different but competitive solutions. For example, consider the stress at the cross-section. It is made up of two components, W/A and M/Z where:

W = direct load
A = area of the cross-section
M = applied bending moment
Z = section modulus.

Let the stress be f, then: $f = \dfrac{W}{A} \pm \dfrac{M}{Z}$

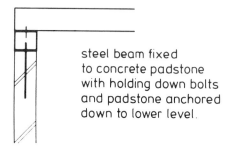

steel beam fixed
to concrete padstone
with holding down bolts
and padstone anchored
down to lower level.

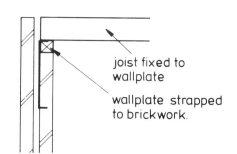

joist fixed to
wallplate

wallplate strapped
to brickwork.

Figure 3.3 Typical strapping details

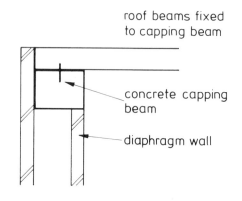

roof beams fixed
to capping beam

concrete capping
beam

diaphragm wall

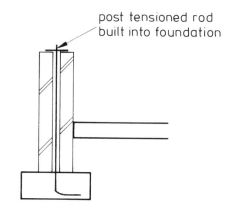

post tensioned rod
built into foundation

Figure 3.4

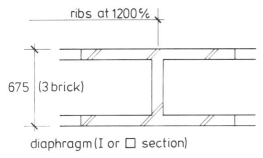 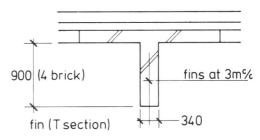

Figure 3.5

It can be seen that to exploit the material compressive stresses rather than tensile stresses are required, i.e. a higher value of W/A than M/Z.

Assume that W and M are fixed values, then it is desirable to have a large Z value relative to the area A of the cross-section, i.e. a large Z/A ratio. This can be achieved by positioning the material to produce an increased section modulus for a similar area, i.e. diaphragm, fin wall or other geometrical section (see Figure 3.5).

This, when compared with, say, a normal cavity wall gives increased Z/A ratios (see comparison in Table 3.1).

would be to artificially increase W. The old method of doing this would be to add mass by increasing the amount of masonry, i.e. thickening the wall. However, an increase in the value of W can also be achieved by post-tensioning (see Figure 3.6).

This method produces extra compressive stresses in the brickwork which must then be cancelled out before any tension can be developed, thereby helping to keep the brick stresses at an acceptable level. Again the post-tensioning is kept simple by using large diameter rods rather than cables and applying the force by means of a simple threaded rod and nut bearing on to an end

Table 3.1 Comparison of properties of various wall types

Wall	Area, A	Section modulus $Z \times 10^{-3}$ (m³) (minimum value for fin wall)	$\dfrac{Z}{A}$	Proportional increase in area	Proportional increase in $\dfrac{Z}{A}$
260 mm cavity wall	0.205	3.50	0.017	1.0	1.0
Diaphragm wall	0.245	52.49	0.214	1.2	12.6
Fin wall	0.307	26.40	0.086	1.5	5.1

(Assuming fin centres at 3.0 m$^c/_c$ and Z value of effective fin only is used.)
Values relate to the equivalent of 1 m length of each wall type.

It can be seen that with little increase in area, above the area of a normal cavity wall construction, massive increases in section modulus and hence Z/A ratio can be achieved. It should be noted that the sections being considered are all simple sections designed to take into account the method of construction and material being used – essential factors if real economy is to be achieved.

Considering again the two components of the stress make-up, i.e. W/A and M/Z, it can be seen that another method of increasing the value W/A

plate. The force is then applied, to the required value, by tightening the nut to the specified torque (see Figure 3.6).

Consider condition 'C' – axial loading

The ability of the wall to support vertical loading is restricted by its tendency to buckle, and again there are a number of ways of overcoming the problem. The most obvious is to thicken up the wall, but in many situations this is not the most economical solution and the engineer should consider other possibilities.

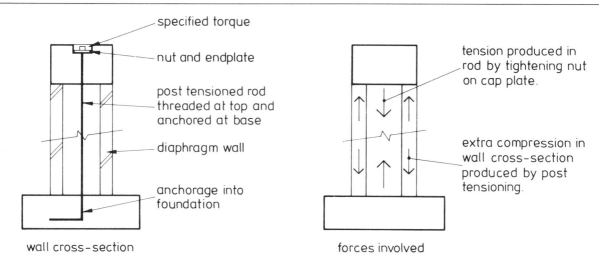

wall cross-section

- specified torque
- nut and endplate
- post tensioned rod threaded at top and anchored at base
- diaphragm wall
- anchorage into foundation

forces involved

- tension produced in rod by tightening nut on cap plate.
- extra compression in wall cross-section produced by post tensioning.

Figure 3.6

The stiffness of the section depends on the ratio of:

$$\frac{\text{effective height or length}}{\text{effective thickness}}$$

or

$$\frac{\text{effective height or length}}{\text{radius of gyration}}$$

i.e. l/t or l/r, depending on how the slenderness ratio or section properties are expressed. As in all struts, the greater the slenderness ratio the weaker the strut, and it is advisable in exploiting the material to reduce the slenderness ratio whenever economically possible. There are two ways, therefore, of reducing this factor. One is to improve the effective thickness or radius of gyration, and again this can be achieved by using a diaphragm or fin form (see Figure 3.5).

Alternatively, a reduction in effective length would have a similar effect and extra restraint from essential building elements should be considered. For example, if a substantial suspended ceiling or a ceiling which could be economically braced to provide restraint were in the close vicinity of the wall, then extra restraint could be provided assuming that the sequence of construction and sequence of loading are co-ordinated, (see Figure 3.7). This would reduce the effective height. A further consideration would be to make use of a reduced effective length in the plan, possibly by changes in the construction of intersecting partitions etc. (see plan in Figure 3.7).

These solutions can often be achieved much more economically than by increasing the thickness of the wall. The above solutions bring in the exploitation of essential building elements which can also help the conditions previously considered as explained below.

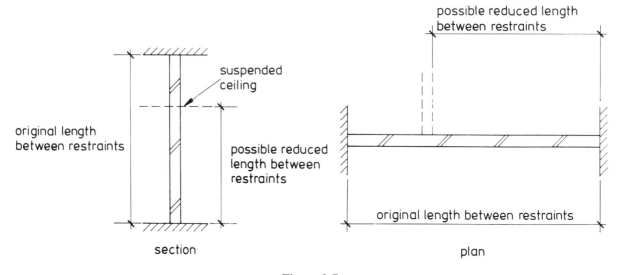

original length between restraints

suspended ceiling

possible reduced length between restraints

section

possible reduced length between restraints

original length between restraints

plan

Figure 3.7

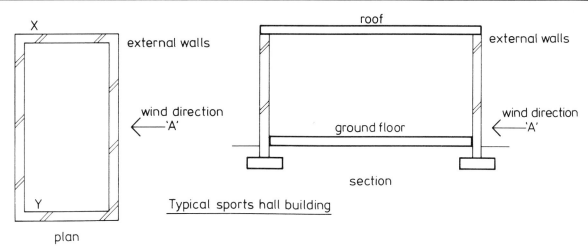

Figure 3.8

3.3 EXPLOITATION OF ESSENTIAL BUILDING ELEMENTS

Consider again condition 'B' and the two components W/A and M/Z. It has already been shown that improving Z relative to A and/or increasing W reduces the tensile stresses, but a further possibility would be to reduce the applied bending moment M. This condition now brings into consideration the overall building stability and how best to exploit the building elements to produce the most economical structure. For example, consider the simple plan shape of the single-storey sports hall building which has an open plan and simple flat roof, shown in Figure 3.8.

Assume that a uniformly distributed wind loading is applied in the direction of arrow A. The walls could be designed as pure cantilevers, with a maximum bending moment of $WL^2/2$. Assume that this was the condition at the cross-section considered under condition 'A' in Figure 3.1.

Methods of exploiting the materials at the cross-sections only, have so far been considered. For economical solutions, there is a need to exploit all essential building elements. In the building shown in Figure 3.8, it can be seen that the walls best able to resist the wind forces from the direction of the arrow A are the end gables X and Y. In addition, by inspecting the section in Figure 3.8, it can be seen that the roof element, if suitably and adequately fixed, could prop the tops of the walls to which the wind is applied and span as a horizontal girder (or plate) between the gables X and Y, transferring its loading as a reaction to the tops of these walls. Walls which resist such reactions are often termed 'shear walls'.

Consider again the applied bending moments in the outside wall, which is now propped at roof level, and it will be found that the maximum bending moment is $WL^2/8$ which is ¼ of the previous value (see Figure 3.9).

This exploitation of essential elements applies to all building forms, particularly when considering wind forces and restraint conditions. Take the case of a multi-storey building subject to wind forces. Again, the elements on which the wind is

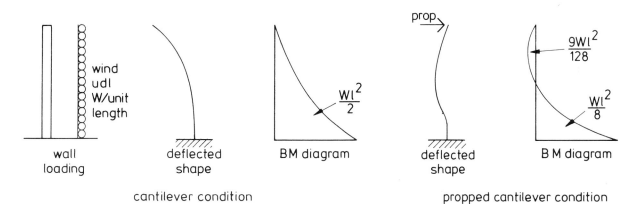

Figure 3.9

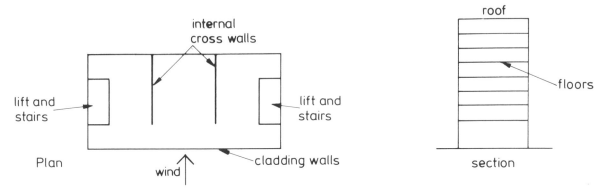

Figure 3.10

directly applied are usually the outer cladding walls, which have their weakest axis at right angles to the wind direction (see Figure 3.10).

It can be seen that the walls best able to resist these forces are the internal crosswalls, the gable walls and the vertical shafts forming the stairs and lifts. In addition, it can be seen that the floor and roof elements could be economically designed to act as horizontal girders spanning between these vertical wall elements and reducing the bending moments on the outer cladding walls to a minimum.

The advantage of using masonry for these forms of construction is that whilst they provide the essential cladding and dividing walls, they can also act as the structure – thus changing elements which would normally be required to be supported on a structural frame into the supporting elements. Moreover, the load from a structural masonry form tends to be more uniformly distributed and, therefore, at foundation level, there is less need to spread the load to reduce the bearing pressure. This makes for economies where strip, pad or raft foundations are employed.

In cases where a piled, or pier and beam, foundation must be used, masonry can also produce advantages since composite action between the ground beam and the masonry can reduce greatly the sizes of foundation beams required (see Figure 3.11).

The ground beam is designed taking into account the interaction between the masonry and the reinforced concrete (rc), which gives the beam a greatly increased lever arm over that of the rc beam acting alone.

Taking a broader view of the approach to design, it is important that engineers are involved in the architectural and planning design layout and proposals at the earliest opportunity. For example, in the case of multi-storey buildings the stiffness of the structure and the distribution of bending stresses depend greatly on the overall layout of the building. In a similar manner to the exploitation of the wall section, the overall plan of a building can often be designed to satisfy both the architectural requirements and provide a high Z/A ratio to resist wind forces.

The use of 'T', 'L' and other plan configurations, can greatly enhance masonry's lateral load resistance (see Figure 3.12), as can the positioning of stairs and lift shafts.

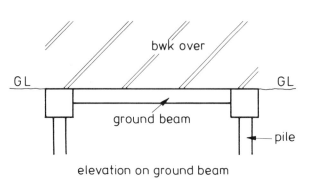

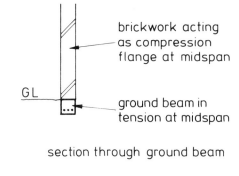

Figure 3.11

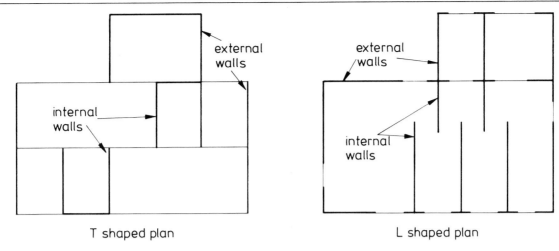

T shaped plan   L shaped plan

Figure 3.12

Whilst, from a structural point of view, the ideal layout for multi-storey structures should repeat on every floor, it is not, as some engineers think, essential if loadbearing masonry is to be used. In the same way that a concrete-framed structure contains beams to pick up walls over, the masonry structure can also accommodate beams supporting loadbearing masonry above. Whilst it should always be the designer's aim to provide repetitive plans for loadbearing walls, the masonry structure can still prove competitive for more flexible layouts.

For some buildings, the plan layout at ground floor level has to be varied from the upper floors – there is often a need for a much more open plan at this level, requiring minimum intrusion from the structural supports. Many engineers would, under such circumstances, rule out loadbearing masonry completely and introduce a framed solution for the whole structure. In practice, however, a podium construction can often be adopted combining, say, a reinforced concrete frame up to first floor level and loadbearing masonry structure from first to roof (see Figure 3.13). This form of construction has proved very

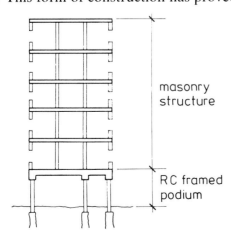

masonry structure

RC framed podium

Figure 3.13

economical for many schemes designed by the authors.

The approach of many engineers to the problem of multi-storey structures is to first consider the most suitable locations for column supports and the most economical material with which to frame up the building. All too frequently, they fail to realise the possibilities of a loadbearing masonry solution.

From the authors' experience, loadbearing masonry is so economical and competitive that this should be considered first, before contemplating a concrete or steel frame solution. But, as was stated at the very beginning of this chapter, it is essential that the proposals are kept simple and practical, and based on the use of masonry as a structural material in its own right. This, after all, is no different to the requirements for other forms of construction. Steel and concrete frames behave differently from masonry, are constructed differently and should be designed differently.

For example, concrete and masonry are similar to the extent that they are both strong in compression and weak in tension. Nevertheless, this provides no grounds for supposing that a similar design approach may be adopted. The two materials are very different:

(a) concrete tends to shrink, whereas some masonry tends to expand
(b) concrete is easily reinforced or pre-tensioned, while masonry is more easily post-tensioned
(c) thin prestressing tendons are usually required for concrete, whereas large diameter post-tensioning rods are more suitable for masonry

(d) concrete behaves similarly in all directions, whereas brickwork behaves differently according to the direction of the stress to the bedding planes.

The variations are enormous. Hence, the design approach must be completely different. In particular, it is important not to overlook the methods of construction, since these constitute a major difference. The authors recommend that unless the reader has a good knowledge of building construction, he or she should do further reading of reliable construction text-books.

More detailed information relating to many aspects of this chapter is covered under appropriate headings in the remaining chapters.

CHAPTER 4

LIMIT STATE DESIGN

At the design stage of a structure, it is not possible to predict accurately the loadings which will act on the structure during its planned life.

Nor is it possible to define precisely the behaviour of a structure under these loadings, or to predict with certainty the strength of the materials which combine to form the structure. It is necessary, therefore, for the engineer to introduce factors of safety when designing a structure in order to ensure that it will be satisfactory for its intended use. In determining the factors of safety to be used, it is necessary to first define the meaning of the word 'satisfactory' when related to the use of the structure.

Obviously, one criterion for the design of the building is that it should not fall down, but there are other conditions which need to be examined. The structure should be readily usable by the occupants, as well as visually acceptable in the long term as well as the short. It should not excessively deflect or crack, and should be sufficiently durable to maintain its initial condition. There are other criteria which may be applicable to certain structures including, for example, resistance to fire, explosion, impact and vibration.

In the past, there have been two basic methods of applying a factor of safety to the design of structures. Both methods have involved the introduction of a single factor to cover all the uncertainties mentioned above, in order to ensure the stability and safety of the structure. The first method, and the one commonly used in the past, is permissible stress design. This involves determining the ultimate stresses for the materials involved, and dividing these by a factor of safety to arrive at a permissible or working stress. The second, is the load-factor method which involves multiplying the working loads by a factor of safety and using the ultimate stress of the materials.

Both these techniques have their faults. The former is based on elastic stresses, and is not strictly applicable to plastic or semi-plastic materials such as masonry. The latter is applied to loads and does not take into account variations in materials. Both have the disadvantage of applying one factor of safety to cover all conditions of materials, loading, workmanship, and use of structure, thus preventing adjustments where one item is, perhaps, of a higher or lower standard than normal.

Limit state design is an attempt to consider each item more closely so as to enable a more accurate factor of safety to be applied in the design, and depends upon the case being considered. This is achieved by breaking down the overall factor of safety used in the design into its various components, and then placing a specific factor – known as a 'partial factor of safety' – on that component for a given condition. It is thus possible to build up an overall, or 'global', factor of safety from these individual factors.

The global factor of safety may be divided into two groups:

Group 1 The factor of safety to be applied to the materials and to the workmanship used in the construction of the structure.

Group 2 The factor of safety to be applied to the loads in the overall structure and the consequences of failure.

Group 1
The factor of safety for materials and workmanship will be based on the probability of the material failing to reach its expected strength. In reinforced concrete, for example, there are two basic components. One is steel, which is manufactured in a factory with a high degree of control over its production, and the probability of an inferior piece of steel is moderately low. How-

ever, concrete is made from natural materials which are mixed together on site, or at a batching plant, where the degree of control is relatively reduced. In addition, the placing of the concrete will affect its overall strength and, since this is done by unskilled or semi-skilled labour, the degree of control is further reduced.

The materials used in basic structural masonry are: (i) the structural units which are manufactured from natural clays or concrete or stone, and (ii) mortar which is made up of cement, sand, water and a plasticiser – usually lime. The quality control of the structural units could be likened to that of steel, with controlled sampling of the product from the production line. It is also possible to check the strength of a randomly selected unit, or group of units, by means of load tests which will give a reasonable indication of the strength of the units actually used. This is an advantage over concrete in that the method of testing concrete is to manufacture a cube which may not be cured in the same manner as the actual concrete used on site. Masonry units are, therefore, products over which there is reasonable control and some limit to the probability of failure.

Mortar presents a different problem. Although it is manufactured from similar materials to concrete, on many sites there is less control over the mixing of the ingredients, which is generally done in smaller batches than concrete, and there is only limited control over the placing of mortar. Therefore, there is a greater probability of failure. However, on larger sites, where the contractor is able to establish a good site organisation, and material testing does not become disproportionately expensive, it is possible to exercise a higher degree of control over the production of the masonry. Thus it is advantageous to vary this partial factor of safety on the design of brickwork from site to site. It becomes necessary, therefore, to assess at an early stage the degree of quality control to be expected on site, and thus the partial factor to be used.

Group 2

The second group of factors is applied to the loadings to be used in the design of the structure. These factors are introduced to take account of inaccuracies in the assumed design loading, errors in the design of the structure, constructional tolerances and the consequences of failure.

The dead weight of the structure may be determined reasonably accurately, but the superimposed loading is much more difficult. It is also not possible to predict accurately the probability of overloading of the superimposed load, and it is therefore necessary to use a higher factor of safety for superimposed loading than dead loading.

It is also advantageous to adjust the factors of safety for different combinations of loadings. It is unlikely, for example, that overloading of the dead, superimposed and wind loads would all occur simultaneously, and thus the factors are adjusted for various combinations of loading.

The assumptions and theories used in the design of the structure will not precisely fit the way in which the building acts. Similarly, inaccuracies during the construction of the structure will vary the actual conditions from those assumed in the design. The consequences of failure of the structure vary considerably, depending upon the type, use and location of the building. For example, the effects of failure of a shopping centre or a grandstand are quite different to those of site temporary works such as shuttering to a concrete beam.

It is thus possible to build up an overall factor of safety from those in Group 1, which are applied to the materials, and those from Group 2, which are applied to the loading used in the design of the structure. However, it is also necessary to vary the safety factors depending upon the condition being investigated and designed. When checking the ultimate failure of the structure, it is obviously necessary to use a higher factor of safety than when checking the deflection of a beam. Thus the partial factors of safety must be adjusted according to the limit state being considered. The following examples should illustrate the point. Consider the following two cases:

Case 1

A skip of loose bricks is to be hoisted over a busy shopping street. The skip weighs 1 kN and can carry approximately 5 kN of bricks, and is to be suspended from a steel rope of ultimate tensile stress of 260 N/mm^2. Since the rope is of mild steel, and to be used in town, it may be assumed that high quality steel is available and the partial factor of safety in this material may be taken as 1.15.

The skip is to be filled with loose bricks and could easily be overloaded and, as it passes over a busy shopping street, the consequences of failure could be serious. The partial factor of safety on the loading should therefore be taken as 1.4 of its dead load, which is not so likely to be increased, and 1.8 on the live load which is liable to increase.

Thus:

Design loading $= (1 \times 1.4) + (5 \times 1.8)$
$= 10.4 \text{ kN}$

Design steel stress $= \dfrac{260}{1.15} = 226 \text{ N/mm}^2$

Area of rope $= \dfrac{10.4 \times 10^3}{226} = 46.0 \text{ mm}^2$

Case 2

A similar problem, but with the bricks in pre-packed units of actual weight of 5 kN and passing over a river. High quality steel is not available. So an increased material factor of safety of 1.5 should be used.

The bricks are in pre-packed units of known weight and overloading is less probable. In addition, the consequences of failure are not serious, merely the loss of the bricks. The partial factor of safety on the loading should be taken as 1.3 on the dead load and 1.5 on the live load.

Thus:

Design loading $= (1 \times 1.3) + (5 \times 1.5)$
$= 8.8 \text{ kN}$

Design steel stress $= \dfrac{260}{1.5} = 173 \text{ N/mm}^2$

Area of rope $= \dfrac{8.8 \times 10^3}{173} = 51 \text{ mm}^2$

CHAPTER 5

BASIS OF DESIGN (1): VERTICAL LOADING

This and the succeeding chapter provide a basic design method based on limit state principles, and explanations of the requirements of BS 5628 as they relate to actual design. In later chapters, detailed examples will be given of the design of elements and complete structures. It will be seen from these examples that BS 5628 does not have universal application or provide a solution to all problems, and that the basis of design set out here must be extended to solve the less straightforward problems.

As a structural material, masonry is required to resist direct compression, bending stresses and shear stresses. Masonry is strongest in compression, and this characteristic will be the starting point in establishing the basis of design. Consideration will first be given to walls and columns acting in compression.

In the introduction to limit state principles (Chapter 4), it was shown that the aim of the design process is to ensure that the design strength of the particular element under consideration is greater than, or equal to, the design load the element is required to resist. The design strength of an element is a function of the characteristic strength of the masonry and the relevant partial factors of safety on the materials. This is expressed mathematically as $f(f_k/\gamma_m)$. The design load to be resisted is a function of the characteristic load and the partial factors of safety applicable to loads. This is expressed mathematically as $f(\gamma_f F_k)$.

Thus the aim of the design process may be expressed by the formula:

$$f\left(\frac{f_k}{\gamma_m}\right) \geq f(\gamma_f F_k)$$

where
f_k = characteristic compressive strength of masonry

γ_m = partial factor of safety on materials
F_k = characteristic loading
γ_f = partial factor of safety on loadings
f is a mathematical function involving the symbols in parentheses.

5.1 COMPRESSIVE STRENGTH OF MASONRY

A wall or column carrying a compressive load behaves like any other strut, and its loadbearing capacity depends on the compressive strength of the materials, the cross-sectional area and the geometrical properties as expressed by the slenderness ratio (see Figure 5.1).

The compressive strength of a wall depends on the strength of the units used, the bricks or blocks, and the mortar. The assessment of the combined strength of these elements will also be

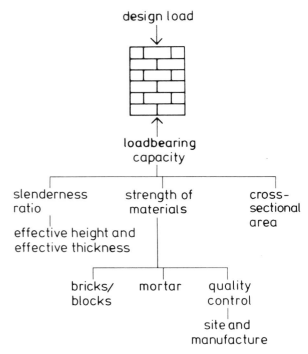

Figure 5.1 Factors affecting the compressive strength of masonry

affected by the degree of quality control exercised in manufacture and construction. The slenderness ratio, in turn, depends upon the effective height (or length, as will be seen later) and the effective thickness of the wall or column.

5.2 CHARACTERISTIC STRENGTH AND CHARACTERISTIC LOAD

The terms 'characteristic strength' and 'characteristic load' have already been used. These occur frequently in limit state design, and should be explained. They derive from the statistical methods used to analyse a number of different results.

If, when considering the compressive strength of a number of bricks, the values of compressive strength are plotted graphically against the number of bricks reaching each specified strength, then a distribution curve is obtained (see Figure 5.2).

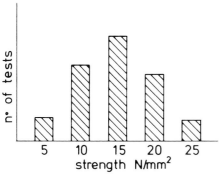

Figure 5.2

This graph shows most of the results to be close to one particular value, with a fewer number of results (samples) at greater or lower strengths. If a sufficiently large sample is taken, then the distribution curve can be approximated to the 'normal' or Gaussian distribution (see Figure 5.3).

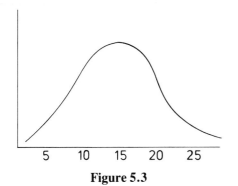

Figure 5.3

The peak of this curve corresponds to the 'mean' strength of the samples, and the area beneath the curve (defined as unity) represents probability. The area under the curve to the left of the vertical line, at any value of compressive strength, represents the probability that any result will have a strength less than that being considered. Figure 5.4 shows that the probability of any brick or block having a strength of less than x is 0.05, i.e. 1 in 20 of the bricks or blocks will have a strength less than x.

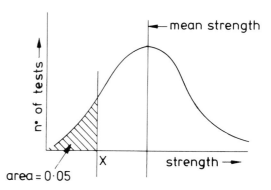

Figure 5.4

The 'standard deviation' of a set of results is the term used to quantify how narrow a range the results cover. Standard deviation, is defined mathematically as:

$$\sigma = \frac{\Sigma(\bar{x} - x_n)^2}{n - 1}$$

i.e. it is the sum of all the differences between each result and the mean, multiplied by itself and divided by the number of differences. The values of standard deviation may be plotted on the normal distribution curve (see Figure 5.5).

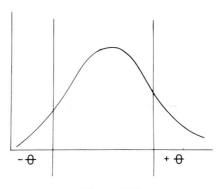

Figure 5.5

Thus to obtain a value of strength above which a certain proportion of the results will probably lie, it is only necessary to take the mean strength and add or subtract a certain fraction of the standard deviation. Therefore, if it is required that not more than 1 in 20 of the results should fail, in order to achieve a required factor of safety, then the strength corresponding to this value can be defined in terms of:

(mean strength) − K × (standard deviation)

where K is the relevant factor corresponding to the 1 in 20 limit, which is, in fact, 1.64.

The value of strength so obtained is termed the *characteristic strength.*

The characteristic strength = (mean strength) − 1.64 × (standard deviation).

Similarly, the values of flexural and shear strength may be obtained.

The characteristic strength of masonry in compression, shear, tension, etc., as defined in BS 5628, is the value of strength below which the probability of test results failing is not more than 5%, i.e. 1 in 20.

Ideally, the dead and imposed loadings should be based on similar statistical concepts. At present, however, insufficient data are available to express loads in statistical terms. In practice, the values given in appropriate British Standards are used as follows:

(a) Characteristic dead load, G_k
The weight of the structure complete with finishes, fixtures and permanent partitions. This is taken as being equal to the dead load, as defined and calculated in accordance with CP 3: Chapter V: Part 1.

(b) Characteristic imposed load, Q_k
The imposed load as defined in and calculated in accordance with CP 3, Chapter V: Part 1.

(c) Characteristic wind load, W_k
The wind load calculated in accordance with CP 3: Chapter V: Part 2.

The characteristic load, as defined in BS 5628, is ideally the load which has a probability of not more than 5% of being exceeded where the load acts favourably. That is, for example, if the load were likely to cause overstressing. Where the load may be providing stability, e.g. restraining a cantilever beam, the load which has a probability of at least 95% of being exceeded (see BS 5628, clause 22).

5.3 PARTIAL SAFETY FACTORS FOR LOADS, γ_f

Inaccuracies are likely to arise in practice due to errors in design assumptions, errors in calculations, the effect of construction tolerances, possible load increases and unusual stress redistribution. Such errors are taken into account by applying a partial factor of safety, γ_f, to the characteristic loads, so that:

Design load = characteristic load × partial factor of safety

$= (G_k, Q_k \text{ or } W_k) \times \gamma_f$

where

G_k = characteristic dead load
Q_k = characteristic imposed load
W_k = characteristic wind load.

The partial factors applied to each type of loading vary according to the accuracy of the estimation of that loading. For example, the characteristic dead load may be assessed more accurately than the characteristic imposed loading. This is reflected in the 1.4 partial safety factor applied to the dead load, and the 1.6 factor applied to the imposed loading. However, it should be noted that where the imposed loading may be more accurately assessed, e.g. a water-retaining structure it would be reasonable to adjust the partial safety factor.

The partial factor of safety takes into account the importance of the limit state being considered,

Table 5.1 Partial factors of safety on loadings for various load combinations (BS 5628, clause 22)

Load combination	Limit state – ultimate		
	Dead	Imposed	Wind
Dead and imposed	$1.4G_k$ or $0.9G_k$	$1.6Q_k$	
Dead and wind	$1.4G_k$ or $0.9G_k$		The larger of $1.4W_k$ or $0.015G_k$
Dead and wind – free-standing walls and laterally loaded panels whose removal would in no way affect the stability of the remaining structure	$1.4G_k$ or $0.9G_k$		The larger of $1.2W_k$ or $0.015G_k$
Dead, imposed and wind	$1.2G_k$	$1.2Q_k$	The larger of $1.2W_k$ or $0.015G_k$

the accuracy of the estimate of the loadings and the probability of the load combination. It does not account for gross errors in design calculations or faulty construction.

The partial factors of safety, given in clause 22 of BS 5628, are shown in Table 5.1. Where alternative values are given, those producing the more severe condition should be selected.

EXAMPLE 1

The characteristic loads (from CP 3) for a floor used for offices are:

(a) characteristic dead load, $G_k = 3.0$ kN/m^2
(b) characteristic imposed load, $Q_k = 2.5$ kN/m^2.
Determine the design load.

$$\begin{aligned} \text{Design load} &= G_k\gamma_f + Q_k\gamma_f \\ &= (3.0 \times 1.4) + (2.5 \times 1.6) \\ &= 4.2 + 4.0 \\ &= 8.2 \text{ kN/m}^2 \end{aligned}$$

Loads on walls from uniform continuous floor and roof slabs are normally calculated on the assumption that the beams and slabs are simply supported by the walls.

5.4 CHARACTERISTIC COMPRESSIVE STRENGTH OF MASONRY, f_k

The characteristic compressive strength of masonry depends upon:

(a) the characteristic strength of the masonry unit
(b) the mortar designation
(c) the shape of the unit
(d) whether the work is bonded or unbonded

(e) the thickness of the mortar joints
(f) the standard of workmanship.

(a) Units
Typical characteristic strengths of masonry units are given in Table 5.2.

(b) Mortar
The designated grades of mortar given in BS 5628, Table 1, are provided in Table 5.3.

Table 5.2 Compressive strength of masonry units

	Compressive strength of unit (N/mm^2)		
	Bricks	Blocks	
	5.0	2.8	
	10.0	3.5	
	15.0	5.0	Typical structural units
	20.0	7.0	
Typically available values	27.5	10.0	
	35.0	15.0	Blocks of these strengths may not be
	50.0	20.0	readily available
	70.0	35.0	
	100.0		

Table 5.3 Mortar designations (BS 5628, Table 1)

Grade	Cement	Lime	Sand	Masonry cement	Sand	Cement	Sand with plasticiser
(i)	1	0–¼	3	—	—	—	—
(ii)	1	½	4–4½	1	2½–3½	1	3–4
(iii)	1	1	5–6	1	4–5	1	5–6
(iv)	1	2	8–9	1	5½–6½	1	7–8

(c) Unit shape

It has been found by experiment that, for a given unit strength, the larger the individual units which comprise the wall or column under consideration, the higher the strength of the resulting construction. This is reasonable when it is realised that the joints in properly designed masonry should be the weakest point of any construction, and the larger the units the fewer the number of mortar joints.

(d) Bond

The values quoted in the tables which follow (5.4, 5.5, 5.6 and 5.7), based on BS 5628, Table 2, are for 'normally bonded masonry'. This is taken to mean stretcher bond for single-leaf walls, and flemish or other equivalent bond for greater than single-leaf walls. It would appear that, provided one of the established bond patterns is adopted, variations in bond have little effect on the comparative strength of the wall or column.

(e) Joint thickness

The main structural role of the mortar in masonry construction is to provide a bedding between the units to ensure the uniform transfer of compressive stress. As stated previously, the mortar joint is the weakest element of masonry construction. Generally, the thicker the joints, the weaker the resulting structure. The conventional joint thickness is 10 mm. Reasonable variations on this thickness will not usually be critical. However, the evenness and line of the mortar joints provide a good guide to the quality of workmanship and, in the case of face work, are of aesthetic importance.

(f) Standard of workmanship

This, perhaps, is the most important factor to be considered because it can affect all the items listed previously. Control of the unit manufacture, under factory conditions, may be fairly sophisticated, but, control of workmanship on site is far more difficult. Account is taken of the possible variations in workmanship in the partial factors of safety. This will be discussed later in this chapter.

5.4.1 Brickwork

Table 5.4 gives the characteristic compressive strength, f_k, of normally bonded brickwork constructed with standard format bricks. The values quoted are for walls constructed under laboratory conditions, tested at an age of 28 days under axial compression in such a manner that the

Table 5.4 Characteristic compressive strength of masonry, f_k: standard format bricks (BS 5628, Table 2(a))

Mortar designation	Compressive strength of unit (N/mm^2)								
	5	10	15	20	27.5	35	50	70	100
(i)	2.5	4.4	6.0	7.4	9.2	11.4	15.0	19.2	24.0
(ii)	2.5	4.2	5.3	6.4	7.9	9.4	12.2	15.1	18.2
(iii)	2.5	4.1	5.0	5.8	7.1	8.5	10.6	13.1	15.5
(iv)	2.2	3.5	4.4	5.2	6.2	7.3	9.0	10.8	12.7

Table 5.5 Characteristic compressive strength of masonry, f_k: narrow walls – standard format bricks

Mortar designation	Compressive strength of unit (N/mm^2)								
	5	10	15	20	27.5	35	50	70	100
(i)	2.9	5.0	6.9	8.5	10.6	13.1	17.3	22.1	27.0
(ii)	2.9	4.8	6.1	7.4	9.1	10.8	14.0	17.4	20.9
(iii)	2.9	4.7	5.8	6.7	8.2	9.8	12.2	15.1	17.8
(iv)	2.5	4.0	5.0	6.0	7.1	8.4	10.4	12.4	14.6

Table 5.6 Characteristic compressive strength of masonry, f_k: narrow walls – modular bricks

Mortar designation	Compressive strength of unit (N/mm^2)								
	5	10	15	20	27.5	35	50	70	100
(i)	3.1	5.5	7.5	9.3	11.5	14.3	18.8	24.0	30.0
(ii)	3.1	5.3	6.6	8.0	9.9	11.8	15.3	18.9	22.8
(iii)	3.1	5.1	6.3	7.3	8.9	10.6	13.3	16.4	19.4
(iv)	2.8	4.4	5.5	6.5	7.8	9.1	11.3	13.5	15.9

Table 5.7 Characteristic compressive strength of masonry, f_k: modular bricks

Mortar designation	Compressive strength of unit (N/mm^2)								
	5	10	15	20	27.5	35	50	70	100
(i)	2.8	4.8	6.6	8.1	10.1	12.5	16.5	21.1	26.4
(ii)	2.8	4.6	5.8	7.0	8.7	10.3	13.4	16.0	20.0
(iii)	2.8	4.5	5.5	6.4	7.8	9.4	11.7	14.4	17.0
(iv)	2.4	3.9	4.8	5.7	6.8	8.0	9.9	11.9	14.0

effects of slenderness (see later) may be neglected. Linear interpolation within the table is permitted, and Figure 5.6(a) may be used for this purpose.

Where solid walls or the loaded inner leaf of a cavity wall are constructed in standard bricks with a thickness equal to the width of a single brick, the values of f_k obtained from Table 5.4 may be multiplied by 1.15. These enhanced values of f_k for so-called narrow brickwork have been computed and are given in Table 5.5. This increase is based on experimental results and relates to the absence of a vertical mortar joint parallel with the face within the thickness of the wall.

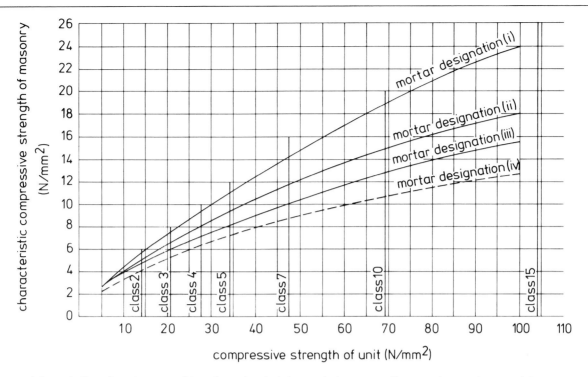

interpolation for classes of loadbearing bricks not shown on the graph may be used for average crushing strengths intermediate between those given on the graph, as described in clause 10 of BS 3921:1974 and clause 7 of BS 187:1978.

Figure 5.6(a) Characteristic compressive strength, f_k, of brick masonry (see Table 5.4)

With walls constructed in modular bricks (see Appendix 1), the values of f_k from Table 5.4 may be multiplied by 1.25 for narrow walls, as noted above, and 1.10 for a greater thickness of wall. The Code quotes these values as being applicable only for 90 mm wide and 90 mm high modular bricks complying with the requirements of BS 1180, or as detailed in DD 34 or DD 59. Designers should check with manufacturers if the units do not meet these requirements. These enhanced values of f_k for modular bricks have been computed and are given in Tables 5.6 and 5.7.

Where bricks are used which do not comply with the standard format or modular requirements, the values of f_k should be obtained from wall tests carried out in accordance with the procedures given in BS 5628.

Guidance on the choice of appropriate mortars is given in Appendix 1.

EXAMPLE 2
Determine the characteristic compressive strength of brickwork constructed with standard format bricks of unit compressive strength 15 N/mm² and a 1:1:6 mortar.

From Table 5.3 the mortar designation is (iii) and referring to Table 5.4 the characteristic compressive strength $f_k = 5.0$ N/mm².

5.4.2 Blockwork
When a wall is constructed in blockwork, the increased size of the units means that there are fewer joints than an equivalent wall built with bricks of the standard format. As the joints are generally the weakest part of a wall, a reduction in the number of joints per unit length or height of wall results in an increased compressive strength for a given strength of unit and mortar. The characteristic compressive strength of blockwork thus depends more on the shape factor of the units, that is the ratio of unit height to least horizontal dimension (see Figure 5.7).

The compressive strength also varies depending on whether the units are solid or hollow (see Appendix 1).

Values for the characteristic compressive strength of walls constructed with blocks having a shape factor of 0.6 are given in Table 5.8. Values for hollow block walls with a shape factor be-

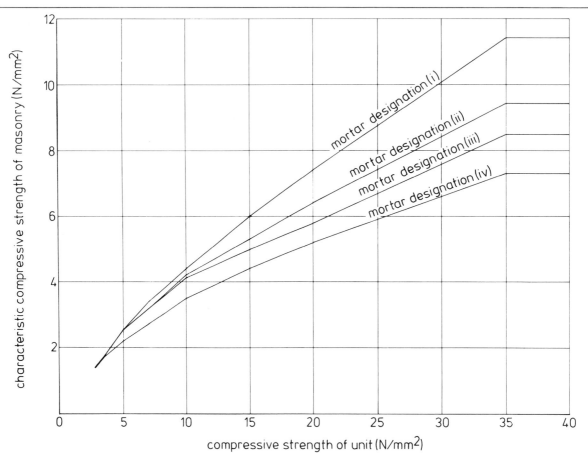

Figure 5.6(b) Characteristic compressive strength, f_k, of block masonry constructed of blocks having a ratio of height to least horizontal dimension of 0.6 (see Table 5.8)

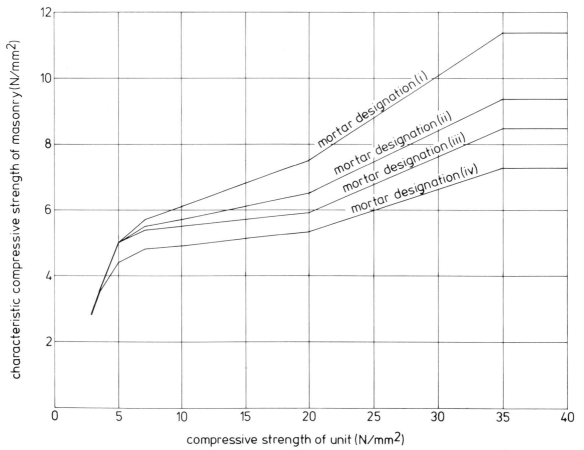

Figure 5.6(c) Characteristic compressive strength, f_k, of block masonry constructed of hollow blocks having a ratio of height to least horizontal dimension of 2.0 and 4.0 (see Table 5.9)

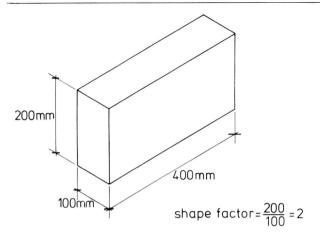

$$\text{shape factor} = \frac{200}{100} = 2$$

Figure 5.7

Table 5.9 Characteristic compressive strength of masonry, f_k: hollow blocks having a ratio of height to least horizontal dimension of between 2.0 and 4.0 (BS 5628, Table 2(c))

Mortar designation	Compressive strength of unit (N/mm^2)							
	2.8	3.5	5.0	7.0	10	15	20	35 or greater
(i)	2.8	3.5	5.0	5.7	6.1	6.8	7.5	11.4
(ii)	2.8	3.5	5.0	5.5	5.7	6.1	6.5	9.4
(iii)	2.8	3.5	5.0	5.4	5.5	5.7	5.9	8.5
(iv)	2.8	3.5	4.4	4.8	4.9	5.1	5.3	7.3

Table 5.10 Characteristic compressive strength of masonry, f_k: solid concrete blocks having a ratio of height to least horizontal dimension of between 2.0 and 4.0 (BS 5628, Table 2(d))

Mortar designation	Compressive strength of unit (N/mm^2)							
	2.8	3.5	5.0	7.0	10	15	20	35 or greater
(i)	2.8	3.5	5.0	6.8	8.8	12.0	14.8	22.8
(ii)	2.8	3.5	5.0	6.4	8.4	10.6	12.8	18.8
(iii)	2.8	3.5	5.0	6.4	8.2	10.0	11.6	17.0
(iv)	2.8	3.5	4.4	5.6	7.0	8.8	10.4	14.6

tween 2.0 and 4.0 are given in Table 5.9, and for solid block walls with similar shape factors in Table 5.10. Linear interpolation within the tables is permitted, and Figures 5.6(b) and (c) may be used for this purpose.

For values of f_k with units of shape factors between 0.6 and 2.0, interpolation should be made between Table 5.8 and 5.10 depending on whether solid or hollow blocks are being used.

Table 5.8 Characteristic compressive strength of masonry, f_k: blocks having a ratio of height to least horizontal dimension of 0.6 (BS 5628, Table 2(b))

Mortar designation	Compressive strength of unit (N/mm^2)							
	2.8	3.5	5.0	7.0	10	15	20	35 or greater
(i)	1.4	1.7	2.5	3.4	4.4	6.0	7.4	11.4
(ii)	1.4	1.7	2.5	3.2	4.2	5.3	6.4	9.4
(iii)	1.4	1.7	2.5	3.2	4.1	5.0	5.8	8.5
(iv)	1.4	1.7	2.2	2.8	3.5	4.4	5.2	7.3

When hollow blocks are used and completely filled with *in situ* concrete with a compressive strength, at the appropriate age, of not less than that of the blocks, they should be treated as solid blocks as noted above. The compressive strength of the hollow block should be based on the net area of the block and not the gross area. It will be seen in the Appendix dealing with materials that the compressive strength of hollow blocks is based on the gross area of the block for blocks in accordance with BS 2028, 1364.

EXAMPLE 3

Determine the characteristic compressive strength f_k of a wall constructed in hollow blocks (as shown in Figure 5.8) of compressive strength 7 N/mm^2, in accordance with BS 2028, 1364, if the blocks are filled with concrete having a 28 day compressive strength equal to that of the blocks and a mortar designation (iii) is used.

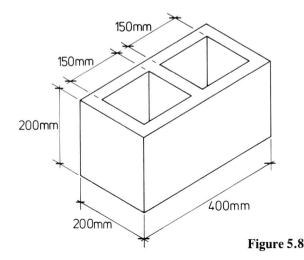

gross area $= 400 \times 200$
$= 80\,000\,\text{mm}^2$
nett area $=$ gross area $-$ openings
$= 80\,000 - (2 \times 150 \times 100)$
$= 50\,000\,\text{mm}^2$

Figure 5.8

Strength of infill concrete:
Unit strength 7 N/mm² based on gross area which in terms of crushing load per block

$$= \text{crushing strength} \times \text{gross area} \times \frac{1}{10^3} \, \text{kN}$$

$$= \frac{7 \times 80\,000}{10^3} = 560 \, \text{kN}$$

Therefore, crushing strength of unit based on net area

$$= \frac{\text{load} \times 10^3}{\text{net area}} \, \text{N/mm}^2$$

$$= \frac{560 \times 10^3}{50\,000} = 11.2 \, \text{N/mm}^2$$

Therefore strength of concrete $\geqslant 11.2$ N/mm², say, grade 15 mix to CP 110. The infilled block thus becomes equivalent to an 11.2 N/mm² solid unit for purposes of determining the f_k value.

$$\text{Shape factor of unit} = \frac{\text{height}}{\text{least horizontal dimension}}$$

$$= \frac{200}{200} = 1$$

f_k is then obtained by interpolation between Tables 5.8 and 5.9.

Assuming a 10 N/mm² unit for simplicity f_k, from Table 5.8, = 4.1 N/mm²; for mortar designation (iii) (shape factor 0.6) and from Table 5.9, f_k = 5.5 N/mm²; for mortar designation (iii) (shape factor 2.0–4.0), therefore for a shape factor of 1:

$$f_k = 4.1 + \frac{(5.5 - 4.1) \times (1.0 - 0.6)}{(2.0 - 0.6)}$$

$$= 4.5 \, \text{N/mm}^2$$

5.4.3 Natural stone masonry and random rubble masonry
BS 5628 recommends that natural stone masonry should be designed on the basis of solid concrete blocks of an equivalent compressive strength. Construction with natural stone masonry of a massive type with large well-dressed stones and relatively thin joints has a compressive strength more closely related to the intrinsic strength of the stone. Working stresses in excess of the maximum value quoted in the previous tables may be allowed in massive stone masonry, as described, provided the designer is satisfied that the properties of the stone and the method of wall construction warrant the increase.

As a guide, the Code suggests that the characteristic strength of random rubble masonry may be taken as being 95% of the corresponding strength of natural stone masonry built with similar materials. However, the structural design of stone masonry and random rubble walling requires great care, and designers should obtain detailed information on the particular design and construction methods for the type of wall being considered before attempting any detailed analysis.

5.4.4 Alternative construction techniques
The values for characteristic compressive strength quoted in the previous sections were for 'normally' bonded masonry and thus do not necessarily apply where the method of construction varies from 'normal'.

For aesthetic or other reasons, structural units may be laid other than on their normal bed faces, e.g. brick-on-edge or brick-on-end construction. The characteristic compressive strength in such an instance is obtained from the tables quoted, but using the compressive strength of unit as determined for the appropriate direction.

With reference to Figure 5.9 the characteristic strength of masonry for brick-on-end construction as in (A) is determined from Table 5.3, using

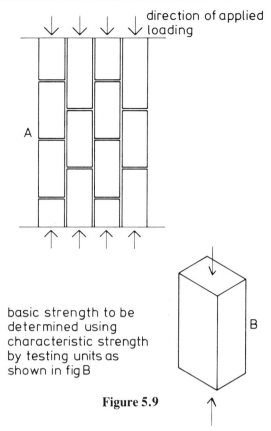

basic strength to be determined using characteristic strength by testing units as shown in fig B

Figure 5.9

the compressive strength of unit obtained by testing the bricks as shown in (B).

As explained in Appendix 1, the compressive strength of hollow blocks and perforated bricks is determined by dividing the load at failure by the gross plan area of the unit. This is done so that the calculation of the strength of walls of solid and hollow units may be carried out in an identical manner, i.e. load × plan area of wall. However, where walls are constructed of hollow units, and sometimes when using perforated bricks, the units are laid with mortar on the two outer edges of the bed face only. This is termed shell bedding (see Figure 5.10).

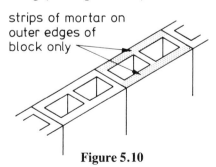

strips of mortar on outer edges of block only

Figure 5.10

In walls constructed in this way, the characteristic compressive strength should be obtained in the normal manner, but the design strength of the wall should be reduced by the ratio of the bedded area to the gross area of block (see Figure 5.11).

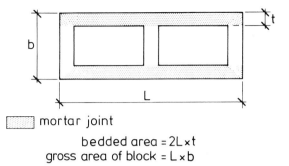

mortar joint

bedded area = 2L×t
gross area of block = L×b

Figure 5.11

The design strength of the wall is thus equal to the f_k obtained from Tables 5.8, 5.9, 5.10, as appropriate, multiplied by a factor equal to the bedded area divided by the gross area.

Thus the characteristic compressive strength of the shell bedded wall

$$= f_k \frac{2l \times t}{l \times b}$$

where f_k = characteristic compressive stress determined in accordance with section 5.1.

5.5 PARTIAL SAFETY FACTORS FOR MATERIAL STRENGTH, γ_m

The degree of care exercised in the control of the manufacture of the units and in the construction of the masonry affects the design strength of the wall, column, etc. The characteristic strength of masonry has to be divided by a partial safety factor to obtain the *design* strength. The partial safety factor depends on the degree of quality control on manufacture and construction. BS 5628 recognises two categories of control, namely 'normal' and 'special'.

5.5.1 Manufacturing control (BS 5628, clause 27.2.1)

(a) Normal category
Used when the supplier can meet the compressive strength requirements of the appropriate British Standard.

(b) Special category
Used when suppliers can meet a specified strength limit (known as the 'acceptance limit') when not more than 2½% (as opposed to the 5% in section 5.2) of the test results will fall below the acceptance limit, and also when the supplier's quality control scheme can satisfy the buyer that the acceptance limit is consistently met. This may be assumed where the unit manufacturer com-

plies with the following two quality control procedures:

(i) The units supplied are to have a specified strength limit termed the 'acceptance limit' for compressive strength, such that the average compressive strength of a sample of units from any consignment selected and tested in accordance with the appropriate British Standard Specification has a probability of not more than 2½% of being below this acceptance limit. Thus for any particular unit, the specified strength limit, i.e. the 'acceptance limit', may be set for example at 7 N/mm². Therefore, in any sample of these units, the average strength must only have a 1 in 40 chance of falling below this acceptance limit.

(ii) A quality control scheme must be operated to show that the requirements of (i) are consistently satisfied. The results of the quality control scheme should be made available to the purchaser or his representative, i.e. the designer.

5.5.2 Construction control

(a) Normal category
Used in normal construction where the requirements of the special category are not satisfied.

(b) Special category
Used when:

(i) Supervision, by frequent site visits by the designer or by a permanent representative on site, ensures that the work is built in accordance with CP 121 and the designer's specification.

(ii) Preliminary and regular mortar compressive strength tests in accordance with Appendix A of BS 5628 show that the mortar complies consistently with the strength requirements of Table 5.3.

The partial safety factors on material strength are tabulated in Table 5.11. These values do not vary for ultimate or serviceability limit state design, but may vary when considering the effects of misuse or accident (see Chapter 8).

Table 5.11 Partial factors of safety on materials (BS 5628, Table 4)

		Category of construction control	
		Special	Normal
Category of manufacturing control	Special	2.5	3.1
	Normal	2.8	3.5

EXAMPLE 4

Determine the *design* strength of the brickwork in Example 3 (section 5.4.2) if the manufacturing and construction controls are normal category.

$$\text{Design strength} = \frac{f_k \text{ (characteristic compressive strength)}}{\gamma_m \text{ (partial safety factor for materials)}}$$

$$= \frac{5.0}{3.5} \text{ N/mm}^2$$

$$= 1.43 \text{ N/mm}^2$$

EXAMPLE 5

Determine the design strength of the brickwork in Example 3 (section 5.4.2), the manufacturing control is normal and the construction control is special category.

$$\text{Design strength} = \frac{5.0}{2.8} \text{ N/mm}^2$$

$$= 1.78 \text{ N/mm}^2$$

5.6 SLENDERNESS RATIO

Slender masonry walls and columns under compressive loading are likely to buckle in the same way as concrete, steel or timber columns in compression. It is, therefore, necessary to determine the masonry wall or column's slenderness ratio in order to relate a failure in buckling to the compressive load carrying capacity of a wall or column.

The slenderness ratio (SR) of masonry walls or columns is defined in BS 5628 as:

$$\frac{\text{effective height (or length)}}{\text{effective thickness}} = \frac{h_{\text{ef}} \text{ or } l_{\text{ef}}}{t_{\text{ef}}}$$

(The term 'effective' is dealt with later.) The effective length should be used where this gives the lesser SR.

EXAMPLE 6

A wall has an effective height of 2.25 m and an effective thickness of 102.5 mm. Determine its SR:

$$SR = \frac{\text{effective height}}{\text{effective thickness}}$$

$$= \frac{2250}{102.5} \text{ mm}$$

$$= 22$$

This concept is satisfactory for walls and other sections which are rectangular on plan. However, it is difficult to apply to other geometrical configurations. The Code gives no guidance for such instances, but, for diaphragm and fin walls, guidance is given in the relevant sections of this book.

The slenderness ratio is a measure of the tendency of a member under compressive loading to fail by buckling before failure by crushing occurs. The greater the slenderness ratio, the greater the tendency for the member to fail by buckling, and thus the lower the loadbearing capacity of the member.

BS 5628 (clause 28.1) generally limits the slenderness ratio to not greater than 27. However, in the case of walls less than 90 mm thick in buildings of more than two storeys, this value should not exceed 20. This exception to the general rule should be interpreted as meaning thin walls which are *continuous* for more than two storeys.

5.7 HORIZONTAL AND VERTICAL LATERAL SUPPORTS

The effective heights and effective lengths used to determine the slenderness ratio of an element are determined from the actual height or length, which is then modified, depending on the restraint conditions, i.e. the manner in which the member is restrained from buckling by adjacent members. In the case of walls, a further factor may be included to allow for the additional stiffening effect of any piers. The support provided by adjacent members will generally be at right angles to the member and is thus termed a 'lateral' support. In the case of the slenderness ratio being determined from the effective height, the lateral supports (such as floors and roofs) at right angles to the member are acting in a horizontal plane, and are thus termed horizontal lateral supports.

In the case of the slenderness ratio being determined from the effective length, the lateral supports at right angles to the member are acting in a vertical plane, and are thus termed vertical lateral supports, as illustrated in Figure 5.12.

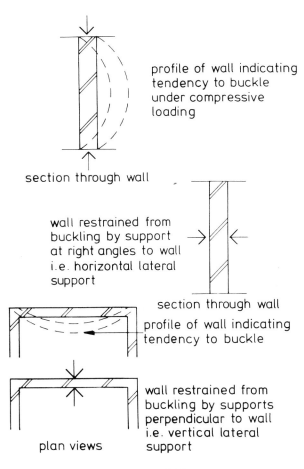

profile of wall indicating tendency to buckle under compressive loading

section through wall

wall restrained from buckling by support at right angles to wall i.e. horizontal lateral support

section through wall

profile of wall indicating tendency to buckle

wall restrained from buckling by supports perpendicular to wall i.e. vertical lateral support

plan views

Figure 5.12

To be considered as a lateral support for the purposes of assessing the effective height, or length, the lateral support should be capable of transmitting to the supporting structure (i.e. the walls or other elements providing lateral stability to the structure as a whole) the sum of the following forces:

(a) the simple static reactions to the total applied design horizontal forces at the line of lateral support, and
(b) 2½% of the total design vertical load (dead plus imposed) that the wall, or other element, is designed to carry at the line of lateral support.

EXAMPLE 7

For the wall shown in Figure 5.13, determine the loading on the support at mid-height for this to be considered as a horizontal lateral support.

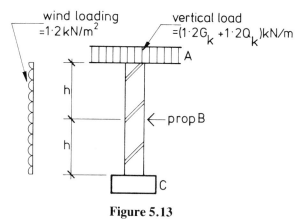

Figure 5.13

Horizontal lateral support must be capable of resisting:

(a) simple static reactions

i.e. $1.2W_k \times h = 1.2W_k h$ kN/m

plus

(b) 2½% of total design vertical load,

i.e. $\dfrac{2.5}{100} \times (1.2G_k + 1.2Q_k + \text{self-weight of wall from A to B})$ kN/m

N.B.

(i) The wind loading is capable of reversal, and thus in practice, both a prop and a tie are required.
(ii) The loading on the prop from the wind must be transferred via floors, walls, etc., to the foundations.

5.7.1 Methods of compliance: walls – horizontal lateral supports

There are many methods of providing horizontal lateral supports, some are given in BS 5628 and will be described later. The Code separates several forms of support which are defined as providing 'enhanced' resistance to lateral movement. The remainder only provide 'simple' resistance to lateral movement. The details suggested in the Code are given in Figure 5.14. Those details providing 'enhanced' restraint are defined as follows and illustrated in Figure 5.15:

(a) Floors or roofs of any form of construction which span onto the wall, or other element, from both sides at the same level.
(b) An *in situ* concrete floor or roof, or a precast concrete floor or roof, giving equivalent restraint, irrespective of the direction of span, having a bearing of at least one half the thickness of the wall, or other element, or 90 mm, whichever is greater.
(c) For houses of not more than three storeys, a timber floor spanning onto a wall from one side only, but having a bearing of not less

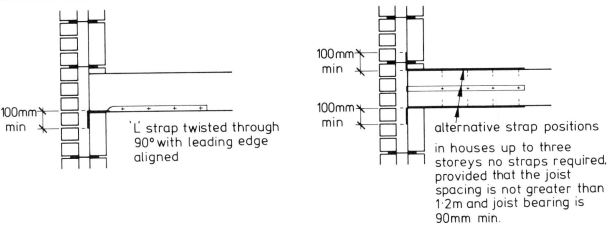

Figure 5.14(a) Timber floor bearing directly onto wall

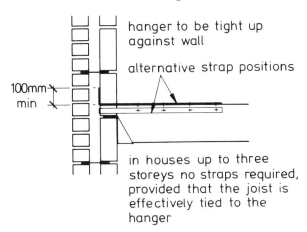

Figure 5.14(b) Timber floor using typical joist hanger

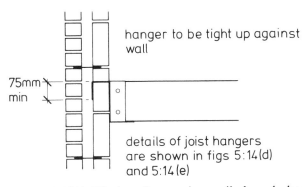

Figure 5.14(c) Timber floor using nailed or bolted joist hangers acting as a tie

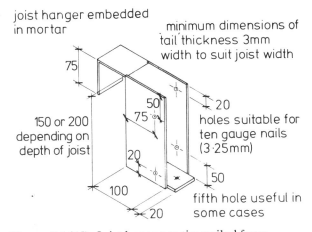

Figure 5.14(d) Joist hanger as tie: nailed form

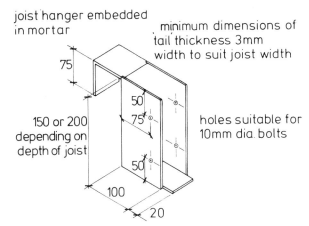

Figure 5.14(e) Joist hanger as tie: bolted form

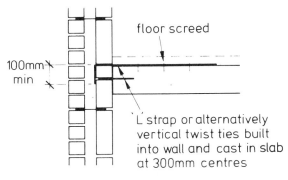

Figure 5.14(f) *In situ* concrete floor abutting external cavity wall

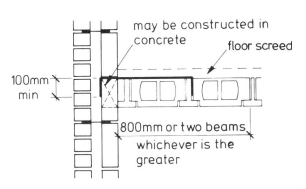

Figure 5.14(g) Beam and pot floor abutting external cavity wall

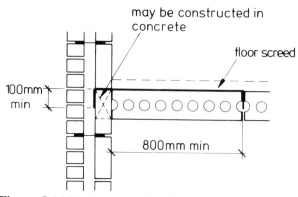

Figure 5.14(h) Precast units abutting external cavity wall

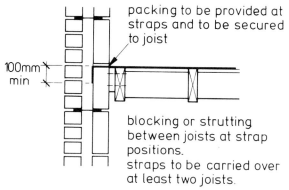

Figure 5.14(j) Timber floor abutting external cavity wall

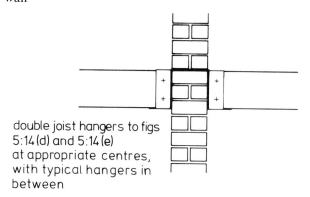

double joist hangers to figs 5:14 (d) and 5:14 (e) at appropriate centres, with typical hangers in between

Figure 5.14(k) Timber floor using double joist hanger acting as tie

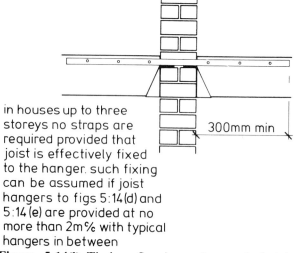

in houses up to three storeys no straps are required provided that joist is effectively fixed to the hanger. such fixing can be assumed if joist hangers to figs 5:14 (d) and 5:14 (e) are provided at no more than 2m% with typical hangers in between

Figure 5.14(l) Timber flooring using typical joist hanger

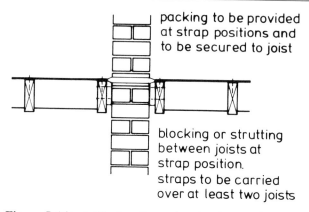

packing to be provided at strap positions and to be secured to joist

blocking or strutting between joists at strap position. straps to be carried over at least two joists

Figure 5.14(m) Timber floor abutting internal wall

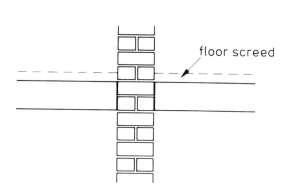

floor screed

Figure 5.14(n) *In situ* floor abutting internal wall

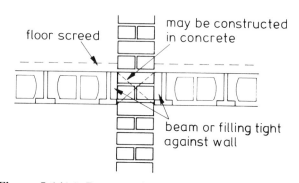

floor screed

may be constructed in concrete

beam or filling tight against wall

Figure 5.14(p) Beam and pot floor abutting internal wall

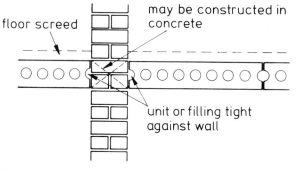

floor screed

may be constructed in concrete

unit or filling tight against wall

Figure 5.14(q) Precast units abutting internal wall

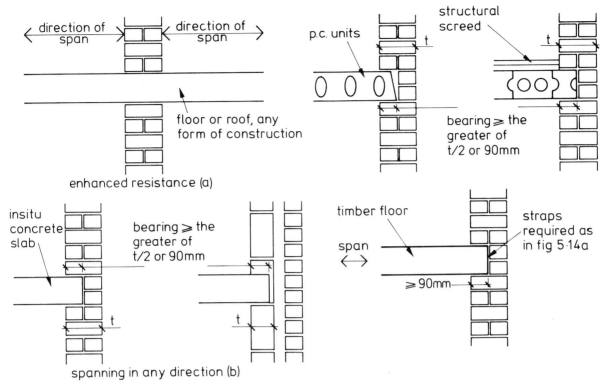

Figure 5.15 Details providing 'enhanced' resistance

than 90 mm. The Code does not state explicitly in this clause that metal restraint straps are required in this situation, but, this is implied and certainly recommended by the authors.

It should be emphasised that, when considering the above conditions, the restraining member, i.e. floor or roof construction, must be capable of resisting the required loading given in 5.7, and transmitting the static reactions to other members providing the overall stability. This is particularly important in the case of timber trussed rafters.

The requirement that a precast concrete floor, or roof, should provide equivalent restraint to that of an *in situ* slab, needs special attention, especially if a 'beam and pot' construction is to be adopted without a structural screed. The precast units must be capable of resisting the required loading and transferring the forces back to the stabilising walls.

5.7.2 Methods of compliance: walls – vertical lateral supports
Vertical lateral supports are, as with horizontal lateral supports, classed as those providing simple resistance and those providing enhanced resistance. Such vertical supports must be capable of resisting the forces defined in (a) and (b) of 5.7, and also of transmitting the static reactions

to suitable foundations. Once again, the Code gives 'rule of thumb' details, and allows that, in other cases, suitable lateral supports may be confirmed by calculation. There appear to be no criteria, however, for verifying that a support is in fact an enhanced support within the terms of the Code. The Code requirements are illustrated in Figures 5.16 and 5.17 and defined as follows:

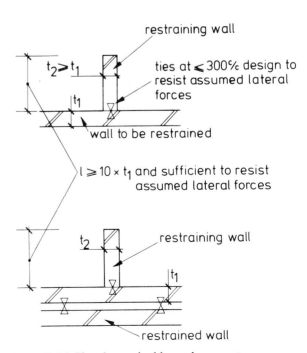

Figure 5.16 Simple vertical lateral supports

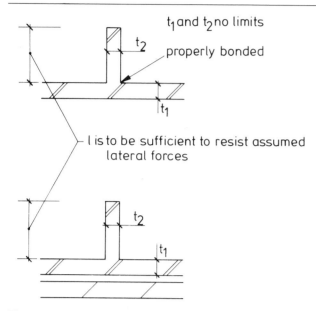

Figure 5.17 Vertical lateral supports providing 'enhanced' resistance

(a) Simple resistance may be assumed where an intersecting or return wall, not less than the thickness of the supported wall, extends from the intersection at least ten times the thickness of the supported wall, and is connected to it by metal anchors (ties) designed to resist the assumed lateral forces (i.e. simple static reactions and 2½% of the vertical load, as defined previously). The Code also stipulates that the ties should be evenly distributed throughout the height, at not more than 300 mm centres. In the case of cavity wall construction, the thickness of the intersecting or return wall should be not less than the thickness of the loadbearing leaf, and should extend from the intersection at least ten times the thickness of the loadbearing leaf. If both leaves are loadbearing, then it is reasonable to assume that the construction is adequate if the requirements are met for one leaf only, or for the thicker leaf if different thicknesses are used.
(b) Enhanced resistance may be assumed where an intersecting or return wall is properly bonded to the supported wall or leaf of a cavity wall. There are no specific requirements with respect to the thickness of the two walls.

On the basis of this and later Code requirements, it would appear that enhanced resistance to lateral movement may be assumed where calculation verifies that a restraint provides the equivalent rotational resistance, or continuity, as would properly bonded construction.

5.8 EFFECTIVE HEIGHT OR LENGTH: WALLS

As stated in 5.7 the effective height or length, used in the determination of the slenderness ratio of an element, is based on the actual height or length which is modified depending on the restraint conditions. Examples of simple and enhanced restraints have been given in the previous section, and the effective height or length may be determined in accordance with BS 5628 from the actual height or length as follows:

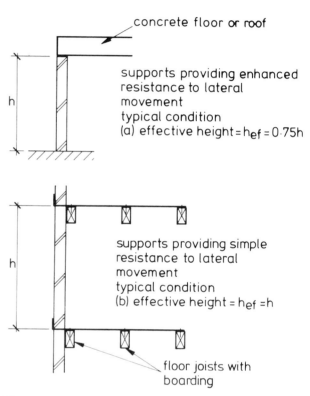

Figure 5.18 Examples of effective heights for walls

Effective height =
either

(a) 0.75 times the clear distance between horizontal lateral supports which provide enhanced resistance to lateral movement

or

(b) the clear distance between horizontal lateral supports which provide simple resistance to lateral movement. (See Figure 5.18.)

Effective length =
either

(a) 0.75 times the clear distance between vertical lateral supports which provide enhanced resistance to lateral movement

$$\text{Effective length} = l_{ef} = 0.75L$$

or
(b) twice the distance between a vertical lateral support which provides enhanced resistance to lateral movement, and a free edge

or

(c) the clear distance between vertical lateral supports which provide simple resistance to lateral movement

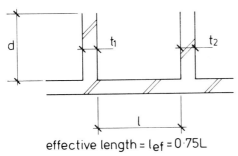

effective length = l_{ef} = 0·75L

Figure 5.19(a) Typical condition (a) supports providing enhanced resistance to lateral movement

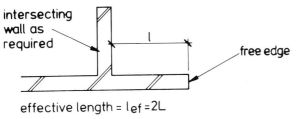

effective length = l_{ef} = 2L

Figure 5.19(b) Typical condition (b) support providing enhanced resistance to lateral movement and a free edge

(d) 2.5 times the distance between a vertical lateral support which provides simple resistance to lateral movement, and a free edge. (See Figure 5.19.)

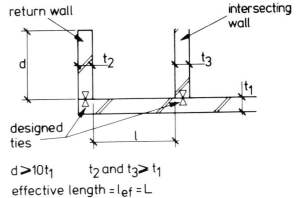

$d > 10t_1$ t_2 and $t_3 > t_1$
effective length = l_{ef} = L

Figure 5.19(c) Typical condition (c) supports providing simple resistance to lateral movement

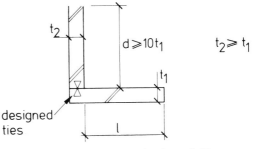

effective length = l_{ef} = 2·5l

Figure 5.19(d) Typical condition (d) support providing simple resistance to lateral movement

EXAMPLE 8

Determine the slenderness ratio for the wall as shown in Figure 5.20, assuming t_{ef} = 102.5 mm.

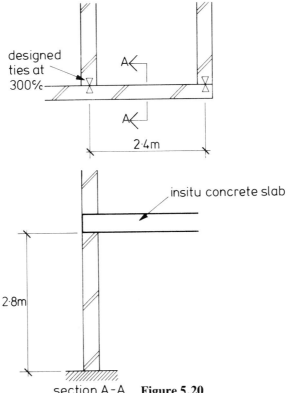

section A-A   **Figure 5.20**

Slenderness ratio, SR, = lesser of $\dfrac{h_{ef}}{t_{ef}}$ and $\dfrac{l_{ef}}{t_{ef}}$.

$h_{ef} = 0.75h = 0.75 \times 2.8 = 2.1$ m
$t_{ef} = 102.5$ mm

Therefore, based on effective height SR = 20.5.

$l_{ef} = L = 2.4$, based on condition (c)

Therefore, based on effective length SR = 23.4

Therefore, use lesser SR = 20.5.

5.9 EFFECTIVE THICKNESS OF WALLS
5.9.1 Solid walls

For a solid wall not stiffened by intersecting or return walls, the effective thickness is equal to the actual thickness. Where a solid wall is stiffened by piers, the effective thickness t_{ef} is increased by an amount depending on the stiffening effect of the piers, i.e. their size and spacing.

$$t_{ef} = t \times K$$

where

   t = the actual thickness of the solid wall
   K = the appropriate stiffness factor as given in Table 5.12.

Table 5.12 (BS 5628, Table 5)

Ratio of pier spacing (centre to centre) to pier width	Ratio: $\dfrac{\text{pier thickness}}{\text{wall thickness}} = \dfrac{t_p}{t}$		
	1	2	3
6	1.0	1.4	2.0
10	1.0	1.2	1.4
20	1.0	1.0	1.0

Note: Linear interpolation is permissible, but not extrapolation.

EXAMPLE 9

Determine the effective thickness of the wall shown in Figure 5.21.

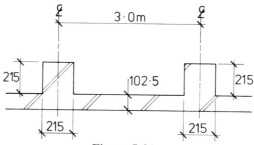

Figure 5.21

$$\frac{\text{Pier spacing c/c}}{\text{Pier width}} = \frac{3000}{215} = 13.95$$

$$\frac{\text{Pier thickness } (t_p)}{\text{Wall thickness } (t)} = \frac{215}{102.5} = 2.1$$

From Table 5.12, K lies between 1.2 and 1.0.

By interpolation

$$K = 1.2 - \frac{(1.2 - 1.0)}{10} \times 13.95$$

$$= 0.92$$

Thus

$$t_{ef} = 0.92 \times 102.5 = 94.3 \text{ mm}$$

Where a solid wall is stiffened by intersecting walls, the stiffening coefficient as obtained from Table 5.12 may be used to calculate the effective thickness, in the same way as for a wall stiffened by piers. For the purposes of calculating the effective thickness, the intersecting walls are assumed to be equivalent to piers whose widths are equal to the thickness of the intersecting walls, and of thickness equal to three times the thickness of the stiffened wall (see Figure 5.22).

EXAMPLE 10

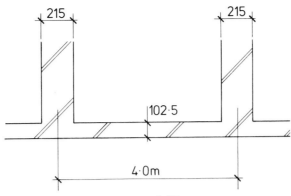

Figure 5.22

$$\frac{\text{Equivalent pier spacing}}{\text{Equivalent pier width}} = \frac{4000}{215} = 18.6$$

$$\frac{\text{Equivalent pier thickness}}{\text{Wall thickness}} = \frac{3 \times 102.5}{102.5} = 3$$

(note this value will be 3 in all cases)

5.9.2 Cavity walls

The addition of another leaf of masonry, joined only with ties to a solid wall, will obviously reduce the tendency of the solid wall to buckle, whether only one or both of the leaves are loadbearing. However, two leaves of masonry separated by a cavity will not produce a wall of equal stiffness to that formed by properly bonding the two leaves together. The effective thicknesses of a cavity wall is thus taken to be two-thirds of the sum of the thickness of the two leaves. Where one of the leaves is thicker than the other, the value for the effective thickness may (on that basis) be less than that of the thicker leaf alone and, in this case, the effective thickness of the cavity wall should be taken as the actual thickness of the thicker leaf.

EXAMPLE 11

Determine the effective thickness of the cavity wall shown in Figure 5.23.

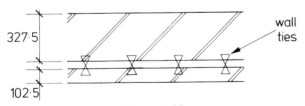

Figure 5.23

Actual thickness of leaves   = 327.5 mm and 102.5 mm
Effective thickness of combined leaves = ⅔ (327.5 + 102.5)
                                       = 286.7 mm

But this combined effective thickness is less than the actual thickness of the thicker leaf alone.

Thus, t_{ef} for SR calculation = 327.5 mm.

Where one of the leaves of a cavity wall is stiffened by piers, the effective thickness of this leaf should be calculated as described in 5.9.1. The effective thickness of the cavity wall is then determined, as set out above, using the increased value for the effective thickness of the leaf stiffened with piers.

EXAMPLE 12

Determine the effective thickness of the cavity wall shown in Figure 5.24.

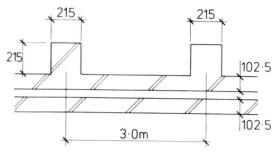

Figure 5.24

From Example 9, the effective thickness of the leaf with piers is t_{ef} = 115 mm.
Thus the effective thickness of the cavity wall = ⅔ (115 + 102.5) = 145 mm.

5.10 LOADBEARING CAPACITY REDUCTION FACTOR, β

In section 5.6, the tendency of structural materials to buckle under compressive loading was considered, and methods for determining the slenderness ratio for various types of wall were given in subsequent sections. The slenderness ratio is a measurement of how slender a wall is, which in turn, is a measure of the tendency to buckle when subjected to compressive loading. The higher the wall's tendency to buckle, the lower the potential strength of the wall, since it will fail by buckling before failing due to crushing of the units or joints.

Even if the characteristic strengths of the masonry in walls (1) and (2) in Figure 5.25 are equal, wall (2) will still have the higher loadbearing capacity as it is less likely to buckle than wall (1).

The design strength of a wall or column must therefore be reduced by a factor, termed the capacity reduction factor β depending on the slenderness ratio of the wall or column. Values for β are given in BS 5628 and reproduced in Table 5.13.

Table 5.13 Capacity reduction factor, β (BS 5628, Table 7, part only)

Slenderness ratio h_{ef}/t_{ef}	β
0	1.00
6	1.00
8	1.00
10	0.97
12	0.93
14	0.89
16	0.83
18	0.77
20	0.70
22	0.62
24	0.53
26	0.45
27	0.40

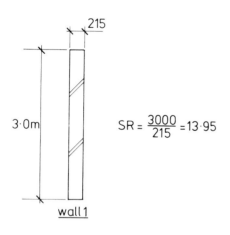

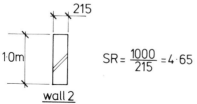

Figure 5.25

It should be noted that these values for β, which are, in fact, maximum or worst case values, are strictly correct only for the central fifth of the member. Over the remaining height of the member, β varies between this value and unity at restraints (see BS 5628, Appendix B).

5.11 DESIGN COMPRESSIVE STRENGTH OF A WALL

Having determined the characteristic strength of masonry, f_k, the relevant partial factor of safety for materials, γ_m, and the capacity reduction factor, β, the design strength of a wall may now be calculated. The design compressive strength is given by the product of the capacity reduction factor and the characteristic compressive strength divided by the partial safety factor for materials. This may be expressed as follows: $\beta f_k/\gamma_m$, which will give the strength as a force per unit area of wall. The design strength per unit length of wall is thus given by the formula:

$$\text{design strength per unit length} = \frac{\beta t f_k}{\gamma_m}$$

which is the expression given in BS 5628 (clause 32.2.1),
where

β = capacity reduction factor (see section 5.10),

f_k = characteristic strength of masonry (see section 5.4),

γ_m = relevant partial safety factor on materials (see section 5.5),

t = actual thickness of the wall.

EXAMPLE 13

A wall has an effective height of 2.25 m and an effective thickness of 102.5 mm (Example 6). The brick strength is 15 N/mm^2 and the mortar mix is 1:1:6 (Examples 2 and 3). The manufacturing control is normal and the construction controls are special.

Determine (a) the design strength of the wall
          (b) the loadbearing capacity of the wall.

$$\frac{f_k}{\gamma_m} = 1.78 \text{ N/mm}^2 \text{ (Example 5)}$$
$$\beta = 0.62 \text{ (Example 6 and Table 5.13)}$$

$$\text{Design strength} = \frac{\beta f_k}{\gamma_m}$$
$$= 0.62 \times 1.78$$
$$= 1.10 \text{ N/mm}^2$$

$$\text{Design load} = \text{stress} \times \text{area}$$
$$= 1.10 \times 102.5$$
$$= 112.75 \text{ kN/m run}$$

EXAMPLE 14

Determine the loadbearing capacity of the wall given in Example 13, when both the manufacturing and construction controls are normal.

$$\frac{f_k}{\gamma_m} = 1.43 \text{ N/mm}^2 \text{ (Example 4)}$$

$$\text{Design strength} = \frac{\beta f_k}{\gamma_m}$$
$$= 0.62 \times 1.43$$
$$= 0.88 \text{ N/mm}^2$$

$$\text{Design load} = 0.88 \times 102.5$$
$$= 90.20 \text{ kN/m run}$$

5.12 COLUMNS

In terms of loadbearing masonry subject to axial compressive loading, a column is only a special case in the design of walls. A column is defined in BS 5628 as an isolated vertical loadbearing member whose width is not more than four times its thickness (see Figure 5.26).

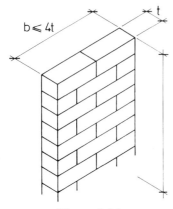

Figure 5.26

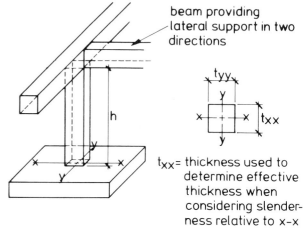

t_{xx}= thickness used to determine effective thickness when considering slenderness relative to x-x

effective height relative to x-x axis = $h_{ef_{xx}}$ = h
effective height relative to y-y axis = $h_{ef_{yy}}$ = h

5.12.1 Slenderness ratio: columns

As an isolated member, a column does not gain from the lateral support provided in the longitudinal direction by the adjacent elements of a wall. The slenderness ratio of a column must, therefore, be checked in two directions, and the worst case used to determine the design strength.

The effective height of a column is defined as the distance between lateral supports, or twice the actual height in respect of a direction in which lateral support is not provided. This is illustrated in Figure 5.27 and Table 5.14.

The effective thickness of a solid column is equal to the actual thickness, t, relative to the direction being considered. The effective thickness of a

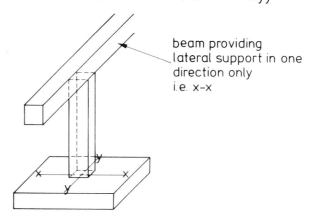

effective height relative to x-x axis = $h_{ef_{xx}}$ = h
effective height relative to y-y axis = $h_{ef_{yy}}$ = 2h

Figure 5.27

Table 5.14

End condition	Type of restraint		Effective height, h_{ef}
Column restrained at least against lateral movement top and bottom		Floor or roof of any construction spanning onto column from both sides at the same level	h in respect of both axes
		Concrete floor or roof, irrespective of direction of span, which has a bearing of at least $\frac{2}{3}t$ but not less than 90 mm	h in respect of both axes
Column restrained against lateral movement at top and bottom by at least two ties 30 × 5 mm min. at not more than 1.25 m centres		No bearing or bearing less than case above	h in respect of minor axis
		Floor or roof of any construction irrespective of direction of span	$2h$ in respect of major axis

cavity column, perpendicular to the cavity, is taken as two-thirds of the sum of the leaf thicknesses, or the actual thickness of the thicker leaf, whichever is the greater. In the other direction, the effective thickness is equal to the plan length of each leaf (see Figure 5.28). As in the case of walls, the slenderness ratio of a column about either axis is restricted to not more than 27.

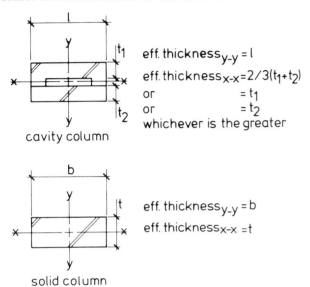

eff. thickness$_{y-y}$ = l

eff. thickness$_{x-x}$ = 2/3(t$_1$+t$_2$)
or   = t$_1$
or   = t$_2$
whichever is the greater

cavity column

eff. thickness$_{y-y}$ = b

eff. thickness$_{x-x}$ = t

solid column

Figure 5.28

In respect of isolated columns, the Code does not specifically provide for any reduction in the effective height if enhanced lateral supports are provided. It would seem reasonable, however, at least in the case of an *in situ* concrete roof or floor slab, or precast concrete slab providing equivalent restraint, that enhanced lateral support may be assumed and the effective height modified accordingly. A similar assumption may be made in certain cases of: 'floors or roofs of any form of construction which span onto the column from both sides at the same level'. However, for this to be valid, the floor or roof construction must have sufficient rigidity perpendicular to the span to provide resistance to the assumed forces as noted in section 5.7. The effective height of a column may also be taken as 0.75 times the clear distance between lateral supports which provide enhanced resistance to lateral movement as shown in Figure 5.27. This, though, does not comply with BS 5628.

5.12.2 Columns formed by openings
Most walls contain door, window or some other form of opening and these are often close together so that a section of the wall between the adjacent openings becomes very narrow. In cases where this section of walling is by definition a column, i.e. width not more than four times its

thickness, the effective height relative to an axis perpendicular to the wall will, due to the reduced restraint offered by the remaining section of wall, be less than that of a completely isolated member but greater than that of a continuous wall.

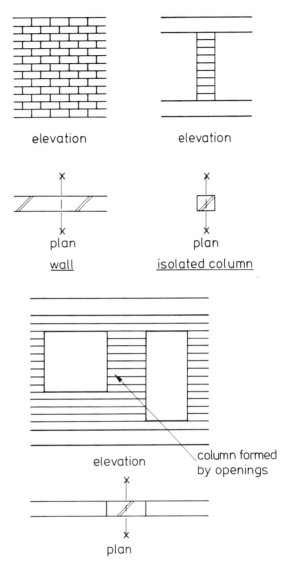

elevation   elevation

plan   plan

wall   isolated column

elevation   column formed by openings

plan

Figure 5.29

This is illustrated in Figure 5.29, from which it can be seen that for consideration of slenderness relative to an axis x–x, the effective height of the column formed by the openings will be greater than for the wall, but less than for the isolated member. The assessment of the effective height will vary depending on the size of columns, etc. Thus each case should be considered separately. A conservative approach would be to treat such a section as an isolated column. The Code BS 5628 gives two general recommendations for the assessment of the effective height as follows:

'(a) Where simple resistance to lateral movement of the wall containing the column is provided, the effective height should be taken as the distance between supports.'

'(b) Where enhanced resistance to lateral movement of the wall containing the column is provided, the effective height should be taken as 0.75 times the distance between the supports plus 0.25 times the height of the taller of the two openings.'

This is illustrated in Figure 5.30 and means that if one of the openings is full height, the effective height will again be equal to the distance between lateral restraints. It is important to note that, even if the wall containing the column is provided with adequate restraint, the column itself may not be, particularly if both the openings continue up to the level of the lateral restraints (see Figure 5.31).

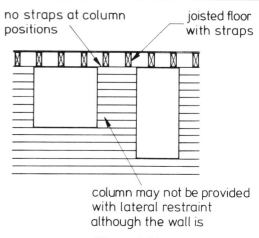

Figure 5.31

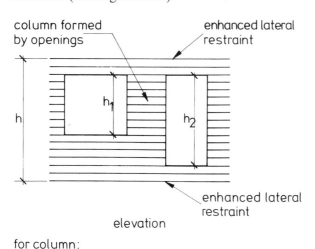

for column:
$$\text{effective height} = h_{ef} = 0.75h + 0.25h_2$$

Figure 5.30

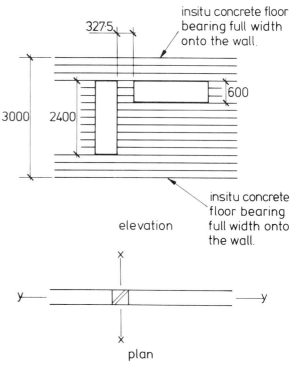

Figure 5.32

EXAMPLE 15

Determine in accordance with BS 5628, the slenderness ratio of the column formed by openings in the wall shown in Figure 5.32.

Effective heights:
$$h_{ef\,xx} = 0.75 \times 3000 + 0.25 \times 2400$$
$$= 2850$$
$$h_{ef\,yy} = 3000$$

Effective thicknesses:
$$t_{ef\,xx} = 327.5$$
$$t_{ef\,yy} = 215$$

Slenderness ratios:
$$SR_{xx} = \frac{2850}{327.5} = 8.7$$
$$SR_{yy} = \frac{3000}{215} = 13.9$$

Thus the slenderness ratio for the column is 13.9 as this is the worst case.

5.12.3 Design strength

The design compressive strength of a column is given by the product of the capacity reduction factor, the column area and the characteristic compressive strength of masonry, divided by the relevant partial safety factor for materials. This may be expressed as follows:

$$\beta \, bt \, \frac{f_k}{\gamma_m}$$

where

b = width of column
t = depth of column

Therefore

bt = cross-sectional area.
β = capacity reduction factor (see section 5.10)
f = characteristic compressive strength (see section 5.4)
γ_m = partial factor of safety for materials (see section 5.5).

See also section 5.12.4.

EXAMPLE 16

Determine the design compressive strength of a column, 440 mm × 440 mm, 4.4 m clear height between concrete floors giving enhanced lateral restraint. The bricks have a compressive strength of 35 N/mm^2, and the mortar is designation (ii). The manufacturing and construction controls are normal.

$$\text{Slenderness ratio} = \frac{h_{ef}}{t_{ef}} = \frac{4.4}{0.44} = 10$$

Therefore $\beta = 0.97$.

Characteristic compressive strength of masonry:

$f_k = 9.4 \, N/mm^2$ (see section 5.12.4),

$\gamma_m = 3.5$

Therefore

$$\text{design strength} = 0.97 \times 440 \times 440 \times \frac{9.5}{3.5} \times 10^{-3}$$

$$= 509 \, kN$$

5.12.4 Columns or walls of small plan area

In a member whose plan area is relatively small, the number of individual units available to support the loading is less than in the case of a wall. A wall may consist of, say, fifty units, of which some will be of greater compressive strength than others, and some of lesser compressive strength, with a reasonable spread of values about the mean value. A column may be required to support a similar load per unit area, but may consist of only, say, four units on plan. If all the four units in the column are of similar strength, which may be comparatively low, the effect on the design strength of the column would be greater than if four units in the wall were of a similarly low strength. The probability of a relatively lower strength of units is greater for the column than the wall. It is necessary, therefore, to adjust the design strength of a column or wall of small plan area, to ensure that the probability of failure is similar to that of a normal wall. Logically, this should be achieved by adjusting the partial factor of safety for materials. The Code provides for a modification factor to be applied to the characteristic strength. The recommendation given in BS 5628 applies to walls or columns whose horizontal cross-sectional area is less than 0.2 m^2, and states that the characteristic compressive strength should be multiplied by a factor given by the following formula:

$$(0.7 + 1.5A)$$

where A = horizontal cross-sectional area of the column or wall in m^2.

EXAMPLE 17

Determine the design strength of the column shown in Figure 5.33, constructed with 20 N/mm² units in mortar designation (iv), normal construction and materials control.

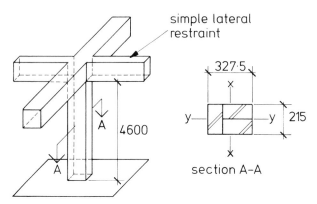

Figure 5.33

Capacity reduction factor:
Slenderness ratios,

$$h_{ef\,yy} = h_{ef\,xx} = 4600$$

$$t_{ef\,xx} = 327.5, \text{SR}_{xx} = \frac{4600}{327.5} = 14$$

$$t_{ef\,yy} = 215, \text{SR}_{yy} = \frac{4600}{215} = 21$$

Therefore β = 0.66.

Characteristic compressive strength:
20 N/mm² units, therefore f_k = 5.2 N/mm² in mortar designation (iv).
But, plan area = 0.3275 × 0.215 = 0.07 m², i.e. < 0.2 m², therefore modification factor applies.

$$\text{Modification factor} = 0.75 + 1.5 \times 0.07$$
$$= 0.855$$

Therefore, modified characteristic compressive strength
$$= 0.855 \times 5.2 \text{ N/mm}^2$$

Partial factor of safety:
Controls normal, γ_m = 3.5

Therefore

$$\text{design strength} = \frac{\beta \times A \times f_k}{\gamma_m}$$

$$= \frac{0.66 \times 0.07 \times 10^6 \times 0.855 \times 5.2}{3.5 \times 10^3}$$

$$= 58 \text{ kN}$$

5.13 ECCENTRIC LOADING

When considering a member subject to compressive loading, it is unlikely that the loading will ever be truly applied concentrically. In most instances, the load will be applied at some eccentricity to the centroid of the member, whether due to construction tolerances, varying imposed loads on adjacent floor spans or other causes. Generally, in the absence of evidence to the contrary, it is assumed that the load transmitted to a wall by a single floor or roof acts at one-third of the length of bearing from the loaded face. That is, as if a triangular stress distribution is assumed under the bearing (see Figure 5.34).

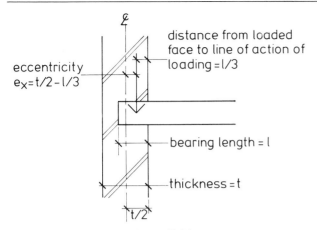

Figure 5.34

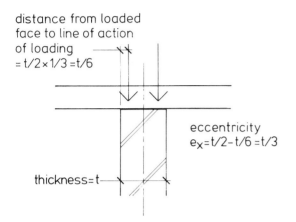

Figure 5.35

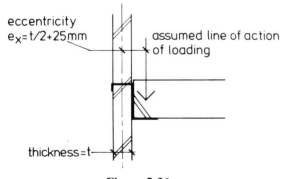

Figure 5.36

Where a uniform floor is continuous over a wall, the Code recommends that each span of the floor should be taken as being supported individually, on half the total bearing area (see Figure 5.35).

Where loads are supported at a distance from the face of the wall, as where joist hangers or continuous bearers are used, the load should be assumed to be applied at a distance of 25 mm from the face of the wall (see Figure 5.36).

The eccentricity has a maximum value, e_x, just under the applied load, and the member must be designed to resist the extra stresses incurred due to this eccentricity. But, the effect of the eccen-

tricity may be assumed to decrease down the height of the member, until its effect is zero at the bottom of the member. Thus the vertical load on a member may be considered as being axial (concentric) immediately above a lateral support (see Figure 5.37).

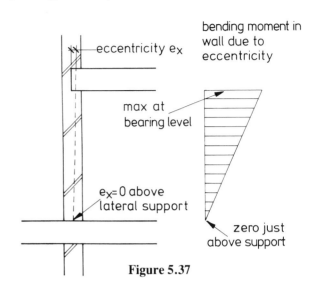

Figure 5.37

In the case of walls, it is not necessary to consider the effects of eccentricity where e_x is less than $0.05t$.

5.14 COMBINED EFFECT OF SLENDERNESS AND ECCENTRICITY OF LOAD

It was seen in 5.10, that the loadbearing capacity of a member was reduced due to the effects of slenderness on the tendency of the member to buckle. The application of an eccentric load will further increase the tendency of the wall to buckle, and thus reduce the load-carrying capacity of the member. This reduction is catered for by reduced values of β, the capacity reduction factor, depending on the ratio of the eccentricity, e_x, to the member thickness.

5.14.1 Walls
Values of the capacity reduction factor, β, for walls are given in Table 5.15 for values of eccentricity, e_x, from 0 to $0.3t$, where t is the thickness of the wall. The values from Table 5.13 are included in this table. Intermediate values may be obtained by linear interpolation between slenderness ratios and eccentricities. As stated in section 5.10, these values of β are maximum values and are strictly only correct for the central fifth of the member (see sections 5.10 and 5.14.2).

Table 5.15 Capacity reduction factor, β (BS 5628, Table 7)

Slenderness ratio h_{ef}/t_{ef}	Eccentricity at top of wall, e_x			
	Up to 0.05t (see note 1)	0.1t	0.2t	0.3t
0	1.00	0.88	0.66	0.44
6	1.00	0.88	0.66	0.44
8	1.00	0.88	0.66	0.44
10	0.97	0.88	0.66	0.44
12	0.93	0.87	0.66	0.44
14	0.89	0.83	0.66	0.44
16	0.83	0.77	0.64	0.44
18	0.77	0.70	0.57	0.44
20	0.70	0.64	0.51	0.37
22	0.62	0.56	0.43	0.30
24	0.53	0.47	0.34	
26	0.45	0.38		
27	0.40	0.33		

Note 1. It is not necessary to consider the effects of eccentricities up to and including 0.05t.
Note 2. Linear interpolation between eccentricities and slenderness ratios is permitted.
Note 3. The derivation of β is given in Appendix B of BS 5628.

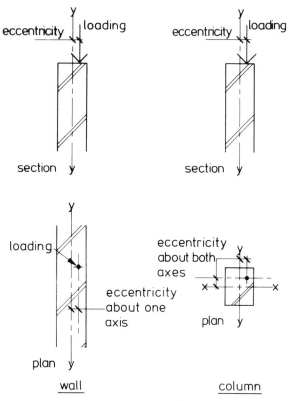

Figure 5.39

EXAMPLE 18

Determine from Table 5.15 the value of β when $h_{ef}/t_{ef} = 16$ for the wall shown in Figure 5.38.

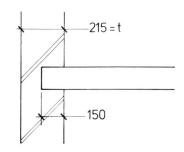

Figure 5.38

$$e_x = \frac{215}{2} - \frac{150}{3} = 57.5$$

Thus in terms of t,

$$e_x = \frac{57.5}{215}t = 0.27t$$

When $e_x = 0.2t$, β = 0.64
    $e_x = 0.3t$, β = 0.44

Therefore for
    $e_x = 0.25t$, β = 0.54

5.14.2 Columns

Because the application of loading to a column may be eccentric relative to two axes, as compared to a wall where the eccentricity is generally related only to an axis in the plane parallel with the centre line, the treatment of eccentricity for columns is necessarily more involved (see Figure 5.39).

The Code defines eccentricity as relative to the major or minor axis of the column. The major axis being defined as the principle axis, about which the member has the larger moment of inertia. The minor axis being perpendicular to the major axis (see Figure 5.40).

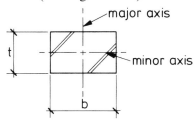

Figure 5.40

The dimensions of the section are then taken as being '*b*' for the side perpendicular to the major axis and '*t*' perpendicular to the minor axis. The values of the capacity reduction factor β for columns are determined in accordance with BS 5628 as follows:

Case 1: Nominal eccentricity both axes
When the eccentricities about major and minor axes at the top of the column are less than $0.05b$ and $0.05t$ respectively, β is taken from the range of values given in Table 5.15 for e_x up to $0.05t$, with the slenderness ratio based on the value of t_{ef} appropriate to the minor axes (see Figure 5.41).

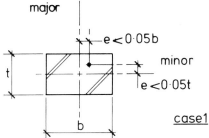

slenderness ratio: S.R. based on:-

$$\frac{\text{effective height} \left(\begin{array}{l}\text{determined relative} \\ \text{to minor axis}\end{array}\right)}{\text{effective thickness (based on }t)}$$

Figure 5.41

Case 2: Nominal eccentricity – major axis, eccentric about minor axis
When the eccentricities about the major and minor axes are less than $0.05b$ but greater than $0.05t$ respectively, β is taken from Table 5.15, using the values of eccentricity and slenderness ratio appropriate to the minor axis (see Figure 5.42).

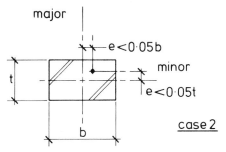

case 2

slenderness ratio: S.R. based on:-

$$\frac{\text{effective height (minor axis)}}{\text{effective thickness (minor axis)}}$$

Figure 5.42

Case 3: Nominal eccentricity – minor axis, eccentric about major axis
When the eccentricities about the major and minor axes are greater than $0.05b$ but less than $0.05t$ respectively, β is taken from Table 5.15 using the value of eccentricity appropriate to the major axis and the value of slenderness ratio appropriate to the minor axis (see Figure 5.43).

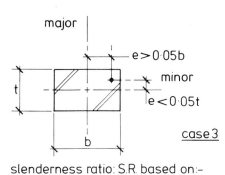

case 3

slenderness ratio: S.R. based on:-

$$\frac{\text{effective height (minor axis)}}{\text{effective thickness (minor axis)}}$$

Figure 5.43

Case 4: Eccentricity about both axes greater than nominal
When the eccentricities about major and minor axes are greater than $0.05b$ and $0.05t$ respectively, β is calculated by deriving additional eccentricities and substituting in the appropriate formula as follows (based on BS 5628, Appendix B):

The eccentricity of applied loading is assumed to vary from e_x at the point of application to zero above the lateral support (see Figure 5.37), and an additional eccentricity considered to allow for slenderness effects, i.e. the tendency of the member to buckle. This additional eccentricity, e_a, is considered to vary linearly from zero at the lateral supports to a value over the central fifth of the member height given by the formula:

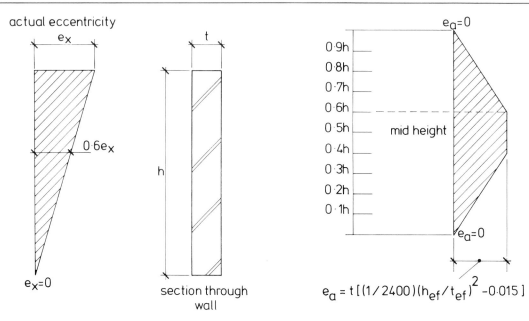

Figure 5.44 Variation of e_a and e_x over height of member

$$e_a = t \left(\frac{1}{2400} \left(\frac{h_{ef}}{t_{ef}} \right)^2 - 0.015 \right)$$

where

t_{ef} = effective thickness of the member
h_{ef} = effective height. (See Figure 5.44.)

The total design eccentricity, e_t, for calculation of the capacity reduction factor is given by the sum of e_x and e_a at the point being considered. When considering the mid-height section, where e_a is maximum, the maximum value of e_t will be:

$$e_t = 0.6e_x + e_a \text{ at mid-height}$$

When considering the top of the member, e_a will be zero and e_x at a maximum. Thus e_t will be equal to e_x:

$$e_t = e_x \text{ at top of member}$$

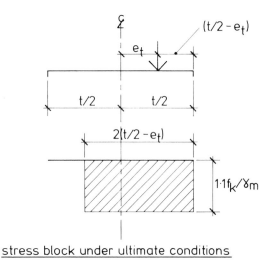

stress block under ultimate conditions

Figure 5.45

The ultimate stress block for an eccentrically loaded section is then assumed to be as given in Figure 5.45.

From which the design vertical loading may be seen to be equal to the area of the stress block multiplied by the design stress, i.e.:

$$2 \left(\frac{t}{2} - e_t \right) \times \frac{1.1 f_k \text{ per unit length}}{\gamma_m}$$

which may be expressed as follows

$$\frac{1.1 \left(1 - \frac{2e_t}{t} \right) t f_k}{\gamma_m}$$

or

$$\frac{\beta t f_k}{\gamma_m}$$

where

$$\beta = 1.1 \left(1 - \frac{2e_t}{t} \right)$$

f_k = characteristic strength of masonry
γ_m = partial factor of safety on materials
e_t = total design eccentricity appropriate to the point under consideration.

Thus for columns in Case 4, the value of β may be calculated for each axis and the minimum design capacity calculated. This method has a more general application and may be used to determine β for any member at any position. The values given in Table 5.15 are strictly only appropriate for the mid-height region of a member and, in some instances, the determination of β may be required at other points.

5.15 CONCENTRATED LOADS

The design compressive stress locally at the bearing of a concentrated load may be greater than the general level of stress within the body of a wall or other member. Such relatively higher stress concentrations occur only over a small area, and are rapidly reduced by dispersion within the body of the member. Typical examples of such concentrated loads are beam bearings which are usually either rigid in themselves, e.g. deep concrete beams, or are provided with spreaders, e.g. padstones. The distribution of stress under such bearings varies according to the particular details being adopted, and various methods are available for analysing the distribution. BS 5628, clause 34, considers that, in general, the concentrated load may be assumed

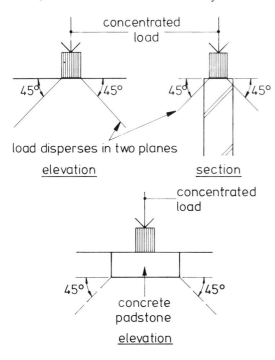

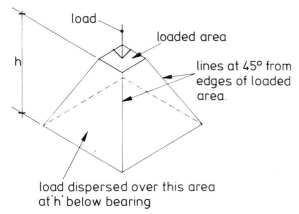

Figure 5.46

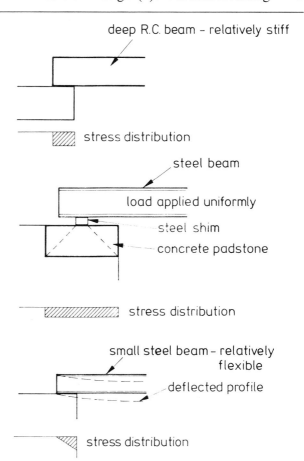

Figure 5.47 Stress distributions under concentrated loads

to be uniformly distributed over the area of the bearing and dispersed in two planes within a zone contained by lines extending downward at 45° from the edges of the loaded area (see Figure 5.46). This, however, tends to conflict with the recommendation in the Code that, when considering the eccentricity of applied loading the load may be assumed to act at one third of the depth of the bearing area from the loaded face (see section 5.13). Each case should, therefore, be considered separately – the rigidity of the beam, lintol, etc. being an important factor. Also, the bearing detail may be such that the load can be applied uniformly (see Figure 5.47).

The design local compressive stresses recommended in the Code vary according to type of bearing being considered. Three types are considered, being designated as types 1, 2 and 3 respectively. Types 1 and 2 are defined in terms of bearing area in relation to the thickness of the member. Type 3 bearing is for the special case of a spreader beam designed in accordance with the elastic theory, located at the end of a wall and spanning in its plane. For this type of bearing, the local stress is not uniformly distributed, but may be calculated from elastic theory. The

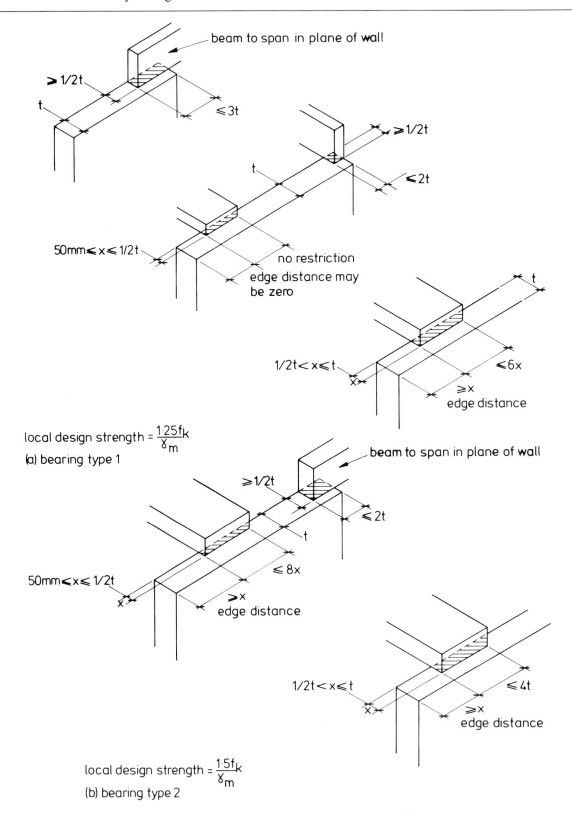

beam to span in plane of wall

$\geqslant 1/2t$

t

$\leqslant 3t$

$\geqslant 1/2t$

t

$\leqslant 2t$

$50mm \leqslant x \leqslant 1/2t$

no restriction
edge distance may
be zero

t

$1/2t < x \leqslant t$

$\leqslant 6x$

$\geqslant x$
edge distance

local design strength $= \dfrac{1\cdot 25 f_k}{\gamma_m}$

(a) bearing type 1

beam to span in plane of wall

$\geqslant 1/2t$

$\leqslant 2t$

t

$50mm < x \leqslant 1/2t$

$\leqslant 8x$

$\geqslant x$
edge distance

$1/2t < x \leqslant t$

$\leqslant 4t$

$\geqslant x$
edge distance

local design strength $= \dfrac{1\cdot 5 f_k}{\gamma_m}$

(b) bearing type 2

Figure 5.48(a–c) Concentrated loads: types of bearings (BS 5628 Figure 4)

appropriate design local compressive stresses for all three types are given in Table 5.16, and each type is illustrated in Figure 5.48.

The Code requires that where concentrated loads are applied to a member, not only should the applied stresses be checked in the immediate

area of the concentrated load, but also the stresses at a distance of $0.4h_b$ below the bearing, where h_b is the height of the bearing relative to the lower support. The applied stresses in the immediate area of the concentrated load will be equal to the sum of the design applied stress due to the concentrated load and the distributed

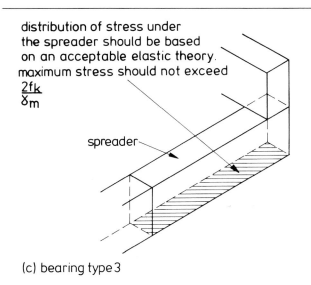

distribution of stress under the spreader should be based on an acceptable elastic theory. maximum stress should not exceed $\frac{2f_k}{\gamma_m}$

spreader

(c) bearing type 3

Table 5.16

Bearing type	Design local compressive stress in masonry
1	$1.25\dfrac{f_k}{\gamma_m}$
2	$1.5\dfrac{f_k}{\gamma_m}$
3	$2.0\dfrac{f_k}{\gamma_m}$

design applied stress existing within the member at that position (see Figure 5.49). The total combined stress must be less than or equal to the design local compressive stress from Table 5.16. Referring to Figure 5.49.

Total stress at x–x

$= \dfrac{\text{design reaction at end of beam}}{\text{bearing area}}$

$+ \dfrac{\text{design load from floor and brickwork above x–x}}{\text{area of wall}}$

i.e. local concentrated stress + uniformly distributed stress

$\leqslant$ design local compressive stress from Table 5.16.

The total applied stress at a distance of $0.4h_b$ below the bearing will be equal to the sum of the design applied stress due to the concentrated load, reduced from its maximum value due to an assumed dispersion at 45°, and the distributed design applied stress existing within the member at $0.4h_b$ below the bearing. The total combined stress must be less than, or equal to, the design

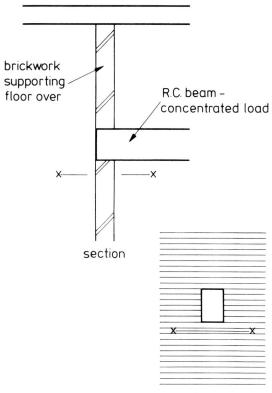

brickwork supporting floor over

R.C. beam – concentrated load

section

Figure 5.49   elevation

strength of the member calculated in accordance with sections 5.11 and 5.12 for walls and columns respectively (see Figures 5.50 and 5.51).

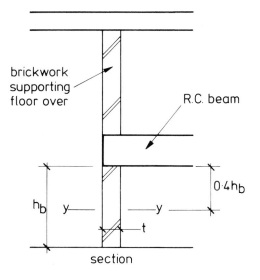

brickwork supporting floor over

R.C. beam

$0.4h_b$

h_b

t

section

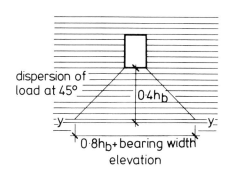

dispersion of load at 45°

$0.4h_b$

$0.8h_b$ + bearing width

elevation

Figure 5.50

Referring to Figure 5.50.

Total stress at y–y

$$= \frac{\text{design reaction at end of beam}}{(\text{bearing width} + 0.8h_b) \times \text{thickness of wall}}$$

$$+ \frac{\text{design load from floor and brickwork above y–y}}{\text{area of wall}}$$

$$\leqslant \frac{\beta t f_k}{\gamma_m}$$

It should be noted that, the value of β, the capacity reduction factor, should be that for the position under consideration (see Figure 5.35 and BS 5628, Appendix B). The design strength of the wall may also require checking at other locations, e.g. mid-height which may not coincide with $0.4h_b$ below the bearing. It is also possible that such concentrated loads may provide a degree of horizontal lateral restraint to a member and thus reduce the height between supports to be considered in determining the effective height.

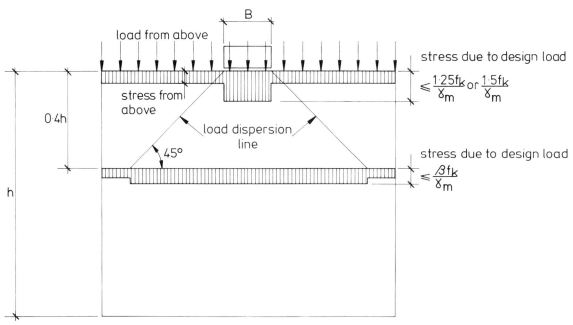

stress due to design load to be compared with design strength as indicated above
(a) load distribution for bearing types 1 and 2

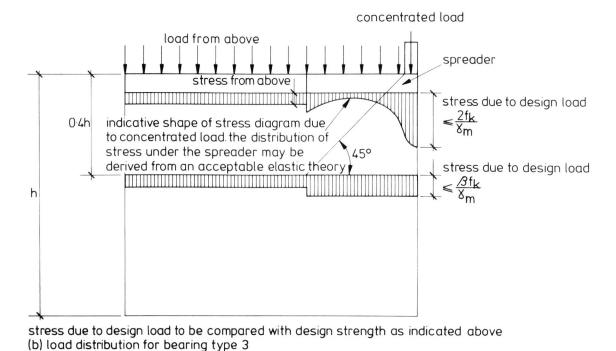

stress due to design load to be compared with design strength as indicated above
(b) load distribution for bearing type 3

Figure 5.51 (BS 5628 Figure 5)

CHAPTER 6

BASIS OF DESIGN (2): LATERAL LOADING – TENSILE AND SHEAR STRENGTH

Having dealt with walls and columns acting in compression, resistance to lateral loads must now be considered. In the walls of a house, for example, in addition to supporting the vertical loads from the roof and floors, masonry is also subject to the pressure of the wind against the outside walls and must, therefore, be designed to resist tensile and shear stresses, as well as the compressive stresses.

When a wall supports a uniformly distributed concentric vertical axial load, every part of the wall is assumed to be subjected to an equal compressive stress at any particular cross-section. When a wall supports a lateral loading, and bends or flexes, the stress at any particular cross-section may vary from being compressive at one face to being tensile at the other face – unless the wall is cracked (see Figure 6.1). As the resulting tensile stress is due to flexure of the wall, it is termed the flexural tensile stress.

When axial loading and lateral loading are combined, the resultant stresses in an uncracked wall, or other geometric section under considera-

tion, are those given by the combination of the stress due to the vertical load (a uniform compressive stress) and the lateral load – i.e. a stress varying from a maximum compressive stress on one face to a maximum flexural tensile stress on the other. Depending on the relative values of the vertical load and the bending moment due to the lateral load, the wall may be subject to entirely compressive stress, as shown in Figure 6.2, case (1), or both compressive and tensile stress, case (2).

If the tensile stresses which develop exceed the tensile resistance of the wall – or other section being considered – the section will crack. Cracking will also occur where no tensile resistance can be developed due, for instance, to the inclusion of a damp proof course. The applied loading is then resisted purely by compression within the section, as shown in Figure 6.2, case (3). The stress diagrams illustrated are for compressive stresses within the elastic range. Stress diagrams appropriate to conditions approaching the ultimate limit state will be dealt with later in this chapter.

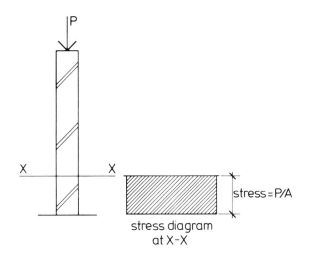

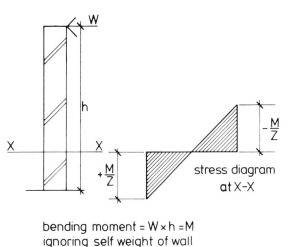

bending moment = W × h = M
ignoring self weight of wall

Figure 6.1

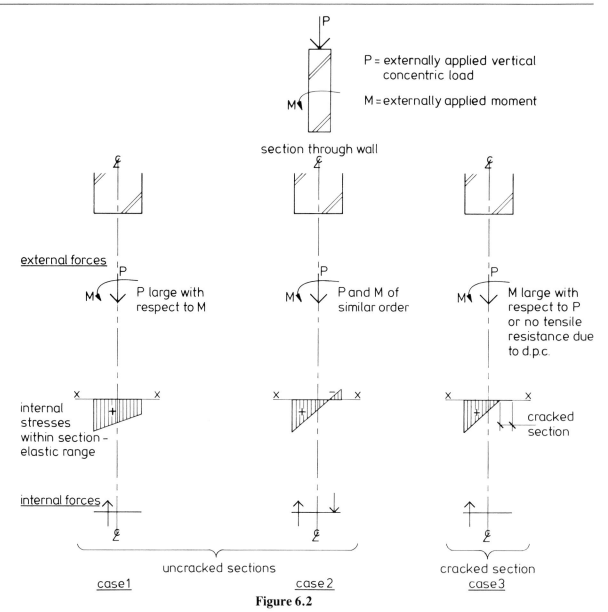

Figure 6.2

Because of the self-weight of masonry, all walls are subject to some vertical loading. In case (1), Figure 6.2, the compressive strength of the wall will govern the design of – in this instance – an external wall in the lower storey of a multi-storey building where the vertical loading due to the roof, floors and walls is large, and the bending moment due to the wind pressure is comparatively small.

In case (2), the tensile strength of the wall must be considered in addition to the compressive strength. This is an external wall in an upper storey of a multi-storey building where the vertical loading from the roof is small, or non-existent due to wind uplift forces, and the wind pressure is at its maximum. As masonry is comparatively weak in tension, it is usually the tensile stress that governs the design in this situation.

In case (3), the tensile strength – if any – of the wall has been exceeded, the section has cracked,

and higher compressive stresses have resulted over a reduced area of the section. Thus in this case, the compressive stress governs the design. This example is typical of the external walls of lightly loaded structures, and all laterally loaded members which include a damp proof membrane incapable of transferring tensile stresses across the joint. Some engineering bricks employed as dpcs are capable of transferring tensile stresses, but reliance on this must be given careful consideration by the designer.

6.1 DIRECT TENSILE STRESS

Tensile stresses due to bending are termed flexural tensile stresses in order to distinguish them from the tensile stress due to the application of a direct tensile force such as that resulting from wind uplift. In view of masonry's comparative weakness in tension, and the many other factors involved, such as workmanship, etc., it is considered unwise to place any reliance on its direct tensile strength.

Nevertheless, BS 5628 does enable designers, at their own discretion, to allow limited direct tensile stresses in two instances. The first is when suction forces arising from wind loads are transmitted to masonry walls. However, some form of restraint straps are generally necessary, and it is always advisable to provide some positive anchorage. The second case where direct tensile stresses may be allowed is when considering the probable effects of misuse or accidental damage (see Chapter 8).

When considering these two cases, the Code limits the tensile stress to half the values of the flexural tensile stress given in Table 6.1, and states that in no circumstances should the combined flexural and direct tensile stresses exceed the values provided.

6.2 CHARACTERISTIC FLEXURAL STRENGTH (TENSILE) OF MASONRY, f_{kx}

Masonry is a brittle material and its resistance to flexural tension depends on the type of masonry unit, the mortar grade and, most importantly, the bond between the mortar and the unit. The correct type of unit and mortar grade can readily be specified.

Much research has been undertaken into the mechanism of the bond between masonry units and mortar. It has been found under laboratory conditions that, when clay bricks are being used, the strength of the bond varies according to their water absorption properties. Thus for clay bricks, BS 5628 provides characteristic flexural strength values for various ranges of water absorption (see Table 6.1). Nevertheless, achieving a good bond between bricks and mortar still depends to a large extent on the degree of skill and care taken during construction.

Failure to provide adequate temporary propping against wind or lateral pressure, or inadequate curing during construction, may result in cracks occurring at a critical section which may invalidate any design assumptions based on flexural tensile resistance. This is not to say that masonry cannot be designed to resist flexural tensile stresses, but the designer's judgement of what is safe and reasonable is crucial, and should be even more critical than when considering other types of loading resistance.

Flexural tension should only be relied on at a dpc if the material has been proved by tests to permit the joint to transmit tension, or if the dpc consists of bricks in accordance with BS 743. Care must also be taken to ensure that the dpc is properly bedded in mortar, since a test on a bonded dpc is useless, if, on site, the dpc is laid dry.

Masonry is not isotropic, i.e. it does not have similar properties in all directions, and, therefore, does not provide the same resistance to bending in both directions. For example, a square wall panel of masonry with only vertical supports on each side will provide a greater resistance to bending due to lateral loading than if only horizontal supports were provided at the top and bottom (see Figure 6.3).

This difference in strength, however, is reduced by the effect of the self-weight of the wall, which will tend to reduce the flexural tensile stresses developed – as was explained at the beginning of this chapter. As the height of a wall is increased, the compressive stress due to the self-weight of the masonry will also increase. The combination of this increased stress and the flexural tensile stresses will mean that, if the vertical loading is

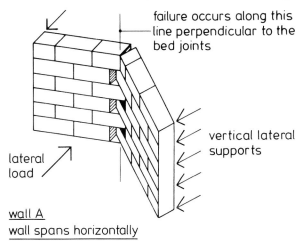

wall A
wall spans horizontally

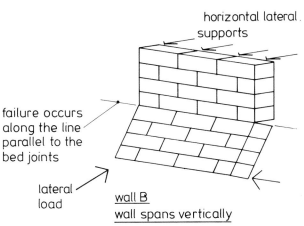

Figure 6.3

significant, the wall could resist a greater lateral loading when spanning between top and bottom supports, e.g. referring to Figure 6.3, wall B, than when spanning between vertical supports at each side, e.g. wall A.

Any other dead or permanent imposed loadings will similarly increase the compressive stress in the wall and improve its resistance to bending, provided that the compressive stresses are within allowable limits. This dead loading is often termed the 'pre-load' on the wall.

Thus masonry subject to little or no vertical loading tends to be stronger when spanning horizontally than when spanning vertically. On the other hand, walls which are subject to large vertical loading tend to be stronger when spanning vertically than when spanning horizontally.

Values for the characteristic flexural tensile strength, f_{kx}, both perpendicular and parallel to the bed joints are given in Table 6.1. These values take no account of any pre-load in the wall.

6.2.1 Orthogonal ratio

As previously explained, masonry is not isotropic and the difference in resistance to bending when spanning vertically and horizontally is defined as the orthogonal ratio, μ. It is used primarily for the calculation of bending moments in panel wall design. However, it is dealt with here because of its relationship to characteristic flexural strength.

The orthogonal ratio, as defined in BS 5628, is the ratio of the values of the respective characteristic flexural strengths when spanning vertically and horizontally. This may be expressed as follows:

Orthogonal ratio, μ

$$= \frac{f_{kx\ par}}{f_{kx\ perp}}$$

where

$f_{kx\ par}$ = characteristic flexural strength parallel to bed joints

$f_{kx\ perp}$ = characteristic flexural strength perpendicular to bed joints.

Table 6.1 Characteristic flexural strength of masonry, f_{kx}, N/mm^2 (BS 5628, Table 3)

Plane of failure parallel to bed joints			Plane of failure perpendicular to bed joints		
Mortar designation					
(i)	(ii) and (iii)	(iv)	(i)	(ii) and (iii)	(iv)
Clay bricks having a water absorption					
less than 7% — 0.7 / 0.5		0.4	2.0	1.5	1.2
between 7% and 12% — 0.5 / 0.4		0.35	1.5	1.1	1.0
over 12% — 0.4 / 0.3		0.25	1.1	0.9	0.8
Calcium silicate bricks — 0.3		0.2	0.9		0.6
Concrete bricks — 0.3		*	0.9		*
Concrete blocks of compressive strength in N/mm^2					
2.8	0.25	0.20	0.4		0.4
3.5			0.45		0.4
7.0			0.6		0.5
10.5			0.75		0.6
14.0 and over			0.90†		0.7†

*Values not at present available, pending research
†When used with flexural strength in parallel direction, assume the orthogonal ratio $\mu = 0.3$

EXAMPLE 1

Determine the orthogonal ratio, μ, for clay bricks having a water absorption of 9%, laid in mortar designation (ii), when no significant vertical load exists within the panel.

From Table 6.1, the bricks are in the range 7–12% for water absorption.

Thus
$$f_{kx\ par} = 0.4\ \text{N/mm}^2$$
and
$$f_{kx\ perp} = 1.1\ \text{N/mm}^2$$
Therefore
$$\mu = \frac{0.4}{1.1} = 0.36$$

The effect of any vertical loading in the member will tend to increase its resistance to bending when spanning vertically, and thus must be taken into account when determining the orthogonal ratio. The stress due to the design vertical load – which may only be the self-weight of the panel – is therefore added to the characteristic strength parallel to the bed joints, and the sum of these two stresses is used to determine the appropriate value of the orthogonal ratio. This may be expressed as follows:

$$\text{orthogonal ratio, } \mu = \frac{f_{kx\ par} + g_d}{f_{kx\ perp}}$$

where g_d = compressive stress due to the design vertical load in N/mm^2.

The Code recommends that the design vertical load should be modified by multiplying by the partial safety factor on materials γ_m. This produces the expression:

$$\mu = \frac{f_{kx\ par} + \gamma_m g_d}{f_{kx\ perp}}$$

However, as the design vertical load is already provided with a factor of safety on loads (i.e. $\gamma_f G_k$), it does not seem logical to further reduce this stress value, which is not a property of the material dependent on workmanship, etc., but generally only due to the dead weight of the construction. The authors thus consider that the design vertical load, $g_d = \gamma_f G_k$, should be used unmodified.

EXAMPLE 2

Determine the orthogonal ratio, μ, at mid-height for the wall illustrated in Figure 6.4, constructed of the bricks and mortar described in Example 1 of density, $\rho = 20\ \text{kN/m}^3$.

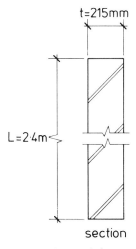

t=215mm

L=2·4m

section

Figure 6.4

Characteristic vertical load, G_k, at mid-height

$$= \text{thickness} \times \frac{\text{height}}{2} \times \text{density/unit length}$$

$$= 0.215 \times \frac{2.4}{2} \times 20$$

$$= 5.16 \, \text{kN/m}$$

Design vertical load at mid-height

$$= \text{partial safety factor on loads} \times \text{characteristic load}$$

$$= \gamma_f \times G_k$$

$$= 0.9 \times 5.16 \, (\text{from Table 5.1, } \gamma_f = 0.9, \text{ i.e. minimum vertical load})$$

$$= 4.6 \, \text{kN/m run}$$

Design vertical stress per m length, g_d,

$$= \frac{\text{design vertical load}}{\text{area}}$$

$$= \frac{4.6 \times 10^3}{215 \times 10^3} \, \text{N/mm}^2$$

$$= 0.02 \, \text{N/mm}^2$$

Therefore

$$\mu = \frac{(0.4 + 0.02)}{1.1}$$

$$= 0.38$$

6.3 MOMENTS OF RESISTANCE: GENERAL

With the exception of panel wall design, the topic of lateral loading is not considered to any great extent in BS 5628. However, the application of structural masonry is not limited only to vertical loadbearing elements and cladding panels to structural frames. Design methods should cover all applications of masonry.

The section which follows is based on the principles laid down in the Code. These principles have been abstracted, and produced in the form given in the Code, as they have more general application than in their original context.

The effects of lateral loading on masonry members were discussed at the beginning of this chapter. The three cases considered were as follows:

1. Entire section in compression – design governed by compressive strength requirements.
2. Limited tensile stresses developing – design generally governed by tensile stress requirements, compressive stresses to be checked (particularly for geometric profile sections).
3. Section cracked, or no tensile resistance due to presence of dpc unable to transmit tension – design governed by compressive stresses.

6.3.1 Moments of resistance: uncracked sections

At sections where flexural tension can develop, that is uncracked sections and those where continuity is not broken by the inclusion of a dpc unable to transfer tensile stresses, the design moment of resistance is given by the product of the design stress which can develop in the section and the section modulus. When considering the moment of resistance to bending about a vertical axis, i.e. a member spanning horizontally, the design stress will simply be the characteristic flexural strength perpendicular to the bed joints, divided by the appropriate partial safety factor for materials. Thus for members spanning horizontally, the design moment of resistance, MR, may be expressed as follows:

$$MR = \frac{f_{kx\ perp}}{\gamma_m} \times Z \text{ (uncracked section)}$$

where

$f_{kx\ perp}$ = characteristic flexural strength perpendicular to bed joints (see Table 6.1)

γ_m = partial factor of safety for materials (see Table 5.11)

Z = elastic section modulus.

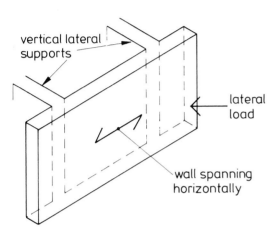

design moment of resistance horizontally

$$MR = \left(\frac{f_{kx}perp}{\gamma_m}\right)z$$

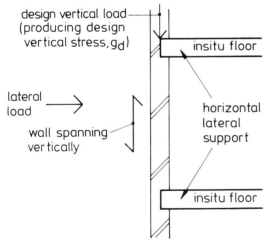

design moment of resistance vertically

$$MR = \left(\frac{f_{kx}\ par}{\gamma_m} + g_d\right)z$$

Figure 6.5

It should be noted that the section modulus for hollow blocks should be based on the geometric properties of the unit, i.e. Z based on the net area of the section, not the gross area (see Figure 6.5).

When considering geometric sections (spanning vertically) other than a rectangular section, e.g. diaphragms or fins, the outstanding length of flange from the face of the rib or fin should be taken as follows:

(a) 4 × the effective thickness of the wall forming the flange, when the flange is unrestrained;
(b) 6 × the effective thickness of the wall forming the flange when the wall is continuous.

In no case should this distance be more than half the distance between ribs or fins. This consideration affects the assessment of the section modulus, and also the other geometric properties of the section. The requirements are discussed further in Chapter 13 which deals with diaphragm and fin walls.

When considering the moment of resistance to bending about a horizontal axis, i.e. a member spanning vertically, the design stress which may be developed is the sum of the design tensile strength parallel to the bed joints and the stress due to the design vertical loading. Thus for members spanning vertically, the design moment of resistance, MR, may be expressed as follows:

$$MR = \left(\frac{f_{kx\ par}}{\gamma_m} + g_d\right) Z \text{ (uncracked section)}$$

(see Figure 6.5)

where

$f_{kx\ par}$ = characteristic flexural stress parallel to bed joints (see Table 6.1)

γ_m = partial factor of safety for materials (see Table 5.11)

g_d = design vertical stress due to dead loads, i.e. $\gamma_f G_k$/area

Z = elastic section modulus.

EXAMPLE 3

Determine the type of brick required, laid in mortar designation (ii), for the wall shown in Figure 6.6, assuming $\gamma_m = 2.5$, and density, $\rho = 18\ \text{kN/m}^3$, if the wall is subject to a lateral characteristic wind load of 0.8 kN/m² and is only supported along the top and bottom edges.

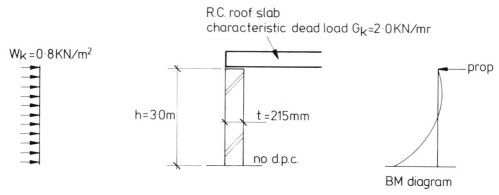

Figure 6.6

Consider 1 m width of wall:

Characteristic wind load, W_k = 0.8 kN per m height
Design wind load = $\gamma_f W_k$ = 1.4 × 0.8 = 1.12 kN per m height
Total design wind load   = 1.12 × 3 = 3.36 kN
Therefore, applied moment,
maximum at base = 3.36 × ⅜ kNm   = 1.26 kN.m
(For the purposes of this example, the wall is treated as a propped cantilever.)

MR ⩾ applied moment

but

$$MR = \left(\frac{f_{kx\ par}}{\gamma_m} + g_d\right) Z$$

Therefore

$$1.26 \leqslant \left(\frac{f_{kx\ par}}{\gamma_m} + g_d\right) Z$$

Thus to calculate $f_{kx\ par}$ required, substitute $\gamma_m = 2.5$.

$$g_d = \frac{\gamma_f G_k + \gamma_f \rho \times h \times t \times \text{unit length}}{t \times \text{unit length}}$$

$$= \frac{0.9 \times 2 \times 10^3 + 0.9 \times 18 \times 10^3 \times 3 \times 0.215 \times 1}{215 \times 10^3}$$

$$= \frac{12249}{215 \times 10^3} = 0.057 \text{ N/mm}^2$$

and

$$Z = \frac{10^3 \times 215^2}{6} = 7.7 \times 10^6 \text{ mm}^3$$

Hence

$$f_{kx\ par} \geqslant \left(\frac{1.26 \times 10^6}{7.7 \times 10^6} - 0.057\right) \times 2.5 \text{ N/mm}^2$$

$$\geqslant 0.267 \text{ N/mm}^2$$

From Table 6.1, with mortar designation (ii), any clay bricks would be suitable. The brick strength must also be checked for compressive stresses, as explained in Chapter 5.

The maximum design compressive stress will occur at the base, and will be equal to the sum of the design compressive stress due to the axial load and the maximum design compressive stress due to bending. The required design compressive strength of the brickwork must be equal to or exceed this stress. This may be expressed as follows:

design compressive strength of brickwork required (see Chapter 5, section 5.10)	design compressive stress due to axial load	maximum design compressive stress due to bending
$\geqslant$		$+$

$$\text{i.e.} \quad \frac{f_k \text{ required}}{\gamma_m} \geqslant \frac{g_d}{\beta} + \frac{M_A}{Z}$$

where

f_k = characteristic design compressive stress required
γ_m = 2.5
g_d = 0.057 N/mm^2
M_A = 1.26 kNm
Z = 7.7 × 10^6 mm^3
β = capacity reduction factor at base level, obtained as follows:

$$\text{slenderness ratio} = \frac{0.75 \times 3000}{215} = 10.5$$

eccentricity at base = zero (see Figure 5.44).

Thus β = 0.96 (see Table 5.15).

Hence

$$f_k \text{ required} \geqslant \left(\frac{g_d}{\beta} + \frac{M_A}{Z} \right) \gamma_m$$

$$\geqslant \left(\frac{0.057}{0.96} + \frac{1.26 \times 10^6}{7.7 \times 10^6} \right) 2.5$$

$$\geqslant 0.56 \text{ N/mm}^2$$

From Table 5.4 any unit with a compressive strength of 5 N/mm^2 or more, laid in mortar designation (ii), will provide a characteristic compressive strength of 0.56 N/mm^2.

It should be noted that the capacity reduction factor, β, has only been applied to the design compressive stress due to axial loading. This is not the design stress, but it is considered that the β is applicable only to the design compressive strength in respect of vertical axial loading, except for bending in the perpendicular plane of the wall where buckling can occur perpendicular to the axis of bending and not to the stress due to bending. The compressive strength of the wall should also be checked where the capacity reduction factor, β, is maximum.

6.3.2 Moments of resistance: cracked sections

At sections where flexural tension cannot be developed, e.g. where the section is already cracked, or where a dpc unable to transmit tensile stresses is provided, the design moment of resistance to lateral loading is provided solely by the gravitational stability moment produced by the self-weight of the member and any net dead load about the appropriate lever arm. This stability moment must, therefore, be sufficient to resist the applied overturning moment due to the lateral loading, and the compressive stresses will govern the design.

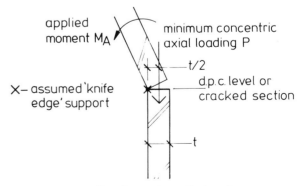

section through wall showing exaggerated action of stability moment

Figure 6.7

This action is illustrated in Figure 6.7. Under the action of the applied moment, M_A, the section is tending to overturn about the pivot which, for the present, is assumed to be a 'knife-edge' support. The tendency to overturning is resisted, and stability is provided, by the action of the concentric vertical axial loading about its lever arm about X, which is equal to approximately half the thickness of the section in this instance. This may be expressed as follows:

$$M_A \leqslant P \times \frac{t}{2}$$

The necessary factors of safety, etc. must, of course, be included – but this is basically the principle involved.

The assumption of a knife-edge support, that is to say non-yielding points of contact, is not correct when dealing with masonry. The compressive stresses at the edge X would be infinitely large. This would cause some local crushing of units or mortar, or squeezing of the damp proof membrane, if one is present, and increase the contact area. The actual width of this contact area, and actual stress distribution over it, is complex. In order to simplify the design analysis, an equivalent stress block is assumed. In BS 5628, a rectangular stress block is assumed, the value of the compressive strength being taken as the characteristic compressive strength of masonry divided by the partial factor of safety. Where this stress block is actually shown in the Code (Appendix B), the compressive strength is increased by a factor of 1.1, which relates the assumed compressive strength for the rectangular section to the actual stress distribution. This factor is included in the following determination of the moment of resistance. The stress block considered is shown in Figure 6.8.

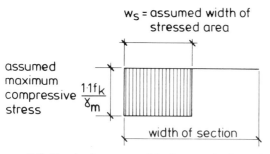

Figure 6.8 Equivalent stress block – cracked section

As the applied stress is of a local concentrated nature, it is not considered relevant to apply the capacity reduction factor for slenderness. The assumed width of the stressed area, w_s, will depend on the vertical loading within the wall. The total upward reaction, i.e. stress multiplied by the area, must be equal to the applied vertical load in accordance with the laws of statics, to maintain the equilibrium of the section. This is illustrated in Figure 6.9.

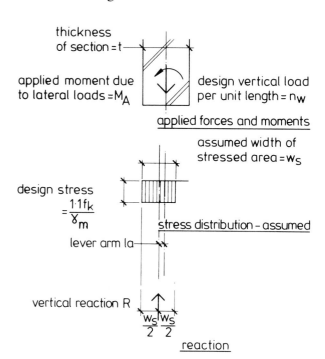

Figure 6.9

From Figure 6.9, for equilibrium:

Vertical forces: applied external vertical load = internal vertical reaction

i.e.
$$n_w = R \qquad (1)$$

Moments: applied external overturning moment about centre line of axial load = internal moment of reaction about centre line of applied axial load, i.e. lever arm.

i.e.
$$M_A = R \times l_a$$

but
$$l_a = \left(\frac{t}{2} - \frac{w_s}{2} \right)$$

therefore
$$M_A = R \times \left(\frac{t}{2} - \frac{w_s}{2} \right) \qquad (2)$$

The upward reaction,
$$R = \text{stress} \times \text{area}$$

$$= 1.1 \frac{f_k}{\gamma_m} \times w_s / \text{unit length}$$

But, from equation (1)
$$R = n_w$$

therefore
$$n_w = 1.1 \frac{f_k}{\gamma_m} w_s$$

i.e.
$$w_s = \frac{n_w \gamma_m}{1.1 f_k}$$

Substituting for R and w_s in equation (2):

$$M_A = n_w \left(\frac{t}{2} - \frac{n_w \gamma_m}{1.1 f_k \times 2} \right)$$

which may be rearranged:

$$M_A = \frac{n_w}{2} \left(t - \frac{n_w \gamma_m}{1.1 f_k} \right) \text{ (cracked section)}$$

This is the expression for equilibrium, and thus the design moment of resistance vertically, MR, is given by the expression:

$$\text{MR} = \frac{n_w}{2} \left(t - \frac{n_w \gamma_m}{1.1 f_k} \right) \text{ (cracked section)}$$

N.B. n_w is the design load, and therefore includes the factor of safety. Note also that the 1.1 factor applied to f_k does not strictly comply with the formula given in BS 5628, clause 36.5.3. However, it is catered for elsewhere in the Code (Appendix B).

EXAMPLE 4

Repeat Example 3 assuming a sheet dpc at base level which is not capable of transmitting tensile stresses.

Applied design moment at base $M_A = 1.26$ kN.m/m run. With a dpc at the base, the section must be designed as a cracked section, and thus the design moment of resistance, MR, must be greater than or equal to the applied design moment.

$$\text{MR} \geqslant M_A, \text{ i.e. } \frac{n_w}{2} \left(t - \frac{n_w \gamma_m}{1.1 f_k} \right) \geqslant M_A$$

where

f_k = characteristic design compressive stress
$\gamma_m = 2.5$

and, from Example 3

$t = 215\,mm$
$n_w = \gamma_f G_k + \gamma_f \times \rho \times h \times t$
$\quad = 0.9 \times 2 + 0.9 \times 18 \times 3 \times 0.215 = 12.25 \text{ kN/m}$
$M_A = 1.26 \text{ kN·m}$

Hence
$$f_k \text{ required} \geqslant \frac{n_w \gamma_m}{1.1 \left(t - \dfrac{2M_A}{n_w} \right)}$$

$$\geqslant \frac{12.25 \times 2.5 \times 10^3}{1.1 \left(215 - \dfrac{2 \times 1.26 \times 10^6}{12.25 \times 10^3} \right) \times 10^3}$$

$$\geqslant 2.99 \text{ N/mm}^2$$

From Table 5.4, units with a compressive strength of 10 N/mm² or more are required, laid in mortar designation (ii). The compressive strength of the wall should also be checked where the capacity reduction factor, β, is a maximum. The characteristic flexural tensile stress required should be determined at the point of maximum bending moment in the span, and at this point will be based on an uncracked section.

6.4   CAVITY WALLS

When cavity walls are subject to lateral loading, usually only one leaf is loaded. For example, in the case of external wind loading (and no internal pressure), it is only the external leaf of the cavity which is loaded. The other leaf can only contribute to the resistance if (a) the two leaves are joined in such a way as to act together (as in the case of diaphragm walls), or (b) the load is transmitted to the other leaf of the wall and is shared in some ratio between the two leaves.

In order to achieve the first option, the connection between the two leaves must be able to transmit both horizontal and vertical internal shear stresses. The standard types of wall ties, i.e. vertical twist, butterfly and double-triangle, manufactured in accordance with BS 1243, are not strong enough to transmit these stresses across a cavity. However, when provided at the spacings recommended in BS 5628 (clause 29.1), will ensure that the applied lateral loading is shared between the two leaves, as illustrated in Figure 6.10.

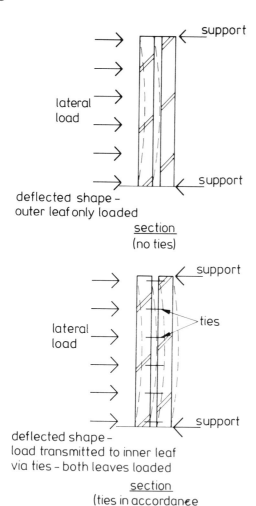

deflected shape –
outer leaf only loaded

<u>section</u>
(no ties)

<u>section</u>
(ties in accordance
with B.S. 5628 cl 29.1)

deflected shape –
load transmitted to inner leaf
via ties – both leaves loaded

Figure 6.10

6.4.1   Vertical twist ties

With vertical twist type ties, or ties of equivalent strength, the loading is considered to be fully transmitted by the ties from the outer leaf to the inner leaf. The two leaves will thus deflect together to support the applied loading.

Under the action of loading, the two leaves must deflect to a similar profile, and the amount of deflection of each leaf must be similar, otherwise they will move apart. If both leaves have the same moment of inertia, I, modulus of elasticity, E, and length, L, the load will be shared equally between the two as the deflection, which is measured in terms of EI/L, will be similar. If one leaf has a greater moment of inertia than the other, a greater load will be required to produce the same deflection as the leaf with the smaller moment of inertia.

Thus the leaf with the greater moment of inertia will resist a greater load than the leaf with the smaller moment. That is to say, the load resisted by each leaf will be in proportion to its stiffness, as measured by the quantity EI/L. But, as E and L are generally similar for each leaf of a cavity wall, the proportion may be based on the I value.

With regard to the cavity wall with differing I values for each leaf shown in Figure 6.11:

If W_1 is the load on (1)

$$\text{deflection, } \delta_1 = \frac{5 \times W_1 \times L_1^3}{384 \times E_1 \times I_1}$$

and W_2 is the load on (2)

$$\text{deflection, } \delta_2 = \frac{5 \times W_2 \times L_2^3}{384 \times E_2 \times I_2}$$

But, deflections must be equal, i.e. $\delta_1 = \delta_2$, and here, $E_1 = E_2$ and $L_1 = L_2$.
Thus

$$\frac{W_1}{I_1} = \frac{W_2}{I_2}$$

But, the total lateral load, $W = W_1 + W_2$

i.e. $W_2 = W - W_1$.

Substituting this in the above equation

$$\frac{W_1}{I_1} = \frac{W - W_1}{I_2}$$

Multiply by $I_1 \times I_2$, then

$$W_1 I_2 = WI_1 - W_1 I_1$$

which becomes

$$W_1 = \frac{WI_1}{I_1 + I_2}$$

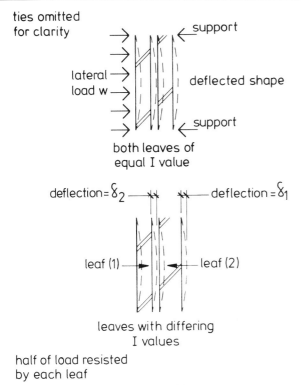

both leaves of
equal I value

leaves with differing
I values

half of load resisted
by each leaf

Figure 6.11

That is to say, the load on leaf (1) is obtained from the total load in proportion to the moment of inertia of leaf (1). Similarly, the load on leaf (2) will be in proportion to its moment of inertia.

When dealing with cavity walls with vertical twist ties, or equivalent, the applied design load may, in accordance with BS 5628, be apportioned between each leaf of the wall in proportion to its moment of inertia, related to the sum of the moments of inertia for each leaf. Each leaf may then be designed to produce a moment of resistance to this loading on the basis of a cracked or uncracked section, as appropriate.

When it is required to check the compressive resistance of the ties, or to assess the strength required for another type of tie to be considered equivalent to a vertical twist tie, the Code states that provided the cavity is no greater than 75

mm, the appropriate value of the load in tension may be used for the compression load. Values for the characteristic strengths of wall ties are given in Table 6.5, later in this chapter.

6.4.2   Double-triangle and wire butterfly ties

When double-triangle or butterfly ties are used, the loading may, in accordance with BS 5628, be apportioned as described in 6.4.1, provided that the ties are capable of transmitting the compressive forces to which they are subjected. The Code states that for double-triangle and wire butterfly ties laid in mortar designation (i), (ii) or (iii), the characteristic compressive resistance may be taken as 1.25 kN and 0.5 kN respectively. These types of wall tie should not be used where the width of the cavity is more than 75 mm.

In cases where these types are not capable of transmitting the necessary compressive loading, the loaded leaf will be required to resist a greater proportion of the applied loading or specially designed ties could be used.

6.4.3   Requirements for ties

The required spacing of ties is given in clause 29.1.4 of the Code and in Table 6.2. The spacing may be varied, provided the number of ties per square metre on elevation is not less than the values given in the table. Additional ties may be necessary around the sides of openings.

The minimum embedment of a tie in a mortar joint should be 50 mm in each leaf. The width of the cavity may vary between 50 mm and 150 mm but, in accordance with the Code, may not be wider than 75 mm where either of the leaves is less than 90 mm in thickness. However, the Code does allow that in special circumstances, with appropriate supervision, the width of the cavity may be reduced below 50 mm.

Where large uninterrupted expanses of cavity walling are constructed, the differential move-

Table 6.2   Spacing of ties (BS 5628, Table 6)

Leaf thickness	Cavity width (mm)	Spacing (mm)		Number of ties per square metre
		Horizontally	Vertically	
Less than 90 mm (but in no case less than 75 mm)	50–75	450	450	4.9
90 mm or more	50–75	900	450	2.5
90 mm or more	75–100	750	450	3.0
90 mm or more	100–150	450	450	4.9

ments due to thermal movement, elastic shortening under load, etc., between the two leaves of a wall may cause loosening of the ties. The heights and lengths of the external cavity walls should, therefore, be limited for this reason, in addition to any movement joint requirements.

The Code recommends that, the outer leaf should be supported at intervals of not more than every third storey, or 9 m, whichever is less. For buildings not exceeding four storeys, or 12 m in height, whichever is less, the Code considers it satisfactory for a wall to be uninterrupted for its full height. These requirements should be carefully considered in conjunction with the requirements for movement joints, as discussed in Appendix 3. These heights are considered by the authors to be an absolute maximum. Greater frequency of support should be provided, where possible.

6.4.4 Double-leaf (collar-jointed) walls

When a wall is constructed of two separate leaves with a vertical joint not exceeding 25 mm wide between them, i.e. a cavity wall with a very narrow cavity, in accordance with the Code it may be designed as a cavity wall, or as a single-leaf effectively 'solid' wall–with an effective thickness equal to actual overall thickness–provided the following conditions are satisfied:

(1) Each leaf is at least 90 mm thick.
(2) For concrete blockwork, the characteristic compressive strength, f_k, (see Chapter 5) should be multiplied by 0.9.
(3) If the two leaves of the wall are of different materials, e.g. one leaf clay bricks and the other concrete blocks, it should be designed for assessment of strength requirements on the assumption that it is entirely constructed of the weaker strength units. The possibility of differential movement between the two differing materials should be considered, and additional joints, etc., provided if required.
(4) The vertical load is applied to both leaves, and the eccentricity of the vertical load does not exceed $0.2t$, where t is the overall thickness of the wall, i.e. two leaves plus the vertical joint thickness.
(5) Flat metal wall ties of cross-sectional area 20 mm × 3 mm are provided at centres not exceeding 450 mm both horizontally and vertically. Alternatively, an equivalent mesh may be provided at the same vertical centres.
(6) The minimum embedment of the ties into each leaf is 50 mm.

(7) The vertical 'collar' joint between the two leaves is solidly filled with mortar as the work proceeds. This, perhaps, is the most difficult requirement to ensure is properly carried out. Solidly filling a narrow vertical joint between two leaves of masonry is not as easy as filling the perpend joints between individual units and, because of the additional time and labour involved, there is always a possibility that it will not be done thoroughly. It is also very difficult to check that the work has been completed satisfactorily. Thus if this type of wall is to be designed and used as a 'solid' wall, particular attention must be paid to the supervision of the work, and the operatives should be made fully aware at the outset of the standard of workmanship that is required.

6.4.5 Grouted cavity walls

In the case of cavity walls with a cavity of between 50 mm and 100 mm filled with concrete, the wall may be designed, in accordance with the Code, as a single-leaf wall – i.e. effectively a solid wall – the effective thickness being taken as equal to the actual overall thickness, subject to the following conditions:

(1) The concrete has a 28 day strength not less than that of the mortar.
(2) Requirements (1), (3), (4), (5) and (6) for collar-jointed walls (see section 6.4.4, above) are complied with.

6.4.6 Differing orthogonal ratios

For cavity walls in which the two leaves have different orthogonal ratios (see section 6.2.1), the Code recommends that the applied lateral load should be shared between the two leaves in proportion to their design moments of resistance. This requirement is subject, of course, to sections 6.4.1 and 6.4.2 regarding the transfer of loading between the two leaves. The orthogonal ratio is mainly used for panel walls – thus design is usually related to uncracked sections. The section modulus, Z, used in the analysis of such sections, is related to the moment of inertia and hence to the relative stiffness of each leaf.

6.5 EFFECTIVE ECCENTRICITY METHOD OF DESIGN

As explained in the introduction to this chapter, vertical loading on a member tends to increase its resistance to bending. As shown in Figure 6.2, case (1), where the vertical loading is sufficiently large, the internal stresses within the section are compressive throughout the section. In such cases, an effective eccentricity may be obtained. The applied moment on the section, due to the lateral loading, may be replaced by the actual axial loading at some eccentricity to the centre line, as illustrated in Figure 6.12.

The section may then be designed, as described in Chapter 5, for an axial load applied at an eccentricity of e_{ef}. The design compressive stress may then be assessed using Table 5.15, based on an eccentricity at the top of the member, e_x equal to e_{ef}, and the compressive stress would be load/area with no increase for M/Z due to the eccentricity as this has been taken into account in the capacity reduction factor, β.

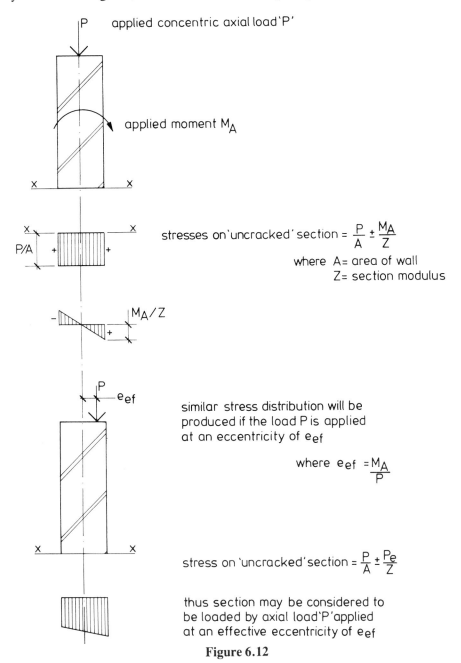

Figure 6.12

EXAMPLE 5

The brick wall shown in Figure 6.13 is subject to a uniformly distributed vertical line loading of 50 kN/m run (design load) applied along the centre line. The wall is simultaneously subject to a design lateral load of 0.8 kN/m².

The end conditions are obviously important in determining the applied bending moment but, for this example, assume that the bending moment due to the lateral load may be taken as $0.125WL^2$ for simplicity of analysis, the wall being considered to span simply supported between top and bottom lateral restraints.

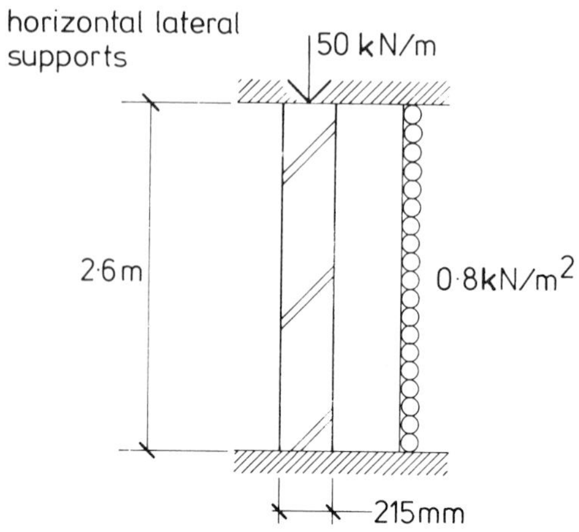

Figure 6.13

Therefore

$$\text{design bending moment} = 0.125 \times 0.8 \times 2.6^2$$
$$= 0.676 \text{ kNm/m length}$$
$$\text{design axial load} = 50 \text{ kN/m length}$$

Therefore, resultant effective eccentricity, e_{ef}

$$= \frac{0.676 \times 10^6}{50 \times 10^3}$$

$$= 13.52 \text{ mm}$$

Eccentricity, as proportion of t, $= \dfrac{13.52}{215}t$, i.e. $0.06t$.

The slenderness ratio of the wall $= \dfrac{h_{ef}}{t_{ef}} = \dfrac{2600}{215}$

$$= 12.1$$

Thus from Table 5.15, the capacity reduction factor, β, for the wall with combined axial and lateral loading will be given as follows:

Slenderness ratio $= 12$
Therefore for eccentricity at top of wall, $e_x = 0 - 0.5t, \beta = 0.93$
and for $e_x = 0.1t$, $\beta = 0.87$
Thus by interpolation, for $e_x = 0.06t$, $\beta = 0.92$.

Determination of the brick strength and mortar designation is then carried out as described in Chapter 5.

6.6 ARCH METHOD OF DESIGN

6.6.1 Vertical arching

As an alternative to the effective eccentricity method for walls and columns under axial loading, another method of design is given in the Code, based on the formation of a vertical arch.

There are various prerequisites to the use of this method relating to the supports, the design load and the dimensions of the panel, which must be fulfilled in each case. It is felt by the authors that there are certain dangers in the use of this design method, unless these requirements are most strictly complied with. In addition, careful thought should be given to the requirements in respect of practical considerations. For example, the development of arch thrusts requires rigid supports. The provision of concrete floors, etc, may afford adequate restraint. However, concrete and masonry move differentially, and cracks may develop at the junction of the two materials which may invalidate design assumptions. Similarly, the vertical axial load available to resist the arch thrust must be carefully analysed. It will generally be the characteristic dead load, but the dispersion of vertical loading to other parts of the structure may reduce the actual load available to resist the arch thrust. Alternatively, the dead load at the time of applying the lateral load may be less than in the final condition if, for instance, a basement retaining wall is back-filled at an early stage and before additional vertical loading is available from the superstructure. All these aspects require very close scrutiny before this method of analysis is considered suitable.

The formula given in the Code is based on the following analysis, for which reference to Figure 6.14 should be made. The member is considered to behave under ultimate conditions in the 'three-pinned arch' mode of failure, under the action of a lateral loading, q_{lat}, and a vertical axial load, n. A small crack develops, and hinges form at positions A, B and C (see Figure 6.14). The member deflects slightly and an arch ABC is formed within its thickness.

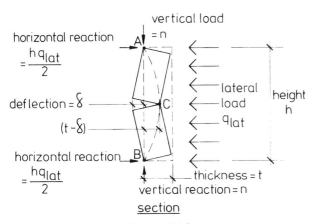

Figure 6.14

Taking moments about C:

$$n \times (t - \delta) + q_{lat} \times \frac{h}{2} \times \frac{h}{4} = \frac{hq_{lat}}{2} \times \frac{h}{2}$$

but, as the deflection, δ, is very small

$$t - \delta = t$$

Substituting and rearranging in the above equation, the lateral loading, q_{lat}, may be expressed as follows:

$$q_{lat} = \frac{8nt}{h^2}$$

A general factor of safety is then applied to this expression to obtain the design lateral strength of the wall. This general factor of safety is taken in the Code to be equal to the appropriate value of

the partial factor of safety for materials. The actual factor of safety, in this instance, is not a partial factor of safety for materials, as no materials strengths are directly involved. It is simply the numerical value which is being used.

As discussed in section 6.3.2, the assumption of a pin, i.e. knife-edge, support is not correct, and the points of contact A, B and C will have some finite width parallel to the thickness of the wall. The width of this bearing will have the effect of marginally reducing the lateral strength, but it can be assumed that this is well catered for within the general factor of safety adopted.

Thus in accordance with BS 5628, the design lateral strength of an axially loaded wall or column may be determined from the following formula:

$$q_{lat} = \frac{8 \times t \times n}{h^2 \gamma_m}$$

where

q_{lat} = design lateral strength per unit area
n = axial load per unit length of wall available to resist the arch thrust. For normal design it should be based on the characteristic dead load. When considering the effects of misuse or accident, it should be the approximate design load (see Chapter 8).
h = clear height of wall or column
t = actual thickness of wall or column
γ_m = factor of safety, taken as equal to the partial factor of safety for materials, see Table 5.11.

This formula may be used provided that the wall is contained between concrete floors affording adequate lateral support and adequate resistance to rotation across the full width of the wall.

The stresses occurring at dpc level, and the effectiveness of the lateral restraints in resisting the horizontal forces, must also be considered. The axial stress due to n, or the appropriate design load, must not be less than 0.1 N/mm², and the h/t ratio must not exceed 25 in the case of narrow brick or block walls, or 20 for all other types of wall.

EXAMPLE 6
Determine the design lateral strength of the wall shown in Figure 6.15

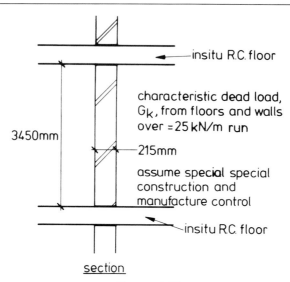

Figure 6.15

From Table 5.11 $\gamma_m = 2.5$
and from Figure 6.15 $n = 25$ kN/m run
                   $t = 215$ mm
                   $h = 3450$ mm

Thus considering 1 m length of wall:

$$\text{design lateral strength, } q_{lat} = \frac{8 \times 0.215 \times 25}{3.45^2 \times 2.5}$$

$$= 1.45 \text{ kN/m height}$$

i.e. design lateral strength of wall = 1.45 kN/m^2 on elevation

Thus the wall is capable of resisting a design lateral loading of 1.45 kN/m^2 on elevation,

i.e. $\gamma_f W_k = 1.45$ kN/m^2.

The strength of the units required and the mortar should then be determined as in Chapter 5, the stresses due to the lateral loading being ignored as far as this part of the design is concerned.

The lateral strength is an inherent property of the wall, relative to the particular vertical loading being considered. On the basis of the formula given in the Code, it should be noted that the lateral strength varies in direct proportion to the characteristic dead load, and in inverse proportion to the square of the height of the member.

6.6.2 Vertical arching: return walls

In situations where walls or columns are considered suitable for design on the basis of vertical arching, see 6.6.1, and are supported by return walls of suitable strength, the design lateral strength may be increased. If such return walls are provided, a proportion of the applied loading will be resisted by horizontal spanning of the member, decreasing the amount to be resisted by vertical spanning. The return walls must, of course, be capable of resisting the horizontal reactions transmitted to them. The Code provides for an increase in the value obtained from the formula for the design lateral strength. The amount of the increase depends on whether return walls are provided on one or both sides,

and on the ratio of the clear height, h, to the length, L, of the member. Values for the appropriate modification factor, k, to be applied to the design lateral strength obtained from the formula in 6.6.1 are given in Table 6.3.

Table 6.3 Design lateral strength with returns = $k \times q_{lat}$ (BS 5628, Table 10)

Number of returns		Value of k			
	$\dfrac{L}{h} =$	0.75	1.0	2.0	3.0
1		1.6	1.5	1.1	1.0
2		4.0	3.0	1.5	1.2

EXAMPLE 7

Determine the design lateral strength of the wall in Figure 6.15, if it is provided with returns, as shown in Figure 6.16.

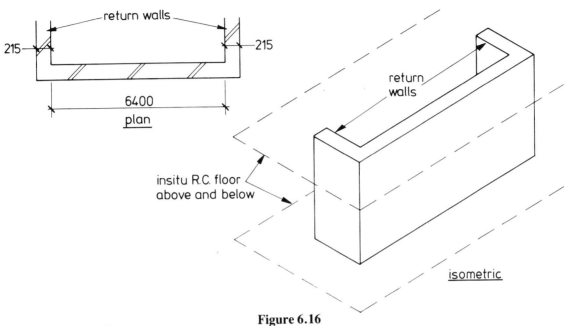

Figure 6.16

Modification factor, k:

$L = 6.4$, $h = 3.45$.

Therefore, $L/h = 1.85$, two returns.

Thus from Table 6.3, the modification factor may be taken as $k = 1.5$.

Thus

$$\text{design lateral strength} = kq_{\text{lat}} = 1.5 \times 1.45$$
$$= 2.18 \text{ kN/m}^2 \text{ on elevation}$$

The return walls should, of course, be checked to ensure that they are capable of resisting the horizontal reactions. The appropriate method of design is given later in this chapter. The loading on the walls may reasonably be assumed to be that on the area of wall on elevation, enclosed by lines drawn at 45° to the corners, as illustrated in Figure 6.17.

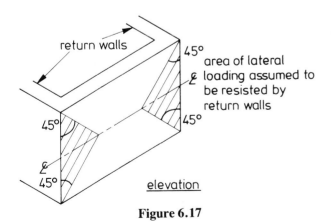

Figure 6.17

6.6.3 Horizontal arching

In the case of walls with minimal axial loading, but which are built 'solidly' between supports capable of resisting an arch thrust, the Code provides a method of design based on the assumption that, under lateral loading, a horizontal arch is developed within the thickness of the wall. A similar assumption could be made when a number of walls are built continuously past supports (see Figure 6.18).

The warnings given in section 6.6.1 are equally applicable when horizontal arch thrusts are being considered. A small change in the length of a wall in arching can considerably reduce the arching resistance. Very careful consideration is required as all masonry will move, either expanding or contracting, due to the effects of temperature, moisture, etc. In practice, it is also difficult to butt masonry tightly to the flanges of steel columns, etc.

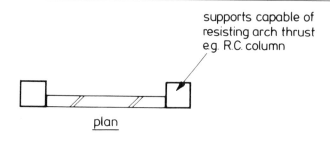

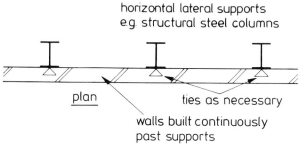

ties as necessary

walls built continuously
past supports

NB the requirements for movement
joints should be carefully considered

Figure 6.18

Longer lengths of masonry should always be provided with joints, and the position of the joints should be considered before this method of analysis is adopted. Designers must make certain that no joints are introduced into walls which have been designed on the basis of arching.

Unlike vertical arching, which is essentially an inherent property of an axially loaded wall, horizontal arching has to be assessed on the basis of the applied lateral loading, the compressive strength of the masonry, and the effectiveness of the junction between the wall and the supports. The effectiveness of the wall/support junction, and the ability of the support to satisfactorily resist the arch thrust – without, for example, excessive deflection, which would invalidate the design assumptions – require careful attention to

detail. In particular, the wall/support junction should be solidly filled with mortar.

The design analysis given in the Code is shown in Figure 6.19 and is similar to that for vertical arching. However, the width of the bearing at the assumed 'pinned' joints, see section 6.6.1, is taken as one tenth of the wall thickness – the method adopted being similar to the design of cracked sections, see section 6.3.2.

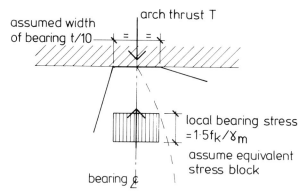

Figure 6.20 Enlarged detail at A, bearing C and B similar

From Figures 6.19 and 6.20:

Taking moments at centre line of bearing C for the external forces

$$q_{\text{lat}} \times \frac{L}{2} \times \frac{L}{2} = T\left(t - \delta - \frac{t}{10}\right) + q_{\text{lat}} \times \frac{L}{2} \times \frac{L}{4}$$

But
arch thrust, T = bearing stress × area

$$= 1.5 \frac{f_k}{\gamma_m} \times \frac{t}{10} \text{ per unit height}$$

Therefore

$$q_{\text{lat}} \frac{L^2}{4} = 1.5 \frac{f_k}{\gamma_m} \times \frac{t}{10}\left(t - \delta - \frac{t}{10}\right) + q_{\text{lat}} \frac{L^2}{8}$$

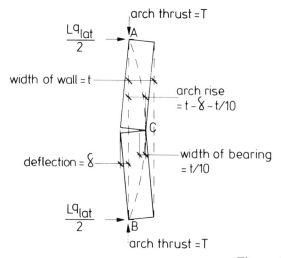

Figure 6.19

The Code recommends that, if the ratio of length to thickness, i.e. L/t, is 25 or less, the deflection under the design lateral load can be ignored. If L/t exceeds 25, then allowance should be made.

Thus if $L/t \leqslant 25$, δ assumed $= 0$:

$$q_{lat}\frac{L^2}{8} = 1.5\,\frac{f_k}{\gamma_m} \times \frac{t}{10}\left(t - \frac{t}{10}\right)$$

$$q_{lat} = 1.5\,\frac{f_k}{\gamma_m} \times 9 \times \frac{t^2}{100} \times \frac{8}{L^2}$$

$$q_{lat} = \frac{108}{100} \times \frac{f_k}{\gamma_m} \times \left(\frac{t}{L}\right)^2$$

Considering the design assumptions, etc., the factor 108/100 is ignored, and this formula is then given in the Code as the design lateral strength:

$$q_{lat} = \frac{f_k}{\gamma_m}\left(\frac{t}{L}\right)^2$$

where

q_{lat} = design lateral strength per unit area of wall

t = overall thickness

f_k = characteristic compressive strength of masonry (see Chapter 5)

L = length of wall

γ_m = partial factor of safety for materials (see Table 5.11).

The Code also states that the supporting structure has to be designed to be capable of resisting the arch thrust with negligible deformation.

EXAMPLE 8

Determine the compressive strength required for bricks in the wall shown in Figure 6.21, to resist a design lateral loading, $\gamma_f W_k$ of 3.3 kN/m². The bricks are to be laid in mortar designation (ii), and γ_m = 2.5 (special/special).

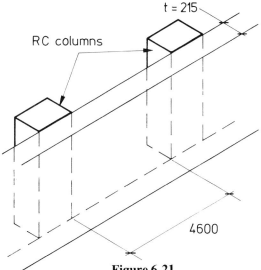

Figure 6.21

Applied design lateral loading = 3.3 kN/m².
Therefore, design lateral strength, q_{lat}, required = 3.3 kN/m².

$L/t = 4600/215 = 21$, i.e. $L/t \leqslant 25$ and therefore

$$q_{lat} = \frac{f_k}{\gamma_m}\left(\frac{t}{L}\right)^2$$

Therefore, knowing q_{lat} required:

$$f_k \text{ required} = \frac{q_{lat} \text{ required} \times \gamma_m}{(t/L)^2}$$

$$= \frac{3.3 \times 2.5 \times 10^{-3}}{(215/4600)^2} = 3.78 \text{ N/mm}^2$$

Thus from Table 5.4, with mortar designation (ii), units with a compressive strength of 10 N/mm² are required.

6.7 FREE-STANDING WALLS

6.7.1 General

Free-standing walls may be external boundary walls, parapet walls or internal walls where no restraint is provided to the top or sides of the wall. They are designed as vertical cantilevers, allowance being made for the stability moment due to the self-weight of the wall.

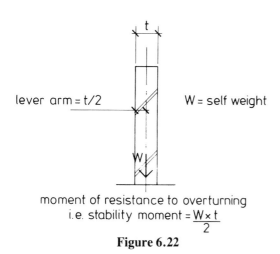

moment of resistance to overturning
i.e. stability moment = $\dfrac{W \times t}{2}$

Figure 6.22

Consider a section of wall as in Figure 6.22. When a lateral loading is applied to the wall, the tendency to overturn will be resisted by a moment due to the product of the self-weight of the wall and the lever arm. The lever arm being the distance from the line of action of the self-weight of the wall and the point about which the wall is tending to overturn. This moment is termed the stability moment (see also section 6.3.2).

The walls are designed to cantilever, either from the top of the foundations or from the point of horizontal lateral restraint – provided that the restraint is capable of resisting the horizontal reaction or shear from the wall. When stiffer elements, such as piers, are introduced into a free-standing wall, the sections of wall between the piers may be designed as panel walls supported on three sides, or spanning horizontally (see later, section 6.9) and the piers themselves designed as cantilevers to resist the reactions from the panel. The Code recommends that, the mortar used for free-standing walls should not be weaker than designation (iii). In addition, it recommends limiting the height of a free-standing wall to twelve times its effective thickness.

6.7.2 Design bending moments

The calculation of the design bending moments of a free-standing wall is based on simple statics,

the design bending moment on a wall being assessed by taking moments about a particular point. As shown in Figure 6.23, when a wall is subjected to a uniformly distributed wind loading of W_k, and a horizontal line loading of Q_k, the bending moment is obtained as follows:

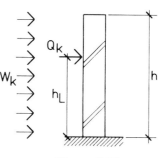

Figure 6.23

Total characteristic wind load, W_k	$= W_k \times h$ per unit length
Applied design wind load	$= W_k h \gamma_f$ per unit length
Applied design bending moment, M_A, about face of wall due to wind	$=$ load $\times$ lever arm $= W_k h \gamma_f \times h/2$
Characteristic imposed lateral load	$= Q_k$
Applied design imposed lateral load	$= \gamma_f Q_k$
Applied design bending moment, M_A, about face of wall due to imposed load	$=$ load $\times$ lever arm $= \gamma_f Q_k \times h_L$
Therefore, total design bending moment, M_A	$= W_k \gamma_f h^2/2 + Q_k \gamma_f h_L$

where

W_k = characteristic wind load (see section 5.2)

γ_f = partial safety factor for loads (see section 5.3; Note: factors only applicable to free-standing walls, Table 5.1.)

h = clear height of wall or pier above restraint, assuming moments taken about this point

Q_k = characteristic imposed load (see section 5.2)

h_L = the vertical distance between the point of application of the horizontal load, Q_k, and the lateral restraint – assuming moments taken about this point.

EXAMPLE 9

A 6 m high internal free-standing wall in a bus depot is to be designed for an internal wind loading of 0.2 kN/m², and an imposed loading of 0.74 kN/m run from a handrail fixed to the wall 1.0 m above floor level (see Figure 6.24).

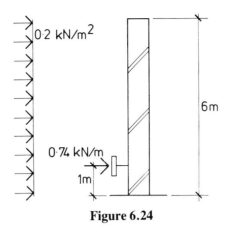

Figure 6.24

$\gamma_f = 1.2$, see Table 5.1

Design wind load	$= 0.2 \times 6 \times 1.2$	$= 1.44$ kN/m run
Design imposed load	$= 0.74 \times 1.2$	$= 0.89$ kN/m run
Design bending moment	$= (1.44 \times 6/2) + (0.89 \times 1.0)$	
	$= 4.32 + 0.89$	$= 5.21$ kN/m run

6.7.3 Design moment of resistance

The design moment of resistance must be assessed on the basis of either an uncracked or a cracked section, in accordance with sections 6.3.1 and 6.3.2, depending on whether a damp proof course capable of transmitting the tensile stresses is included. In external boundary walls, the damp proof course often consists of two courses of engineering bricks, which would probably be capable of transmitting the tensile stresses. In order to achieve the best use of materials, free-standing walls are often built in geometrical configurations other than the usual rectangular section. Fin walls and diaphragm walls are suitable, as also are chevron, curved and zig-zag walls. The relevant section properties should be used, therefore, when deriving the design moment of resistance for each particular section being considered.

6.8 RETAINING WALLS

Retaining walls are generally considered to be free-standing walls retaining earth, liquid or stored material, on one side. The design procedure is similar to that for free-standing walls, as discussed in section 6.7. According to BS 5628, the earth pressure should be treated as an imposed load, and the characteristic value taken as the active pressure in accordance with the relevant Code of Practice for earth-retaining structures.

6.9 PANEL WALLS

In section 6.7, free-standing walls were considered – a free-standing wall generally being considered as a wall with no restraint at the top and sides, and little or no imposed vertical loading. Such walls behave as vertical cantilevers. In section 6.6, the lateral strength of axially loaded walls was discussed, and from this the increased lateral resistance of a wall, due to the application of a vertical imposed loading, was demonstrated.

In buildings, walls are seldom free-standing, being usually continuous past floors, and with return walls, columns or piers providing restraint to the sides. Thus the wall is no longer a simple vertical cantilever, and may be continuous in one or two perpendicular directions. If such a wall also performs a supporting function, in addition to its enclosing or cladding function, the increased lateral resistance due to the vertical loading will mean that the panel is much stiffer in the vertical direction – providing most resistance in this direction and spanning vertically. If, on the other hand, the wall does not perform a loadbearing function by supporting the vertical loading, and acts merely as a cladding, it is often termed a 'panel wall'. The term is usually applied to a non-loadbearing cladding wall supported by a structural frame. Walls of this kind are subjected to mainly lateral loading, with little or no vertical loading other than their self-weight.

Basically, however, this is structurally wasteful of the high capacity of masonry to support vertical loading.

6.9.1 Limiting dimensions

The Code, BS 5628, recommends various limiting dimensions to ensure that panel walls are not too slender. The recommended heights and lengths vary according to the support conditions at the panel edges. Figures 6.25–6.27 show typical examples and provide relevant values for masonry set in mortar designations (i) to (iv) designed in accordance with the Code.

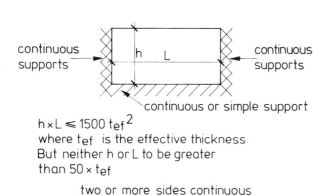

$h \times L \leqslant 1500\, t_{ef}^{2}$
where t_{ef} is the effective thickness.
But neither h or L to be greater
than $50 \times t_{ef}$

two or more sides continuous

$h \times L \leqslant 1350\, t_{ef}^{2}$
neither h or L to be greater
than $50 \times t_{ef}$

all other cases

(a) Panel supported on three edges

Figure 6.25

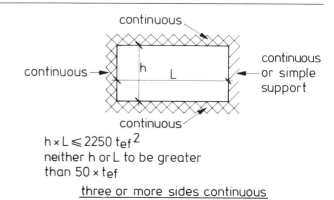

$h \times L \leqslant 2250\, t_{ef}^{2}$
neither h or L to be greater
than $50 \times t_{ef}$

three or more sides continuous

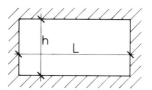

$h \times L \leqslant 2025\, t_{ef}^{2}$
neither h or L to be greater
than $50 \times t_{ef}$

all other cases

(b) Panel supported on four edges

Figure 6.26

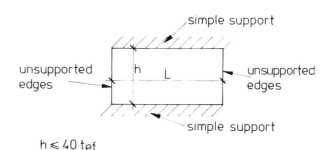

$h \leqslant 40\, t_{ef}$

(c) Panel simply supported top and bottom

Figure 6.27

The bending moments and shear forces capable of being resisted by panel walls vary with the edge support conditions. Panels may be simply supported, fully continuous or free, depending on the support condition at any particular edge. Supports are generally formed vertically by piers, intersecting or return walls, or steel or concrete columns. Lateral supports are provided by roofs, floors and foundations.

A simple support may be assumed where a panel is adequately tied to the supporting structure with metal wall ties, or similar. 'Tied', in this context, means a connection capable of resisting the tensile or compressive load, depending on the direction of loading which, in the case of

wind load, may be in both directions, as wind loading is reversible. The supporting structure must obviously be able to resist the applied loading, and the wall ties or other fixings should be designed to transfer this loading (see section 6.9.2). A simple support may also be generally assumed where a dpc occurs, although the effectiveness of the support provided should be checked by calculation.

Continuity may be assumed where masonry is provided with return ends, or is continuous past and tied to a column or beam. (Again, in all cases, the supporting structure must be capable of resisting the applied loading.) In the case of cavity walls, only one leaf need be continuous, provided wall ties are used (see section 6.4.3) between the two leaves and between each section of the discontinuous leaf and the support. Where the leaves are of differing thickness, the thicker leaf should be continuous in accordance with the Code.

Typical examples of support conditions are illustrated in Figures 6.28 and 6.29.

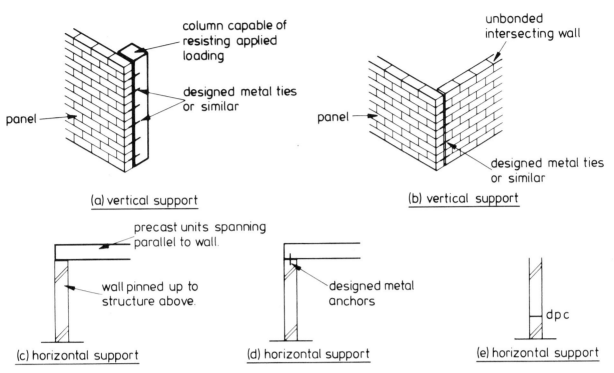

Figure 6.28 Details providing simple support conditions

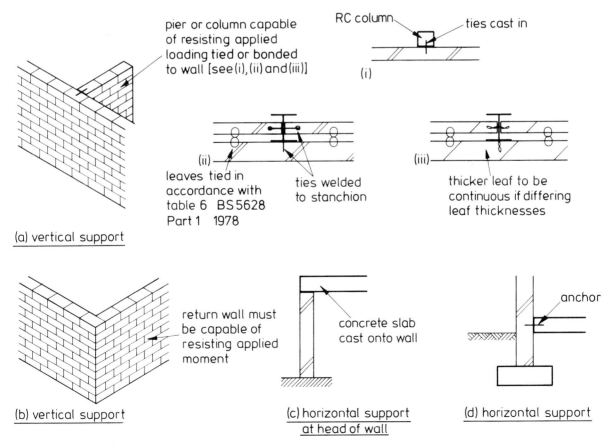

Figure 6.29 Details providing continuity at supports

6.9.2 Design methods

As with many other structural elements or forms, walls subject to mainly lateral loads are not capable of precise calculation. The Code provides two approximate methods which may be used to assess the strength of such walls:

(a) as a panel supported on a number of sides
(b) as an arch spanning between supports.

Walls of irregular shape, or those with openings, require careful consideration and neither of the above methods can be used directly. Although some guidance is given in Appendix D of the Code, each case should be studied carefully by the designer to assess reasonable assumptions to be used in the design. (For further information and examples, see Chapter 11.)

The alternative method (b) is described in section 6.6.3. Method (a) is dealt with in the sections which follow, 6.9.3–6.9.5.

6.9.3 Design bending moment

The design bending moments vary with the design load, the vertical and horizontal spans, the orthogonal ratio, μ, and the relevant design bending moment coefficient, α. The orthogonal ratio was discussed in section 6.2.1. The bending moment coefficient, α, depends on the edge support conditions, i.e. continuous, discontinuous or free, and the position of the section under consideration.

The design bending moment per unit height of a panel in the horizontal direction may, in accordance with the Code, be expressed as follows:

Applied design bending moment,

$$M_A = \alpha W_k \gamma_f L^2$$

when the plane of bending is perpendicular to the bed joints,

where

α = the bending moment coefficient from Table 6.4
γ_f = the partial safety factor for loads (see Table 5.1)
L = the length of the panel between supports
W_k = the characteristic wind load per unit area (see section 5.2).

The Code states that, at a damp proof course, the bending moment coefficient, α, may be taken as for an edge over which full continuity exists when there is sufficient vertical load on the dpc to ensure that its flexural strength is not exceeded. This means that full continuity may be assumed for the determination of α, provided the dpc can transmit the tensile stresses or, if not, that there is sufficient vertical loading to ensure that no tensile stresses are developed across the section, i.e. as in case (1) Figure 6.2.

The applied design moment in the vertical direction (M_A) is given by the following formula:

$$M_A = \mu \, \alpha \, W_k \gamma_f L^2$$

when the plane of bending is parallel to the bed joints, provided that the edge conditions justify treating the panel as partially fixed, where μ = the orthogonal ratio modified for the vertical loading as appropriate (see section 6.2.1). The remaining terms being as for the horizontal moment.

Values of the bending moment coefficient, α, are given in Table 6.4 for various values of the orthogonal ratio, μ, modified for vertical loading as necessary (see section 6.2.1), various edge conditions, and various ratios of height to length. This table is based on Table 9 of BS 5628.

Table 6.4 Bending moment coefficients in laterally loaded wall panels (BS 5628, Table 9)

Note 1. Linear interpolation of μ and h/L is permitted.

Note 2. When the dimensions of a wall are outside the range of h/L given in this table, it will usually be sufficient to calculate the moments on the basis of a simple span. For example, a panel of type A having h/L less than 0.3 will tend to act as a free-standing wall, whilst the same panel having h/L greater than 1.75 will tend to span horizontally.

Key to support conditions

———— denotes free edge

⁄⁄⁄⁄⁄⁄ simply supported edge

⨯⨯⨯⨯⨯ an edge over which full continuity exists.

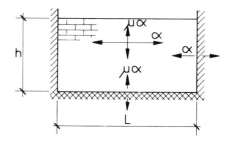

	μ	Values of α h/L 0.30	0.50	0.75	1.00	1.25	1.50	1.75
A	1.00	0.031	0.045	0.059	0.071	0.079	0.085	0.090
	0.90	0.032	0.047	0.061	0.073	0.081	0.087	0.092
	0.80	0.034	0.049	0.064	0.075	0.083	0.089	0.093
	0.70	0.035	0.051	0.066	0.077	0.085	0.091	0.095
	0.60	0.038	0.053	0.069	0.080	0.088	0.093	0.097
	0.50	0.040	0.056	0.073	0.083	0.090	0.095	0.099
	0.40	0.043	0.061	0.077	0.087	0.093	0.098	0.101
	0.35	0.045	0.064	0.080	0.089	0.095	0.100	0.103
	0.30	0.048	0.067	0.082	0.091	0.097	0.101	0.104
B	1.00	0.024	0.035	0.046	0.053	0.059	0.062	0.065
	0.90	0.025	0.036	0.047	0.055	0.060	0.063	0.066
	0.80	0.027	0.037	0.049	0.056	0.061	0.065	0.067
	0.70	0.028	0.039	0.051	0.058	0.062	0.066	0.068
	0.60	0.030	0.042	0.053	0.059	0.064	0.067	0.069
	0.50	0.031	0.044	0.055	0.061	0.066	0.069	0.071
	0.40	0.034	0.047	0.057	0.063	0.067	0.070	0.072
	0.35	0.035	0.049	0.059	0.065	0.068	0.071	0.073
	0.30	0.037	0.051	0.061	0.066	0.070	0.072	0.074
C	1.00	0.020	0.028	0.037	0.042	0.045	0.048	0.050
	0.90	0.021	0.029	0.038	0.043	0.046	0.048	0.050
	0.80	0.022	0.031	0.039	0.043	0.047	0.049	0.051
	0.70	0.023	0.032	0.040	0.044	0.048	0.050	0.051
	0.60	0.024	0.034	0.041	0.046	0.049	0.051	0.052
	0.50	0.025	0.035	0.043	0.047	0.050	0.052	0.053
	0.40	0.027	0.038	0.044	0.048	0.051	0.053	0.054
	0.35	0.029	0.039	0.045	0.049	0.052	0.053	0.054
	0.30	0.030	0.040	0.046	0.050	0.052	0.054	0.055
D	1.00	0.013	0.021	0.029	0.035	0.040	0.043	0.045
	0.90	0.014	0.022	0.031	0.036	0.040	0.043	0.046
	0.80	0.015	0.023	0.032	0.038	0.041	0.044	0.047
	0.70	0.016	0.025	0.033	0.039	0.043	0.045	0.047
	0.60	0.017	0.026	0.035	0.040	0.044	0.046	0.048
	0.50	0.018	0.028	0.037	0.042	0.045	0.048	0.050
	0.40	0.020	0.031	0.039	0.043	0.047	0.049	0.051
	0.35	0.022	0.032	0.040	0.044	0.048	0.050	0.051
	0.30	0.023	0.034	0.041	0.046	0.049	0.051	0.052
E	1.00	0.008	0.018	0.030	0.042	0.051	0.059	0.066
	0.90	0.009	0.019	0.032	0.044	0.054	0.062	0.068
	0.80	0.010	0.021	0.035	0.046	0.056	0.064	0.071
	0.70	0.011	0.023	0.037	0.049	0.059	0.067	0.073
	0.60	0.012	0.025	0.040	0.053	0.062	0.070	0.076
	0.50	0.014	0.028	0.044	0.057	0.066	0.074	0.080
	0.40	0.017	0.032	0.049	0.062	0.071	0.078	0.084
	0.35	0.018	0.035	0.052	0.064	0.074	0.081	0.086
	0.30	0.020	0.038	0.055	0.068	0.077	0.083	0.089
F	1.00	0.008	0.016	0.026	0.034	0.041	0.046	0.051
	0.90	0.008	0.017	0.027	0.036	0.042	0.048	0.052
	0.80	0.009	0.018	0.029	0.037	0.044	0.049	0.054
	0.70	0.010	0.020	0.031	0.039	0.046	0.051	0.055
	0.60	0.011	0.022	0.033	0.042	0.048	0.053	0.057
	0.50	0.013	0.024	0.036	0.044	0.051	0.056	0.059
	0.40	0.015	0.027	0.039	0.048	0.054	0.058	0.062
	0.35	0.016	0.029	0.041	0.050	0.055	0.060	0.063
	0.30	0.018	0.031	0.044	0.052	0.057	0.062	0.065

	Values of α						
	h/L						
μ	0.30	0.50	0.75	1.00	1.25	1.50	1.75
G							
1.00	0.007	0.014	0.022	0.028	0.033	0.037	0.040
0.90	0.008	0.015	0.023	0.029	0.034	0.038	0.041
0.80	0.008	0.016	0.024	0.031	0.035	0.039	0.042
0.70	0.009	0.017	0.026	0.032	0.037	0.040	0.043
0.60	0.010	0.019	0.028	0.034	0.038	0.042	0.044
0.50	0.011	0.021	0.030	0.036	0.040	0.043	0.046
0.40	0.013	0.023	0.032	0.038	0.042	0.045	0.047
0.35	0.014	0.025	0.033	0.039	0.043	0.046	0.048
0.30	0.016	0.026	0.035	0.041	0.044	0.047	0.049
H							
1.00	0.005	0.011	0.018	0.024	0.029	0.033	0.036
0.90	0.006	0.012	0.019	0.025	0.030	0.034	0.037
0.80	0.006	0.013	0.020	0.027	0.032	0.035	0.038
0.70	0.007	0.014	0.022	0.028	0.033	0.037	0.040
0.60	0.008	0.015	0.024	0.030	0.035	0.038	0.041
0.50	0.009	0.017	0.025	0.032	0.036	0.040	0.043
0.40	0.010	0.019	0.028	0.034	0.039	0.042	0.045
0.35	0.011	0.021	0.029	0.036	0.040	0.043	0.046
0.30	0.013	0.022	0.031	0.037	0.041	0.044	0.047
I							
1.00	0.004	0.009	0.015	0.021	0.026	0.030	0.033
0.90	0.004	0.010	0.016	0.022	0.027	0.031	0.034
0.80	0.005	0.010	0.017	0.023	0.028	0.032	0.035
0.70	0.005	0.011	0.019	0.025	0.030	0.033	0.037
0.60	0.006	0.013	0.020	0.026	0.031	0.035	0.038
0.50	0.007	0.014	0.022	0.028	0.033	0.037	0.040
0.40	0.008	0.016	0.024	0.031	0.035	0.039	0.042
0.35	0.009	0.017	0.026	0.032	0.037	0.040	0.043
0.30	0.010	0.019	0.028	0.034	0.033	0.042	0.044

Key to support conditions

——— denotes free edge

///// simply supported edge

XXXXX an edge over which full continuity exists

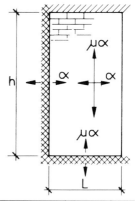

	Values of α						
	h/L						
μ	0.30	0.50	0.75	1.00	1.25	1.50	1.75
J							
1.00	0.009	0.023	0.046	0.071	0.096	0.122	0.151
0.90	0.010	0.026	0.050	0.076	0.103	0.131	0.162
0.80	0.012	0.028	0.054	0.083	0.111	0.142	0.175
0.70	0.013	0.032	0.060	0.091	0.121	0.156	0.191
0.60	0.015	0.036	0.067	0.100	0.135	0.173	0.211
0.50	0.018	0.042	0.077	0.113	0.153	0.195	0.237
0.40	0.021	0.050	0.090	0.131	0.177	0.225	0.272
0.35	0.024	0.055	0.098	0.144	0.194	0.244	0.296
0.30	0.027	0.062	0.108	0.160	0.214	0.269	0.325

K	1.00	0.009	0.021	0.038	0.056	0.074	0.091	0.108
	0.90	0.010	0.023	0.041	0.060	0.079	0.097	0.113
	0.80	0.011	0.025	0.045	0.065	0.084	0.103	0.120
	0.70	0.012	0.028	0.049	0.070	0.091	0.110	0.128
	0.60	0.014	0.031	0.054	0.077	0.099	0.119	0.138
	0.50	0.016	0.035	0.061	0.085	0.109	0.130	0.149
	0.40	0.019	0.041	0.069	0.097	0.121	0.144	0.164
	0.35	0.021	0.045	0.075	0.104	0.129	0.152	0.173
	0.30	0.024	0.050	0.082	0.112	0.139	0.162	0.183
L	1.00	0.006	0.015	0.029	0.044	0.059	0.073	0.088
	0.90	0.007	0.017	0.032	0.047	0.063	0.078	0.093
	0.80	0.008	0.018	0.034	0.051	0.067	0.084	0.099
	0.70	0.009	0.021	0.038	0.056	0.073	0.090	0.106
	0.60	0.010	0.023	0.042	0.061	0.080	0.098	0.115
	0.50	0.012	0.027	0.048	0.068	0.089	0.108	0.126
	0.40	0.014	0.032	0.055	0.078	0.100	0.121	0.139
	0.35	0.016	0.035	0.060	0.084	0.108	0.129	0.148
	0.30	0.018	0.039	0.066	0.092	0.116	0.138	0.158

Table 6.5 Characteristic strengths of wall ties used as panel supports (BS 5628, Table 8)

Type	Characteristic strengths of ties engaged in dovetail slots set in structural concrete	
	Tension (kN)	Shear (kN)
Dovetail slot types of ties		
(a) Galvanized or stainless steel fishtail anchors 3 mm thick, 17 mm min. width in 1.25 mm thick galvanized or stainless steel slots, 150 mm long, set in structural concrete	4.0	5.0
(b) Galvanized or stainless steel fishtail anchors 2 mm thick, 17 mm min. width, in 2 mm thick galvanized or stainless steel slots, 150 mm long, set in structural concrete	3.0	4.5
(c) Copper fishtail anchors 3 mm thick, 17 mm min. width, in 1.25 mm copper slots, 150 mm long, set in structural concrete	3.5	4.0

	Characteristic loads in ties embedded in mortar			
	Tension (kN)			Shear* (kN)
	Mortar designations			Mortar designation
	(i) and (ii)	(iii)	(iv)	(i), (ii) or (iii)
Cavity wall ties[†]				
(a) Wire butterfly type: Zinc coated mild steel or stainless steel	3.0	2.5	2.0	2.0
(b) Vertical twist type: Zinc coated mild steel or bronze or stainless steel	5.0	4.0	2.5	3.5
(c) Double-triangle type: Zinc coated mild steel or bronze or stainless steel	5.0	4.0	2.5	3.0

* Applicable only to cases where shear exists between closely abutting surfaces.
[†] See BS 1243: 1978.

6.9.4 Design moments of resistance

The design moments of resistance are those based on an uncracked section, as given in section 6.3.1

6.9.5 Design of ties

Where wall ties are used to provide simple support, they should be checked to ensure that they can adequately resist the applied loadings.

Values for the characteristic strengths of wall ties used as panel supports are given in Table 6.5. The value of the partial factor of safety, γ_m, for use with ties used as supports, should be 3.0. However, when considering the probable effects of misuse or accidental damage, this value may be halved. The reaction along an edge of a wall due to the design load may normally be assumed to be uniformly distributed for the purposes of designing the means of support. The load to be transmitted from a panel to its support may, in the case of simple supports, be taken by ties to one leaf only, provided that there is adequate connection between the two leaves.

EXAMPLE 10

Design suitable ties for the wall load shown in Figure 6.30.

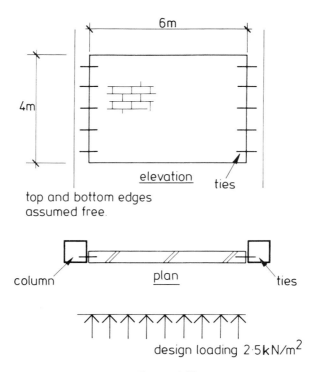

Figure 6.30

Total design load = $2.5 \times 4 \times 6$ = 60 kN

Reaction at each edge = $\dfrac{60}{2}$ = 30 kN

Try ties type (a), Table 6.5:

Characteristic load in shear = 5.0 kN

Design load in shear = $\dfrac{5.0}{\gamma_m} = \dfrac{5.0}{3}$ = 1.67 kN

Number of ties required = $\dfrac{30}{1.67}$ = 17.96, i.e. 18

Therefore, provide type (a) ties at $\dfrac{4000}{18}$ = 222 mm ‰, i.e. vertically at 225 mm ‰.

6.10 ECCENTRICITY OF LOADING IN PLANE OF WALL

In crosswall construction, and other similar structural forms, the walls are subject to loading in the plane of the wall. This is usually the result of wind loading on the building elevation. The walls then act as vertical cantilevers, stability and the resistance moment being provided by the vertical loading in the wall (see Figure 6.31).

The resultant eccentricity of the combined lateral and vertical loading is calculated from consideration of statics. All the stresses in the walls are then calculated by an elastic analysis, and the increased compressive stress due to the wind loading in the plane of the wall should be checked against the wall's tendency to buckle. An example is given in Chapter 11.

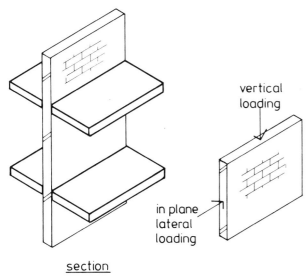

section

Figure 6.31

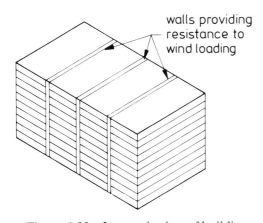

Figure 6.32   Isometric view of building

Where, as in Figure 6.32, several walls provide resistance to wind loading, the horizontal force should be distributed between the walls in proportion to their flexural stiffness at right angles to the direction of the force. Thus in unreinforced masonry, where Young's modulus will not vary from wall to wall, the horizontal force is distributed between the walls in relation to their respective moments of inertia. The connections between the various walls must be capable of transmitting the necessary loadings.

It is important to check that the floors, roof, etc., which are generally used to distribute the loading from the elevations to the crosswalls, are capable

of distributing them in the proportion determined from the respective I values of the walls.

6.10.1 Design of walls loaded eccentrically in the plane of the wall

In the case of crosswalls, and walls acting in a similar fashion, the intensity of loading at any particular position should, in accordance with the Code, be assessed on the basis of the load distribution shown in Figure 6.33. The strength of the wall should then be determined in accordance with sections 5.13 and 5.14.

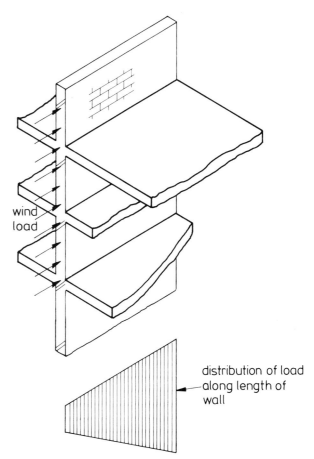

Figure 6.33 Load distribution from loading eccentric to plane of wall

6.11 WALLS SUBJECTED TO SHEAR FORCES

So far, consideration has been given to masonry subject to flexural bending stresses. However, when members are subjected to bending, they are also required to resist shear forces, and the resulting shear stresses must generally be investigated.

6.11.1 Characteristic and design shear strength

The characteristic shear strength of masonry greatly depends on the axial stress or pre-load on the wall, or other member, and the mortar designation. The Code allows for the increase in shear strength due to vertical loading by allowing the characteristic shear stress to be increased by 0.6 times the stress due to vertical loading up to certain limiting values – both dead and imposed loading being considered to give the worst conditions.

In accordance with the Code, for mortar designations (i), (ii) and (iii), the characteristic shear strength for walls may be taken as follows:

$$f_v = (0.35 + 0.6g_A) \text{ N/mm}^2$$

with a limit of 1.75 N/mm², where g_A is the design vertical load per unit area of the wall cross-section due to vertical dead loads, and imposed load when the imposed load is permanent, calculated from the appropriate loading condition.

For mortar designation (iv), the characteristic shear strength for walls may be taken as:

$$f_v = (0.15 + 0.6g_A) \text{ N/mm}^2$$

with a maximum of 1.4 N/mm².

The design shear strength is then obtained by reducing the characteristic strength by an appropriate partial factor of safety, and is given by the formula, f_v/γ_{mv}, where γ_{mv} is the partial factor of safety.

For mortar not weaker than designation (iv), γ_{mv} = 2.5. When considering the probable effects of misuse or accident, γ_{mv} may be reduced to 1.25.

6.11.2 Resistance to shear

Provision against the ultimate limit state of shear being reached may be assumed where the shear stress due to the horizontal design load is less than, or equal to, the design shear strength:

$$v_h \leqslant \frac{f_v}{\gamma_{mv}}$$

where v_h is the shear stress produced by the horizontal design load calculated as acting uniformly over the horizontal cross-sectional area of the wall.

EXAMPLE 11

Check the shear stress at the base of the wall shown in Figure 6.34. Assume mortar designation (ii).

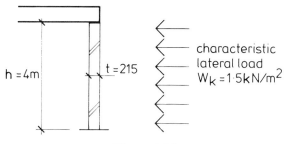

insitu slab
characteristic dead load G_k = 5kN/m

h = 4m   t = 215

characteristic lateral load W_k = 1·5kN/m²

Figure 6.34

Consider 1 m length of wall:

Applied design lateral load at base, V, (assume, for this example, half applied load resisted at base) $= \gamma_f W_k \times \dfrac{h}{2}$ per unit length

Appropriate γ_f, from Table 5.1 $= 1.4$

Therefore, $V = 1.4 \times 1.5 \times \dfrac{4}{2}$ $= 4.2$ kN/m run

Therefore, applied design shear stress, v_h, (assumed to act uniformly over horizontal cross-section of wall)

$$= \frac{4.2 \times 10^3}{10^3 \times 215}$$
$$= 0.019 \text{ N/mm}^2$$

Characteristic shear stress, f_v $= (0.35 + 0.6g_A) \text{ N/mm}^2$

$$g_A = \frac{\gamma_f G_k}{\text{cross-sectional area}}$$

Appropriate γ_f, from Table 5.1 $= 0.9$

Thus $g_A = \dfrac{0.9 \times 5 \times 10^3}{10^3 \times 215}$ $= 0.021 \text{ N/mm}^2$

Therefore, $f_v = 0.35 + 0.6 \times 0.021$ $= 0.36 \text{ N/mm}^2$

Therefore design resistance to shear stress $= \dfrac{f_v}{\gamma_{mv}} = \dfrac{0.36}{2.5}$

$= 0.14 \text{ N/mm}^2$

Thus $v_h < \dfrac{f_v}{\gamma_{mv}}$ and is therefore satisfactory.

CHAPTER 7

STRAPPING, PROPPING AND TYING OF LOADBEARING MASONRY

In all structures, it is essential that the design engineer, in making assumptions regarding the behaviour of the structure, ensures that all such assumptions are soundly based, practical to achieve, and adequately catered for in the structural details. There are few aspects of the design/detail process more involved in this problem than the strapping, propping and tying requirements of loadbearing masonry.

Assumptions regarding restraint, end support conditions, resistance to uplift forces, etc., must all be carefully considered and detailed. These points are just as important in small single-storey structures as they are in tall heavy engineering projects. In fact, modern lightweight cladding and partitions have meant that more and more building types are sensitive to relatively small changes in loading conditions. For example, when lightweight roof decking is used, stress reversal occurs at a much lower wind suction than for more heavy forms of roof cladding (see Figure 7.1).

In the past, reliance for stability, particularly on single-storey buildings, was often put upon traditional room-dividing walls without the need for calculations. However, since the introduction of lightweight non-loadbearing partitions, the stability of such structures has become critical for relatively small wind loadings (see Figure 7.2).

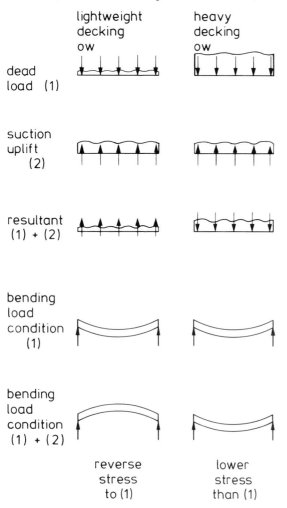

Figure 7.1

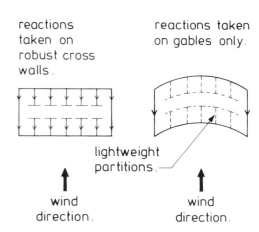

Figure 7.2 Plans on typical single-storey building

In the above examples, the need for adequate connections to transfer wind uplift and shear loads from the various elements to which they are applied into the elements on which they are to be resisted, is absolutely critical to the overall stability of the building. Wall restraints, uplift straps, ties and seating connections are far more

critical than many engineers and architects realise and, in masonry structures in particular, these
elements often become the items which determine the completed building's life. They are all
too often the ill-conceived weak link of the
structure which can be such that repair or replacement is more costly than demolition and
rebuilding. It is essential that more consideration
is given to corrosion and the life expectancy of
these elements for the conditions in which they
are built than has been given in the recent past.

For centuries, masonry structures have proved to
be very adaptable and have been handed down
from one generation to the next, and successfully
modified or altered to meet changing needs and
patterns of use. The economic and environmental advantages of this fact can scarcely be over-
emphasised. In recent years, however, the rate of
change in the approach to design has not kept
pace with the rapid changes in materials and
construction techniques. Numerous failures have
already occurred and, if the traditional virtues of
masonry construction are to be maintained, it is
important that the pitfalls of recent trends be
corrected in the next generation of buildings.

Extra attention to the function and expected life
of the small but essential elements already referred to will go a long way towards reducing the
number of future problems. The forces in these
elements, the practicability of the construction,
the environment in which they must survive and
the materials of which they are made, are most
important considerations.

The importance of both the overall stability and
the detailed connection of element to element
cannot be over-emphasised. Many building failures have resulted from the lack of an engineering check on overall stability and the connection
of element to element, and this is particularly
highlighted by hybrid structures where elements
of different materials have been designed by
different specialists and no overall stability check
has been made.

In high-rise construction, compliance with the
accidental damage requirements (see Chapter 8),
plus the greater awareness of stability problems
with tall buildings tend to lead engineers to a
more suitable and robust structure. This is not so
with low-rise structures, such as domestic dwell-
ings up to three storeys, certain schools, hostels,
old peoples homes, etc. The tendency here is to
stick to so called 'traditional' details and a 'tradi-

tional' approach to design. But modern construction methods and materials have produced 'non-
traditional' buildings and thus a greater need for
engineering design. Surely, engineering design is
as much to do with the joining of individual
elements to form a robust structure, as it is to do
with the design of the individual elements of
which it is composed.

It is this aspect of design, namely the strapping of
walls to floors and roofs, with which this chapter
is concerned and, in the broadest terms, the two
objectives could be summarised as:

(i) to identify situations where tying problems
    exist,
(ii) to suggest specimen details which would
    form the basis of a solution for each situa-
    tion.

Here, as in all engineering design, there are no
special details which are universally applicable
and the engineer must select his own solution
for the individual requirements of structural
adequacy. Whilst it is right and proper for codes
and building regulations to insist on the placing
of straps in certain situations, and to insist on a
minimum standard, the varied nature of tying
problems forbids placing too firm restrictions on
the engineer's freedom to deal with each case on
its merits. Nevertheless, the general guidance
given in this chapter should prove useful.

7.1 STRUCTURAL ACTION

Before highlighting the problem areas, it may be
useful to review quickly the way in which a
typical low-rise masonry structure would be
made to work. From this exercise, the reason for
the inclusion of ties in the design should become
evident and certain problem areas should
emerge.

The design of a solid wall under eccentric loading
can be undertaken using the basis of design in
Chapters 5 and 6. The strength of the material
and the geometric property of slenderness ratio
are the basis for the design. Assumptions do
need to be made concerning the support of the
wall at its top and at its base, in order that the
slenderness ratio can be calculated. It follows
that the design of such walls implicitly places on
the engineer the responsibility to ensure that
such edge conditions present in the completed
structure are as assumed in the design. The
easiest way and, indeed, in many cases the only
way, is to tie the wall to a floor or roof.

As an example, consider a two-storey house which is built unrestrained except for the roof which is carried by the wall. Such a wall (Figure 7.3), will have a possible failure mode similar to that shown. This is an inherently weak situation when viewed as a stability problem, and an extremely thick wall would be required on this basis.

It must be remembered, of course, that this presupposes that the wall has no returns. In the majority of domestic construction, returns are present if only because of the traditional approach to house building. Certainly, following the aftermath of Ronan Point, the multi-storey flats failure which drew attention to accidental damage and progressive collapse, the unreturned type of wall is no longer common, since good practice now suggests that a loadbearing panel should be strengthened with at least one substantial return wall. Having said this, however, the mode of failure shown in Figure 7.3 is possible towards the free edge of a three-sided panel, or in the central region of a four-sided one of large length. If the wall is restrained at floor level, the failure mode would change to the S shape shown in Figure 7.3, giving a smaller slenderness ratio which would allow a wall of less thickness to be justified in this case. Clearly then, there is a

distinct advantage to be gained in structural efficiency by restraining the deflection at all floor levels, as well of course at the roof level. Such restraint of horizontal movement can be of two main types.

Figure 7.4(A), shows schematically a pinned horizontal support. The connections between wall and floor do prevent horizontal movement but are not sufficiently rigid to afford any resistance to rotation on the wall at the floor levels. In Figure 7.4(B), however, the floors are carried through the walls. If the floors are of sufficient stiffness they afford a degree of resistance to rotation which results in more advantageous effective length, and hence slenderness ratio.

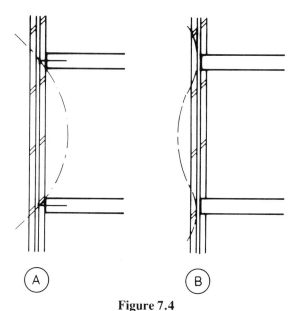

Figure 7.4

There is another stability requirement, however, regarding strapping, namely the workmanship and inbuilt 'out of balance' factor should the wall be built with an initial out of balance, i.e. as shown in Figure 7.5. Here a force P would be required to hold the wall stable against the moment effect of the applied dead and live load, W, which is now eccentric with respect to the base of the wall.

The moment equation is written:

$$W \times e = P \times h$$

neglecting the self-weight of the wall.

where

W = the vertical load
P = the required tie force
e = the eccentricity
h = the height (see Figure 7.5).

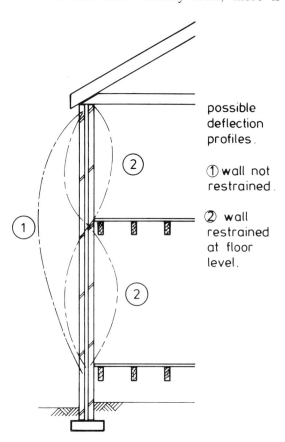

possible deflection profiles.

① wall not restrained.

② wall restrained at floor level.

Figure 7.3

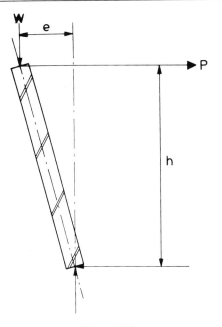

Figure 7.5

P is therefore related to the maximum load, W, to which the wall is likely to be subjected, and the eccentricity e.

If P is assumed to be $n\%$ of W then:

$$W \times e = \frac{nW}{100} h$$

This suggests that the eccentricity, e, is $n\%$ of h. At present, a figure of 2½% of the maximum applied weight is recommended for P which is sufficient to restrain a wall of 2.54 m storey height when it is built out of balance by 63 mm. Generally speaking, there is a consensus of opinion that 2½% is a figure of the right order which, furthermore, is common to concrete and steel

Codes of Practice and which allows a certain added factor to cover other aspects of workmanship detail and eccentricities due to deflections. It is certainly a figure which by previous practice has been shown to be both workable and adequate.

At this stage, the wall is restrained at the floor level and any out of balance will be resisted by the tie. Any wind loading, whether positive pressure or negative suction, will also need to be carried by the tie arrangements. It is necessary to consider the distribution of these forces when they have been transferred via the strapping into the masonry.

Consider a simple box type structure as shown in Figure 7.6

In section, the structure has a degree of stiffness based on the cantilever action of the walls. Nevertheless, it is basically a mechanism (unless some form of movement joint could be introduced) and, on application of wind load, the structure would sway through the position shown dotted on the figure as it collapsed. This mode of failure can be prevented by restraining the sway using walls A and B. If the floor is strong enough to act as a stiff plate the two walls C and D can be held vertical if the wind force is transferred into walls A and B. Both these walls are extremely stable and stiff when loaded with in-plane forces and made to act as a shear wall. To obtain this action, some form of shear key must be present, or introduced, between the floor plate and walls A and B at the floor level.

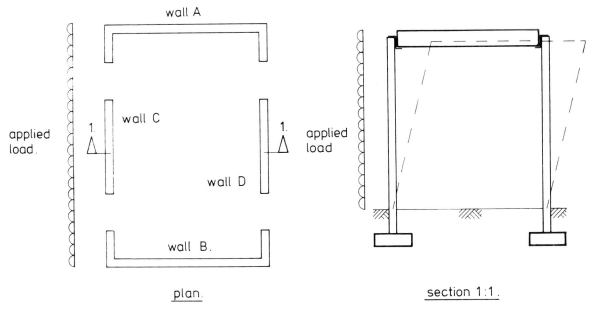

plan.   section 1:1.

Figure 7.6

Similarly the converse should apply. When a wind force is considered to act on walls A or B, they may not have sufficient returns to give total structural stability and, consequently, the wind forces when transferred into the floor plate must be taken via a shear key on to walls C and D. These two walls, acting as shear walls (together with any contributions from the four returns) provide the stability in the longitudinal direction of the building.

There are, therefore, two main actions to be considered. The first is to restrain the wall from horizontal movement at floor level, whilst the second is to provide a means of shear transfer between the floor and the wall. Each of these is considered in more detail below.

7.2 HORIZONTAL MOVEMENT

Floor bearing onto walls

Many possible arrangements exist to ensure that floors bear onto walls. These can range from the floor being *in situ* concrete which is carried in bearing on the full width of the wall, to the floor being of timber joists and boarding which is 'suspended' between the walls on joist hangers. A possible mid-range solution is shown in Figure 7.7, where the joists are carried part way into the inner leaf.

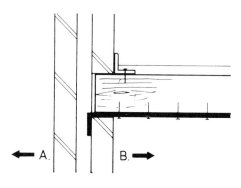

Figure 7.7

To be confident that the wall will have no detrimental movement, it must be strapped to the floor. This need not be at every joist, but must be at a spacing which can be justified by calculations. To stop the wall moving in direction A, an L-type strap could be used as shown. This would then act as a tie. Should the wall attempt to move in direction B relative to the floor, the tie will not restrain it and some other arrangement must be introduced to prohibit such movement. This could be an angle stud affixed to the top of the beam, as indicated, or a batten, positioned where the angle is shown to be,

running along the length of the wall and securely fixed to the joists.

Floors abutting walls

Where floors abut the wall, but are not spanning onto it, a similar type of strapping arrangement could be used (see Figure 7.8).

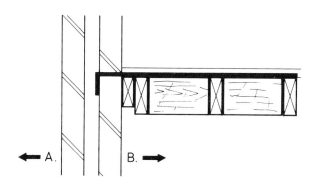

Figure 7.8

The L-shaped strap will restrain the wall from movement in direction A by acting as a tie. Should there be movement in direction B, however, the floor must be made to act as a stop. This can be achieved by packing between the wall and the first joist. The joist, however, is not particularly stiff in the situation where it may bend about its minor axis. To give this external joist added stiffness the first two or three joists could be braced either by diagonal strutting or by blocking as shown on the plan form. Such an arrangement will effectively restrain the wall from any horizontal deflection. Without the packing piece between wall and joist, it would be possible for the tie to either buckle or slip in the mortar joint and it is for this reason that is should be included in the joint detail.

Internal walls with floors solidly abutting on either side will have the same effect, in general, of restraining wall movements.

The foregoing details rely, however, on the assumption that the floor remains horizontally undeflected with respect to the rest of the structure, otherwise a 'house of cards' type of total collapse may ensue.

7.3 SHEAR KEYING BETWEEN WALL AND FLOORS

The final arrangement of ties to restrain all walls from horizontal movement is shown in Figure 7.9.

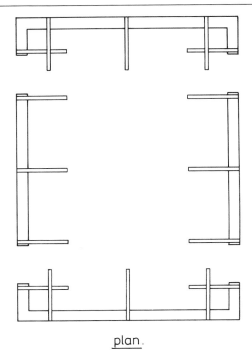

plan.

Figure 7.9

It can be seen that, while one set of straps is acting in tension (or the floor acting as a compression plate), the other set of straps, orthogonally placed with respect to the first set, will act as a shear key. In many cases, this amount of shear keying, combined possibly with other shear connections which are also present, such as joists carried into walls on joist hangers, will suffice. Nevertheless, there may be situations where the forces to be transmitted from the floor into the wall are of such magnitude that special fixings will need to be incorporated which are capable of resisting the shear forces involved.

7.4 HOLDING DOWN ROOFS SUBJECT TO UPWARD FORCES

Thus far, the need for connecting floors to walls has been due to horizontal forces acting on the wall.

Another major area of concern is the connection between the roof and the wall. All that has previously been said about strapping masonry to floors is still relevant to roofs. In many cases, however, there is an added problem of holding down a roof on which suction is liable to act. This is more relevant to flat roofs and pitched roofs of shallow slope. In the situation shown schematically in Figure 7.10, not only must the normal ties and shear keys be present (although they have been omitted for reasons of clarity from the figure), but also the roof must be sufficiently well connected to enough of the masonry to ensure

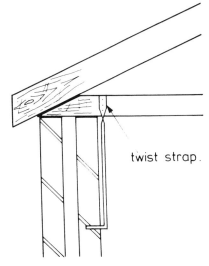

twist strap.

Figure 7.10

that it is still 'held down' when the worst suction acts on the roof. Generally speaking, this is achieved by connecting the roof to several courses of brickwork or blockwork at the top of the wall. The exact number of courses picked up by the holding down strap will depend on the amount of uplift, and will be calculated on the basis that the total weight of the masonry plus roof is greater than the total uplift forces by a suitable safety margin.

In practice, the roof is rarely tied down directly to the masonry. A timber wall plate is positioned on top of the wall and the roof is then tied down to the wall plate (see Figure 7.11). Many prop-

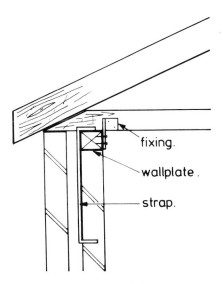

fixing.

wallplate.

strap.

Figure 7.11

rietary brands of fixing are available which are quite adequate for this job and, in general, there is little difficulty in holding the rafter or truss down to the wall plate. This cannot be said of the problem of holding down the wall plate, and its accumulated uplift on the wall, and it is this aspect of the detail which is predominant. It must be remembered that the use of a wall plate also reopens the question of shear keys, and the shear strength of the timber/mortar interface between the wall plate and the wall. In general, where shear is a problem, the holding down details should be appraised to check the adequacy of shear resistance.

7.5 AREAS OF CONCERN

From the foregoing structural considerations, three major areas of concern can be seen to exist:

(i)   the tension/compression connections of walls to floors and roofs,
(ii) the connections of floors and roof plates to shear walls,
(iii) the tying down of wall plates to the walls.

These connections require careful, calculated, and detailed design.

It should be remembered, of course, that these connection requirements are not unique to load-bearing masonry. The design of many structures would require similar careful detailing of connections, the prime example of which is a structure using precast concrete loadbearing panels. It is essential, however, to be both aware of the limitations of the materials, and be ready to exploit any advantageous properties which they may possess. Because of this, the final detail of masonry connections may differ considerably from those used in other forms of construction, although the fundamental considerations for incorporating them into the design are the same for all types of structures.

7.6 OTHER FACTORS INFLUENCING THE DETAILS OF CONNECTIONS

Besides the overriding consideration of structural adequacy, other factors emerge which not only influence the final detail of the strap arrangements, but are also open to debate and could be considered controversial.

Consider the strength of the strap
At present, the current Code suggests that 5 mm thick galvanised mild steel straps are used. The need for such a thickness of metal is questionable on grounds of strength but, in many situations, justifiable on grounds of durability. If it is assumed that durability is not a problem and a 3 mm thickness is considered, this has certain

advantages when considering 'buildability' highlighted by the possible avoidance of rebating the straps into joists.

Considering the strength of a 3 mm strap in tension, the lateral forces on two- and three-storey housing generally vary, depending on location between approximately 0.8 and 2 kN/m run. Assuming 1.8 m centres the strap load lies between 1.44 and 3.6 kN. A tie of 3 mm × 30 mm cross-section with two 6 mm diameter holes gives a net area of 54 mm². The maximum working stress would be approximately 70 N/mm², which is acceptable for mild steel. A 3 mm thickness for the ties should prove more than adequate in pure tension on strength grounds. It is possible that an L strap, bent down into the cavity, would be pulled straight by the range of forces which could act on it. In the absence of the results of laboratory tests, an attempt can be made to analyse the problem using stress blocks.

Consider crushing of the brick
Assume that only the top one-third of the 100 mm turn down of the strap bears on the wall when the strap is tensioned by the 3.6 kN force. The bearing force on the masonry, assuming the triangular distribution shown in Figure 7.12, is then:

$$2 \times \frac{3.6 \times 1000}{30 \times 33} = 7.3 \, \text{N/mm}^2$$

This is an acceptable value for common bricks and many blocks.

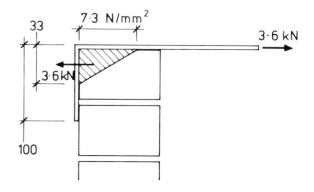

Figure 7.12

Consider the bending value of the strap
Again assume the stress block is 33 mm deep.

The lever arm of the force about the right angle bend is:

$$\frac{33}{3} = 11 \, \text{mm}$$

$$\text{Moment} = 11 \times 3.6 \times 10^3 = 39\,600 \text{ Nmm}$$

$$Z = \frac{30 \times 32}{6} = 45 \text{ mm}^3$$

$$\text{Stress in strap} = \frac{39\,600}{45} = 880 \text{ N/mm}^2$$

This is clearly an unacceptable stress. It may not, however, be a governing factor since a small deflection will ensue at this stress level which, in turn, will affect the bearing area which, in turn, will reduce the moment. The system should therefore reduce to stability, assuming the masonry does not begin to crush. It is this interplay of parameters that makes this topic more amenable to a laboratory investigation than a theoretical analysis. Nevertheless, if the right-angle bend could be strengthened against a bending failure mode, the performance of the strap would be greatly enhanced. One way of achieving this would be to use slot indentations in the steel strap in the region of the right-angle bend. The suggestion of using 100 mm turn-downs into the cavity is a compromise between the requirements of catching more than one brick or block and yet limiting the maximum lever arm of the force about the right-angled bend. The 100 mm value is purely arbitrary. A series of tests, however, may prove useful in validating or modifying this arbitrary choice.

There are many similarities here with joist hangers. While production strength checks are usually made by manufacturers, these tend to be carried out in steel test rigs. Hence, only the action of the steel component is verified. Little practical testing of joist hangers would appear to have been done with a joist hanger positioned in a mortar joint between bricks or blocks. There seems scope here for a similar practical investigation to establish such criteria as load v. deflection curves, eccentricity v. load curves, and load limits for certain mortar brick or block combinations using some of the more commonly available proprietary hangers.

Where more than one solution would fulfil the basic requirement of structural adequacy, the deciding factors are likely to be concerned with ease of construction (buildability), and with its related parameter, cost. One example of such considerations is the question of whether to screw, bolt or nail floor beams into joist hangers. On both grounds nailing is preferable.

Assuming the approximate figure of 2 kN/m run tie force as the maximum of the range with joist hangers spaced at 600 mm centres, the force per joist hanger is 1.2 kN. Assuming a J3 timber group, without allowing any reduction factors, five 10 gauge or four 9 gauge nails would suffice. In many practical situations four 11 gauge nails would be required. Clearly this is a more economical solution than bolting.

Another example on 'buildability' arises where connections have to fulfil several functions. This occurs where a roof must not only be strapped to a wall, but must also be held down to it while, at the same time, the wall plate must also transfer shear load from the wind load on the gable ends. In such situations, it is not uncommon to find this already complex detail further complicated by the positions in which the nail plates have been used on the trusses.

Where precast floor beams have been used spanning in one direction, it is possible to find the screed damaged at the centre of the span against abutting walls, caused by the difference between the deflection of the beams at centre span and the inability of the tie to deflect at the point where it emerges through the mortar course of the wall. This type of irritating performance failure requires careful consideration to be given to likely deflection contours of two-, three- or four-sided panels. Possibly a more stringent deflection limit, with consideration given to the value to be adopted, needs to be introduced for such situations.

In certain situations, for example large walls of more than normal height, there may be concern for the way in which the outer leaf is tied to the inner leaf at the region where the inner leaf is effectively restrained against lateral movement at the top. In situations like this, there is an argument for placing an extra number of wall ties in the immediate vicinity of the strapping.

BS 5628 includes in its Appendix C details and requirements for connections to floors and roofs by means of metal anchors and joist hangers capable of resisting lateral movement. The Code requires that the effective cross-section of anchors and of their fixings should be capable of resisting the appropriate design loading which is the sum of:

(a) the simple static reactions to the lateral applied design horizontal forces at the line of the lateral support, and

(b) 2.5% of the total design vertical load that the wall or column is designed to carry at the line of lateral support; the elements of construction that provide lateral stability to the structure as a whole need not be designed to support this force.

The Code also states that the designer should satisfy himself that loads applied to lateral supports will be transmitted to the elements of construction providing stability, e.g. by the floors or roofs acting as horizontal girders or plates.

The stress assumed for design purposes is to be the characteristic yield strength (or its equivalent), as laid down in the appropriate BS, divided by $\gamma_m = 1.15$. Anchors should be provided at intervals of not more than 2 m in houses of not more than three storeys, and not more than 1.25 m for all storeys in all other buildings. Galvanised mild steel anchors having a cross-section of 30 mm × 5 mm may be assumed to have adequate strength in buildings of up to six storeys in height.

The Code also gives other types of buildings where simple resistance or enhanced resistance may be assumed, provided certain conditions of tying are incorporated (see Chapters 5 and 6, Basis of Design).

The following illustrated examples of strapping and tying should help the engineer in selecting a suitable detail. The design examples which follow the illustrations will give guidance on the assessment of the forces involved.

The typical details which follow show methods of holding down roofs, stabilising walls at floor and roof levels, and shear transfer from roofs and floors to walls. It should be noted that the dimensions are only typical, and each detail must be designed to cater for the load and the environmental conditions.

7.7 ILLUSTRATED EXAMPLES OF STRAPPING AND TYING

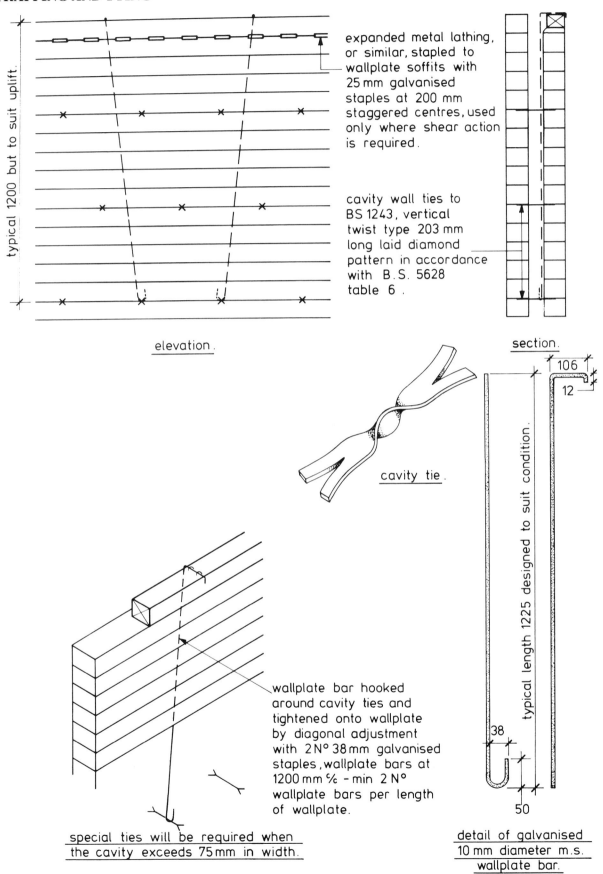

typical 1200 but to suit uplift.

elevation.

expanded metal lathing, or similar, stapled to wallplate soffits with 25 mm galvanised staples at 200 mm staggered centres, used only where shear action is required.

cavity wall ties to BS 1243, vertical twist type 203 mm long laid diamond pattern in accordance with B.S. 5628 table 6 .

section.

cavity tie.

106
12

typical length 1225 designed to suit condition.

38
50

wallplate bar hooked around cavity ties and tightened onto wallplate by diagonal adjustment with 2 N° 38 mm galvanised staples, wallplate bars at 1200 mm ℅ – min 2 N° wallplate bars per length of wallplate.

special ties will be required when the cavity exceeds 75 mm in width.

detail of galvanised 10 mm diameter m.s. wallplate bar.

METHOD OF SECURING WALLPLATE TO AN EXTERNAL LOADBEARING WALL. (UPLIFT TO BE CATERED FOR.)   **Figure 7.13**
BRICK OR 100 mm BLOCKWORK.

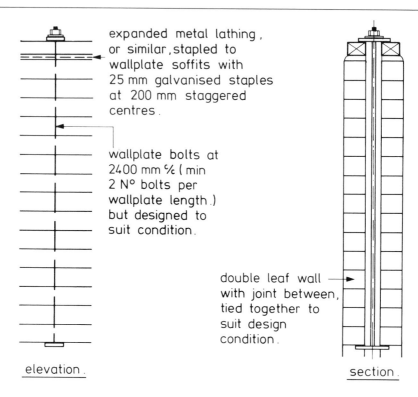

expanded metal lathing,
or similar, stapled to
wallplate soffits with
25 mm galvanised staples
at 200 mm staggered
centres.

wallplate bolts at
2400 mm ℅ (min
2 N° bolts per
wallplate length.)
but designed to
suit condition.

double leaf wall
with joint between,
tied together to
suit design
condition.

elevation.

section.

expanded metal lathing, or similar,
used only where shear action is
required.

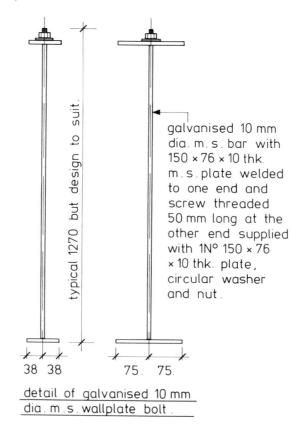

typical 1270 but design to suit.

galvanised 10 mm
dia. m.s. bar with
150 × 76 × 10 thk.
m.s. plate welded
to one end and
screw threaded
50 mm long at the
other end supplied
with 1N° 150 × 76
× 10 thk. plate,
circular washer
and nut.

38 38

75. 75.

detail of galvanised 10 mm
dia. m.s. wallplate bolt.

METHOD OF SECURING WALLPLATE TO INTERNAL
LOADBEARING WALLS. (UPLIFT TO BE CATERED FOR.)
BRICKWORK.

Figure 7.14

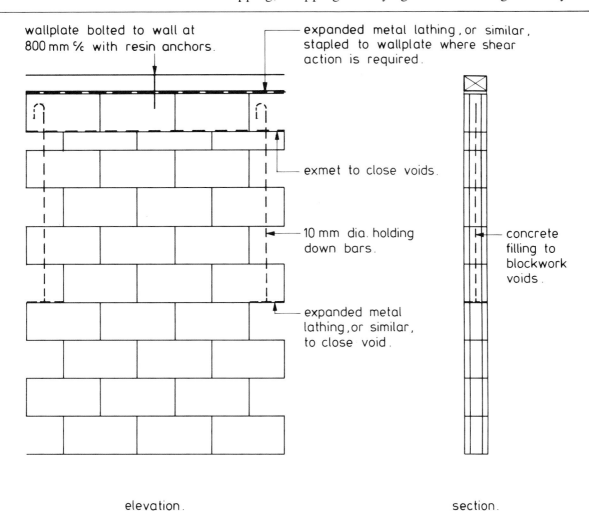

wallplate bolted to wall at
800 mm ℅ with resin anchors.

expanded metal lathing, or similar,
stapled to wallplate where shear
action is required.

exmet to close voids.

10 mm dia. holding
down bars.

expanded metal
lathing, or similar,
to close void.

concrete
filling to
blockwork
voids.

elevation.   section.

typical detail.
10 mm dia. m.s holding down bars 1200 mm long hooked one end placed at 1200 mm ℅
and surrounded in concrete.
minimum 2 N° bars per wall length but design for actual conditions.
concrete to be 1:2:4 mix with 8 mm aggregate.
voids to be filled in depths not exceeding 800 mm as work proceeds.
this detail is only suitable for block walls - 140 mm thk and over.

METHOD OF SECURING WALLPLATE TO HOLLOW BLOCK
INNER LEAF OF CAVITY WALL AND HOLLOW BLOCK
INTERNAL WALLS. (UPLIFT TO BE CATERED FOR.)

Figure 7.15

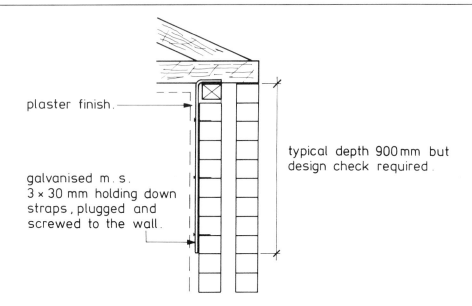

plaster finish.

galvanised m.s.
3 × 30 mm holding down
straps, plugged and
screwed to the wall.

typical depth 900 mm but
design check required.

this is an alternative detail to those in figs. 7:13 and 7:15.

METHOD OF SECURING WALLPLATE.

THIS DETAIL IS PERMISSIBLE WHERE THE WALLS ARE PLASTERED.
(UPLIFT TO BE CATERED FOR.)

Figure 7.16

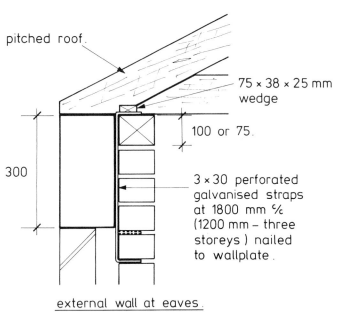

pitched roof.

75 × 38 × 25 mm
wedge

100 or 75.

300

3 × 30 perforated
galvanised straps
at 1800 mm ℀
(1200 mm – three
storeys) nailed
to wallplate.

external wall at eaves.

this detail shows the fixing of timber
wallplates to the top of external and
internal walls under pitched roofs.
straps are galvanised steel straps at
1800 mm ℀ 325 mm or 425 mm long
according to coursing.

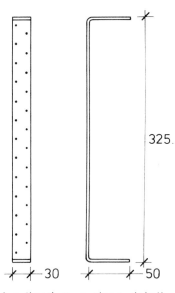

325.

30   50

galvanised m.s. strap detail.

wallplates are 100 × 100 mm where
200 mm nominal high blocks are
used and 100 × 75 mm where
225 mm nominal high blocks and
brickwork are used.
all wallplates bedded in mortar.

METHOD OF SECURING WALLPLATE.

Figure 7.17

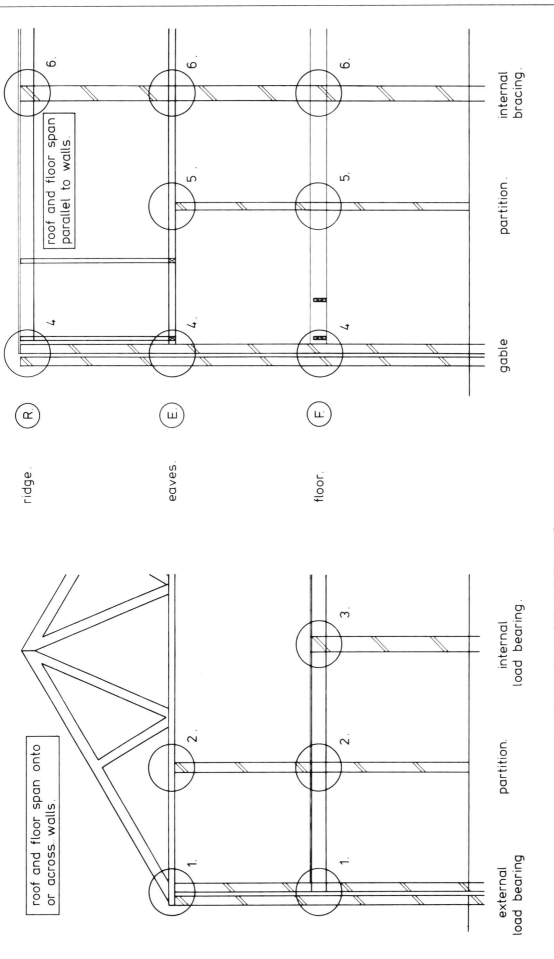

LOCATION OF DETAILS. TIMBER PITCHED ROOF. TIMBER FLOOR.

Figure 7.18

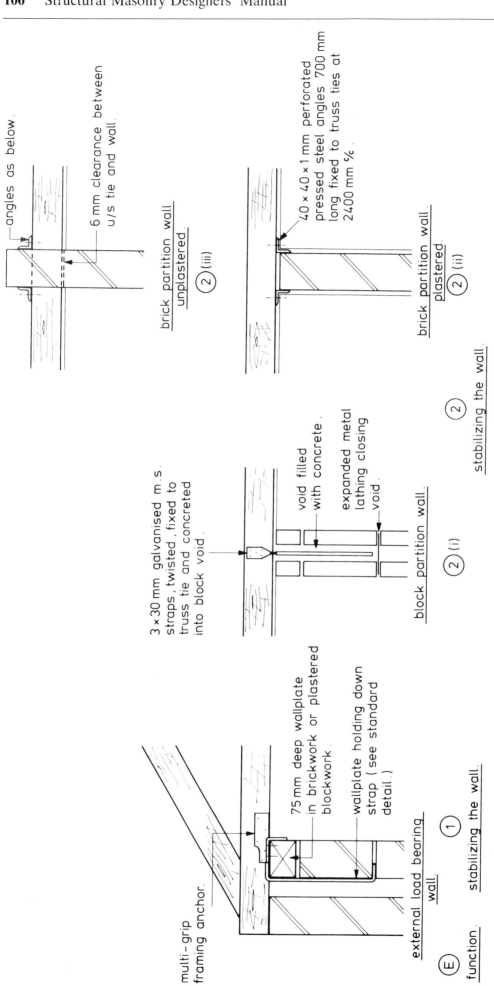

angles as below.

6 mm clearance between u/s tie and wall.

brick partition wall unplastered.
② (iii)

40 × 40 × 1 mm perforated pressed steel angles 700 mm long fixed to truss ties at 2400 mm %.

brick partition wall plastered.
② (iii)

stabilizing the wall.
②

3 × 30 mm galvanised m.s. straps, twisted, fixed to truss tie and concreted into block void.

void filled with concrete.

expanded metal lathing closing void.

block partition wall.
② (i)

multi-grip framing anchor.

75 mm deep wallplate in brickwork or plastered blockwork.

wallplate holding down strap (see standard detail.)

external load bearing wall.
①

function.

stabilizing the wall.
①

Ⓔ

Ⓔ ① ② SEE FIG. 7.18 FOR LOCATION.

TIMBER PITCHED ROOF DETAILS. AT

Figure 7.19

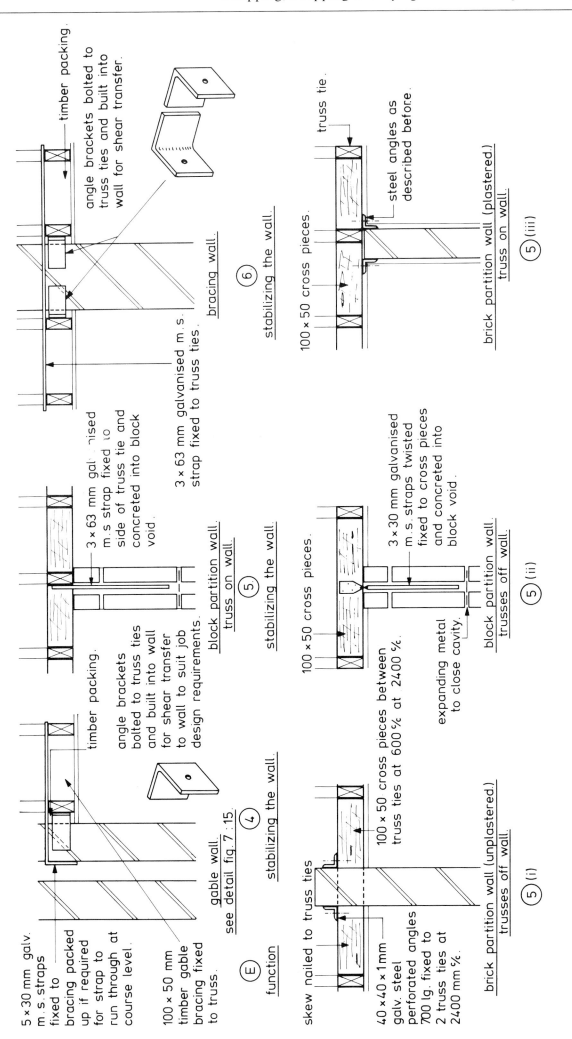

5 × 30 mm galv. m.s. straps fixed to bracing packed up if required for strap to run through at course level.

100 × 50 mm timber gable bracing fixed to truss.

timber packing.

angle brackets bolted to truss ties and built into wall to suit job design requirements.

gable wall.
see detail fig. 7.15.

Ⓔ   stabilizing the wall.

function

timber packing.

angle brackets bolted to truss ties and built into wall for shear transfer.

3 × 63 mm galvanised m.s. strap fixed to side of truss tie and concreted into block void.

block partition wall.
truss on wall.

⑤   stabilizing the wall.

100 × 50 cross pieces.

100 × 50 cross pieces between truss ties at 600 ℅ at 2400 ℅.

expanding metal to close cavity.

3 × 30 mm galvanised m.s. straps twisted fixed to cross pieces and concreted into block void.

block partition wall.
trusses off wall.

⑤(ii)

skew nailed to truss ties

40 × 40 × 1mm galv. steel perforated angles 700 lg. fixed to 2 truss ties at 2400 mm ℅.

100 × 50 mm cross pieces at truss ties at 2400 mm ℅.

brick partition wall (unplastered.)
trusses off wall.

⑤(i)

timber packing.

angle brackets bolted to truss ties and built into wall for shear transfer.

3 × 63 mm galvanised m.s. strap fixed to truss ties.

bracing wall.

⑥   stabilizing the wall.

truss tie.

steel angles as described before.

100 × 50 cross pieces.

brick partition wall (plastered.)
truss on wall.

⑤(iii)

TIMBER PITCHED ROOF DETAILS. AT Ⓔ ④ ⑤ ⑥ SEE FIG 7.18 FOR LOCATION.

Figure 7.20

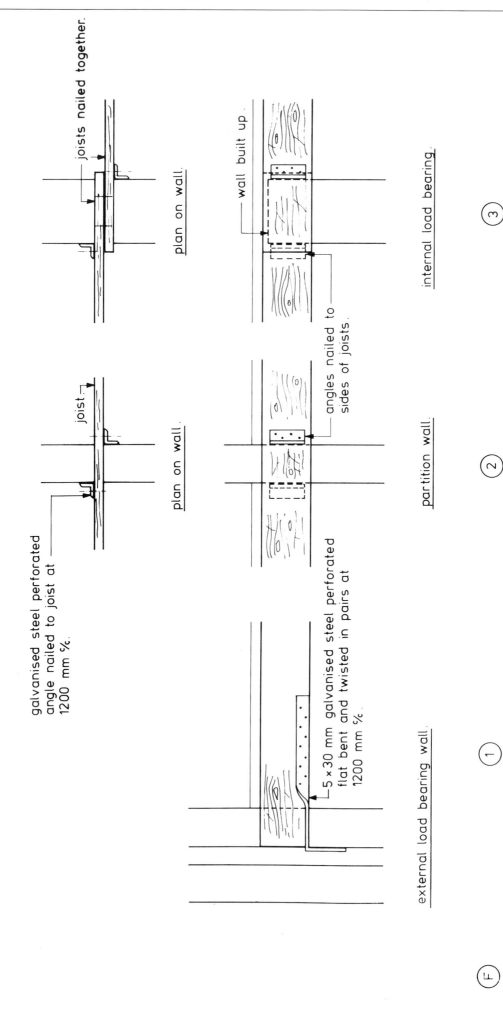

joists nailed together.

plan on wall.

joist.

galvanised steel perforated angle nailed to joist at 1200 mm %.

plan on wall.

wall built up.

angles nailed to sides of joists.

5 × 30 mm galvanised steel perforated flat bent and twisted in pairs at 1200 mm %.

external load bearing wall.

partition wall.

internal load bearing.

① stabilizing the wall.

② stabilizing the wall.

③ stabilizing the wall.

Ⓕ function

① stabilizing the wall.

TIMBER FLOOR DETAILS. AT Ⓕ ① ② ③ SEE FIG. 7 : 18 FOR LOCATION.

Figure 7.21

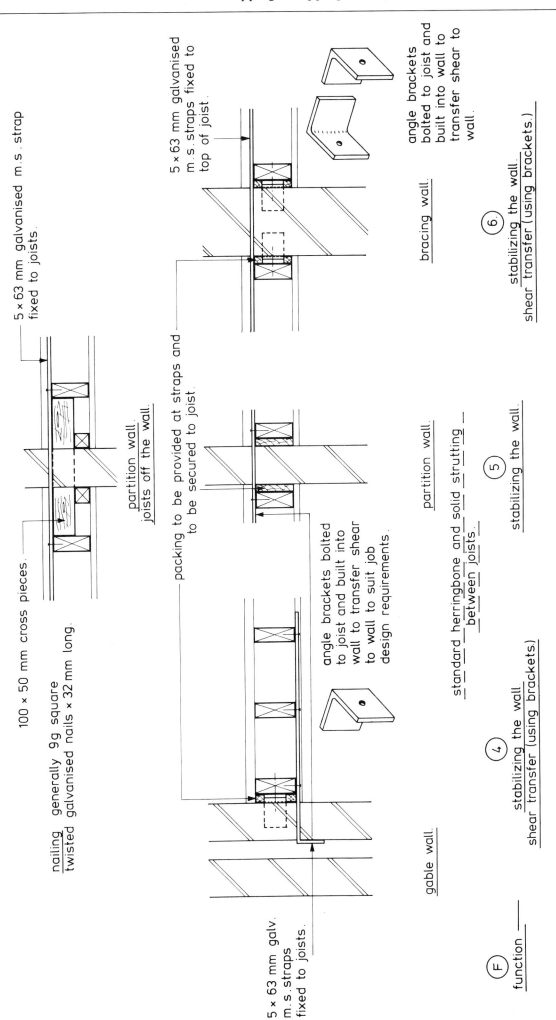

5 × 63 mm galvanised m.s. strap fixed to joists.

5 × 63 mm galvanised m.s. straps fixed to top of joist.

angle brackets bolted to joist and built into wall to transfer shear to wall.

bracing wall.

⑥

stabilizing the wall.
shear transfer (using brackets.)

100 × 50 mm cross pieces.

partition wall.
joists off the wall.

nailing generally 9g square
twisted galvanised nails × 32 mm long.

packing to be provided at straps and to be secured to joist.

partition wall.

standard herringbone and solid strutting between joists.

⑤

stabilizing the wall.

5 × 63 mm galv. m.s. straps fixed to joists.

angle brackets bolted to joist and built into wall to transfer shear to wall to suit job design requirements.

gable wall.

Ⓕ

function

④

stabilizing the wall
shear transfer (using brackets.)

TIMBER FLOOR DETAILS. AT Ⓕ ④⑤⑥ SEE FIG. 7:18 FOR LOCATION.

Figure 7.22

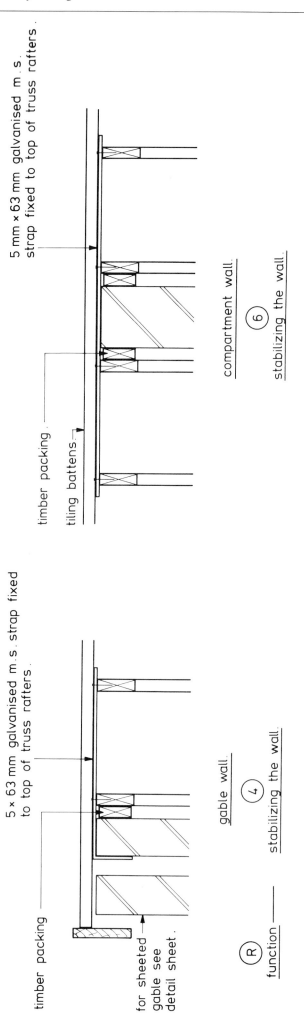

TIMBER PITCHED ROOF DETAILS. AT Ⓡ ④ ⑥ SEE FIG. 7·18 FOR LOCATION.

Figure 7.23

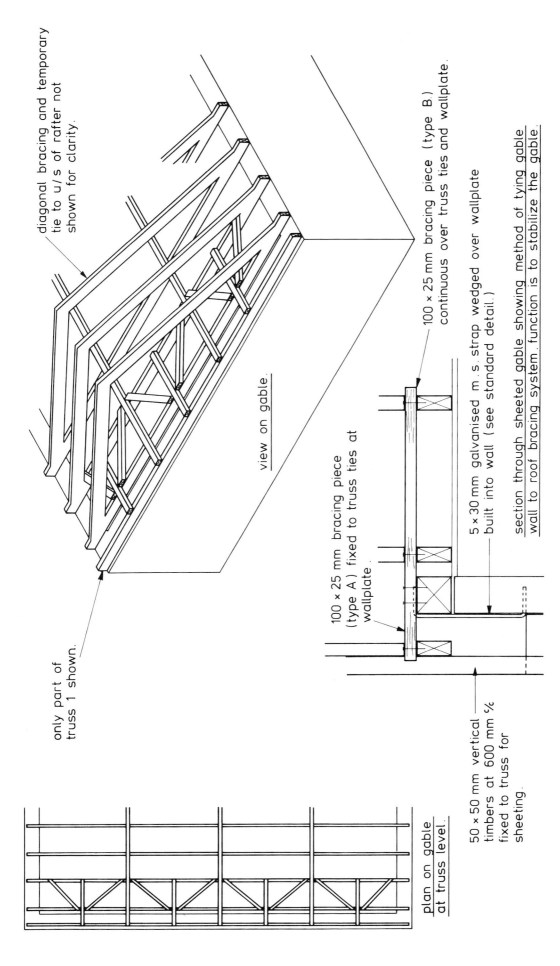

diagonal bracing and temporary tie to u/s of rafter not shown for clarity.

view on gable

only part of truss 1 shown.

100 × 25 mm bracing piece (type B.) continuous over truss ties and wallplate.

100 × 25 mm bracing piece (type A) fixed to truss ties at wallplate.

5 × 30 mm galvanised m.s. strap wedged over wallplate built into wall (see standard detail.)

section through sheeted gable showing method of tying gable wall to roof bracing system. function is to stabilize the gable.

plan on gable at truss level.

50 × 50 mm vertical timbers at 600 mm ℅ fixed to truss for sheeting.

TIMBER PITCHED ROOF. DETAILS OF SHEETED GABLE.

Figure 7.24

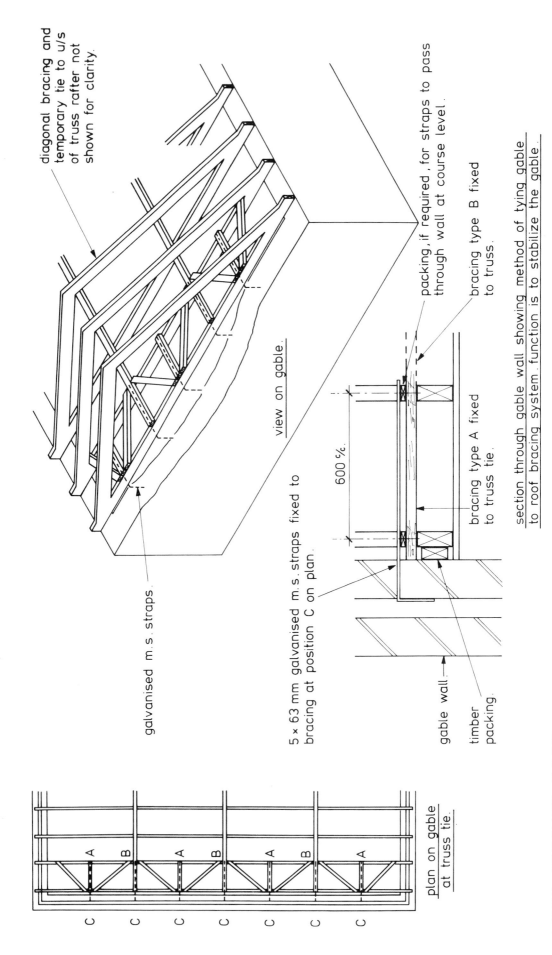

diagonal bracing and temporary tie to u/s of truss rafter not shown for clarity.

packing, if required, for straps to pass through wall at course level.

bracing type B fixed to truss.

view on gable.

galvanised m.s. straps.

5 × 63 mm galvanised m.s. straps fixed to bracing at position C on plan.

600 %.

bracing type A fixed to truss tie.

section through gable wall showing method of tying gable to roof bracing system. function is to stabilize the gable.

gable wall.

timber packing.

plan on gable at truss tie.

TIMBER PITCHED ROOF. DETAILS OF BRICK GABLE WALL.

Figure 7.25

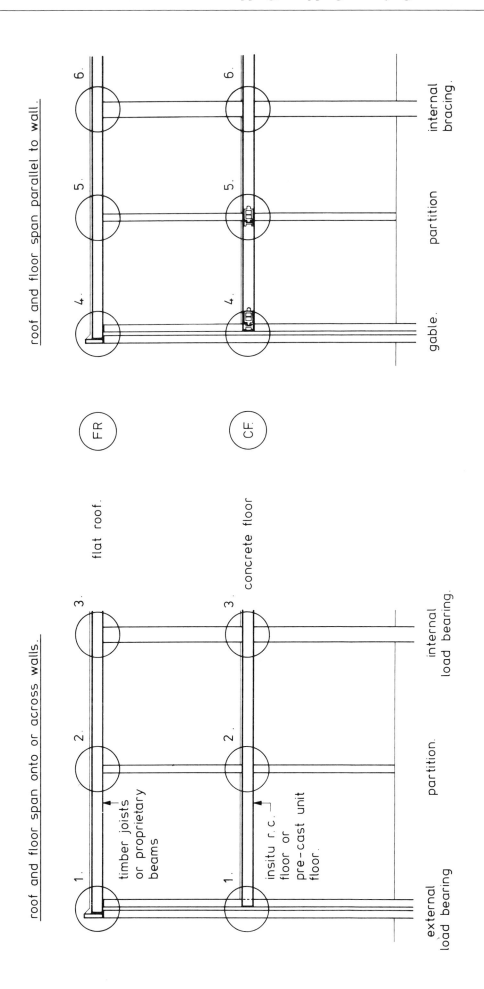

LOCATION OF DETAILS. TIMBER FLAT ROOF. CONCRETE FLOOR.

Figure 7.26

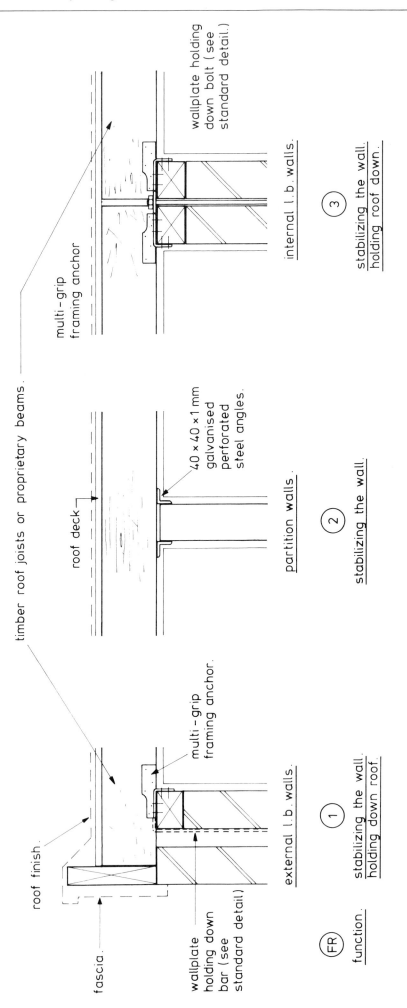

TIMBER FLAT ROOF DETAILS. AT (FR) (1) (2) (3) SEE FIG. 7:26 FOR LOCATION.

Figure 7.27

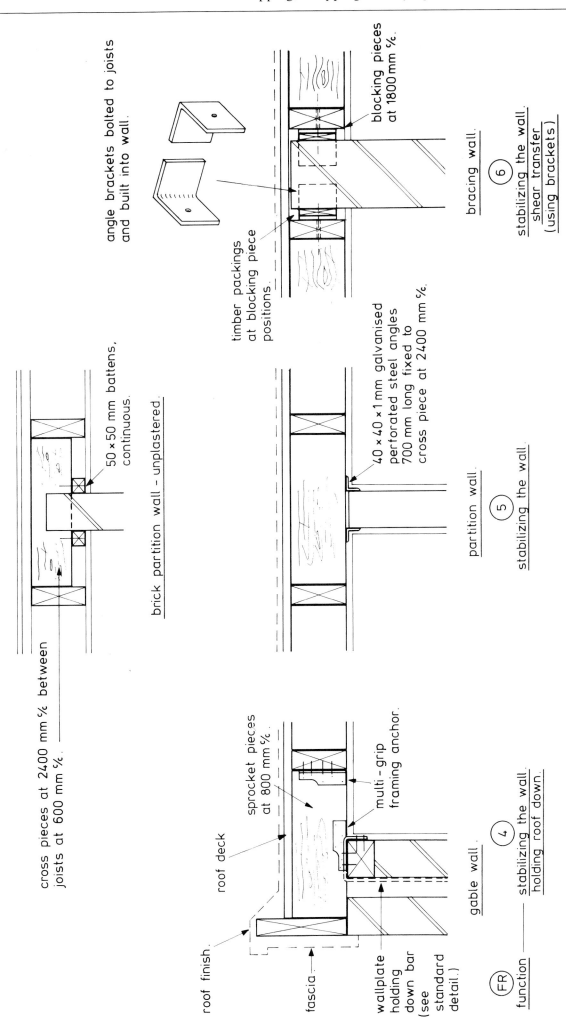

angle brackets bolted to joists and built into wall.

blocking pieces at 1800mm ‰.

timber packings at blocking piece positions.

bracing wall.

(6)

stabilizing the wall. shear transfer (using brackets)

cross pieces at 2400 mm ‰ between joists at 600 mm ‰.

50 × 50 mm battens, continuous.

brick partition wall – unplastered.

40 × 40 × 1mm galvanised perforated steel angles 700 mm long fixed to cross piece at 2400 mm ‰.

partition wall.

(5)

stabilizing the wall.

roof finish.

fascia.

roof deck

sprocket pieces at 800 mm ‰.

multi-grip framing anchor.

wallplate holding down bar (see standard detail.)

gable wall.

(4)

stabilizing the wall. holding roof down.

(FR)

function

TIMBER FLAT ROOF DETAILS. AT (FR) (4) (5) (6)   SEE FIG. 7:26 FOR LOCATION.

Figure 7.28

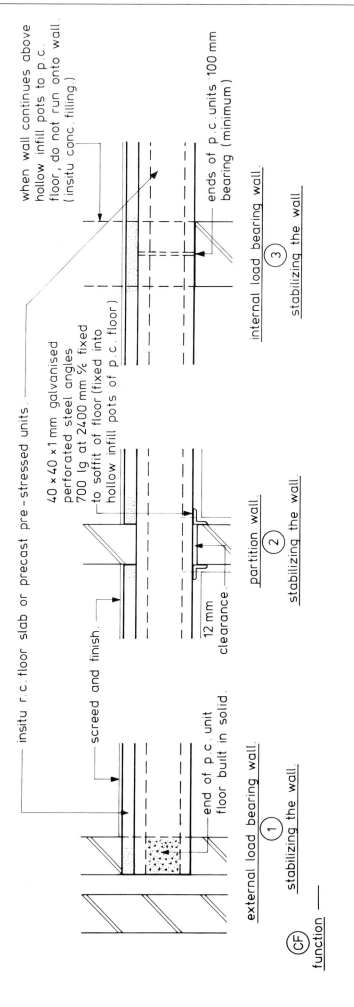

when wall continues above hollow infill pots to p.c. floor, do not run onto wall. (insitu conc. filling.)

ends of p.c. units·100 mm bearing (minimum)

insitu r.c. floor slab or precast pre-stressed units.

40 × 40 × 1 mm galvanised perforated steel angles 700 lg at 2400 mm ‰ fixed to soffit of floor (fixed into hollow infill pots of p.c. floor)

screed and finish.

12 mm clearance.

end of p.c. unit floor built in solid.

internal load bearing wall.

③

stabilizing the wall.

partition wall.

②

stabilizing the wall.

external load bearing wall.

①

stabilizing the wall.

Ⓒ Ⓕ
function

CONCRETE FLOOR DETAILS. AT ⒸⒻ ① ② ③ SEE FIG.7·26 FOR LOCATION.

Figure 7.29

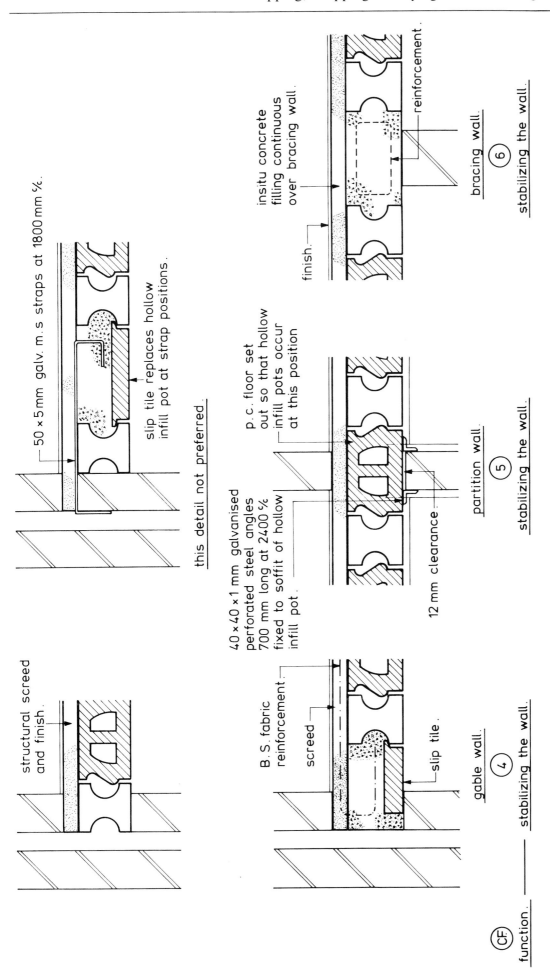

CONCRETE FLOOR DETAIL. AT CF ④ ⑤ ⑥ SEE FIG. 7 : 26 . FOR LOCATION .

Figure 7.30

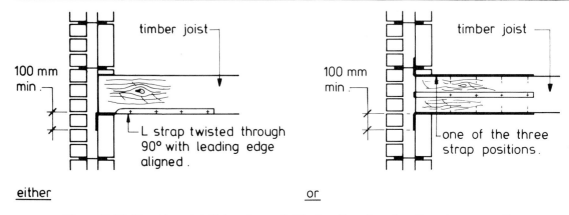

Figure 7.31 Function stabilising the wall. Timber floor bearing directly onto wall

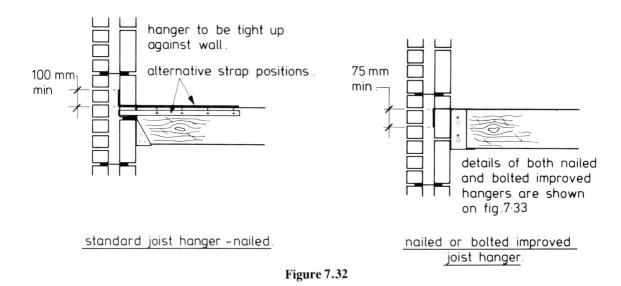

Figure 7.32

Figure 7.33

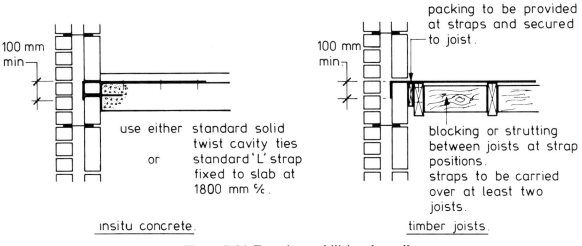

insitu concrete.

use either standard solid twist cavity ties

or standard 'L' strap fixed to slab at 1800 mm %.

packing to be provided at straps and secured to joist.

blocking or strutting between joists at strap positions.
straps to be carried over at least two joists.

timber joists.

Figure 7.34 Function stabilising the wall

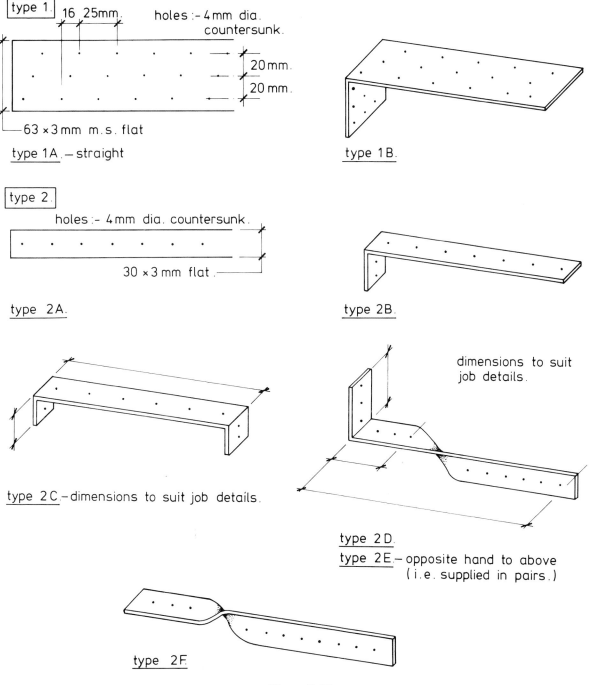

type 1

16 25mm.

holes :- 4mm dia. countersunk.

20mm.

20mm.

63 × 3mm m.s. flat

type 1A.— straight

type 1B.

type 2

holes :- 4mm dia. countersunk.

30 × 3mm flat.

type 2A.

type 2B.

type 2C.—dimensions to suit job details.

dimensions to suit job details.

type 2D.

type 2E.— opposite hand to above (i.e. supplied in pairs.)

type 2F.

Figure 7.35

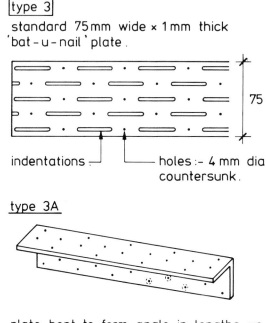

type 3

standard 75 mm wide × 1 mm thick 'bat - u - nail' plate .

indentations —— —— holes :- 4 mm dia countersunk .

type 3A

plate bent to form angle in lengths up to 700 mm .

Figure 7.36

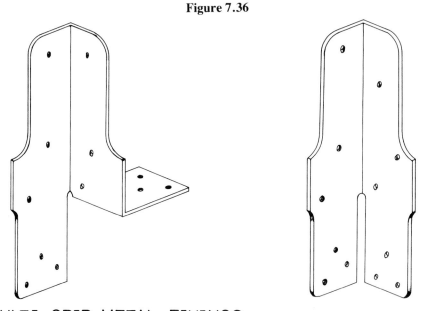

TYPICAL MULTI-GRIP METAL FIXINGS

Figure 7.37

7.8 DESIGN EXAMPLE: STRAPS AND TIES FOR A THREE-STOREY MASONRY BUILDING

A typical three-storey building is to be built using loadbearing brick and block walls, precast concrete floor units and gang nail roof trusses.

The following calculations are for a number of typical elements and illustrate the procedures adopted and demonstrate the use of different strap details. For this reason, the calculations are not necessarily complete in themselves, being designed to be illustrative by nature. The designs are also based on working load design limits.

Roof construction
Roof: Tiles on battens and felt on gang nail trusses at 600 mm centres.
Ceiling: 12 mm plasterboard. Fibre glass between trusses.

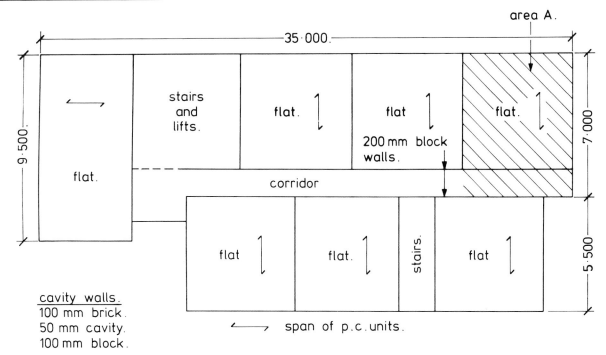

Figure 7.38 Typical floor plan

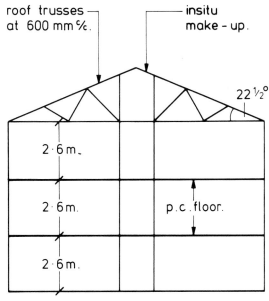

Figure 7.39 Cross-section

Floor construction

Floor: Precast concrete floor units. 65 mm finishing screed.

Ceiling: 12 mm plasterboard on battens.

Loadings (CP3, Chapter V: Part 1, Residential Buildings)

Roof	*Dead loads*	*Floors*	*Dead loads*
Tiles	$= 0.45 \text{ kN/m}^2$	Screed and finishes	$= 1.60 \text{ kN/m}^2$
Felt and battens	$= 0.10 \text{ kN/m}^2$	Precast units	$= 2.20 \text{ kN/m}^2$
Trusses	$= 0.15 \text{ kN/m}^2$	Plasterboard and battens	$= 0.25 \text{ kN/m}^2$
Plasterboard	$= 0.20 \text{ kN/m}^2$	Services	$= 0.10 \text{ kN/m}^2$
	0.90 kN/m^2		4.15 kN/m^2

	Imposed load		*Imposed load*
	0.75 kN/m^2		1.50 kN/m^2

Wind loading (from CP3: Chapter V: Part 2):

Table 1, Building location – Wirral – Merseyside
Basic wind speed, $V = 46$ m/s
cl. 5.1. Design wind speed, $V_S = V \times S_1 \times S_2 \times S_3$
Table 2 $S_1 = 1.0$
Table 3, Group 3, Class B $S_2 = 0.75$ kN/m^2
                                           $H = 10.4$ m
cl. 5.6 $S_3 = 1.0$
Therefore $V_S = 46 \times 1 \times 0.75 \times 1$ $= 34.5$ m/s
Table 4, Dynamic pressure, $q = 0.73$ kN/m^2

The calculation of this pressure has been based on a height of 10.4 m (ridge level) from the ground. The actual wind pressure will decrease as the height of the section under consideration decreases. However, to simplify the analysis, this value will be used throughout.

The wind forces on a building are divided into two categories:

(i) The forces from the wind pressure acting normal to the surfaces of the building.
(ii) The forces parallel to the surface under consideration. These forces can be ignored since, in accordance with clause 7.4, when d/h or d/b are less than 4, frictional drag need not be considered. This is the case in this example.

Wind pressure on walls:

Table 7, $\dfrac{h}{w} = \dfrac{7.8}{12.5} = 0.62$ i.e. $\dfrac{1}{2} < \dfrac{h}{w} < \dfrac{3}{2}$

$\dfrac{l}{w} = \dfrac{35.0}{12.5} = 2.8$ i.e. $\dfrac{3}{2} < \dfrac{l}{w} < 4$

Therefore maximum $C_{pe} = +0.7$ and -0.5 (front and rear)

(except local values) $= +0.7$ and -0.7 (gables)

Appendix E. In a building of this nature there is a negligible probability of a dominant opening occurring during a storm.

Therefore, $C_{pi} = +0.2$ and -0.3

Pressure on walls:

Applied pressure $= q(C_{pe} - C_{pi})$
Maximum positive value $= 0.73\,(+0.7 - (-)\,0.3)$
i.e. inward pressure $= 0.73$ kN/m^2
Maximum negative value $= 0.73\,(-0.7 - (+)\,0.2)$
i.e. outward pressure
or suction $= -0.66$ kN/m^2 on gable wall
and $= 0.73\,(-0.5 - (+)\,0.2)$
                            $= -0.51$ kN/m^2 on front and rear wall

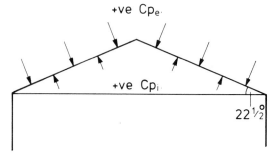

Figure 7.40 Wind pressure on roof

Table 8, $\dfrac{h}{w} = \dfrac{7.8}{12.5} = 0.62$ i.e. $\dfrac{1}{2} < \dfrac{h}{w} < \dfrac{3}{2}$

Roof angle 22½°

Wind on main elevation:

Windward slope	Leeward slope
C_{pe} -0.6	-0.5

Wind on gables, maximum $C_{pe} = -0.8$

C_{pi} for both cases as before $= +0.2$ and -0.3.

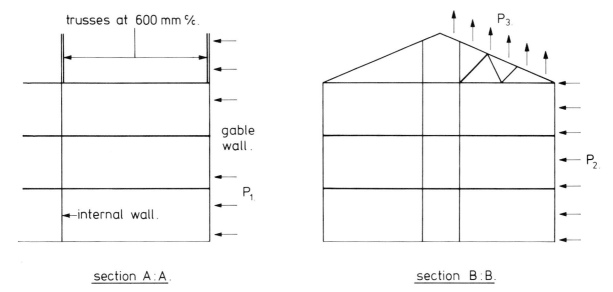

section A:A.   section B:B.

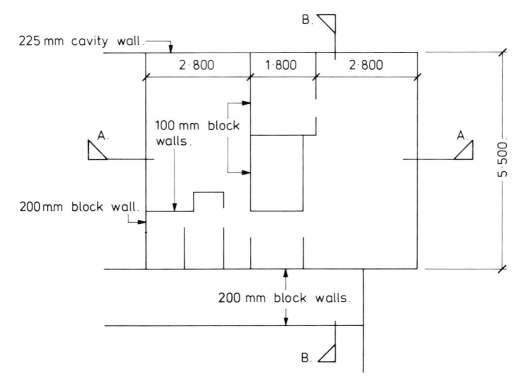

enlarged plan on area 'A'

Figure 7.41

When the wind is normal to the main elevations and C_{pe} values are such that there is a net horizontal force applied to the roof, the value of this force is the difference between the C_{pe} values multiplied by the dynamic wind pressure and this value resolved into the horizontal direction.

Net horizontal force $= (0.6 - 0.5)\,0.73 \times 5 \sin 22\tfrac{1}{2}° = 0.028\ \text{kN/m}^2$

This value is sufficiently small for it to be assumed, without further calculations, that the normal truss fixings will dissipate the load into the lateral walls.

It will therefore be assumed that the pressure applied to the roof is uniform over the roof and has a maximum value of:

$0.73\,(-0.8 - (+)\,0.2) = 0.73\ \text{kN/m}^2$

Resolving this pressure into the vertical direction then

Pressure $= -0.73 \times \cos 22\tfrac{1}{2}° = -0.67\ \text{kN/m}^2$

Design of elements to resist wind loading:

Consider area A shown shaded on Figure 7.38.

Forces applied to walls and roof:

Pressure on gable wall $P_1 = +0.73\ \text{kN/m}^2$ and $-0.66\ \text{kN/m}^2$
Pressure on rear elevation $P_2 = +0.73\ \text{kN/m}^2$ and $-0.51\ \text{kN/m}^2$
Pressure on roof $P_3 = -0.67\ \text{kN/m}^2$

These wind forces applied to the various parts of the building must be transmitted through the walls down to the ground level.

For wind pressure on the gable walls, the corridor, front and rear external walls acting as shear walls will best transfer this loading to ground level. For wind on the front and rear elevations, the gable wall will transfer the forces down to ground level together with walls internal to Flat A which are parallel to the gable wall.

It is therefore necessary to check that these walls are capable of taking this loading to ground floor level.

(1) Consider the gable wall
Since the gable wall has no support from internal walls, it must be designed to span vertically between floors. The floors must, therefore, be capable of transferring this load to the corridor and external walls.

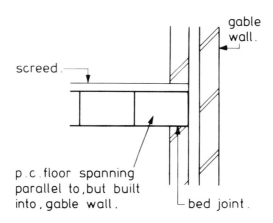

Figure 7.42

The horizontal force at floor level due to wind load is:

$$F = 0.73 \times 2.6 = 1.9 \text{ kN/m}^2$$

Calculate the axial load on the inner leaf assuming density of blockwork = 18 kN/m³. The worst condition will be at first floor level, maximum load in the wall is then

$$3 \times 2.6 \times 18 \times 0.1 = 14.04 \text{ kN/m}^2$$

From BS 5628: Part 1: 1978: Section 4, clause 28.2.1, the lateral restraint must resist the horizontal forces plus 2½% of the axial load.

$$\text{Force to be resisted} = 1.9 + \left(14.0 \times \frac{25}{1000} \right) = 2.25 \text{ kN/m}^2$$

From BS 5628, clause 25:

$$\text{Permissible shear} = \frac{0.35 + 0}{3.5} = 0.10 \text{ N/mm}^2 \quad \left(\text{from: } \frac{f_v + 0.6g_A}{\gamma_m} \right)$$

$$\text{Actual shear stress} = \frac{2.25 \times 10^3}{100 \times 10^3} = 0.023 \text{ N/mm}^2$$

Therefore detail satisfactory.

A precast concrete floor with screed, will be adequate to transfer these forces to the corridor and external walls by diaphragm, or plate action.

At eaves and at roof level, the forces must also be transferred. This must be done by the roof acting as a plate and in order to do this it may need to be braced.

The masonry will tend to act as a propped cantilever and thus:

$$F_1 = \frac{3}{8} h \times \text{wind pressure}$$

$$F_2 = \frac{5}{8} h \times \text{wind pressure} + \text{pressure from wall below}$$

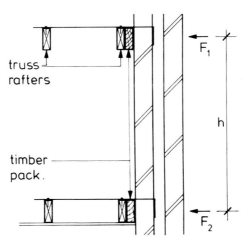

Figure 7.43

The maximum height, H, and thus maximum values of F_1 and F_2 occur at the ridge and are:

$$F_1 = \frac{3}{8} \times 2.6 \times (+) 0.73 \qquad = +0.71 \text{ kN/m}^2$$

and $\dfrac{3}{8} \times 2.6 \times (-)\,0.66 \qquad = -0.64\ \text{kN/m}^2$

$F_2 = \left(\dfrac{5}{8} + \dfrac{1}{2}\right)\ 2.6 \times (+)\,0.73 = +2.14\ \text{kN/m}^2$

and $\left(\dfrac{5}{8} + \dfrac{1}{2}\right)\ 2.6 \times (-)\,0.66 = -1.93\ \text{kN/m}^2$

The positive (inward) forces may be transferred into the roof by introducing a tight pack between the wall and the truss. However the negative (outward) forces require additional measures.

Therefore, in accordance with Appendix C of BS 5628, introduce metal straps of minimum cross-sectional area of 30 × 5 mm at 1.200 m centres at both roof and eaves level.

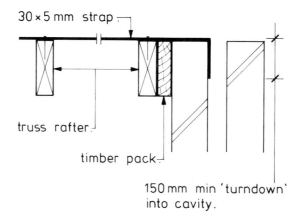

Figure 7.44

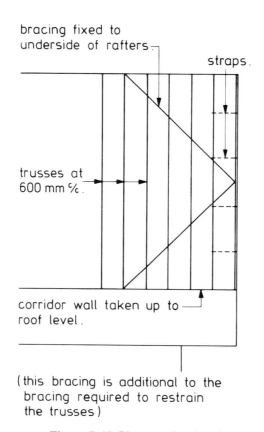

Figure 7.45 Plan at rafter level

The strap should be designed as follows:

(i)   To resist direct tensile force from the wind plus 2½% of the axial load.
(ii) To resist bending at the return down the cavity face.
(iii) To avoid any local crushing of the masonry.
(iv) With suitable fixings to connect the strap to the trusses.

Having transferred the wind loading into the roof structure, it is now necessary to design the roof and any bracing necessary to transmit this loading to the resisting walls.

Introducing bracing as shown, the bracing should be fixed to the underside of the rafters.

Both the bracing and its fixings should be designed to resist the compressive and tensile forces from pressure and suction on the gable wall. Similarly, bracing should be introduced at eaves level to transmit the forces to the lateral walls

Consider the external wall to the rear elevation
The external wall, see Figure 7.46, has a series of intersecting internal walls at a maximum of 2.8 m centres, these intersecting walls are built directly off the precast units. It is not advisable, for reasons of different deflections, to bond these walls into the external walls. Metal anchors or wall ties may be introduced in the bed joints to afford these non-loadbearing intersection walls some degree of lateral support. These ties also offer the external cavity wall lateral support. To accommodate differential deflection butterfly ties or similar should be used in every bed joint of the blockwork. The external walls can then be made to span horizontally between the internal walls. The internal walls are not continuous to ground level, and this necessitates the forces being transferred at each floor into the main dividing walls, which will then transfer the forces to ground level.

Consider a metre strip of wall, the maximum force on the internal wall is:

$$(2.8 + 1.8) \times 0.73 = 1.7 \text{ kN/m height.}$$

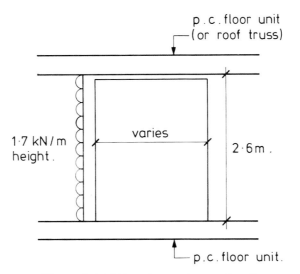

Figure 7.46 Elevation on internal wall

The internal wall must be designed to resist:

(i) overturning in the plane of the wall,
(ii) horizontal shear failure on the bed joints.

Having transferred the force into the floor then, as for the gables, the floor must be designed to transfer this loading into the resisting walls.

Consider wind forces on the roof
Roof loading (from pages 00 and 00)
Dead weight = 0.9 kN/m²
Imposed load = 0.75 kN/m²
Wind loading = −0.67 kN/m² (uplift)

The dead weight of the roof must be sufficiently heavy to provide a factor of safety of 1.4 over the uplift from the wind. If this factor is not achieved the roof must be strapped down to prevent the possibility of it lifting off. Thus the factor of safety = 0.9/0.67 = 1.34 which is inadequate.

Therefore uplift to be designed for $=\dfrac{0.9}{1.4} - 0.67 = 0.03$ kN/m²

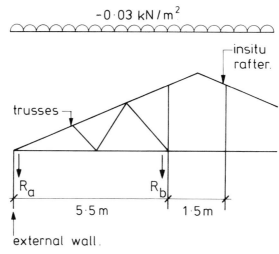

Figure 7.47

Net upward force $R_A = \dfrac{0.03 \times 6.25}{5.5}\left(\dfrac{6.25}{2} - 0.75\right) = 0.08$ kN/m

At the corridor wall = 0.11 kN/m

The roof trusses bear onto a wall plate which then sits on the wall. The fixing of the trusses to the wall plate should be designed to resist the uplift force.

The force on each truss = 0.11 × 0.6 = 0.07 kN
(trusses at 600 mm %)

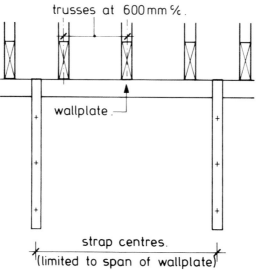

Figure 7.48 Elevation on straps

Thus if trusses are skew nailed to the wall plate and using the withdrawal value given in CP 112: Part 2 1971, Table 23, for 8 SWG nails in J3 timber:

Resistance to withdrawal using two nails = 2.98 N/mm penetration

Penetration length required $=\dfrac{0.07 \times 10^3}{2 \times 2.98} = 11.7$ mm

To ensure this penetration is achieved use 75 mm long nails. Alternatively, use proprietary fixings which can be justified.

It is necessary to tie down the wall plate to the blockwork to resist the net upward forces. The centres of the straps must be limited to the maximum distance the wall plate can span to transfer the uplift from each truss to the strap positions.

The strap should be designed to resist the full tensile force due to uplift and the fixings likewise designed to transfer the force into the blockwork. For this purpose the strap can be assumed to be connected to triangular areas of blockwork (as shown in Figure 7.49) enclosed by a 45° line (see Figure 7.50).

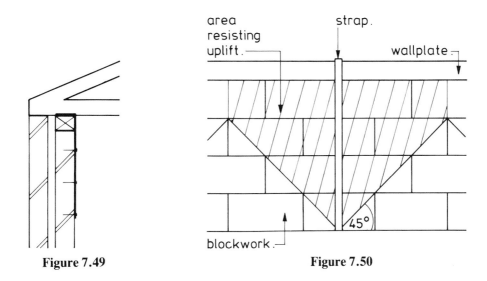

Figure 7.49   **Figure 7.50**

Note: Where straps are close together (or very long) the area from two adjoining straps may overlap. Obviously the same area of blockwork cannot be used to resist the uplift on both straps.

CHAPTER 8

STABILITY, ACCIDENTAL DAMAGE AND PROGRESSIVE COLLAPSE

During recent years, there has been an increasing number of collapses of different types of structures built in a variety of structural materials. Apart from the ever-present causes of mistakes in design, erection and construction, the situation has been aggravated by changes in:

(a) the methods of design and construction,
(b) the increased use of specialist subcontractors,

(c) the lack of awareness, by some designers, that sound structural elements can be inadequately connected and thus form unstable structures,
(d) the increase in the hazard of accidental forces.

(a) Methods of design and construction
The continuing decrease in the factors of safety (with the accompanying increase in stresses),

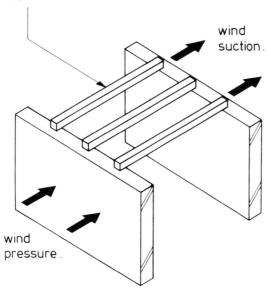

roof beams or trusses sitting on wall assumed to provide lateral support to wall.

wind suction.

wind pressure.

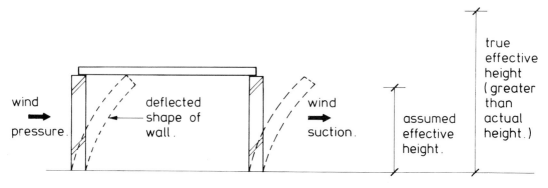

Figure 8.1(a)

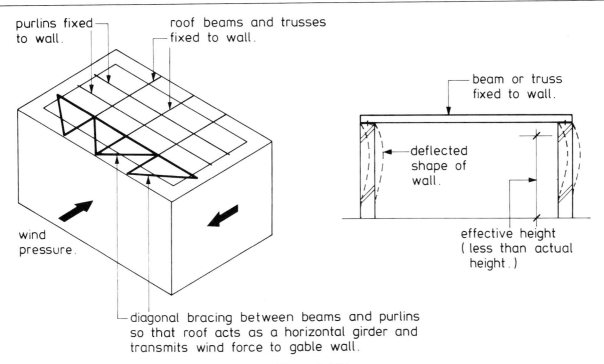

purlins fixed to wall.

roof beams and trusses fixed to wall.

wind pressure.

diagonal bracing between beams and purlins so that roof acts as a horizontal girder and transmits wind force to gable wall.

beam or truss fixed to wall.

deflected shape of wall.

effective height (less than actual height.)

Figure 8.1(b)

together with the continuing pressure to cut costs by reducing the amount of material and labour in construction, have led to the almost total disappearance of traditional massive structural forms. Buildings are far less robust than they used to be, and are now relatively flimsy. The introduction of new and comparatively untried materials, which are not standing up to the test of time, has not helped the situation. Structural design is becoming more and more complex and sophisticated, increasing the possibility of errors being made by young engineers whose education is becoming more theoretical and less practical.

(b) Increased use of specialist sub-contractors

It is not uncommon, for example, to find that a piling contractor has designed and formed the foundations for steel columns, designed and erected by a steelwork fabricator, used to support a prestressed concrete deck designed and erected by another specialist, on which has been erected a glued laminated timber frame designed by yet another specialist. Each element of this hybrid structure is normally perfectly structurally adequate in itself, but, since there has been no overall engineering supervision and responsibility, the resulting complete structure may be structurally unsound.

(c) Inadequate connection of structural elements

It is not uncommon for designers to consider a wall as being restrained top and bottom, and then fail to check that it is in fact so restrained. Designers sometimes fail to appreciate that re-

turns on the ends of walls, or other restraint or stiffening, may be necessary. A number of single-storey factories, and similar structures, have the main loadbearing wall inadequately restrained by the roof trusses, so that the effective height is greater than the designer has allowed for (see Figure 8.1). Further increasing the risk, the non-axial loaded gable wall, subject to wind loading, is not only unrestrained at roof level but movement joints are positioned between it and the side, or return, walls. The gable wall has thus no returns on its ends and is unstable on its horizontal axis and, lacking restraint from the roof, it is not robust on its vertical axis (see Figure 8.2). It is hardly surprising that collapses of gable walls in factories, etc., are not uncommon. Indeed, it is surprising that more of them do not fail.

Similar faults occur in multi-storey structures, where floors and roofs are not adequately tied or strapped to the walls, and where walls lack edge restraint.

In the authors' opinion, this particular aspect of design and construction is of very great importance and, for that reason, the whole of Chapter 7 was devoted to the subject.

(d) Risk of accidental damage

The decrease in robustness, together with an increase in the possibility of blast and impact forces from chemical plants, gas explosions, traffic accidents, etc., and changes in the use of

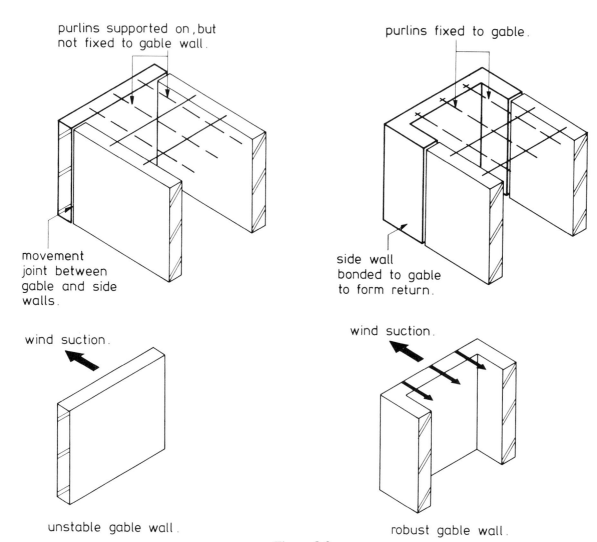

purlins supported on, but not fixed to gable wall.

purlins fixed to gable.

movement joint between gable and side walls.

side wall bonded to gable to form return.

wind suction.

wind suction.

unstable gable wall.

robust gable wall.

Figure 8.2

structures leading to the chance of overloading, have combined to make all structures – in all materials – more susceptible to accidental damage, less stable, and more prone to the risk of progressive collapse.

Having outlined some of the major causes of failures, it has to be admitted that, statistically, the risks are not very great. For example, from surveys by the Construction Industry Research and Information Association, and others, it has been estimated that the risk of a fatality due to a structural accident are 1:3 000 000 – as against 1:8000 from a car accident. Or again, the financial losses due to fires are about 200 times more than the losses due to structural accidents. However, this does not give the designer the right to take risks – low though they may be. People want to be as 'safe as houses', and the emotional, social and political impact and implications of a structural accident far outweigh a rational acceptance of the risk.

8.1 PROGRESSIVE COLLAPSE

'Progressive collapse' is the term used to describe the behaviour of a structure when local failure of a structural member (beam, column, slab, wall, etc.) occurs due to an accident of limited magnitude, destroying or removing the member and causing adjoining structural members to collapse. A chain reaction – 'house of cards' or 'domino' effect – can then spread throughout the whole or the major part of the structure, causing it to become unstable and collapse. Thus the final major damage is out of all proportion to the initial and minor cause.

After the disaster at the Ronan Point concrete system-built high-rise block of flats in May 1968, when the removal of only one wall panel caused widespread damage to the structure and the deaths of four people, it became mandatory for engineers to check certain structures to ensure that the removal of any structural element would

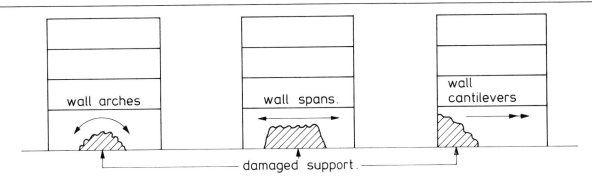

Figure 8.3

not cause a spread of collapse. The factor of safety required against progressive collapse is low, being only 1.05. It should be appreciated that, although the structure might not collapse, it may be unserviceable due to excessive cracking and deflection, and be left with an unacceptably low load-carrying capacity. It may well need extensive repair or even rebuilding. The provision in the design is, therefore, limited to the immediate safety of the occupants.

To keep this problem in proportion – experience has shown that masonry structures have an innate ability to withstand shock and are resistant to progressive collapse. Masonry structures have the capacity to arch, span and cantilever over openings caused by the removal of a structural wall or other support (see Figure 8.3).

Many masonry structures were seriously damaged by bombing and fire during the Second World War, and did not collapse. There have been numerous incidents since, when masonry structures have suffered incompetent demolition during alterations, impact from heavy lorries, blast from explosions, removal of support floors and roofs due to fire damage, etc., without collapsing.

8.2 STABILITY

Note that all new structural codes for all materials will pay attention to this problem. Structural masonry is no exception.

To limit the effects of accidental damage and to preserve structural integrity, BS 5628, clause 20.1, provides general recommendations which may be interpreted as follows:

(1) The designer responsible for the overall stability of the structure should ensure that the design, details, fixings, etc., of elements or parts of the structure are compatible, whether or not the design and details were made by him. This means, for example, not only checking that a specialist roof design can carry its own load, but also ensuring that it is properly tied to the walls and can transfer, say, wind forces from them.

(2) The designer should consider the plan layout of the structure, returns at the ends of walls, interaction between intersecting walls, slabs, trusses, etc., to ensure a stable and robust design.

The Code does not define or quantify 'robustness', but most designers would probably assume it to mean a structure's ability not to suffer a major collapse due to minor accidental damage. To obtain a consensus of opinion on an acceptable degree or quantity of robustness has proved difficult. Probably because any one designer's opinion depends not only on his experience, but also on his confidence and daring, or his caution and apprehension. Figure 8.4 shows plan layouts of masonry structures with differing degrees of robustness.

(3) The designer should check that lateral forces acting on the whole structure are resisted by the walls in the planes parallel to those forces, or are transferred to them by plate action of the floors, roofs, etc., or that the forces are resisted by bracing or other means.

The interaction between the walls, floors and roof affects the robustness of the structure. The lateral forces on the walls can be transferred to shear walls (walls parallel to the line of action of the lateral forces) by the floors or roof acting as horizontal diaphragms or wind girders – this being known as 'plate action'. The plate action of a roof was shown in Figure 8.1(b). Floors, however, vary in their ability to provide plate action. A two-way spanning *in situ* reinforced concrete slab bedded into the external

unstable structure, longitudinally, due to lack of restraint.

stable structure but liable to some distortion if floors are not properly tied to walls to give restraint.

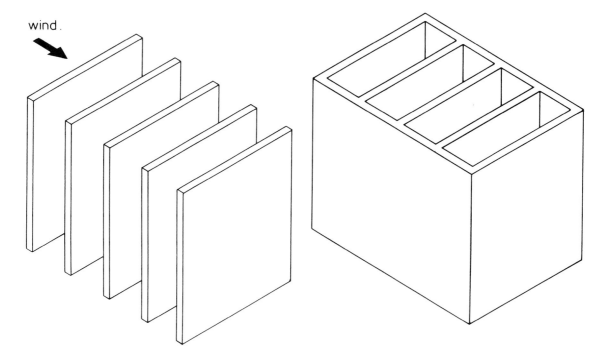

wind.

(a) <u>unstable</u> structure

(b) (possible) <u>stable</u> structure.

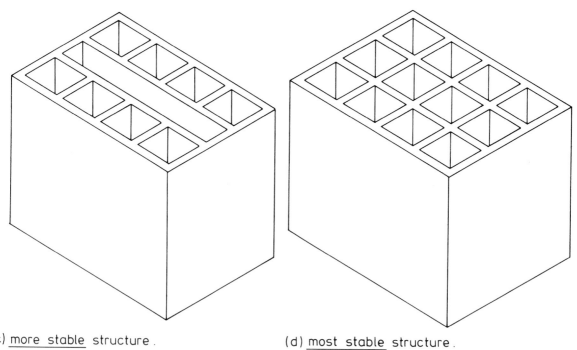

(c) <u>more stable</u> structure.

(d) <u>most stable</u> structure.

Figure 8.4

wall and continuous over the internal walls (see Figure 8.5), is obviously stiffer than precast concrete planks simply supported on the walls and not provided with lateral ties to the edge walls or continuity over the internal walls (see Figure 8.6).

Figure 8.7 shows flooring alternatives for a multi-storey masonry structure and indicates the variation in the degree of robustness imparted to the structure.

(4) Structures should be designed to resist a uniformly distributed horizontal load of

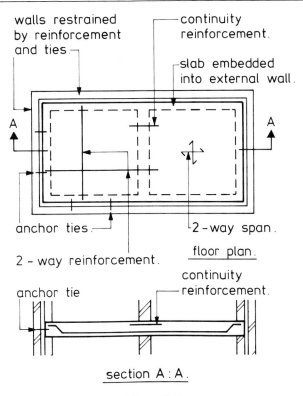

Figure 8.5

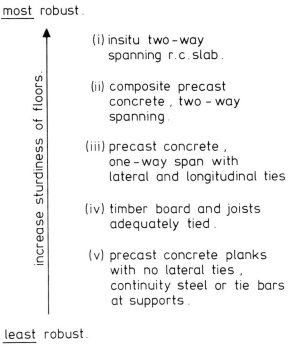

Figure 8.7

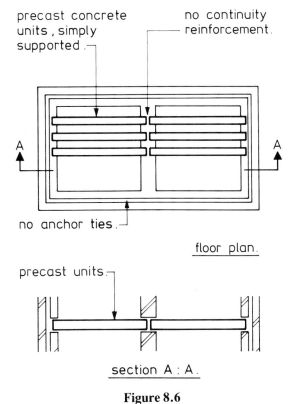

Figure 8.6

1.5% of the total characteristic dead load ($0.015G_k$) above any level for the load conditions (i) dead plus wind load, (ii) dead plus imposed plus wind load.

(5) Adequate strapping or tying (see Chapter 7) should be provided, where appropriate at floors and roofs.

8.3 ACCIDENTAL FORCES (BS 5628, CLAUSE 20.2)

No structure can be expected to resist excessive loads due to an extreme cause, such as a large plane crashing into it, but it should be able to withstand the impact of a lorry, or the possible mis-use of temporary and slight overloading, without collapsing completely. Nor should it suffer damage disproportionate to such a cause. Only the column, wall or slab, etc., subject to the excessive load should be damaged, and *not* the adjoining structural elements.

When because of the use or position of a structure there is a potential hazard such as a fuel dump or a chemical plant (e.g. the Flixborough disaster where an explosion caused severe damage to buildings in the vicinity) the design should ensure that, in the event of an accident, there is an acceptable probability that despite serious damage the structure will remain standing.

A fairly typical case is that of a bus garage where the columns supporting the lintols over the large door openings are liable to vehicle impact. The danger can be reduced by protecting the columns with bollards and designing the lintols as a continuous beam. In the event of a column being removed, the beam should be able to span over the greater length with a factor of safety of 1.05 (see Figure 8.8).

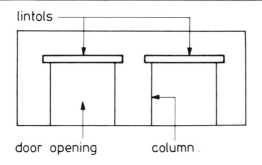

door opening   column .

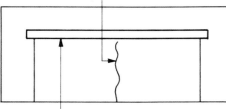

column destroyed .

lintol changed to continuous beam
with f/s of 1·05 i.e. here acting as
a simply supported beam .

Figure 8.8

8.4 DURING CONSTRUCTION

Although it is the contractor's normal contractual responsibility to maintain the safety of the works during construction – i.e. be responsible for the design and erection of all temporary works, to ensure adequate temporary bracing, shores, etc. – the designer should nevertheless consider whether special precautions or special temporary propping are needed to ensure the stability of the structural elements and the overall stability of the structure. If the designer considers that special precautions, etc., are necessary, he should inform the contractor, in writing, that such measures are advisable. Since, unfortunately, some contractors are eager to be absolved of their contractual responsibility to maintain the safety of the works, care must be taken by the designer in specifying the special precautions. If such care is not taken, the designer could find himself responsible for the design of all temporary works.

8.5 EXTENT OF DAMAGE

When a column or wall collapses due to accidental damage, known as 'primary' damage, the beams or slabs supported by the column or wall will suffer 'consequential' or 'secondary' damage. The structure must have adequate resi-

dual stability not to collapse completely, and the Code further advises that the designer should satisfy himself that '. . . collapse of any significant portion of the structure is unlikely to occur'. What is 'significant' is not defined, and is left to the engineer's judgement. The collapse of a carpet warehouse which damages a few carpets, is not so serious as the collapse of a school assembly hall killing or maiming a hundred children.

In the absence of any specific guidance, some designers look to the Building Regulations. These, however, provide no recommendations for buildings of four storeys or less – other than that the structure should be 'robust'. For buildings of five storeys or more, the following guidance is given:

(a) Horizontal area of collapse
Must not exceed 70 m^2 or 15% of the plan area, whichever is the lesser. (The designer can take action to prevent the spread of horizontal damage by tying and connecting the floors.)(see Figure 8.9(a).)

(b) Vertical area of collapse
Failure is acceptable provided that it occurs within the storey in which the incident leading to accidental damage took place, and may involve the storeys above and below. Vertical damage is normally limited in masonry structures by the

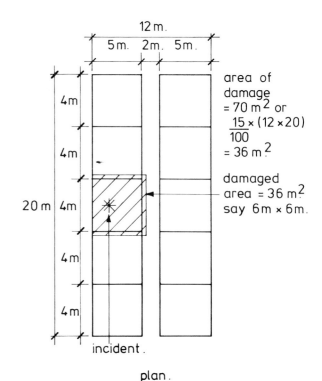

area of
damage
= 70 m^2 or
$\dfrac{15 \times (12 \times 20)}{100}$
= 36 m^2

damaged
area = 36 m^2
say 6m × 6m .

incident .

plan .

Figure 8.9(a)

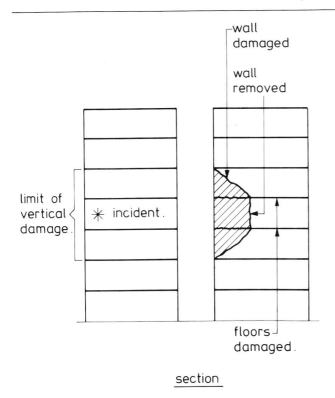

wall damaged
wall removed
limit of vertical damage
incident.
floors damaged.
section

Figure 8.9(b)

walls, etc., arching or spanning over the damaged area, by the interaction of floors and walls, and by vertical tying of the walls and columns, if necessary.

In the seven-storey crosswall structure shown in Figure 8.9(b), an accident has taken place on the third floor. The external wall and two crosswalls have been blown out, and the third and fourth floors damaged. The Building Regulations state that the designer must cater for the removal of any one element at a time. This is reasonable in that when one wall is destroyed, it is likely to reduce the pressure on the other walls. The second floor may be subject to debris loading. This is the limit of allowable damage and the remaining structure must have adequate residual stability, as stated in BS 5628.

8.6 DESIGN FOR ACCIDENTAL DAMAGE

8.6.1 Partial safety factors

The partial safety factor for design load, γ_f, is reduced as follows:

Design dead load $= 0.95G_k$ or $1.05G_k$

Design imposed load $= 0.35Q_k$ (except that, in the case of buildings used predominantly for storage, or where the imposed load is of a permanent nature, $1.05Q_k$ should be used)

Design wind load $= 0.35W_k$

The partial safety factors for material strength, γ_m, may be halved when considering the effects of accidental damage. For a wall or column with a high axial load subjected to a lateral design load, and treated as a 'protected' member (see later for definition), γ_m may be further reduced to 1.05.

EXAMPLE 1

For simplicity, let G_k and Q_k be unity for the two-span continuous slab shown in Figure 8.10. Determine the end reaction and the characteristic strength required for the end walls ($\gamma_m = 2.5$).

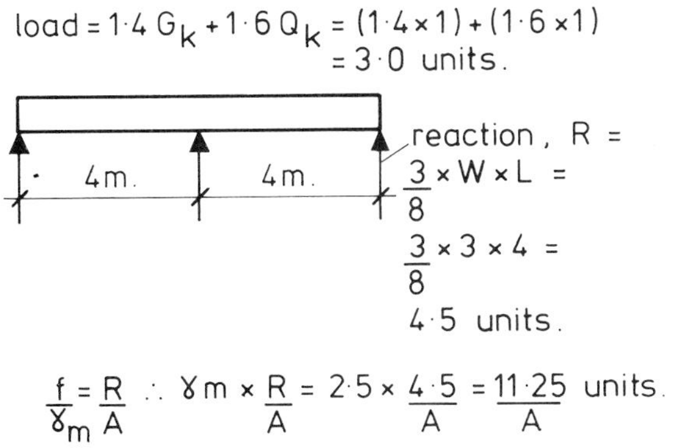

load = 1·4 Gk + 1·6 Qk = (1·4×1) + (1·6×1)
= 3·0 units.
4m.
4m.
reaction, R =
3 × W × L / 8 =
3 × 3 × 4 / 8 =
4·5 units.
f / γm = R / A ∴ γm × R / A = 2·5 × 4·5 / A = 11·25 / A units.

Figure 8.10

If the central support is removed by accident, determine the characteristic strength required for the end walls, $\gamma_m = 1.25$, (see Figure 8.11).

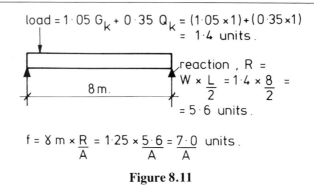

Figure 8.11

Note: Slenderness considerations of the walls have been omitted to simplify demonstration of the principle.

This very simplified example serves to emphasise that when checking a designed structure for accidental damage, the loads are decreased and the allowable stress increased. It should be noted that although the end walls in the example may stand up, the floor slab may have such an excessive deflection as to be unserviceable.

8.6.2 Methods (options) of checking

An experienced designer can check by eye to establish if a designed structure is robust and capable of withstanding accidental damage. But, since robustness – like beauty – is a subjective assessment (see also section 8.2), there may be differences of opinion between the designer and the checking authority.

The multi-storey cellular structure shown in Figure 14.37, having short spans, two-way continuous rc floor slabs with good tying to the outer leaf of the external cavity wall, numerous sturdy partitions bonded into the loadbearing walls, etc., would be unlikely to need checking for accidental damage. Similarly, a deep diaphragm wall of moderate height, capped by a continuous rc beam with a rigidly braced and stiff-sheeted roof firmly fixed to it, is hardly likely to collapse completely if a heavy lorry crashes into it. Where a detailed check is necessary, Table 12 of the Code (see Table 8.1) provides the designer with a number of options.

For buildings up to four storeys (Category 1), provided the plan form and the construction give robustness, and there is interaction of components and containment of spread of damage, no additional check is required. Oddly enough, a piloti base does not count as a storey – thus the building shown in Figure 8.12 is a Category 1 building.

Table 8.1 Detailed accidental damage recommendations (BS 5628, Table 12)

Category 1 *All buildings of four storeys and below*	Plan form and construction to provide robustness, interaction of components and containment of spread of damage (see clause 20)			

		Additional detailed recommendations for category 2		
Category 2		**Option (1)**	**Option (2)**	**Option (3)**
All buildings of five storeys and above	Plan form and construction to provide robustness, interaction of components and containment of spread of damage (see clause 20)	Vertical and horizontal elements, unless protected, proved removable, one at a time, without causing collapse	*Horizontal ties* Peripheral, internal and column or wall in accordance with 37.3 and table 13. *Vertical ties* None or ineffective Vertical elements, unless protected, proved removable, one at a time, without causing collapse	*Horizontal ties* Peripheral, internal and column or wall in accordance with 37.3 and table 13. *Vertical ties* In accordance with 37.4 and table 14

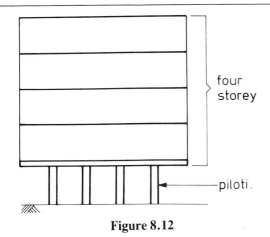

Figure 8.12

Table 8.2 Loadbearing elements (BS 5628, Table 11)

Type of loadbearing element	Extent
Beam	Clear span between supports or between a support and the extremity of a member
Column	Clear height between horizontal lateral supports
Slab or other floor and roof construction	Clear span between supports and/or temporary supports or between a support and the extremity of a member
Wall incorporating one or more lateral supports (note 2)	Length between lateral supports or length between a lateral support and the end of the wall
Wall without lateral supports	Length not exceeding 2.25h anywhere along the wall (for internal walls) Full length (for external walls)

Note 1: Temporary supports to slabs can be provided by substantial or other adequate partitions capable of carrying the required load.

Note 2: Lateral supports to walls can be provided by intersecting or return walls, piers, stiffened sections of wall, substantial non-loadbearing partitions in accordance with (a), (b) and (c) of 37.5, or purpose-designed structural elements

Option 1

Category 2 buildings, that is buildings with five or more storeys, require additional checks. The experienced designer would probably use Option 1 – the removal of loadbearing elements (except 'protected' members, see later) one at a time, without causing collapse.

To check each and every loadbearing element's removal would be an excessively lengthy and monotonous business, and is not normally necessary. On the other hand, it is good experience for a young designer to carry out this exercise on his first project to get the feel of the process. Generally, it will be found that the removal of parts of internal walls and the supports to horizontally tied slabs does not lead to structural collapse. It will also be found that many walls are, in effect, so precompressed by their vertical loads that they are quite capable of carrying the lateral loads required for protected members and, by implication, may be considered as such. A protected member must be capable of resisting an accidental damage design load of 34 kN/m², applied from any direction.

In many masonry structures, it is usually the checking or protection of the external vertical members (walls and columns), and the need for some horizontal tying, which needs investigating.

8.6.3 Loadbearing elements

These are defined in Table 11 and clause 37.5 of the Code (see Table 8.2).

It will be noted from the table that the length of a wall which forms a vertical loadbearing element depends upon the position of the wall and the presence of lateral supports. For example, the complete length of an external wall, without lateral supports, must be removed for an accidental damage check. On the other hand, for an internal wall, only a length 2.25 × clear storey height need be considered to be removed. Thus in the case of an internal wall in a block of flats with a clear height between lateral supports of 2.5 m, only a length of 2.25 × 2.5 m = 5.625 m need be considered removed at any one time (see Figure 8.13).

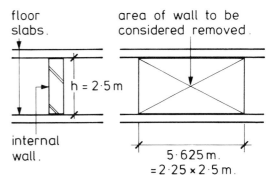

Figure 8.13

The designer must check the worst condition for structural collapse, i.e. where to remove the 5.625 m length of the internal wall to create the most critical overall design condition.

For both internal and external walls, the provision of vertical bracing by lateral supports reduces the length of wall to be examined for the effects of removal. What constitutes a 'lateral support' is defined in clause 37.5 of the Code as:

(a) an intersecting or return wall,
(b) a pier or stiffened section of the wall,
(c) a substantial partition

Detailed descriptions of these are as follows:

(a) An intersecting or return wall tied to the wall for which it provides lateral support, with connections (bonding, wall ties, straps) capable of resisting a force of the lesser of 60 kN or $(20 + 4N_s)$ kN, where N_s is the number of storeys including the ground floor and the basement of the building, per metre height of the wall. The value 60 kN or $(20 + 4N_s)$ kN is the basic horizontal tie force F_t. The intersecting or return wall must have a length of $h/2$ without openings, be at right angles to

the supporting wall, and have an average mass of not less than 340 kg/m³ (see Figure 8.14).

(b) A pier or stiffened section of the wall, not more than 1 m in length, and capable of resisting a force the lesser of 90 kN or $(30 + 6N_s)$ kN, per metre height of the wall (see Figure 8.15).

(c) A substantial partition at right angles to the wall, having an average mass of not less than 150 kg/m³, and tied with connections capable or resisting 30 kN or $(10 + 2N_s)$ kN per metre height of the wall. The partition need not be in a straight line but should, in effect, divide the bay into two compartments (see Figure 8.16).

Since many buildings have return walls, intersecting walls and substantial partitions it is not often necessary to add further lateral supports to a wall. On the few occasions when it is necessary to add lateral supports, it is generally preferable to add masonry piers – rather than to introduce steel or rc columns – so as to limit the number of trades, operations, etc., on the site.

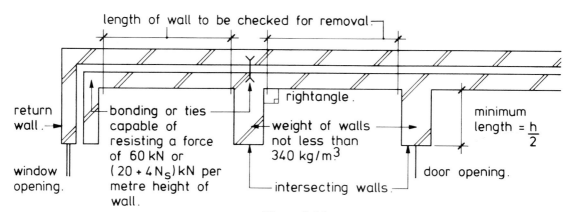

Figure 8.14

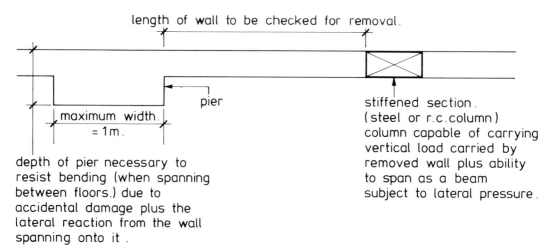

Figure 8.15

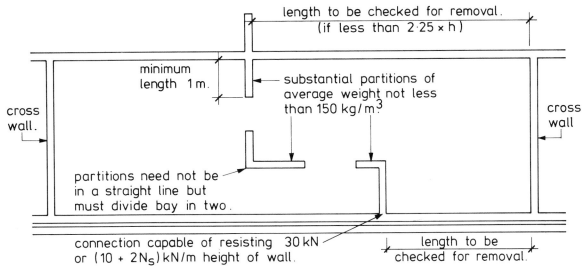

Figure 8.16

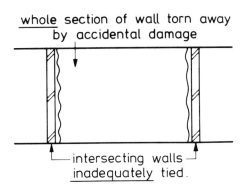

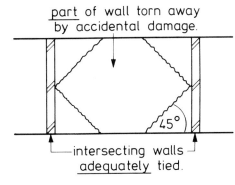

Figure 8.17

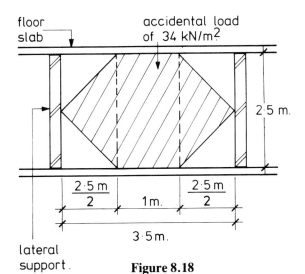

Figure 8.18

The reason for tying the lateral supports to the wall is to prevent large areas tearing off, and to assist a yield line type failure (see Figure 8.17).

For a clear storey height of, say, 2.5 m and a distance of 3.5 m between lateral supports – a typical case in a block of flats – the load on the ties can be calculated approximately (see Figure 8.18).

The accidental load on the shaded area in Figure 8.18

$$= \left(2.5 \times 1 + \frac{2 \times 2.5 \times 2.5}{2} \times \frac{1}{2}\right) \times 34 \text{ kN/m}^2$$

$$= \left(2.5 + \frac{2.5^2}{2}\right) \times 34 \text{ kN/m}^2$$

The tying resistance/m height

$$= \frac{\left(2.5 + \frac{2.5^2}{2}\right)}{2.5} \times 34 \text{ kN}$$

$$= (1 + 1.25) \times 34 \text{ kN}$$

$$= 76.5 \text{ kN}$$

This compares reasonably with the required force of 30 kN in type (c) support, 90 kN in type (b) and 60 kN in type (a). The reason for the comparatively low force of 30 kN in type (c) is that a substantial partition dividing a bay into two compartments assumes that it limits the accidental damage to one compartment only.

8.6.4 Protected member
Some structural members are so vital to the stability of a structure that removal by damage

would cause extensive collapse or damage to the whole structure. An extreme case is depicted in Figure 8.19, where it is obvious that the removal of the column or cantilever would result in the collapse of the structure.

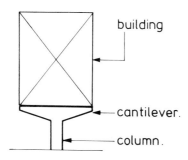

Figure 8.19

Such members, and their connections, must be designed to resist the full accidental loading of 34 kN/m², applied from any direction to the member directly together with the reaction from contiguous (connected) building components (i.e. sheeting, walling, etc.) also subjected to the same accidental loading. In practice, most building components would have a much lower ultimate lateral resistance and would fail way below a loading of 34 kN/m², and thus transmit relatively low reactions to the protected member.

The protected member must, as stated above and shown in Figure 8.20, be able to withstand an accidental loading in any direction.

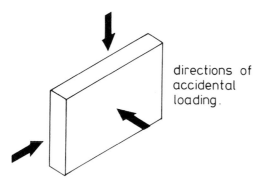

Figure 8.20

The ability of masonry walls or piers to withstand an accidental force of 34 kN/m² applied vertically seldom creates problems. However, problems do arise when the force is applied laterally to a wall that does not have a sufficiently high axial load to provide the precompression necessary to counteract the flexural tensile stress and reduce it to an acceptable level. The acceptable level is that strength multiplied by half the partial factor of safety for material strength, γ_m.

Conversely, it can be considered that a wall which by virtue of its axial load can withstand the lateral accidental loading, with an acceptable partial safety factor, is of itself a protected member. In practice, it is quite common to find that there is sufficient precompression in loadbearing masonry to counteract the flexural tensile stress at about the third, or sometimes the fourth, storey down from the roof. It follows that it is not always necessary to check most walls on the lower storeys of buildings more than five storeys high, particularly internal walls. However, to reiterate a point made earlier, the authors feel that designers lacking practice in checking should check all walls, etc., on their first projects, until they have gained sufficient experience to dispense with checking by calculations and, using their engineering judgement, check by eye.

Option 2

Under this option (see Table 8.1), the removal analysis applies to vertical members only, and does not apply to floor and roof slabs – for which, horizontal tying is required.

The provision of lateral supports to walls, plus the provision of a robust plan form, will obviously reduce the amount and the spread of damage to the vertical loadbearing elements. To reduce the spread of horizontal damage, and the repercussion of secondary damage to the vertical elements, the floors must be able to span or cantilever over the damaged area.

The normal secondary, or distribution, reinforcement in continuous *in situ* rc slabs is usually sufficient to allow the floor to span or cantilever over a damaged, failed or removed, internal vertical support. With floors constructed of simply supported precast concrete units, extra reinforcement is usually necessary. Similarly, timber floors normally need tying round their periphery, and some two-directional tying internally. The basic horizontal tie force, F_t, to be used in determining the amount of tying required is the same as that for the vertical elements, mentioned above, namely $F_t = 60$ kN or $(20 + 4N_s)$ kN, whichever is the lesser of the two values, where $N_s =$ the number of storeys including the ground floor and any basements.

The Code requirements for peripheral (external), internal, and column and wall ties are provided in Table 8.3.

Table 8.3 Requirements for full peripheral, internal and column or wall ties (BS 5628, Table 13)

Type of tie	Unit of tie force	Size of design tie force	Location of tie force (arrowed)	Fixing requirements and notes
A. *Peripheral*	kN	F_t	Around whole perimeter	Ties should be: (a) placed within 1.2 m of edge of floor or roof or in perimeter wall; (b) anchored at re-entrant corners or changes of construction.
B. *Internal* (both ways)	kN/m width	F_t or $\dfrac{F_t(G_k + Q_k)}{7.5} \times \dfrac{L_a}{5}$ whichever is the greater F_t F_t or $\dfrac{F_t(G_k + Q_k)}{7.5} \times \dfrac{L_a}{5}$ whichever is the greater	*One way spans* (i.e. in crosswall or spine construction) (i) in direction of span (ii) in direction perpendicular to span *Two way spans* (in both directions) 	(a) Internal ties should be anchored to perimeter ties or continue as wall or column ties. (b) Internal ties should be provided: (1) uniformly throughout slab width; or (2) concentrated in beams (6 m max. horizontal tie spacing); or (3) within walls 0.5 m max. above or below the slab at 6 m max. horizontal spacing: (4) in addition to peripheral ties spaced evenly in perimeter zone. (c) Calculation of tie forces should assume: (1) $(G_k + Q_k)$ as the sum of average characteristic dead and imposed loads in kN/m²; (2) L_a as the lesser of: the greatest distance in metres in the direction of the tie, between the centres of columns or other vertical loadbearing members whether this distance is spanned by a single slab or by a system of beams and slabs; or 5 × clear storey height, h.
C. *External column*	kN	$2F_t$ or $(h/2.5)F_t$ whichever is the lesser where h is in metres		(a) Corner columns should be tied in both directions (b) Tie connection to masonry may be based on shear strength or friction (but not both). (c) Wall ties (where required) should be (1) spaced uniformly along the length of the wall; or (2) concentrated at centres not more than 5 m apart and not more than 2.5 m from the end of the wall. (d) External column and wall ties may be provided partly or wholly by the same reinforcement as perimeter and internal ties.
D. *External wall*	kN/m length of loadbearing wall			

Basic horizontal tie force = F_t = 60 kN or $(20 + 4 N_s)$ kN, whichever is the lesser of the two values; N_s is the number of storeys (including ground and basement).

At first sight, the table may appear rather formidable. However, on better acquaintance, it proves to be reasonably straightforward.

In clause 37.3, the Code states that horizontal ties should be provided at each floor level and at roof level. When the roof is of lightweight construction, no ties need be provided at that level. Such roofs are defined as 'roofs comprising timber or steel trusses, flat timber roofs, or roofs incorporating concrete or steel purlins with asbestos or wood-wool decking'. The Code also states that 'horizontal ties may be provided in whole or in part by structural members which may already be fully stressed in serving other purposes'. It cites the example, mentioned earlier, of reinforcement in *in situ* slabs, and also masonry in tension. If the masonry's tensile strength is used, it must not be perpendicular to the bed joints (see Figure 8.21).

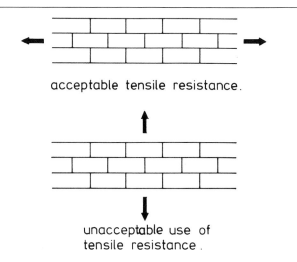

acceptable tensile resistance.

unacceptable use of
tensile resistance.

Figure 8.21

According to the Code, 'ties should be positioned to resist most effectively accidental damage'. Assistance on this topic is provided in the final column of Code Table 13.

EXAMPLE 2

An eight-storey hall of residence with 102.5 mm brick crosswalls at 4 m centres, supporting continuous *in situ* rc floor slabs, has a lightweight roof. Determine the peripheral, internal and external ties required (see Figures 8.22(a) and (b)).

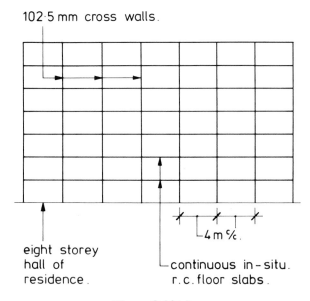

Figure 8.22(a)

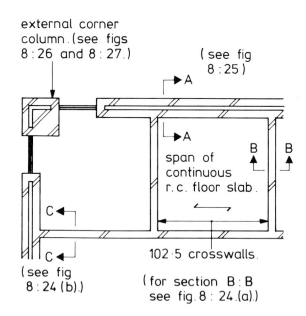

Figure 8.22(b)

Basic tie force = 60 kN or (20 + 4 × 8) = 52 kN. Use 52 kN.
Ties are required at each floor slab, but not at the roof since it is of lightweight construction. (The roof may need vertical straps to act as ties to resist wind suction.)

(a) Peripheral ties

$$F_t = 52 \text{ kN/m}$$

$$A_s = \frac{52 \times 10^3}{410} = 126 \text{ mm}^2$$

Use one No. 16 mm HY bar/m (see Figure 8.23).
In effect, peripheral ties act like a splint running around the outer face of the building, tying it together (see Figures 8.23(a) and (b)).

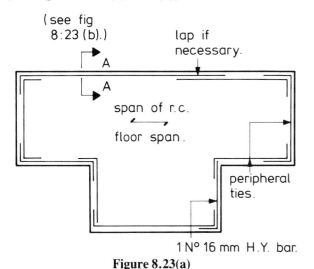

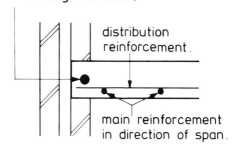

1 N° 16 mm H.Y. bar.

Figure 8.23(a)

section A : A (see fig 8:23 (a))

Figure 8.23(b)

(b) Internal horizontal ties (in direction of span)

$$F_t = 52 \text{ kN or } \frac{F_t(G_k + Q_k)}{7.5} \times \frac{L_a}{5} \text{ kN/m}$$

where

$G_k + Q_k$ = sum of average characteristic dead and imposed loads in kN/m^2
$\qquad\quad$ = 7.5 kN/m^2 for domestic buildings
$L_a \qquad$ = the lesser of the span (in direction of ties) or five times the clear storey height.

Let h = 2.5 m, then L_a = the lesser of 4 m or 5 × 2.5 m
$$= 4 \text{ m}$$
Therefore

$$\frac{F_t(G_k + Q_k)}{7.5} \times \frac{L_a}{5} = \frac{52 \times 7.5}{7.5} \times \frac{4}{5}$$

$$= 41.6 \text{ kN/m}$$

Since this is less than F_t = 52 kN, use 52 kN/m in design of ties, i.e. same as for peripheral ties. The continuity reinforcement would normally be adequate. If it is not, all that is necessary is to add extra reinforcement (see Figures 8.24(a) and (b)).

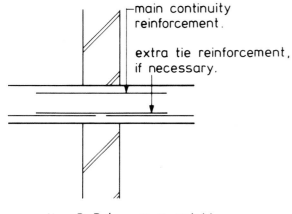

section B : B (see fig 8 : 22 (b))

Figure 8.24(a)

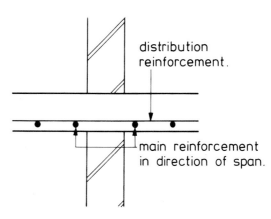

section C : C (see fig 8 : 22 (b))

Figure 8.24(b)

(c) Internal transverse ties (ties parallel to crosswalls)

$$F_t = 52 \text{ kN/m}$$

check that distribution reinforcement is adequate. If not, add extra reinforcement (see Figures 8.24(a) and (b)).

(d) External wall ties

Tie force $= 2F_t$ or $\dfrac{h}{2.5}F_t$ whichever is the lesser.

Since $h = 2.5$, $\dfrac{h}{2.5}F_t = 52 \text{ kN/m}$.

The only practical method of resisting this tie force is in shear between the slab and the masonry (see Figure 8.25).

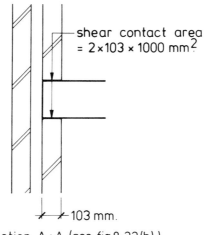

section A : A (see fig 8·22(b))

Figure 8.25

$$\text{Shear contact area} = 2 \times 103 \times 1000 \text{ mm}^2$$

$$\text{Shear resistance} = \frac{f_v}{\gamma_{mv}} \times \text{shear area}$$

Taking the worst case of no increase due to axial loading

f_v $\quad= 0.35 \text{ N/mm}^2$ for brickwork in 1:1:6 mortar
γ_{mv} $= 1.25$ for accidental damage

therefore

$$\text{shear resistance} = \frac{0.35}{1.25} \times 2 \times 103 \times \frac{1000 \times 1}{10^3} \text{ kN/m}$$

$$= 58 \text{ kN/m}$$

Note that lower down the building, the friction against the tie force between the concrete slab and the wall could be used, assuming the coefficient of friction $= 0.6$.

(e) External column tie (see Figure 8.26)
Assume for planning reasons the gable and the main side wall are stopped short of the external corner, and the support is provided by a masonry column:

(i)   the column should be tied in both directions;
(ii) the tie connection may be based on shear strength or friction, but not both.

$$\text{Contact area} = \frac{\text{shear force}}{\text{shear stress}} \times \frac{73.5 \times 10^3}{2 \times 0.35/1.25} = 131\,000 \text{ mm}^2$$

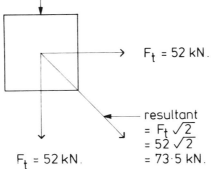

Figure 8.26

Provide 360×360 mm bearing minimum (see Figure 8.27).

Determine F_t for the spine wall structure shown in Figure 8.28.

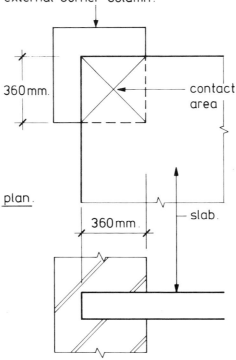

Figure 8.27

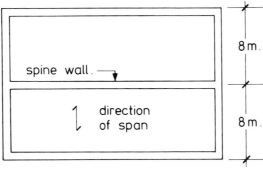

Figure 8.28

F_t = lesser of 60 kN or [20 + (4 × 6)], i.e. 44 kN
L_a = lesser of 8 m or 5 × 3 m, i.e. 8 m

Therefore

$$\frac{F_t(G_k + Q_k)}{7.5} \times \frac{L_a}{5} = \frac{44 \times 20}{7.5} \times \frac{8}{5}$$

$$= 187.7 \text{ kN}$$

Use F_t = 187.7 kN.

This example shows that the internal tie force is affected by the span and the loading of the floor.

The second option, Option 2, is likely to prove most popular with checking authorities until more experience is gained in design for accidental damage. The two options, Option 1 and Option 2, show that little needs adding to a typical robust masonry structure to make it accident resistant.

Option 3
Under this option (see Table 8.1) neither vertical nor horizontal elements may be removed, and both horizontal and vertical tying are required. If the structure is tied (as prescribed in the Code) in three directions, as shown diagrammatically in Figure 8.29, there should only be limited damage due to accidental forces, and the structure should have adequate residual stability.

The vertical tie reinforcement is either placed in *in situ* concrete pockets in a brick wall, or threaded through cellular blocks, and then grouted up (see Figure 8.30).

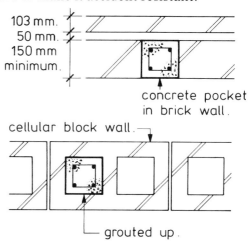

Figure 8.30

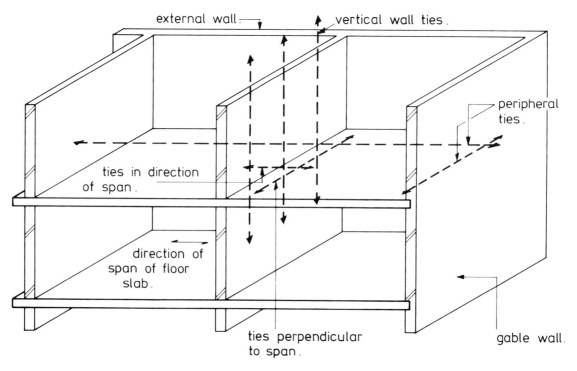

Figure 8.29

Both techniques are clumsy and time-consuming. This, together with the fact that the minimum thickness of a solid wall, or the loadbearing leaf of a cavity wall, must be at least 150 mm thick – thus restricting the use of half brick, 102.5 mm, walls – is likely to make this option unattractive to designers. Its main use is likely to be in buildings subject to high blast forces from chemical works and the like. When the option is used, the vertical ties form, in effect, rc columns which could perform an alternative path load-carrying capacity. This increases the ability of the structure to arch, span and cantilever over damaged areas.

Clause 37.4 of the Code states that vertical tying is effective only when horizontal tying capable of resisting a horizontal force of F_t kN/m width is also present, and that the floor is of precast or *in situ* concrete or other heavy flooring units.

The same clause goes on to state: 'the wall should be contained between concrete surfaces or other similar construction, excluding timber, capable of providing resistance to lateral movement and rotation across the full width of the wall'. The ties should extend from the roof down to either the foundations or to the level where the wall, by virtue of compression due to dead load, may be considered to be protected. The ties should be

continuous, but should be anchored separately and fully at each floor level (see Figure 8.31).

This method reduces the risk of the walls above and below the damaged wall being torn out of position due to it being blasted out by the accidental force. Code Table 14 gives the requirements for vertical ties (see Table 8.4).

Table 8.4 Requirements for vertical ties (BS 5628, Table 14)

Minimum thickness of a solid wall or one loadbearing leaf of a cavity wall	150 mm
Minimum characteristic compressive strength of masonry	5 N/mm²
Maximum ratio h_a/t	20
Allowable mortar designations	(i), (ii), (iii)
Tie force	$\dfrac{34A}{8000}\left(\dfrac{h_a}{t}\right)^2$ N or 100 kN/m length of wall or per column, whichever is the greater
Positioning of ties	5 m centres, max. along the wall and 2.5 m, max. from an unrestrained end of any wall

Note: A = the horizontal cross-sectional area in mm² of the column or wall including piers, but excluding the non-loadbearing leaf, if any, of an external wall of cavity construction
h_a = the clear height of a column or wall between restraining surfaces
t = the thickness of column or wall.

The tie force formula is based on the ability of a wall to arch vertically (if restrained top and bottom) when subject to a lateral load:

$$\text{Tie force} = \frac{34A}{8\,000}\left(\frac{h_a}{t}\right)^2 \text{N}$$

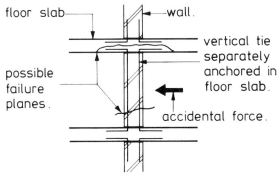

floor slab

wall.

vertical tie separately anchored in floor slab.

possible failure planes.

accidental force.

Figure 8.31

EXAMPLE 3

Determine the area of tying reinforcement required in pockets at 5 m centres (note maximum spacing) in the 150 mm inner loadbearing leaf of a cavity wall (note minimum width). The clear height of the wall is 3 m.

$A = 150 \times 5 \times 1000 = 750\,000 \text{ mm}^2$
$t = 150 \text{ mm}$
$h_a = 3000 \text{ mm}$

$$\text{Vertical tie force} = \frac{34 \times 7.5 \times 10^5}{8 \times 10^3}\left(\frac{3 \times 10^3}{1.5 \times 10^2}\right)^2 \text{N}$$
$$= 128 \times 10^4 \text{ N}$$
$$= 1280 \text{ kN}$$

$$\text{Minimum tie force} = 100 \text{ kN/m} \times 5 \text{ m}$$

$$= 500 \text{ kN}$$

$$A_s = \frac{1280 \times 10^3}{410}$$

$$= 3120 \text{ mm}^2$$

Use four No. 32 mm HY bars per pocket.

8.6.5 General notes

(1) The basic tie force $F_t = 60$ kN or $(20 + 4N_s)$ kN for buildings of various heights is given in the table below.

No. of storeys (N_s)	4	5	6	7	8	9	10	10+
F_t in kN			0	40	44	48	52	56	60	60

At the time of publication, no agreement has been reached for the treatment of accidental damage for buildings of four storeys or less. The designer may consider no extra requirements, other than robustness for four-storey flats, but might consider checking buildings used as schools, hospitals, etc.

(2) Detailed design for accidental damage is an unfamiliar concept to structural designers. It seems likely that, with experience, the rules will improve.

(3) It is possible that a designer's first attempt at checking and designing for accidental damage may be slow and laborious – as indeed, any other first design usually is. But experience shows that what, at first sight, appears complex is really comparatively simple.

(4) With experience, designers will tend to build-in robustness and check for accidental damage using Option 1. If the vertical and horizontal elements are not removable, one at a time, or protected, most designers would probably adjust the structural layout, or add horizontal tying from Option 2. Some designers may adopt Option 2 and check the vertical elements for removal or protection. If the vertical elements are not removable or protected, the designer should either adjust the structural layout or add some vertical tying from Option 3.

(5) In the design of a wide range of structures, particularly those with *in situ* rc floors, the designer will quickly find that all he needs to do is to provide some horizontal ties and check the unprotected external walls on the upper three or possibly four storeys.

CHAPTER 9

STRUCTURAL ELEMENTS AND FORMS

Perhaps the most rewarding part of designing in masonry is forming the various materials and elements into interesting, efficient and useful structural forms. In order to simplify the approach to structural forms, it is probably best to first consider the various structural elements that can be formed with masonry units, and then to consider the ways in which these elements can be put together.

STRUCTURAL ELEMENTS

9.1 SINGLE-LEAF WALLS

A single-leaf wall may be of any thickness provided the masonry units are bonded together (see Figure 9.1). Single-leaf walls are mainly used for internal loadbearing walls, boundary walls and retaining walls.

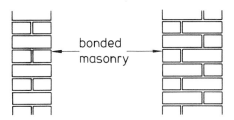

Figure 9.1 Single-leaf walls

9.2 DOUBLE-LEAF COLLAR-JOINTED WALLS

This type of wall is often used internally and externally where a double-leaf thickness is re-quired structurally, but a stretcher bonded face is required for architectural reasons (see Figure 9.2).

Due to the weakness of the collar joint in wall type 1, it is necessary to design the wall in a similar manner to that of a cavity wall. However, by the use of metal ties or mesh reinforcement through the joint, the condition can be improved to a minimum standard (see type 2 in Figure 9.2) which can then be considered as a solid wall. The main improvement is the ability of the joint to take vertical shear forces. Special care must be taken to see that the wall is constructed as specified, since this is critical to the wall's behaviour under load and hidden from view after construction.

9.3 DOUBLE-LEAF CAVITY WALLS

Double-leaf cavity walls are mainly used for external walls, the cavity being incorporated to prevent damp penetration, and the two leaves are tied together with metal ties (see Figure 9.3).

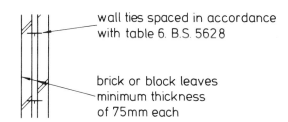

Figure 9.3 Double-leaf cavity wall

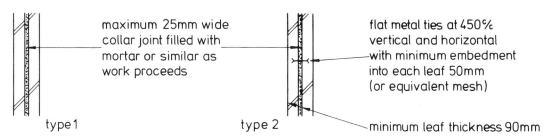

Figure 9.2 Double-leaf collar-jointed walls

Normal ties are considered capable of transferring some horizontal forces across the cavity from one leaf to the other, but not capable of transferring any significant vertical shear forces across the cavity. The wall is, therefore, designed as two separate leaves, each leaf carrying the vertical loads applied directly to it, the only assistance provided by the opposite leaf being increased resistance to buckling and the ability to transfer horizontal loads such as wind across the cavity.

Specially designed cavity walls can be assumed to transfer vertical shear forces, but only when wall ties specifically designed to transfer them without significant distortions are incorporated at suitable centres for the loading conditions involved.

The cavities of double-leaf walls are sometimes filled with non-loadbearing material such as insulation quilt to improve the overall thermal insulation qualities of the wall.

9.4 DOUBLE-LEAF GROUTED CAVITY WALLS

A grouted cavity wall can be designed as a solid wall, provided that the leaves are spaced a minimum of 50 mm apart, suitably tied with metal wall ties, and grouted with concrete of a strength at least equivalent to that of the mortar (see Figure 9.4). This form of construction is often used in similar conditions to the double-leaf collar-jointed wall and designed in a similar manner (see section 9.2).

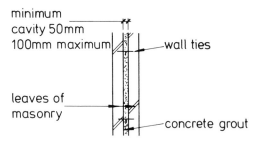

Figure 9.4 Double-leaf grouted cavity wall

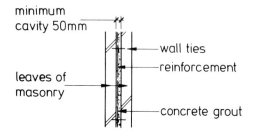

Figure 9.5 Reinforced grouted cavity wall

Grouted cavity walls may also be reinforced (see Figure 9.5). The addition of the reinforcement assists the wall in counteracting tensile forces, and is particularly useful when lateral loads are to be resisted. It should be noted that filling the cavity with grout has a detrimental effect on the wall's ability to resist damp penetration.

9.5 FACED WALLS

Faced walls consist of two different masonry units, bonded together in a manner which provides a particular facing to the wall. They are used where a solid wall is necessary, but where the facing material has to have properties not required for the backing. A typical example is indicated in Figure 9.6 where a 327 mm faced

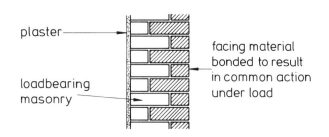

Figure 9.6 Faced wall

wall provides a special facing brick on one side, but incorporates a more economical inner unit to which plaster may be applied. Attention must be given, when selecting different units which are to be bonded together, to see that the shrinkage, thermal and other movements of the units are compatible. The design procedure for a faced wall is similar to that of any other solid wall, but it is generally assumed that the full thickness of the wall is constructed in the weaker unit.

9.6 VENEERED WALLS

Veneered walls have a facing which is attached to the backing, but not bonded to it in such a way as to induce composite action under load (see Figure 9.7).

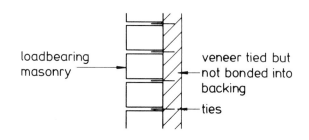

Figure 9.7 Veneered walls

This type of wall is often used where an expensive facing is required and/or the structural qualities of the veneer are of little assistance to the loadbearing capacity of the wall. In other cases, veneered walls are used when the facing veneer is likely to need replacement within the life of the structure.

The design of a veneered wall must take account of the dead weight of the veneer. However, the structural effect of the veneer should be neglected. As in the case of faced walls, the possibility of differential vertical movements from shrinkage, thermal and other effects must be considered to make sure that loosening of the ties and/or buckling of the veneer, etc., will not occur.

9.7 WALLS WITH IMPROVED SECTION MODULUS

Previous chapters have already discussed the problems relating to masonry walls which have to resist large bending moments at positions of low gravitational loads. As mentioned earlier, the greatest problem relates to the tensile stresses.

Consider the calculation involved in determining these stresses:

$$\text{Maximum tension stress} = \frac{W}{A} - \frac{M}{Z}$$

(see Chapter 3)

where

W = vertical load
A = cross-sectional area
M = applied bending moment
Z = section modulus

To improve these stress conditions without changing the mass of the wall, it is necessary to improve the section modulus, Z, without changing the area, A. This can be achieved by a redistribution of the material to locate the majority of it at a greater distance from the neutral axis of the section.

For example consider the two sections (1) and (2) shown in Figure 9.8.

The areas of both sections are equal.
i.e. section (1) area = $2 \times 2 = 4$
section (2) area = $(2 \times 2.5) - (1 \times 1) = 4$

But the Z value of (2) is greater than that of (1),

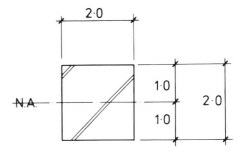

section 1

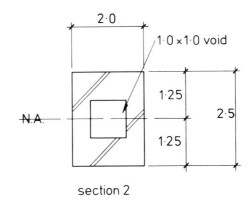

section 2

Figure 9.8

i.e. section (1): $Z = \dfrac{2 \times 2^2}{6} = 1.333$

section (2): $Z = \dfrac{1}{1.25}\left(\dfrac{2 \times 2.5^3}{12} - \dfrac{1 \times 1^3}{12}\right)$

$= 2.013$

This method of improving the section modulus of the wall, when large bending moments are to be resisted, is the main advantage achieved with the wall sections which follow.

9.7.1 Chevron or zig-zag walls

The chevron or zig-zag wall (Figure 9.9) is particularly useful for free-standing walls and other walls required to resist large bending moments. It achieves its extra stiffness and higher Z/A ratio from the changes in direction on plan, and the shape of the wall also results in a very pleasing appearance which has been very successfully used for the external walls of churches, boundary walls, etc.

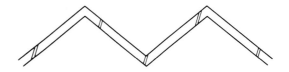

Figure 9.9 Chevron or zig-zag wall

9.7.2 Diaphragm walls

The diaphragm wall (Figure 9.10) is basically a wide cavity wall with cross ribs bonded or specially tied to the leaves of masonry to provide

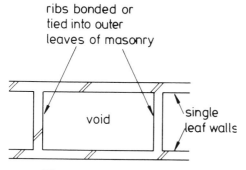

Figure 9.10 Diaphragm wall

suitable vertical shear resistance at the junctions. This type of wall is particularly suitable for tall single-storey buildings enclosing large open areas. The width between the leaves is increased to suit the particular design condition, and large Z/A ratios can be achieved making economical use of the masonry. Diaphragm walls have been successfully used on sports and drama halls, factories, etc., and a typical dimension for such use would be in the region of 550–1 100 mm overall width.

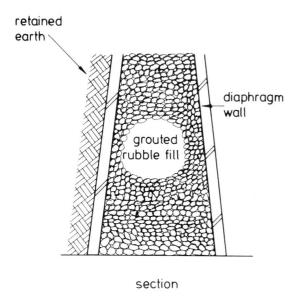

section

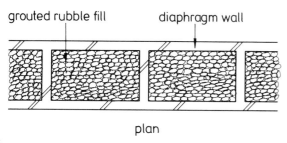

plan

Figure 9.11 Typical mass filled diaphragm

9.7.3 Mass filled diaphragms

The diaphragm wall can also be used for retaining walls and is sometimes mass filled (see Figure 9.11). This form of diaphragm is constructed in lifts, and filled with rubble or other material to provide mass for stability. In some cases the rubble is grouted to form a monolithic mass. For this type of construction, the ribs should always be properly bonded.

9.7.4 Piered walls

A piered wall is a wall stiffened by piers bonded into the wall, at regular centres (see Figure 9.12).

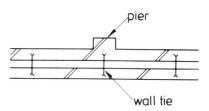

Figure 9.12 Piered wall

The use of piers is mainly suited to local stiffening of loadbearing walls at high concentrated load locations, and in the external walls of single-storey buildings in the range of 3–5 m height which are required to resist some lateral loading. The piers have the effect of increasing the effective thickness of the wall, thus reducing the slenderness ratio and enabling it to carry higher compressive loading. The most complicated form of design tends to be that of a piered cavity wall with combined lateral and vertical loading, where the piers are bonded to the inner or outer leaf and the two leaves tied in the normal manner using standard wall ties.

The design considers the pier and a portion of the leaf bonded to it as a T section and the opposite leaf as a second member tied to it. It is considered that the ties across the cavity are capable of transmitting some horizontal forces but are unable to transmit vertical shear. Due to the unsymmetrical geometry of the section, the pier has greater moment of resistance in one direction than in the other, and for bending moments which can occur in either direction the shape is not ideal.

In cases where the forces involved would demand large piers the fin wall, diaphragm wall and/or post-tensioned wall should be considered.

9.7.5 Fin walls

A fin wall is basically a piered wall in which the pier has been extended to more slender proportions and has taken on the major role in resisting

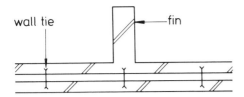

Figure 9.13 Fin wall

lateral load. The fin is the main structural element and is designed as a T section bonded into the intersecting leaf of masonry. The boundary between piers and fins is a rather grey area, but this should not be allowed to confuse the designer since the structural behaviour and design considerations at this boundary are basically the same. Like the diaphragm, the fin wall profile results in a large Z/A ratio, achieving economical use of the masonry in resisting bending moments. Due to the unsymmetrical geometry of the section, the fin has a greater moment of resistance in one direction than in the other and, whilst this is sometimes a slight disadvantage, its attractive form often compensates in its selection for a project. Fin wall construction has been successfully used for retaining walls, sports and drama halls, factories and multi-storey buildings (Figure 9.13).

9.8 REINFORCED WALLS

Reinforced walls have developed from the need for masonry walls to resist tensile stresses in excess of the normal permissible tensile stresses for masonry acting alone.

Walls are often constructed with reinforcement contained in the cavity, although sometimes the reinforcement is located in vertical ducts or holes through the masonry (see Figure 9.14). The void

around the reinforcement is either completely grouted up or completely filled with mortar, as the work proceeds, in order to provide a suitable bond between the reinforcement and the surrounding masonry. The reinforcement is located in the most suitable position to resist the applied moment or tensile force, i.e. in the tension face of the wall. The design of the wall is similar to that of reinforced concrete, using the masonry in the compression zone to resist the compressive stresses, and the reinforcement in the tensile zone to resist the tensile stresses.

Reinforced walls are mainly used for retaining walls where large lateral loads are to be resisted, but they can also be used in any location where the gravitational loads are small compared with the lateral or uplift forces involved. Care should be taken when using reinforced walls to see that suitable and adequate protection against corrosion is provided by the quality of the masonry and mortar and the cover to the reinforcement.

In order to improve the effectiveness of reinforced walls resisting large bending moments, consideration should be given to shapes other than single- or double-leaf walls – for example, the lever arm of the reinforcement can be improved by using fins, diaphragms or piered sections. One method is to construct a pier or fin with a central void, to reinforce this void, and then to grout up the reinforcement. Alternatively, a post-tensioned rod could be used within the void (see Figure 9.15). These solutions have the effect of reducing the required size of the pier or fin and increasing the height to which such walls can be used.

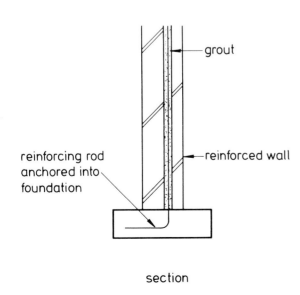

section

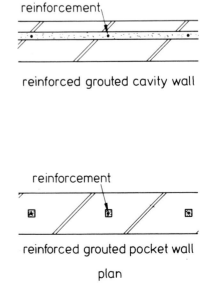

plan

Figure 9.14 Reinforced walls

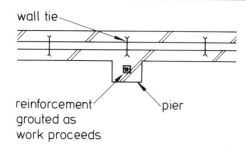

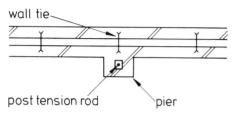

Figure 9.15 Reinforced and post-tensioned pier

Working details should always be as simple as possible, keeping the number of trades and the sequence of operations to a minimum. Grouting can be carried out in short lifts, using a liquid grout. Air vents and weep holes must be provided to allow the air to escape during grouting (but not the grout itself, since this would stain the face of the wall) and to keep a check that the grout has reached the various levels within the void.

9.9 POST-TENSIONED WALLS

Inducing precompression is yet another way of making large improvements in a masonry wall's ability to resist lateral or uplift loads.

The aim of post-tensioning is to induce compressive stresses into the brickwork prior to the application of lateral or uplift forces. The induced compressive stresses must, therefore, be cancelled out by the loading condition before any tensile stress can be developed in the masonry. Thus it is possible to calculate the precompression needed to prevent any tensile stress developing, and to induce this compression by post-tensioning high tensile steel rods which apply their reaction forces onto the panel being considered. The tension is applied to a threaded steel rod by tightening a nut against a cap plate which, in turn, induces compression in the wall. The steel rod is generally anchored into the foundation.

Having established the precompression required, it is then possible to determine the tensile force which must be developed within the post-tensioning rod, the rod diameter required, the torque needing to be applied and the maximum compressive stress induced in the brickwork. A check should be made to ensure that the rod is suitably anchored and that the reaction is adequately catered for (see Figure 9.16).

Post-tensioned walls are often used for spandrel panels below long window openings, retaining walls, tall diaphragm and fin walls, and for other conditions where the gravitational forces involved are small compared with the uplift or lateral forces needing to be resisted.

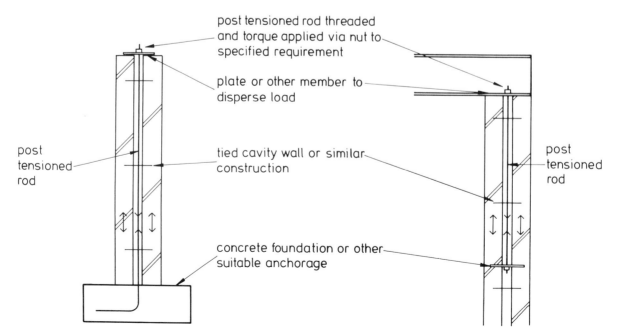

Figure 9.16

It is important, when using post-tensioning rods, to see that suitable and adequate protection is given to all the steel components in order that corrosion does not occur within the required design life of the building. It is also important to ensure that the required torque is suitably noted on any working drawings and details of the wall, and that adequate instructions are given to the contractor. To prevent the possibility of the torque (and thus the prestress) reducing, a locking device should be provided by the use of a lock nut or by grouting up solid the nut's seating plate.

9.10 COLUMNS

Basically a column is a very short length of wall, and is defined as an isolated vertical loadbearing member whose width is not more than four times its thickness (see Figure 9.17). Strictly, this would only apply to rectangular columns, but many other shapes can obviously be utilised provided that adequate bonding of the masonry can be achieved.

Columns are generally used where large, open, uncompartmented areas are required, and the

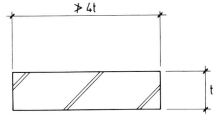

Figure 9.17 Plan on column

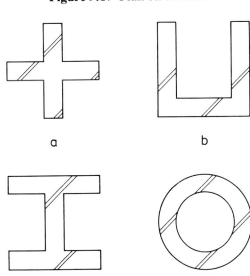

a) cruciform; b) channel; c) I or H section
d) circular tube

Figure 9.18 Typical column section

shapes are often determined by economic, aesthetic or other physical requirements. For example, columns a, b, c and d in Figure 9.18, could be used for either aesthetic appeal, structural suitability or accommodation of ducts, flues, etc., depending on the various design considerations.

The design of columns is dealt with in Chapters 10 and 11 and involves the determination of the slenderness ratio, the reduction factors for area, and the consideration of shear and other stresses.

Columns, like walls, can be built with cavities, can be solid masonry, double-leaf collar-jointed, grouted cavity, faced, veneered, post-tensioned or reinforced. The design considerations for columns are similar to those for walls, as outlined earlier, with additional problems involving the determination of the effective thickness and consideration of the small cross-sectional area.

Since the cross-sectional area of a column is small, the probability of a given proportion of the masonry having a lower than average strength is greater than would be the case for a member of large cross-sectional area. This, of course, also applies to conventional walls of small cross-sectional area. Here, again, it is necessary to introduce a reduction factor related to the small area.

9.11 ARCHES

The arch is one of the most efficient methods of forming a support with materials which have good compressive resistance and low tensile resistance, because its configurations can produce an equilibrium condition made up of compression forces (see Figure 9.19).

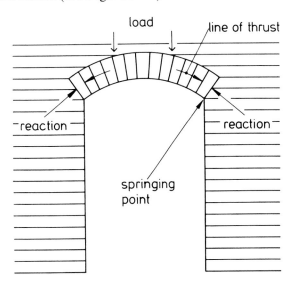

Figure 9.19 Arch

It is one of the most visually attractive structural forms. However, although vast numbers were built during the industrial revolution, arches are, unfortunately, rarely used to-day. Corresponding with the decline in their use has been a decline in the number of craftsmen experienced in this form of construction. For that reason, it seems unlikely that a speedy revival in arch construction will take place – despite the fact that for certain structures the arch offers the best solution.

An interesting past use is to be found in the brick arch floors of many 19th century dock warehouses. The arches were topped off with a weak concrete or other levelling material, and were sometimes constructed above a cast iron framework (see Figure 9.20).

The extra weight on the frame and foundations, when compared with more modern forms of construction, has rendered this type of flooring uneconomical. Contemporary design techniques could, of course, reduce both the weight and the economic disadvantages.

Another past use of interest was in foundations constructed with inverted arches, the loads from columns or walls being dispersed via the arch from a point or knife-edge load onto a more uniform loading on the sub strata (see Figure 9.21).

Although, as noted earlier, it seems unlikely that we shall see a speedy and widespread revival in arch construction, it should certainly not be written off. Conditions have changed greatly since it went out of fashion, and the application of modern techniques of design and construction could well bring this attractive and efficient structural form back into rather more common use.

9.12 CIRCULAR AND ELLIPTICAL TUBE CONSTRUCTION

The arch and the inverted arch can be combined to form a completely circular or elliptical cross-section that is particularly useful for the construction of shafts, tunnels, chimneys, etc. In the case of shafts and tunnels, the critical loading conditions tend to be those of external pressures acting

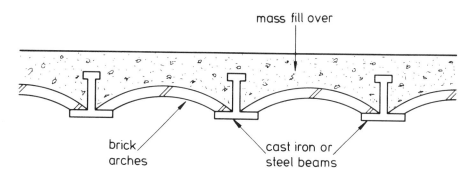

Figure 9.20 Brick arch floor

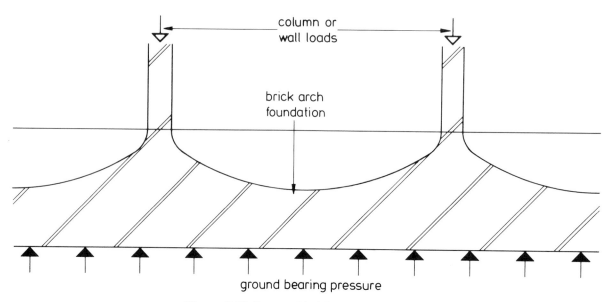

Figure 9.21 Inverted brick arch foundation

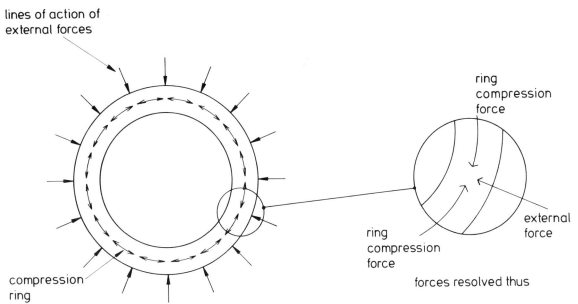

Figure 9.22 Circular tube construction

through lines which radiate from the centre (see Figure 9.22), and which can be resolved into a compression ring of forces ideally suited to masonry construction.

In addition to this property, the tubular shape has good resistance to longitudinal bending moments, and this is exploited in chimney construction.

9.13 COMPOSITE CONSTRUCTION

As mentioned previously, other materials can be combined with masonry to give greater resistance to bending moments, etc. This has often been done unintentionally. For example, reinforced concrete beams are frequently designed to carry large panels of masonry and, in reality, the assumed compressive stresses calculated as existing in the concrete do not in fact occur, because the masonry above the beam resists the bending compression forces (see Figure 9.23).

The condition indicated by the above example has been simplified for clarity. However, it should be noted that existing knowledge on composite action is insufficient for full exploitation and, for that reason, design stresses tend to be lower than in more conventional designs. It is also essential to take into account the vertical stresses occurring in the wall, since these have an effect upon the bending resistance of the composite panel. Thus the design procedure involves calculating the vertical stress, prior to determining the permissible bending stresses. In addition, openings through the wall will reduce the composite action. Nevertheless, quite large openings can be accommodated without destroying the advantages of this action, and the reduced effect can be taken into account in the design.

It is also necessary to design the reinforced concrete beam to support a height of 'wet' uncured masonry for the temporary condition during construction. The lift of wet masonry is also

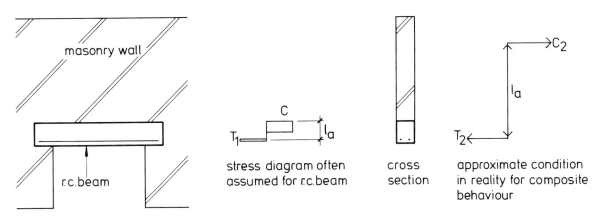

Figure 9.23 Composite construction

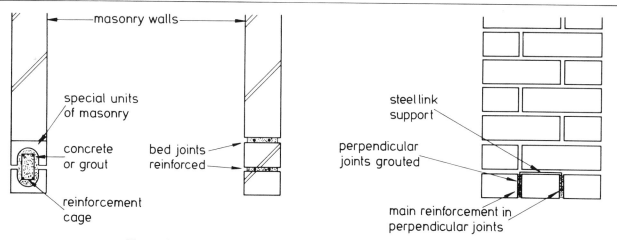

Figure 9.24 Typical sections through reinforced masonry beams

determined from the anticipated height of masonry likely to require support prior to the mortar having achieved sufficient strength to act compositely with the beam.

9.14 HORIZONTALLY REINFORCED MASONRY

As mentioned in 9.4, when combined with reinforcement, masonry can be made to resist much greater bending moments. Hence, reinforced masonry can be used for beams and slabs to span and cantilever over quite large openings. The reinforcement is located in holes in specially manufactured units, or in the perpendicular joints of the masonry (see Figure 9.24).

The design of reinforced masonry is similar in basic priciples to the design of reinforced concrete, the reinforcement being located in the tensile face of the combined section. The design procedure is dealt with in detail in Chapter 15. Special attention is needed when using reinforced masonry in external or exposed conditions to ensure that adequate protection is provided to the reinforcement.

STRUCTURAL FORMS

9.15 CHIMNEYS

During the industrial revolution, the need to remove smoke and other gases to a high level in the atmosphere, to prevent excessive local pollution, brought masonry chimneys onto the industrial scene. The engineers found that the problems caused by wind loading and temperature stresses had taken on new proportions.

Like other chimneys, masonry chimneys are designed as cantilevers resisting horizontal wind loading, and require checking for increased stress conditions caused by the temperature gradient. The effect of wind around chimneys and related oscillation has become better understood – though still not fully – because of the experience gained from this period in engineering history.

The chimney is generally tapered to give increased resistance to bending, as the applied bending itself increases. Consideration must be given to the possibility of sulphate attack, particularly at and near the top of the chimney, and suitable masonry and mortar must be chosen and specified for the expected condition. At low level, allowance must be made for flue openings, etc., and local stiffening is often necessary in this location to facilitate the rapid increase in stresses due to the reduced section at the point of maximum bending (see Figure 9.25).

The use of reinforcement can greatly improve the resistance of the masonry to the applied bending

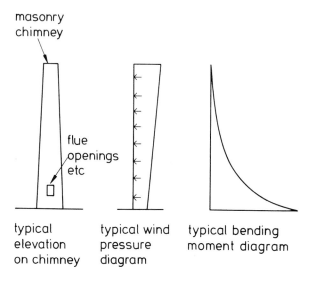

Figure 9.25 Chimneys

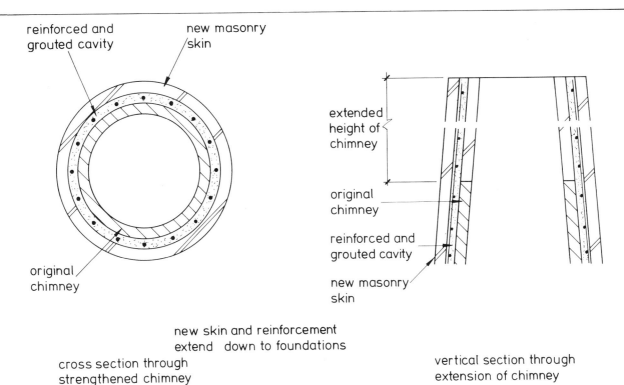

cross section through
strengthened chimney

new skin and reinforcement
extend down to foundations

vertical section through
extension of chimney

Figure 9.26 Chimney extension

moments, and/or reduce the amount of masonry needed.

Reinforced masonry can also be valuable in extending the height of existing chimneys. This can be done by adding an extra outer skin of masonry leaving a reinforced cavity between the old and the new wall and extending the height of the structure as in Figure 9.26.

Chimneys are not necessarily restricted to a circular or square cross-section. Clover leaf, elliptical, hexagonal, triangular and many other cross-sections can be used to accommodate varying numbers of flues and to give a pleasant appearance (see Figure 9.27).

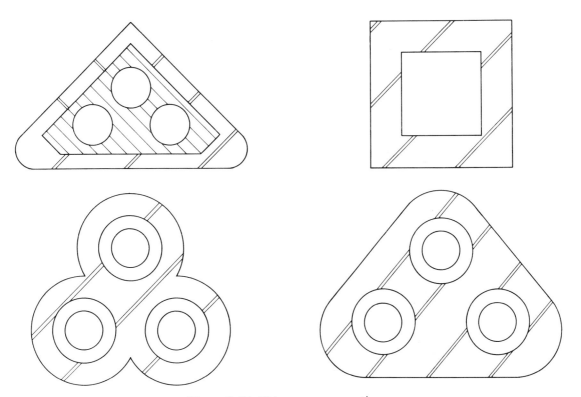

Figure 9.27 Chimney cross-sections

Particular care must be taken when using complicated shapes to see that temperature stresses do not become excessive.

9.16 CROSSWALL CONSTRUCTION

In most buildings, the wall layouts are mainly dictated by functional requirements. Often, however, with a little more consideration of the structural implications, a plan can be developed to suit both the functional and the structural requirements. For example, a multi-storey hostel block containing numerous bedsitters with only a few basic layouts can frequently be planned to have the same types one above the other, and all the room-dividing walls can then line through from the bottom to the top of the building (see Figure 9.28).

The dividing walls can thus be used for the structure of the building, supporting the floor and roof slabs, and resisting lateral wind loads from the main elevations. The corridor walls then support the corridor floors and resist the wind loads acting on the gable ends. The resulting structure forms a very stiff construction, and the stairs and lift shafts add even greater stiffness for resisting wind loads, the slabs being designed as plates to transfer the lateral loads to the main crosswalls in the rooms and staircase areas. An example of the economical cross-section of such walls is a recent nine-storey hostel constructed using only 102 mm thick brick walls for the full height – the stresses in the main crosswalls being within normal acceptable limits without the need for any other structural framework. In designing crosswall buildings it is necessary to ensure that the wall thicknesses required for sound and fire resistance are incorporated, since often, in staircase areas, etc., these requirements demand thicker walls than do the structural considerations.

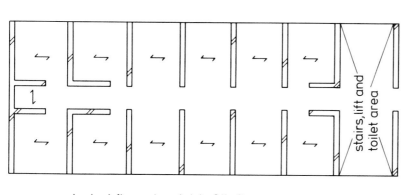

typical floor plan 1st to 9th floors   vertical section
                                                    through crosswalls

Figure 9.28 Crosswall construction

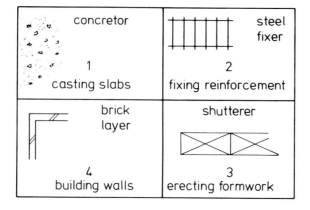

typical building plan showing key to activities

Figure 9.29 Spiral methods of construction

The speed of construction of such buildings is very impressive, particularly if the plan form and size of the building allows it to be constructed in quarters, using the 'spiral' method whereby the trades can follow each other around the building from one quarter to the next completing their section of the work (see Figure 9.29).

From the stage indicated in the figure, the bricklayers on completion of Bay 4 would move up to the next floor and start work in Bay 1. The other trades, i.e. shutterers, steel fixers and concretors would all move on one bay – the construction continuing to 'spiral' up the building keeping all trades constantly employed.

In addition to the normal consideration of wind, superimposed and dead loading, a multi-storey building over stipulated heights must also be designed to prevent progressive collapse due to accidental damage, and to take account of the possible differential movements of the inner and outer leaves of the external cavity walls. These conditions demand extra consideration from the designer, particularly in the detailing and design of floor slabs and the use of alternative support. For example, in order to overcome the critical effects of the vertical differential movement on the outside walls, it is necessary to support the outer leaf of brickwork at intervals of height up the building (see Figure 9.30), and to incorporate special details which allow the movement to take place without detrimental effects to the building.

The requirements to be satisfied with regard to both progressive collapse and differential vertical movement are dealt with in more detail in Chapter 8 and Appendix 3. Some buildings require larger room sizes than those needed for hostel accommodation and larger areas of natural light. An example of this is school classrooms (see Figure 9.31).

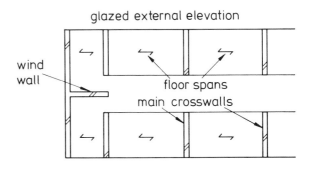

Figure 9.31 Crosswall classroom blocks

With some thought in planning, a floor plan that repeats on all floors can often be achieved, and the main crosswalls can be used as loadbearing elements. It is important in this form of construction to see that a sufficient length of walls at right angles to the main crosswalls is provided to resist wind loading normal to the gables. The floors again span between the main crosswalls, transferring lateral loading to the wind walls, and vertical and lateral loads to the crosswalls. They provide the necessary restraint to the walls at each floor

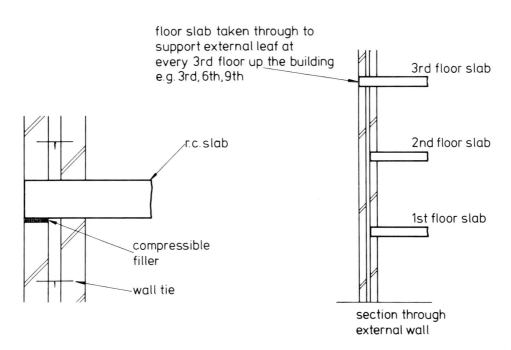

Figure 9.30 Accommodating differential construction

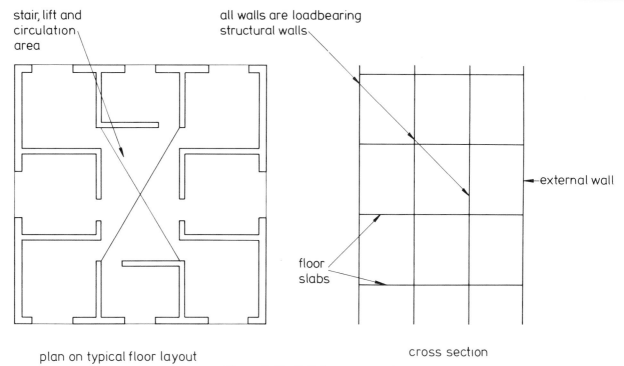

plan on typical floor layout   cross section

Figure 9.32 Cellular construction

level. As in the previous example, the normal room-dividing walls provide the main structure and eliminate the need for a separate structural frame.

9.17 CELLULAR CONSTRUCTION

Another suitable form for multi-storey buildings with small rooms is cellular construction. Generally, this has been used for domestic buildings but it is also suitable for small office accommodation, etc. The rooms in this case form a number of cells, and again the aim is to use all the separating walls for the main structure and to line up the walls from the bottom to the top of the building (see Figure 9.32).

In the majority of cases all the walls are loadbearing with the exception of toilet partitions and other minor room-dividing walls. In all other aspects the wall in cellular construction is similar to the crosswall, except that it is easier to achieve similar stiffness in all wind directions because of the cellular arrangement.

9.18 COLUMN AND PLATE FLOOR CONSTRUCTION

For buildings requiring large open areas, widely spaced columns can be used in a column and plate construction. In this case (see Figure 9.33), the columns are designed to carry the vertical loads and, if sufficient walls can be located to resist the wind forces, such as those around stair

and lift enclosures, gable walls, etc., the floors can be used as plates which span horizontally, transferring the wind forces from the external cladding to the walls.

Some buildings are of such a layout that the crosswalls required to resist wind cannot be accommodated. In these cases the columns can be T, cruciform or channel shapes designed to resist the horizontal reactions from the wind – each individual column resisting its own local area wind reactions (see Figure 9.34).

Columns can also be reinforced or post-tensioned in order to accommodate large bending moments in a neater, smaller and more economical section.

9.19 COMBINED FORMS OF CONSTRUCTION

In many buildings the required layout demands both large open areas and smaller enclosed rooms, and combinations of the forms already mentioned can be used (see Figure 9.35).

The example shown indicates a successful combination of column and plate and crosswall construction. Many variations can be made, always provided that sufficient masonry is accommodated to deal with the loading conditions. The structural forms shown in Figure 9.35, are combined on each floor. However, different forms

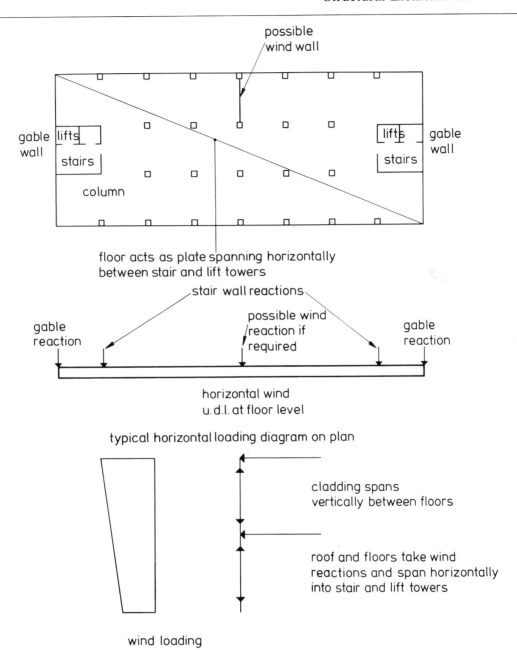

possible wind wall

gable wall

lifts

stairs

gable wall

lifts

stairs

column

floor acts as plate spanning horizontally
between stair and lift towers

stair wall reactions

gable reaction

possible wind reaction if required

gable reaction

horizontal wind
u.d.l. at floor level

typical horizontal loading diagram on plan

cladding spans
vertically between floors

roof and floors take wind
reactions and span horizontally
into stair and lift towers

wind loading

typical horizontal loading diagram on section

Figure 9.33

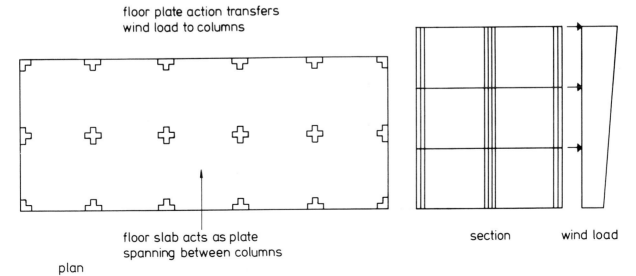

floor plate action transfers
wind load to columns

floor slab acts as plate
spanning between columns

plan

section

wind load

Figure 9.34

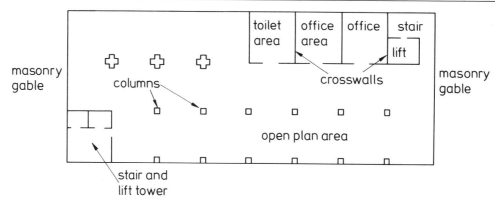

Figure 9.35 Horizontal combined forms

can be successfully combined in a vertical condition as, for example, the podium construction shown in Figure 9.36.

For this condition to be successfully achieved, however, detailed planning is necessary to obtain

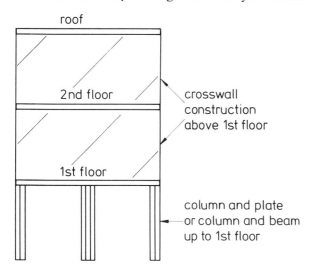

Figure 9.36 Vertical combined forms (podium construction)

the most economical solution, particularly with regard to the locations of the loadbearing walls and columns. The construction is basically a concrete or steel frame up to first floor level and loadbearing masonry above. This construction ties in nicely with the greater flexibility of use often demanded at ground floor level.

9.20 DIAPHRAGM WALL AND PLATE ROOF CONSTRUCTION

Diaphragm wall and plate roof construction is mainly suitable for tall single-storey buildings enclosing large open areas such as sports halls, gymnasia, swimming pools and industrial buildings. Buildings can be constructed using diaphragm external walls, as outlined in 9.7.2, and the roof forms a horizontal plate which is used to prop the walls against lateral loading (see Figure 9.37). In order to transfer the reactions from the wall into the roof diaphragm, a capping beam is often provided which can be used as the seating

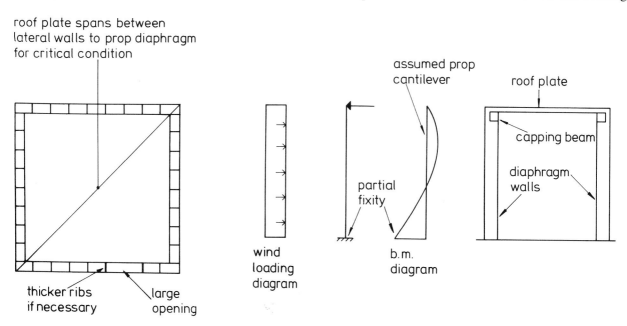

Figure 9.37 Diaphragm and plate roof construction

for fixing the roof beams, to resist the uplift forces and, if necessary, as a boom member of the roof girder.

In Figure 9.37, for the wind direction shown, the roof plate spans from gable to gable and transfers the forces, via the ring beam and roof plate, into the gable wall where the wall stiffness for that direction is greatest. For a wind loading condition on the gable, the plate would span between the main elevations and transfer the loads via the ring beam into the main elevation walls. In many cases, the roof decking material is suitable for use as a plate. However, in situations where this is not so, a horizontally braced girder, or similar, can be used incorporating the ring beam as a boom member to transfer the propping force to the transverse walls of the building. For more detailed information see Chapter 13.

9.21 FIN WALL AND PLATE ROOF CONSTRUCTION

Fin wall and plate roof construction is an alternative to the diaphragm and plate roof. The form is again mainly used on tall single-storey buildings and the main difference is in the type of external wall construction (see Figure 9.38).

The basic design is similar to that of the diaphragm, and plate and is dealt with in more detail in Chapter 13. Again, the walls are designed as propped cantilevers and use is made of the roof deck as a plate for propping the tops of the walls. However, in this situation, bracing is usually needed in the roof to achieve adequate plate action.

9.22 MISCELLANEOUS WALL AND PLATE ROOF CONSTRUCTION

Possible variations of the outside wall configuration for the building types mentioned in 9.20 and 9.21 are numerous, and the main aim should be to achieve a high Z/A ratio and to take maximum advantage of the gravitational forces involved, thus giving a wide scope for imaginative shapes and configurations using the masonry suitably dispersed around the neutral axis of the section (see Figure 9.39).

9.23 SPINE WALL CONSTRUCTION

Spine walls are suitable for more flexible open-plan arrangements where a number of main walls such as corridor walls, stair walls, lift shafts, toilets, services, can be in a fixed location as for example, in office buildings (see Figure 9.40).

Spine walls, together with the external face walls, are used as the main loadbearing walls supporting the floor loads and should line up vertically through the building. It is also most important to provide sufficient wind walls at right angles, as in stair and lift areas (see Figure 9.40), in order to be able to resist the wind loading on all elevations. The principal advantage of this form of construction is the added flexibility of room arrangements, since all dividing partitions can be of a temporary nature and supported on the main floors. Floors are again used as the horizontal plate members distributing the wind reactions from the external elevations to the nearest structural wall normal to the wind direction.

note:-loading and b.m. diagrams
and plate action similar to
diaphragm, see fig. 9.37

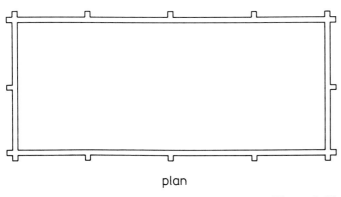

plan

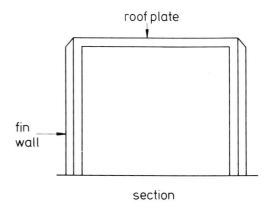

section

Figure 9.38

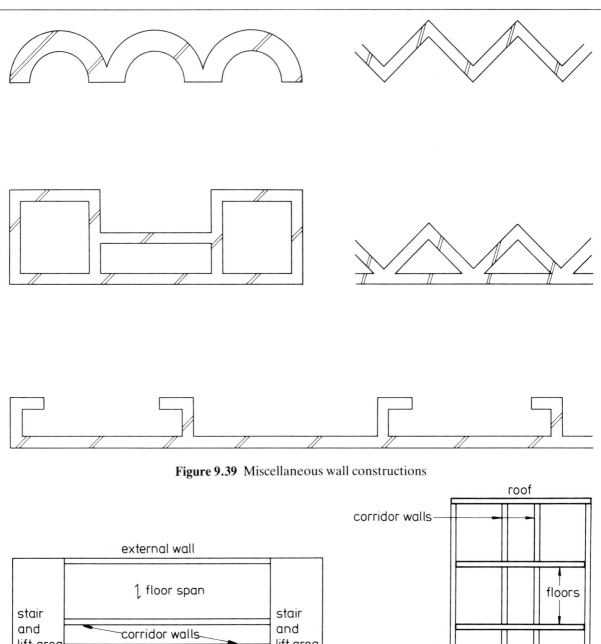

Figure 9.39 Miscellaneous wall constructions

Figure 9.40 Spine wall construction

9.24 ARCH AND BUTTRESSED CONSTRUCTION

In the past, the beautiful form of arch and buttress construction has, been adopted for many structures, for example bridges, industrial buildings, warehouses, churches etc. (see Figure 9.41). The aim is to keep all forces in the masonry in compression.

The thrust from the arch ring is transferred to the foundations by the propping action of the buttress, and the shape and size of both the arch and the buttress are proportioned to produce equilibrium within the form, using compressive forces only. Buttressed arches are still as attractive and useful as ever, and it is a pity that many designers do not consider them, especially when designing churches. The possible combination of the arch and buttress with diaphragm, fin, reinforced and post-tensioned walls opens up for the designer a wonderful opportunity to make masonry church buildings a most attractive development using

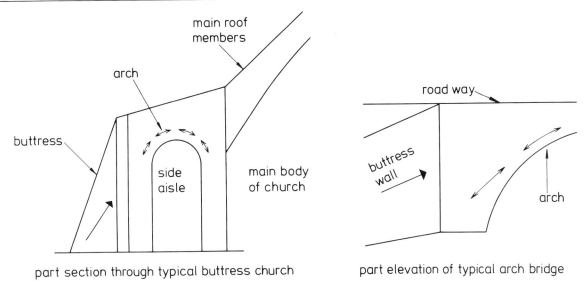

Figure 9.41 Arch and buttress construction

modern forms of construction. The use of this form is, however, not only suited to churches. In the years to come, it is likely that a better understanding of the structural behaviour of the arch and buttress, along with further developments, will produce structures of more economical proportions, making better use of masonry, and possibly combining them with other materials to produce buildings which will compete with those of the past for their visual effect and low maintenance costs.

9.25 COMPRESSION TUBE CONSTRUCTION

Many structures, particularly those to be constructed underground, have to resist external pressures which might crush them. Because of masonry's good resistance to compressive forces, such structures can be designed in a similar

manner to the arch and buttress, keeping all the forces in compression, thereby avoiding or limiting undesirable tensile or bending stresses. The use of circular or elliptical tube forms can give the desired result (see Figure 9.42). This type of construction is often used for tunnels, shafts, sewers, etc.

In the case of the vertical shaft, the design pressures are in balance and the forces resolve into a circular compression ring. The shaft is constructed in sections, working from the top and progressing in short lifts as the shaft is excavated (see Figure 9.43).

In the case of sewers, the theoretical forces are not often equal on all sides, and some bending could occur in a circular form. Variations on the shape of the cross-section will reduce or increase

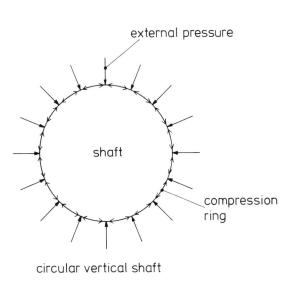

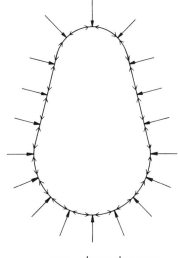

Figure 9.42

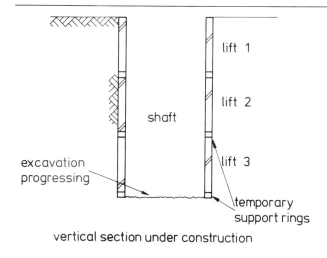

vertical section under construction

Figure 9.43

these bending moments, and the aim should be to produce a shape suitable for the use and keeping the masonry thickness to a minimum – the ideal shape being that which produces only direct compressive forces.

This chapter has outlined some of the elements and forms in structural masonry that are possible, practical and economically advantageous. The design and application of the more common elements and forms are dealt with in detail in the chapters which follow. Doubtless, there are many other possibilities. Certainly, there is room for experienced and imaginative designers to develop their own solutions and to break fresh ground.

CHAPTER 10

DESIGN OF MASONRY ELEMENTS (1): VERTICALLY LOADED

The basis of design of plain masonry has been examined in Chapters 3–6. In this and the following chapter, the recommendations of BS 5628 will be applied to specific design problems.

10.1 PRINCIPLE OF DESIGN

The principle of the design is to satisfy the equation

$$\frac{\beta t f_k}{\gamma_m} \geqslant n_w \text{ (see 5.11 and BS 5628, clause 32.2.1)}$$

in which, for walls, a unit length (linear metre) of wall is considered.

The required wall thickness or column size or, when applicable, the correct choice of geometric profile for a particular element will, initially, as with any other structural material, be unfamiliar and the guidance which follows should be of some help to the inexperienced designer. As in most structural design, the approach is based on trial and error. Experience and familiarity with the materials and components available will lead to more accurate initial assessments.

10.2 ESTIMATION OF ELEMENT SIZE REQUIRED

In general, the mechanism of failure of a wall or column is that of buckling under vertical loading imposed from walls and floors over. Buckling is directly proportional to the stiffness of an element, and the stiffness can be expressed as I/L where I is the second moment of area of the element, and L is the effective length of the element.

For a given effective length, the second moment of area (I) of an element would need to be increased as the loading was increased to contain the buckling tendency to the same degree of safety. For solid walls, this is done by simply increasing the wall thickness and thus the I value. Conversely, for a given loading, the second moment of area (I) of the element would again need to be increased as the effective length was increased.

The capacities of various solid wall thicknesses to carry loads over differing effective lengths can only become familiar to the designer with time and application of the design process. The expression used to measure the tendency of the element to buckle is 'slenderness ratio' and is written as h_{ef}/t_{ef} where h_{ef} is the effective height (or length) of the element, and t_{ef} is its effective thickness. An element with a high slenderness ratio has very little capacity to carry loading due to the tendency to buckle at relatively low stress, whereas an element with a low slenderness ratio has more reserve to carry loading because of its lower tendency to buckle.

However, the designer should not limit his thoughts to simply increasing the wall thickness to provide a greater second moment of area. There are often economies to be achieved by using the material in another geometric profile, in exactly the same way as concrete or steel I, T or box beams were developed. The case for such a choice of element would generally be more applicable to a heavily loaded element of large effective length where the direct stresses and the buckling tendency are extremely high. By the same logic, rectangular rather than square column sections would be a more sensible choice where lateral restraint, capable of preventing buckling of the column, is provided at mid-height and to the minor axis of the element (see Figure 10.1).

The designer should be encouraged to think more in terms of the radius of gyration and second moment of area properties of an element, rather than 'effective thickness' as this will lead

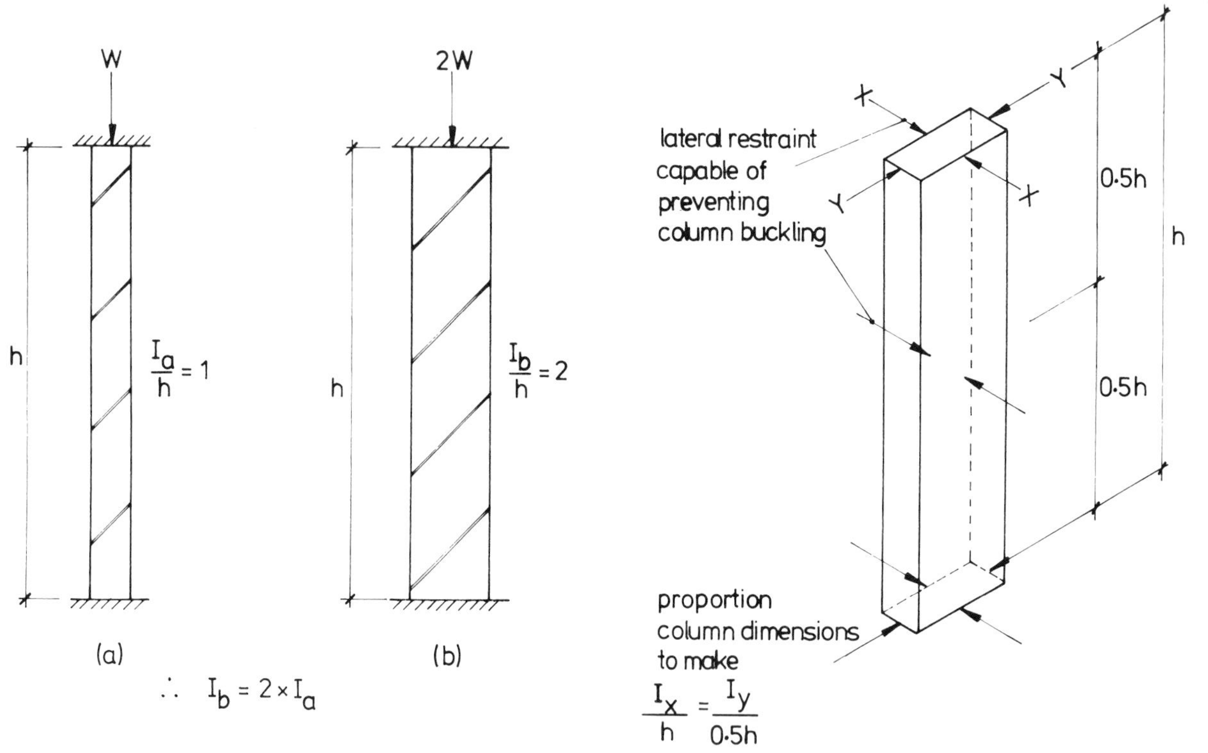

Figure 10.1

to inventiveness and ingenuity in overcoming the more difficult problems.

10.3 SEQUENCE OF DESIGN

It should now be evident that the design sequence can be written as a four-stage operation:

Stage 1: Calculate the characteristic load and design load.

Stage 2: Estimate wall thickness (or column size or geometric profile).

Stage 3: Calculate the design strength required for the wall (or element) to support the design load as calculated in stage 1.

Stage 4: Amend wall thickness or geometric profile if necessary and determine the required brick and mortar strengths.

Stage 2 of the design sequence has been dealt with in 10.2 and stages 1, 3 and 4 will be considered, in practical terms, by examining design examples.

10.4 DESIGN OF SOLID WALLS

As the second moment of area of a solid wall is directly proportional to its thickness, the effective thickness (which for solid walls is the actual thickness) will be used in the calculations which follow.

EXAMPLE 1

Design the internal brick wall in the ground-floor storey of the building shown in Figure 10.2. The wall is plastered both sides and forms part of a large building project where extensive testing of materials and strict site supervision will be implemented.

The loading may be assumed as:

characteristic dead loads	roof =	4.00 kN/m^2
	floors =	5.00 kN/m^2
characteristic superimposed loads	roof =	1.50 kN/m^2
	floors =	3.00 kN/m^2
density of brickwork for own weight		$= 18.00 \text{ kN/m}^3$

The secondary effects of wind stresses due to the possibility of the element concerned acting as a shear

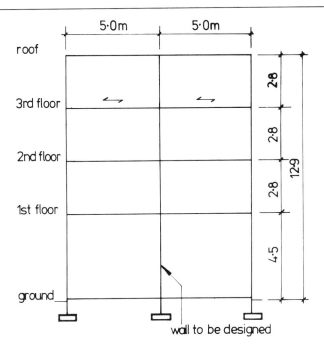

Figure 10.2 Cross-section

wall to provide overall stability should be ignored for the purposes of this design example, but will be investigated in Chapter 11.

Stage 1: Calculate design load on wall n_w

The design load is obtained from the characteristic loads which are increased by the appropriate partial safety factor γ_f to allow for the type of loading and load combination being considered, i.e. dead loading, superimposed loading and wind loading. The degree of partial safety factor applicable to each of these loading types takes account of the degree of accuracy of that particular load. Values of γ_f are given in BS 5628, clause 22. For the combination of loading being considered in this example, dead plus superimposed only, case (a) factors for γ_f are applicable, requiring partial safety factors of 1.4 and 1.6 to be applied to G_k (characteristic dead load) and Q_k (characteristic superimposed load) respectively (see Table 5.1).

(a) Characteristic dead loads, G_k:

$$\text{roof} = \frac{5+5}{2} \times 4 \qquad = 20.00$$

$$3 \text{ floors} = \frac{5+5}{2} \times 5 \times 3 = \underline{75.00}$$
$$= 95.00 \text{ kN/m}$$

Estimation of wall own weight should take account of the possibility of reducing the wall thickness in the upper storeys of the building and, for this purpose, it will be assumed that for this example the wall over can be sensibly reduced to a half brick wall at first floor level.

Wall own weight:

12 mm plaster both sides $= 2 \times 0.012 \times 21 \times 12.9 = \quad 6.00$
102.5 brick wall $\qquad = 0.1025 \times 18 \times 8.4 \quad = \quad 15.50$
215 brick wall $\qquad = 0.215 \times 18 \times 4.5 \quad = \underline{\quad 17.42}$
$$= 38.92 \text{ kN/m}$$

Total characteristic dead load, G_k $\qquad = \quad 38.92 + 95.00$
$$= 133.92 \text{ kN/m}$$

(b) Characteristic superimposed loads, Q_k (super reductions from CP3: Chapter V: Table 2, have been ignored for simplicity):

$$\text{roof} = \frac{5+5}{2} \times 1.5 \quad = \quad 7.50$$

$$3 \text{ floors} = \frac{5+5}{2} \times 3 \times 3 = \underline{45.00}$$
$$Q_k = \overline{52.50} \text{ kN/m}$$

Design load on wall, n_w:

As stated previously, from BS 5628, clause 22(a), for this combination of loading, dead plus superimposed:

$$\gamma_f = 1.4 \times G_k \text{ (characteristic dead loads)}$$

and

$$\gamma_f = 1.6 \times Q_k \text{ (characteristic superimposed loads)}$$

Therefore, design load on wall:

$$n_w = (1.4 \times 133.92) + (1.6 \times 52.50)$$
$$= 271.5 \text{ kN/m}$$

Stage 2: Estimate wall thickness

The wall is required to support its loading over quite a large effective height and, therefore, a reasonably low value of slenderness ratio will be required to limit the buckling tendency. BS 5628 permits a maximum slenderness ratio of 27 (clause 28.1) and for estimation purposes it can be expected that, to provide adequate reserve in the allowable stresses to carry the moderately heavy loads, a limit of around 16 on the slenderness ratio would be required.

Assessed slenderness ratio (SR) = 16

$$\text{Slenderness ratio} = \frac{h_{ef}}{t_{ef}}$$

therefore

$$t_{ef} = \frac{h_{ef}}{\text{SR}}$$

$$t_{ef} = \frac{4.5 \times 0.75}{16}$$

$$\text{Estimated } t_{ef} = 0.211 \text{ m}$$

But, to suit standard brick dimensions, try 215 mm thick wall.

(Note: The factor of 0.75 applied to the actual height to determine the effective height in the equation will be dealt with in stage 3 of the design process following.)

By inspection, it is noted that the design of a 102.5 thick (half brick) wall would produce a slenderness ratio of 32 which is in excess of the maximum permissible value of 27 and, therefore, for standard bricks, a 215 mm thick wall is the minimum thickness appropriate to this effective height.

Stage 3: Calculate design strength of wall

Design strength $= \dfrac{\beta t f_k}{\gamma_m}$ per linear metre (BS 5628, clause 32.2.1).

(a) Determine capacity reduction factor, β:

For the purpose of this example, it is assumed that any intersecting crosswalls are at such centres as to offer little lateral support to the wall. The slenderness ratio of the wall is therefore determined by its effective height rather than its effective length.

Therefore, slenderness ratio (SR) $= \dfrac{\text{effective height}}{\text{effective thickness}} = \dfrac{h_{ef}}{t_{ef}}$

Effective height: The horizontal lateral supports which dictate the effective height of this element are provided by the ground and first-floor rc slabs and, as these slabs span onto the wall, the contribution of their support can be considered to provide enhanced resistance to lateral movement. Clause 28.3.11 of BS 5628 allows an effective height of 0.75 times the clear distance between the lateral supports for this support condition.

Therefore
$$\text{effective height, } h_{\text{ef}} = 0.75 \times (4500-150)$$
$$= 3262.5 \text{ mm}$$

It is perhaps worth considering the logic of this allowance of $0.75 \times h$ by inspecting the deflected shape of the wall at the point where it is about to buckle. The implication of enhanced resistance to lateral movement (as opposed to the alternative simple resistance to lateral movement quoted in BS 5628) is that rotation of the wall at that support position is certainly limited if not completely eliminated. The junction becomes the equivalent of a partially fixed end with the deflected shape as indicated in Figure 10.3.

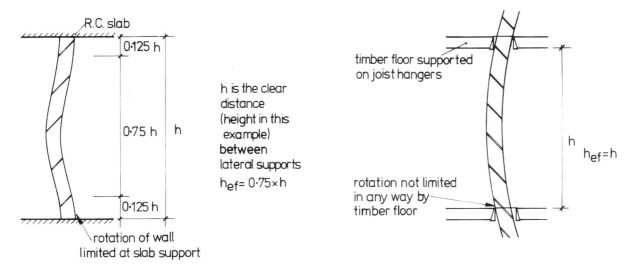

(a) enhanced resistance to lateral movement   (b) simple resistance to lateral movement

Figure 10.3

Effective thickness: The effective thickness t_{ef} of this element, as defined in BS 5628, clause 28.4.1 is the actual thickness and has already been estimated in stage 2 of the design process as 215 mm thick.

Therefore
$$\text{slenderness ratio (SR)} = \frac{h_{\text{ef}}}{t_{\text{ef}}} = \frac{3262.5}{215} = 15.2$$

Eccentricity of loading: The majority of the load supported by the element under consideration is in the loadbearing wall immediately above the first floor slab, and is the accumulation of the loads from the floors, roof and walls over. The proportion of the total load on this element accruing from the first floor slab alone is relatively small. It is assumed that a half brick wall will later be proved to be adequate for the loadbearing wall element immediately above first floor level and the detail at this junction is shown in Figure 10.4.

W_1 = loading in loadbearing wall over.
W_2 = slab loading from left-hand side.
W_3 = slab loading from right-hand side.
W_1 can be assumed to be applied concentrically whereas W_2 and W_3 must be applied at a position equal to $t/6$ from the loaded face (see BS 5628, clause 31).

For this element under full dead plus superimposed loadings, $W_2 = W_3$ because the loads and spans are identical and the resultant of these two loads is concentric on the 215 thick wall under. Consideration should be given, however, to the possibility of one of the floor spans being completely relieved of its superimposed loading, in which case the resultant of W_2 and the reduced W_3 would be

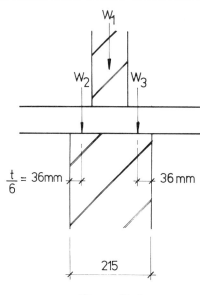

Figure 10.4

eccentric on the 215 thick wall under. The effect of the reduced total load, but applied with an eccentricity, may be a more critical design condition than the axially applied full load. The same imbalance of load, resulting in its eccentric application, would also result from different span lengths from each side of the wall.

For the purpose of this particular example the effects of this eccentricity will be ignored but will be investigated later in the chapter.

Capacity reduction factor, β: From BS 5628, Table 7 (see Table 5.15) and for eccentricities of loading of between 0 and $0.05t$ a β value of 0.854 for a slenderness ratio of 15.2 (as calculated earlier) can, by interpolation, be read off.

(b) Determine partial safety factor, γ_m:
The categories of both constructional control and manufacturing control of the structural units must be selected at the discretion of the designer after careful consideration of the relevant factors applicable to a particular project. From the information supplied for this design example, control of both items could be described as 'special' (BS 5628, Table 4 – see Table 5.11) and a materials partial safety factor γ_m of 2.5 would apply.

(c) Calculate characteristic strength required f_k:

From the basic equation
$$n_w \geqslant \frac{\beta t f_k}{\gamma_m}$$

Rearranged this gives
$$f_k \geqslant \frac{n_w \gamma_m}{\beta t}$$

in which (as previously calculated or assessed) there is:

design load, n_w $\qquad = 271.5 \text{ kN/m}$
partial safety factor, γ_m $\quad = 2.5$
capacity reduction factor, $\beta = 0.854$
wall thickness, t $\qquad = 215 \text{ mm}$

therefore characteristic strength required, $\qquad f_k = \dfrac{271.5 \times 2.5}{0.854 \times 215}$

$$= 3.7 \text{ N/mm}^2$$

The element being considered is a 215 thick wall in which the width of the wall is equal to the length of the brick. For narrow brick walls (referred to earlier as half brick walls) in which the thickness of the wall is equal to the width of the brick a shape factor of 1.15 as defined in BS 5628, clause 23.1.2 is applicable. The application of this shape factor reduces the characteristic strength required of the masonry thus:

$$f_k = \frac{n_w \gamma_m}{\beta t \times 1.15}$$

and will be applied in a later design example.

Stage 4: *Determine required brick and mortar strengths*
It has now been determined that, to support the calculated design loads on this particular element using a 215 thick brick wall, the characteristic strength required of the masonry is 3.7 N/mm².

From BS 5628, Table 2(a) (see Table 5.4) a brick of 10 N/mm² compressive strength set in a mortar designation (iv) is just not adequate (see Table 5.4), therefore use bricks of 15 N/mm² compressive strength set in a designation (iv) mortar.

It can be seen that this is an extremely low strength requirement, but the wall thickness of 215 mm is a minimum for the storey height and therefore no adjustment will be required. This design example has been specifically related to the design of a brick wall. It is evident that, for such a low strength requirement, a concrete block wall of say 200 mm thickness would have been acceptable and perhaps a more structurally economical alternative. This will be considered in a later design example.

EXAMPLE 2
Using the same criteria, design the internal brick wall between first and second floors.

Stage 1: *Calculate design load, n_w*
(a) Characteristic dead loads, G_k:

$$\text{roof} = \frac{5+5}{2} \times 4 \quad = 20.00$$

$$2 \text{ floors} = \frac{5+5}{2} \times 5 \times 2 = 50.00$$
$$= \overline{70.00} \text{ kN/m}$$

Wall own weight:

$$\text{plaster} = 2 \times 0.012 \times 21 \times 8.4 = 4.23$$
$$102.5 \text{ wall} = 0.1025 \times 18 \times 8.4 = 15.50$$
$$= \overline{19.73} \text{ kN/m}$$

$$\text{Total characteristic dead load, } G_k = 19.73 + 70.00$$
$$= 89.73 \text{ kN/m}$$

(b) Characteristic superimposed loads, Q_k:

$$\text{roof} = \frac{5+5}{2} \times 1.5 \quad = 7.50$$

$$2 \text{ floors} = \frac{5+5}{2} \times 3 \times 2 = 30.00$$
$$Q_k = \overline{37.50} \text{ kN/m}$$

Design load on wall, n_w:

$$\left.\begin{array}{l}\text{Dead } n_w = 1.4 \times G_k \\ \text{Superimposed } n_w = 1.6 \times Q_k\end{array}\right\} \text{ as Example 1}$$
$$\text{Design load } n_w = (1.4 \times G_k) + (1.6 \times Q_k)$$
$$= (1.4 \times 89.73) + (1.6 \times 37.50)$$
$$= 185.62 \text{ kN/m}$$

Stage 2: Estimate wall thickness
Assessed slenderness ratio = 21

Therefore wall thickness required, $\qquad t_{ef} = \dfrac{2.8 \times 0.75}{21}$

$$= 0.10 \text{ m}$$

Try brick wall 102.5 mm thick.

Stage 3: Calculate design strength of wall
(a) Determine β from slenderness ratio:

Effective height, h_{ef}	$= (2800 - 150) \times 0.75 = 1987.5$ mm	
Effective thickness, t_{ef}	$=$ actual thickness	$=$ 102.5 mm
Slenderness ratio, SR $= \dfrac{h_{ef}}{t_{ef}} = \dfrac{1987.5}{102.5}$		$=$ 19.4

The possibility of eccentricity of loads due to imbalance of dead and superimposed loads will, for simplicity, again be ignored for this design example. Therefore, from Table 7 of BS 5628, by interpolation (see Table 5.15): $\beta = 0.721$ for SR $= 19.4$ and $e_x = 0$ to $0.05t$.

(b) The partial safety factor, γ_m will remain, as for Example 1, as 2.5 for special categories.

(c) Calculate characteristic strength required, f_k.

It is hoped to use a 102.5 thick brick wall for this element and BS 5628, clause 23.1.2 permits the application of a shape factor of 1.15.

Therefore $\qquad\qquad f_k = \dfrac{n_w \gamma_m}{\beta t \times 1.15}$

$$= \dfrac{185.62 \times 2.5 \times 10^3}{0.721 \times 102.5 \times 1.15 \times 1000}$$

$$= 5.46 \text{ N/mm}^2$$

Stage 4: Determine required brick and mortar strengths
By inspection of Table 2(a) of BS 5628 (see Table 5.4) it is necessary to provide a brick of compressive strength 27.5 N/mm² set in a mortar designation (iv) (6.2 N/mm² provided) or a lower strength brick of 20 N/mm² set in a designation (iii) mortar.

The compatibility of the differing brick strengths and mortar grades required for the elements in the adjacent storey heights of Examples 1 and 2 conflicts with good design practice, as will be discussed in another chapter. The designer may conclude that the lower strength 215 thick wall should extend to the underside of the second-floor slab where a more reasonable reduction could be employed.

EXAMPLE 3

Design the concrete block internal walls shown in Figure 10.5 to support the loading shown from the rc storage slab over.

Loadings on rc slab:

characteristic dead load $\qquad = 6.5 \text{ kN/m}^2$
characteristic superimposed load $= 12.5 \text{ kN/m}^2$

The concrete blocks will individually measure 400×200 on elevation and should be assumed to have a density of 12 kN/m³ and are solid blocks.

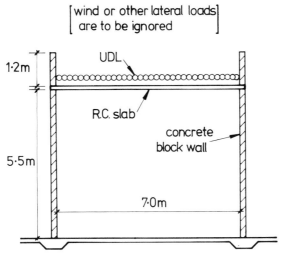

Figure 10.5

Stage 1: Calculate design load, n_w

(a) Characteristic dead loads, G_k:

$$\text{storage slab} = \frac{7.0}{2} \times 6.5 \qquad = 22.75$$

$$\text{block wall} = 12 \times 0.19 \times 6.7 = \underline{15.28}$$
$$= \overline{38.03} \text{ kN/m}$$

(b) Characteristic superimposed loads, Q_k:

$$\text{storage slab} = \frac{7.0}{2} \times 12.5 \qquad = 43.75 \text{ kN/m}$$

For this combination of loading (dead plus superimposed) partial safety factor values for γ_f should be taken as 1.4 and 1.6 respectively for characteristic dead and superimposed loads.

Design load on wall:
$$\begin{aligned} n_w &= (1.4 \times G_k) + (1.6 \times Q_k) \\ &= (1.4 \times 38.03) + (1.6 \times 43.75) = 54.24 + 70.0 \\ &= 124.24 \text{ kN/m} \end{aligned}$$

Stage 2: Estimate wall thickness

The fairly high walls are required to support a moderately heavy load from the storage slab. As the slab spans onto the walls from one side only, eccentricity of loading will influence the capacity reduction factor and, therefore, should be taken into account when assessing the slenderness ratio for wall thickness estimation:

Assessed slenderness ratio = 22

Therefore, estimated wall thickness, $\qquad t_{ef} = \dfrac{5.5}{22} \times 0.75$

$$= 0.1875 \text{ m}$$

Try 190 thick concrete block wall.

Stage 3: Calculate design strength of wall

(a) Determine β from slenderness ratio:

$$\text{Effective height, } h_{ef} = 5.5 \times 0.75 \qquad = 4.125 \text{ m}$$

$$\text{Effective thickness, } t_{\text{ef}} = \text{actual thickness} = 190 \text{ mm}$$

$$\text{Slenderness ratio} = \frac{h_{\text{ef}}}{t_{\text{ef}}} = \frac{4.125}{0.19} = 21.71$$

Eccentricity of load: from BS 5628, clause 31, the eccentricity of the load for this example may be assumed to be applied at one-third the depth of the bearing area from the loaded face (see Figure 10.6).

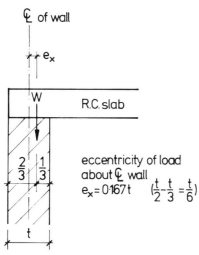

Figure 10.6

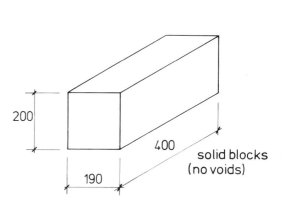

Figure 10.7

From Table 7 of BS 5628 (see Table 5.15): with SR = 21.71 and e_x = 0.167t, by interpolation, β = 0.485

(b) The partial safety factor, γ_m, for materials and workmanship will, for this example, be assumed to be governed by the same conditions as in Example 1, hence γ_m = 2.5.

(c) Characteristic strength required, f_k:

$$f_k = \frac{n_w \gamma_m}{\beta t} = \frac{124.24 \times 2.5 \times 10^3}{0.485 \times 190 \times 1000}$$

$$= 3.37 \text{ N/mm}^2$$

Stage 4: Determine required block and mortar strengths
The applicable section of Table 2, BS 5628, is dependent on the shape of the individual block units, and is related to the ratio of their height to least horizontal dimension. The block to be used, in this example, for the estimated thickness of 200 mm, has the dimensions shown in Figure 10.7.

$$\text{Shape ratio} = \frac{200}{190} = 1.05.$$

Interpolation between Tables 2(b) and 2(d), BS 5628 is necessary to find the block and mortar strength. Try solid blocks with a compressive strength of 7.0 N/mm² set in mortar designation (iv).

Table 2(b), f_k = 2.8 By interpolation for ratio 1.05,
Table 2(d), f_k = 5.6 f_k = 3.696 N/mm², f_k > 3.37 N/mm² required.

Use solid concrete blocks measuring 400 long × 200 high × 190 wide with a compressive strength of 7.0 N/mm² and set in mortar designation (iv).

10.5 DESIGN OF CAVITY WALLS

Cavity walls are of two basic types (see 9.3 and 9.4 and Figure 10.8):

(a) ungrouted cavity walls (the more common type)

(b) cement grouted cavity walls.

10.5.1 Ungrouted cavity walls
The great majority of ungrouted cavity walls are used on external elevations, and the vertical loading in such situations invariably results from

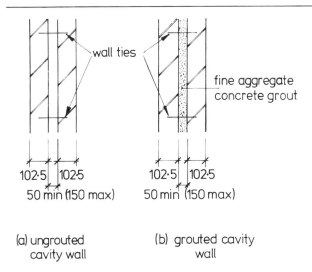

(a) ungrouted
cavity wall

(b) grouted cavity
wall

Figure 10.8

floors and roofs spanning onto the inner leaf only. In addition, wind pressures and suctions impose lateral loading on the wall and this must also be considered in the design. This latter aspect of loading will be considered in Chapter 11. For certain arrangements and situations, cavity walls are used internally and are often loaded with floor and roof loads on both leaves. Examples of internal cavity walls occur where a plan area extends below, say, first floor level, and the line of the external cavity wall over is extended through the ground floor storey as will be seen later in Example 6. Further examples are in the use of cavity party walls for flat developments where the cavity construction is employed for sound insulation of the party wall common to both properties, and at movement joints in a building.

The stiffness of cavity walls ignores the wall ties from the point of view of transferring flexural shears across the cavity, but utilises the wall ties in that each of the two leaves has the effect of helping to prop the other. Allowance is given for this propping effect in the calculation of the effective thickness of cavity walls which is given, in clause 28.4.1 of BS 5628, as equal to two-thirds the sum of the actual thicknesses or the actual thickness of the thicker leaf, whichever is the greater.

EXAMPLE 4

For the building and loading information given for Example 1, design the ungrouted external cavity wall in the bottom storey height.

Stage 1: Calculate design load, n_w
(a) Characteristic dead loads, G_k:

$$\text{roof} = \frac{5}{2} \times 4 \qquad\qquad = 10.00$$

$$3 \text{ floors} = \frac{5}{2} \times 5 \times 3 \qquad = 37.50$$

$$12 \text{ mm plaster} = 0.012 \times 21 \times 12.9 = 3.25$$

$$\text{own weight of wall} = 0.1025 \times 18 \times 12.9 = 23.80$$

Note: inner leaf only considered in ow wall calculation. Outer leaf is self-supporting.

Total characteristic dead load, $G_k = 74.55$ kN/m

For this design example, the effect of wind loading on the wall will be ignored as this aspect will be dealt with in Chapter 11.

(b) Characteristic superimposed load, Q_k:

$$\text{roof} = \frac{5}{2} \times 1.5 \quad = 3.75$$

$$3 \text{ floors} = \frac{5}{2} \times 3 \times 3 = \underline{22.50}$$

Therefore $Q_k = 26.25$ kN/m

Design load on wall: $n_w = (1.4 \times G_k) + (1.6 \times Q_k)$
$$= (1.4 \times 74.55) + (1.6 \times 26.25)$$
$$= 146.37 \text{ kN/m}$$

Stage 2: Estimate wall thickness

The two most common configurations of brick cavity walls are as shown in Figure 10.9.

The internal wall designed in Example 1 was shown to require an extremely low strength brick for a 215 thick brick wall, but a 102.5 thick brick wall could not be used as it exceeded the maximum slenderness ratio of 27. The external leaf of the cavity wall will stiffen the loadbearing inner leaf and therefore, for this example, try a 255 cavity wall.

Check maximum slenderness ratio:

$$SR = \frac{h_{ef}}{t_{ef}}$$

$$= \frac{4.5 \times 0.75 \times 10^3}{\tfrac{2}{3}(102.5 + 102.5)}$$

$$= 24.7$$

which is less than the maximum permissible SR of 27.

Therefore, try a 255 cavity wall.

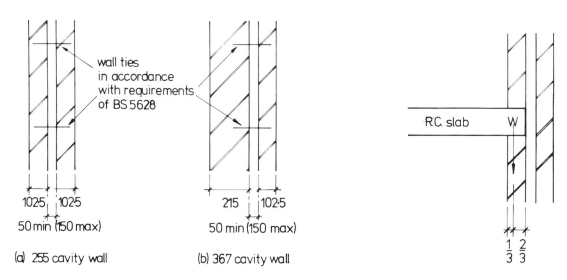

(a) 255 cavity wall   (b) 367 cavity wall

Figure 10.9   **Figure 10.10**

Stage 3: Calculate design strength of wall

(a) Determine β from slenderness ratio:

$$\text{Effective height, } h_{ef} = (4.5 - 0.15) \times 0.75 = \quad 3.2625 \text{ m}$$

$$\text{Effective thickness, } t_{ef} = \tfrac{2}{3}(102.5 + 102.5) \quad = 136.67 \text{ mm}$$

$$\text{Slenderness ratio, } \frac{h_{ef}}{t_{ef}} = \frac{3.2625 \times 10^3}{136.67} \quad = 23.87$$

Eccentricity of load (see Figure 10.10) (from BS 5628, clause 31):

As in Figure 10.6 the eccentricity of the load on the loadbearing inner leaf of the cavity wall, $e_x = 0.167t$ (see Figure 10.10).

From Table 7 of BS 5628 (see Table 5.15), with SR = 23.87 and $e_x = 0.167t$, by interpolation $\beta = 0.389$.

(b) Partial safety factor for materials and workmanship, $\gamma_m = 2.5$.

(c) Characteristic strength required, f_k:

$$f_k = \frac{n_w \gamma_m}{\beta t \times 1.15} = \frac{146.37 \times 2.5 \times 10^3}{0.389 \times 102.5 \times 1.15 \times 1000}$$

$$= 7.98 \text{ N/mm}^2$$

Stage 4: Determine brick and mortar strengths required
From Table 2(a) of BS 5628 (see Table 5.4) with $f_k = 7.98$ N/mm², use bricks with a compressive strength of 32 N/mm² set in a designation (iii) mortar.

Clearly this brick and mortar strength is not compatible with the internal 215 thick brick wall designed in Example 1, and the designer may wish to utilise a 215 thick inner leaf of low strength bricks for the external cavity walls.

EXAMPLE 5
Repeat Example 4 using 215 thick brick inner leaf.

Stage 1: Calculate n_w
(a) Characteristic dead loads, G_k:

$$
\begin{aligned}
\text{roof, as Example 4} &= 10.00 \\
\text{3 floors, as Example 4} &= 37.50 \\
\text{12 mm plaster, as Example 4} &= 3.25 \\
\text{ow wall} = 0.215 \times 18 \times 12.9 &= 49.92
\end{aligned}
$$

Therefore $G_k = 100.67$ kN/m

(b) Characteristic superimposed loads, Q_k:

As Example 4 $Q_k = 26.25$ kN/m

Design load on wall: $n_w = (1.4 \times 100.67) + (1.6 \times 26.25)$
                                     $= 182.94$ kN/m

Stage 3: Calculate design strength of wall
(a) Determine β:

$$h_{ef} = 0.75 \times (4.5 - 0.15) = 3.2625 \text{ m}$$
$$\left. \begin{aligned} t_{ef} = \tfrac{2}{3}(215 + 102.5) &= 211.67 \\ \text{or} = 215 \end{aligned} \right\} \text{ use 215 for } t_{ef}$$
$$\text{SR} = \frac{h_{ef}}{t_{ef}} = \frac{3.2625 \times 10^3}{215} = 15.2$$

Eccentricity, $e_x = 0.167t$ (as Example 4).

From Table 7 of BS 5628 (see Table 5.15), with SR = 15.2 and $e_x = 0.167t$, $\beta = 0.696$.

(b) $\gamma_m = 2.5$ (see Example 4).

(c) Characteristic strength required, f_k:

$$f_k = \frac{n_w \gamma_m}{\beta t} = \frac{182.94 \times 2.5 \times 10^3}{0.696 \times 215 \times 1000}$$

$$= 3.06 \text{ N/mm}^2$$

Stage 4: Determine brick and mortar strength required
From Table 2(a) of BS 5628 (see Table 5.4), use bricks with a compressive strength of 10 N/mm² set in a designation (iv) mortar.

10.5.2 Grouted cavity walls
Internal cavity walls, can be designed, sometimes more economically, by grouting the cavity as has been previously explained. The most common situation for this condition to exist is, perhaps, where a building plan area increases below say first floor level as described in 10.5.1. To maintain standard room sizes in hostel-type buildings,

the cavity wall thickness is often extended through the ground floor storey, and considera- tion could be given in such a situation to the use of a grouted cavity wall.

EXAMPLE 6

Design the internal fairfaced brick wall in the ground floor storey of the building shown in Figure 10.11 as a grouted cavity wall, γ_m can be taken as 2.5 for the purpose of this example, the brickwork density is 18 kN/m³ and the loadings are given as:

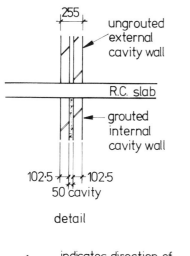

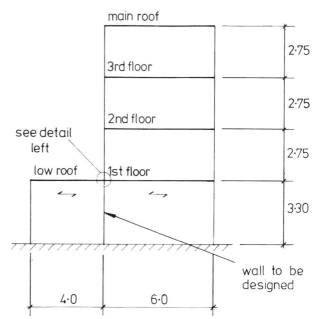

Figure 10.11

	Main roof	Low roof	Floors
Characteristic dead loads (kN/m²)	4.8	6.0	6.0
Characteristic superimposed loads (kN/m²)	1.5	3.0	4.0

Stage 1: Calculate n_w

(a) Characteristic dead loads, G_k:

For this example, the exact position of each of the component dead loads will be considered in order to establish the eccentricity of the loading system on the wall under. The characteristic dead loads will therefore be subdivided thus:

G_{k1} = low wall   = 18 × 0.255 × 3.3   = 15.15
              and   18 × 0.1025 × 2 × 8.25 = 30.44
G_{k1} total   = 45.59 kN/m

G_{k2} = low roof   = $6 \times \dfrac{4}{2}$ = 12.00 kN/m

G_{k3} = main roof = $4.8 \times \dfrac{6}{2}$ = 14.40

          3 floors   = $3 \times 6 \times \dfrac{6}{2}$ = 54.00
G_{k3} total   = 68.40 kN/m

Total characteristic dead load   = $G_{k1} + G_{k2} + G_{k3}$
                                        = 45.59 + 12.00 + 68.40 = 125.99 kN/m

(b) Characteristic superimposed loads, Q_k:
As for the characteristic dead loads, the characteristic superimposed loads will be subdivided thus:

$$Q_{k1} = \text{low roof} = 3 \times \frac{4}{2} \qquad = \quad 6.0 \ \text{kN/m}$$

$$Q_{k2} = \text{main roof} = 1.5 \times \frac{6}{2} \qquad = \quad 4.5$$

$$\begin{aligned} 3 \ \text{floors} \quad &= 4.0 \times \frac{6}{2} \times 3 = 36.0 \\ Q_{k2} \ \text{total} \qquad &= \overline{40.5} \ \text{kN/m} \end{aligned}$$

Total characteristic superimposed load $= Q_{k1} + Q_{k2}$
$$= 6.0 + 40.5 = 46.5 \ \text{kN/m}$$

Design load on wall:

$$\begin{aligned} n_w &= (1.4 \times G_k) + (1.6 \times Q_k) \\ &= (1.4 \times 125.99) + (1.6 \times 46.5) \\ &= 250.79 \ \text{kN/m} \end{aligned}$$

Now consider the resultant position of the design load from the eccentricities of the various components:

Position of resultant design load, n_w, must take account of partial safety factors (γ_f) for loadings as shown below.

Position of resultant: taking moments about left-hand face of wall as shown in Figure 10.12:

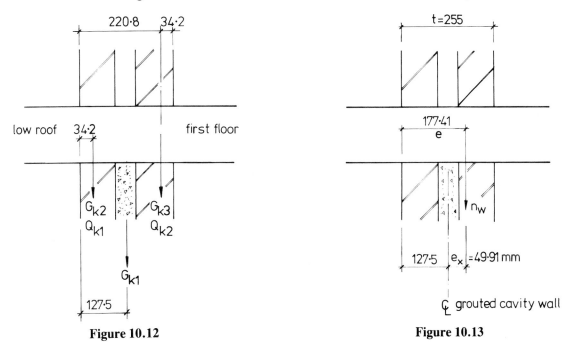

Figure 10.12　　　　　　　　**Figure 10.13**

$$n_w e = (G_{k2} \times 1.4 \times 34.2) + (Q_{k1} \times 1.6 \times 34.2) + (G_{k1} \times 1.4 \times 127.5) +$$

$$(G_{k3} \times 1.4 \times 220.8) + (Q_{k2} \times 1.6 \times 220.8)$$

$$= (574.56) + (328.32) + (8\,137.82) + (21\,143.81) + (14\,307.84)$$

$$= 44\,492.35$$

Therefore $\qquad e = \dfrac{44\,492.35}{250.79} = 177.41 \ \text{mm (see Figure 10.13)}$

$$e_x = 177.41 - 127.50 = 49.91 \ \text{mm}$$

$$= 0.196t$$

In practice, an experienced designer would tend to 'guesstimate' this eccentricity, rather than rely on such a theoretical analysis which is difficult to justify.

Stage 2: Estimate wall thickness
The wall thickness is assumed to be dictated by the external cavity wall over as 255 mm for planning requirements.

Stage 3: Calculate design strength of wall
(a) Determine β:
The effective thickness of a grouted cavity wall, as defined in clause 29.6 of BS 5628, can be taken as the actual overall thickness. Careful supervision of the grouting is essential to ensure compliance with this definition.

$$\text{SR} = \frac{h_{\text{ef}}}{t_{\text{ef}}} = \frac{0.75 \times 3.3 \times 10^3}{255} = 9.71$$

e_x = as calculated previously = $0.196t$.

From Table 7 of BS 5628 (see Table 5.15), by interpolation: $\beta = 0.67$.

(b) γ_m as quoted for this example = 2.5.

(c) Characteristic strength required for wall, f_k:

$$f_k = \frac{n_w \gamma_m}{\beta t} = \frac{250.79 \times 2.5 \times 10^3}{0.67 \times 255 \times 1000}$$
$$= 3.67 \text{ N/mm}^2$$

Stage 4: Determine brick and mortar strengths required
From Table 2(a) of BS 5628 (see Table 5.4), use bricks with a compressive strength of 10 N/mm^2 set in a designation (iv) mortar together with concrete grout to give a 28 day cube strength of 10 N/mm^2 and wall ties in accordance with BS 5628.

10.5.3 Double-leaf (or collar-jointed) walls
The most common use of double-leaf walls is likely to be found where a wall thicker than a half brick wall is required for either functional or design purposes, but where the architect requires stretcher bond and fairfaced work on both wall faces. The design given in Example 1 would be exactly the same for the same total thickness of double leaf wall, provided that the detail conditions as specified in clause 29.5 of BS 5628 are satisfied.

10.6 DESIGN OF WALLS WITH STIFFENING PIERS

The possibility of a half brick wall buckling under axial loads can be significantly reduced by the introduction of piers placed at regular, specified centres, and fully bonded into the wall itself. The use of stiffening piers to increase the second moment of area of a wall section is, of course, not limited to half brick walls and can be applied to any solid wall thickness as well as to cavity walls. It is considered, however, that if a 215 thick brick wall does not have an adequate slenderness ratio to withstand a particular loading condition, the design calls for the selection of a geometric shape best suited to provide the necessary second moment of area. The most common occurrence of this situation would occur in an extremely high wall which is required to support heavy axial loading. The design philosophy for such an element should be based upon second moment of area and radius of gyration, rather than slenderness ratio as traditionally calculated from effective thickness, and a diaphragm or fin wall profile is generally the most suitable geometric form. The design philosophy will be dealt with in more detail later in section 10.8.

EXAMPLE 7

Design the internal wall given for Example 1 as a half brick wall adequately stiffened by the introduction of brick piers.

The design for Example 1 resulted in a 215 thick brick of extremely low strength merely to provide for the maximum permissible slenderness ratio. The best use is not being made of brickwork's natural compressive strength in this design and the pier-stiffened half brick wall is an obvious alternative choice, as will be shown in this example.

Stage 1: Calculate n_w
The design load will be taken as the same as for Example 1:

$$n_w = 271.5 \text{ kN/m}$$

Stage 2: Estimate wall/pier configuration
There are no simple and realistic guidelines that can be applied to the selection of the size and spacing of stiffening piers. The trial and error approach related to the objective of achieving a reasonable slenderness ratio will eventually lead to the designer becoming more familiar with the benefits gained from the introduction of stiffening piers. For this example, we will select a wall/pier profile as shown in Figure 10.14 and check its suitability.

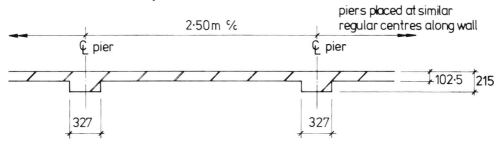

Figure 10.14

Stage 3: Calculate design strength of wall
(a) Determine β:
Effective thickness is improved by the introduction of the stiffening piers and is the product of the actual thickness of the wall (102.5 mm) and the stiffening coefficient K obtained from BS 5628, Table 5 (see Table 5.12):

Stiffened wall properties:

$$\frac{\text{Pier spacing}}{\text{Pier width}} = \frac{2.500}{0.327} = 7.64$$

$$\frac{\text{Pier thickness}, t_p}{\text{Wall thickness}, t} = \frac{215}{102.5} = 2.1$$

By interpolation from BS 5628, Table 5, $K = 1.36$.

$$\text{Effective thickness}, t_{ef} = K \times t = 1.36 \times 102.5$$
$$= 139.4 \text{ mm}$$

$$\text{Slenderness ratio, SR} = \frac{h_{ef}}{t_{ef}} = \frac{0.75 \times (4.5 - 0.15) \times 10^3}{139.4}$$

As for Example 1, $e_x = 0$. Thus from BS 5628, Table 7 (see Table 5.15) $\beta = 0.557$.

(b) Partial safety factor, $\gamma_m = 2.5$, as Example 1.

(c) Calculate characteristic strength, f_k required:
As well as stiffening the wall, the piers are quite capable of supporting some of the axial load and the loadbearing area of the piers should be added to that of the wall in determining the required characteristic strength. The equivalent thickness of solid wall per metre length with allowance for the pier area for this example is calculated as follows:

$$\frac{\text{wall area + pier area}}{\text{length}} = \frac{(102.5 \times 2500) + (112.5 \times 327)}{2500}$$

i.e. equivalent solid thickness = 117.22 mm

$$f_k = \frac{n_w \gamma_m}{\beta \times t \times 1.15} = \frac{271.5 \times 2.5 \times 10^3}{0.557 \times 117.22 \times 1000 \times 1.15}$$

$$= 9.04 \text{ N/mm}^2$$

Stage 4: Select brick and mortar strength required
From Table 2(a) of BS 5628 (see Table 5.4), by interpolation, use a brick of compressive strength 33 N/mm² set in a designation (ii) mortar. The strength requirements for this wall should be compared for compatibility with the external cavity wall designed in Example 4.

EXAMPLE 8

Now reconsider the external ungrouted cavity wall designed in Example 4 to investigate the effect of stiffening the inner leaf with piers.

Stage 1
The design load on the wall will be taken as 146.37 kN/m as for Example 4.

Stage 2
The same configuration of stiffening piers as was used for Example 7 will be considered. The wall profile to be designed is therefore shown in Figure 10.15.

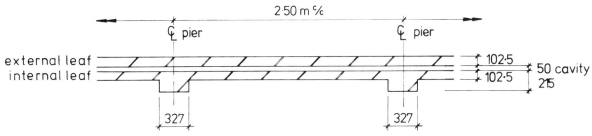

Figure 10.15

Stage 3: Design strength of wall
(a) Determine β:
From BS 5628, clause 28.4.2, Figure 2, the effective thickness of a pier-stiffened cavity wall is given as the greatest of:

(a) ⅔ (102.5 + 102.5*K*) where *K* is the stiffening coefficient for the internal leaf,
(b) 102.5,
(c) *K* × 102.5.

The stiffening coefficient *K* for the internal leaf of the cavity wall is calculated in exactly the same manner as in Example 7 and, as the selected pier spacing and configuration are identical, *K* = 1.36.

$$\text{Effective thickness, } t_{ef} = \tfrac{2}{3}(102.5 + 1.36 \times 102.5)$$
$$= 161.27 \text{ mm}$$

$$\text{Slenderness ratio, SR} = \frac{h_{ef}}{t_{ef}} = \frac{0.75 \times (4.5 - 0.15) \times 10^3}{161.27}$$
$$= 20$$

As for Example 4, $e_x = 0.167t$.

By interpolation from BS 5628, Table 7 (see Table 5.15) β = 0.553.

(b) Partial safety factor for materials (γ_m) will be taken as 2.5 as was used for Example 4.

(c) Characteristic strength required:

$$f_k = \frac{n_w \gamma_m}{\beta \times t \times 1.15}$$

The equivalent thickness of the inner leaf, which is supporting the load, is increased to allow for the piers in the same way as in Example 7:
Equivalent thickness = 117.22 mm (as Example 7).

Therefore
$$f_k = \frac{146.37 \times 2.5 \times 10^3}{0.553 \times 117.22 \times 1000 \times 1.15}$$
$$= 4.91 \text{ N/mm}^2$$

Stage 4: From BS 5628, Table 2(a) (see Table 5.4)
Use a brick of compressive strength 15.0 N/mm^2 set in a designation (iii) mortar.

10.7 MASONRY COLUMNS

The two most common forms of columns generally encountered in design are: (a) the simple rectangular column for the full storey height of a building, and (b) the columns formed by adjacent window or door openings in walls. Other more complex forms are often encountered where an architectural feature is required and these are discussed later in this chapter.

EXAMPLE 9: Simple solid brick columns

It is proposed that the ground floor storey to the building designed in Example 1 should be made 'open plan' by replacing the central spine wall with a series of brick columns placed at 3.0 m % supporting reinforced concrete beams which carry the floors and walls over.

Stage 1: Calculate design load on columns
The design load per metre length of the wall calculated in Example 1 was 271.5 kN/m, therefore, the design load per column spaced at 3.0 m centres = 3 × 271.5 = 814.5 kN.

Stage 2: Estimate column size
The lateral restraint is afforded by the rc first floor slab and beams and, therefore, a square column section is the most suitable profile to provide an equal slenderness ratio to both axes.

The considerable loading will require such a large column section, simply in consideration of the load, that this is likely to provide adequate stability against the tendency to buckle. This aspect will, therefore, have little effect on the column size selection. A simple load/area calculation shows that a 327 square brick column will require an extremely high strength brick and mortar combination. It is therefore decided to use a larger column size of lower unit strength, and a 440 square solid brick column is selected for trial purposes.

Stage 3: Design strength required
(a) Determine β:

$$\text{Slenderness ratio} = \frac{h_{ef}}{t_{ef}}$$

The effective height for columns should be taken as the clear distance between lateral supports and is shown in Figure 10.16 as:

clear distance between lateral supports = 4.5 − 0.5 = 4.0
i.e. effective height, h_{ef} = 4.0 m
effective thickness, t_{ef} = actual thickness   = 0.44 m

$$\text{SR} = \frac{h_{ef}}{t_{ef}} = \frac{4.0}{0.44} = 9$$

The eccentricity of loading on the column can again be taken as 0 to 0.05t. Therefore, from BS 5628, Table 7 (see Table 5.15) for SR = 9 and $e_x = 0$, β = 0.985.

(b) The partial safety factor for materials (γ_m) will again be taken as 2.5 as for Example 1.

(c) Calculate characteristic strength required, f_k:

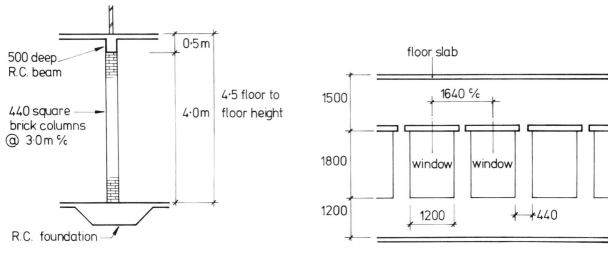

Figure 10.16   **Figure 10.17**

Columns are often subject to clause 23.1.1 of BS 5628 in which the loaded area of the element is considered. The area reduction factor which takes account of the possibility of a below strength unit (brick or block) being included in a small plan area, the single unit therefore representing a significant proportion of the loadbearing element, is applicable where the horizontal loaded cross-sectional area is less than 0.2 m². For this example, the column has a cross-sectional area of $0.44 \times 0.44 = 0.194$ m², the area reduction factor is therefore applicable and is calculated as:

$0.7 + (1.5 \times A) = 0.7 + (1.5 \times 0.194) = 0.991$.

Therefore
$$f_k = \frac{814.5 \times 2.5 \times 10^3}{0.985 \times 0.991 \times 440 \times 440}$$
$$= 10.77 \, \text{N/mm}^2$$

Stage 4: By interpolation from BS 5628, Table 2(a) (see Table 5.4)
Use bricks with a compressive strength of 42 N/mm² set in a designation (ii) mortar for 440 square solid brick columns at 3.0 m ℅.

EXAMPLE 10: Columns formed by adjacent openings

Adjacent door and window openings invariably leave a column of brickwork which is required to support increased load intensity resulting from the lintol loads, in addition to the basic load in that length of wall.

The external cavity wall considered in Example 4 is to be punctured with windows measuring 1200 wide and 1800 high placed at 1640 centres. A typical elevation of the wall is shown in Figure 10.17. Wind loading will be ignored for this example but will be considered in Chapter 11.

Stage 1: Calculate design load, n_w
(a) Characteristic dead loads, G_k:

From Example 4, G_k $\qquad\qquad$ = 74.55 kN/m
Characteristic dead load on column, G_k = 74.55×1.64
$\qquad\qquad\qquad\qquad\qquad$ = 122.26 kN

(b) Characteristic superimposed load, Q_k:

From Example 4, Q_k $\qquad\qquad$ = 26.25 kN/m
Characteristic super load on column, Q_k = 26.25×1.64
$\qquad\qquad\qquad\qquad\qquad$ = 43.05 kN
Design load on column, n_w $\qquad\qquad$ = $(1.4 \times G_k) + (1.6 \times Q_k)$
$\qquad\qquad\qquad\qquad\qquad$ = $(1.4 \times 122.26) + (1.6 \times 43.05)$
$\qquad\qquad\qquad\qquad\qquad$ = 240.0 kN

Stage 2: Estimate column thickness
Based upon a maximum slenderness ratio of 27 and an effective height of 4.350 (being the clear height between supports):

$$\text{minimum } t_{\text{ef}} = \frac{4.350}{27} = 161 \text{ mm}$$

It is clear that thickenings are required between the windows and a 215 thick inner leaf will be adopted.

Stage 3: Calculate design strength of column
By inspection of the elevation in Figure 10.17, it is clear that the columns weakest axis is that tending to buckle perpendicular to the elevation. The column axis parallel to the elevation will therefore not require calculation.

(a) Determine β from slenderness ratio:

$$h_{\text{ef}} = \text{clear height between supports} = 4.350$$

$$t_{\text{ef}} = 0.215 \text{ or } \tfrac{2}{3}(215 + 102.5) \therefore t_{\text{ef}} = 0.215$$

$$\text{Slenderness ratio, SR} = \frac{4350}{215} = 20$$

Eccentricity of loading:
Figure 10.18 shows the bearing detail of the window lintol.

Therefore, from Table 7, BS 5628 (see Table 5.15), with SR = 20 and $e_x = 0.186t$ by interpolation $\beta = 0.5282$.

(b) Partial safety factor, $\gamma_m = 2.5$ (as Example 4).

(c) Area reduction factor
The area reduction factor $= 0.7 + (1.5 \times A) = 0.7 + (1.5 \times 0.215 \times 0.44) = 0.842$.

(d) Characteristic strength required, f_k:

$$f_k = \frac{240 \times 2.5 \times 10^3}{0.528 \times 0.842 \times 0.215 \times 0.440 \times 10^6}$$

$$= 14.3 \text{ N/mm}^2$$

Stage 4: From BS 5628, Table 2(a) (see Table 5.4)
Use bricks with a compressive strength of 47 N/mm² set in a designation (i) mortar. Clearly, this

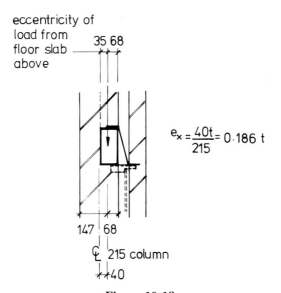

Figure 10.18

extremely high strength brick and mortar is the result of the large window to wall proportions, and the client could be advised to accept smaller windows or thicker columns between the windows.

EXAMPLE 11: Feature columns

Example 9 will be reconsidered, to add interest, by making the columns cruciform-shaped as shown in Figure 10.19.

Column properties:

Area, $A = 0.2311 \text{ m}^2$
$I_{xx} = I_{yy} = 0.00517 \text{ m}^4$

radii of gyration
$r_{xx} = r_{yy} = 0.1496 \text{ m}$

$I_{vv} = I_{uu} = 0.00613 \text{ m}^4$

radii of gyration
$r_{vv} = r_{uu} = 0.1629 \text{ m}$

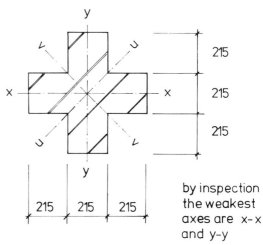

by inspection the weakest axes are x–x and y–y

Figure 10.19

Stage 3: Design strength required
(a) Determine β from slenderness ratio:

$$\text{Slenderness ratio} = \frac{h_{ef}}{t_{ef}}$$

$$h_{ef} = 4.0 \text{ m}$$

t_{ef} = calculate equivalent solid column to give equal radius of gyration

$$\text{Radius of gyration} = \sqrt{(I/A)} \qquad = 0.1496 \text{ m}$$

$$0.1496 = \sqrt{\left(\frac{bt^3/12}{bt}\right)} \quad = \sqrt{\frac{t^2}{12}}$$

$$t = \sqrt{(12 \times 0.1496^2)} = 0.518 \text{ m}$$

$$\text{Slenderness ratio} = \frac{4.0}{0.518}$$

$$= 7.72$$

Local stability of the cruciform column must also be checked by considering the possibility of buckling in the outstanding legs of the profile (see Figure 10.20), thus:

$$\text{Slenderness ratio, SR} = \frac{\text{effective length}}{\text{effective thickness}} = \frac{2 \times 215}{215} = 2$$

Therefore, the previously calculated slenderness ratio value of 7.72 will be used in the determination of β. Eccentricity of loading will be taken as for the previous example $= 0$ to $0.05t$. This can be assumed, for this example, as the rigid floor slab applies the load concentrically into the column. For other loading arrangements the eccentricity should be carefully analysed.

Therefore, from BS 5628, Table 7 (see Table 5.15), for SR = 7.72 and $e_x = 0$ to $0.05t$, β = 1.0.

(b) Partial safety factor γ_m, will again be taken as 2.5.

(c) Area reduction factor is not applicable as area of column exceeds 0.2 m².

(d) Characteristic strength required, f_k:

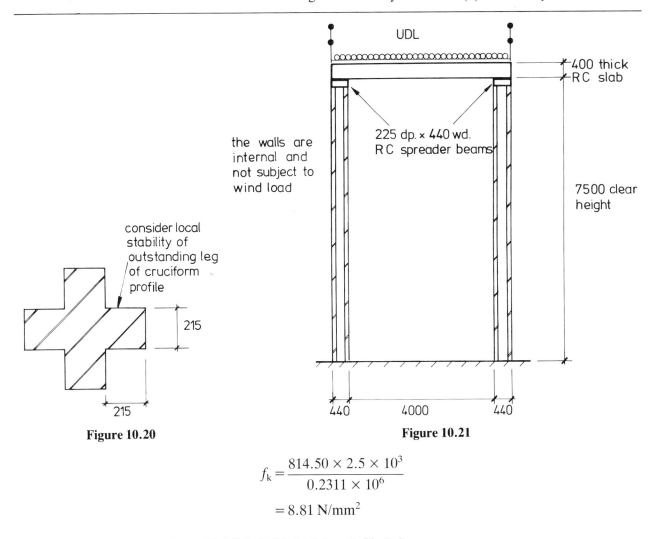

Figure 10.20

Figure 10.21

$$f_k = \frac{814.50 \times 2.5 \times 10^3}{0.2311 \times 10^6}$$

$$= 8.81 \text{ N/mm}^2$$

Stage 4: By interpolation from BS 5628, Table 2(a) (see Table 5.4)
Use bricks with a compressive strength of 32 N/mm² set in a designation (ii) mortar.

10.8 DIAPHRAGM WALLS

The diaphragm wall is mostly used in tall single-storey buildings where its function is primarily to provide stability against wind loading. It can replace the steel or concrete structural frame which may otherwise be required for this purpose. This aspect of the diaphragm wall, and other geometric profiles, are covered in Chapter 13.

However, diaphragm walls can be successfully and economically used to support heavy axial loading, particularly where the load has to be supported at a considerable height. The goemetry of the diaphragm profile provides increased resistance to buckling owing to its large I value and, therefore, the capacity reduction factor β does not reduce the characteristic compressive strength by as much as would be applicable to an equivalent solid wall.

EXAMPLE 12: Design of diaphragm wall under axial loading

An overhead loading platform is shown in Figure 10.21. The superimposed loading on the rc slab is 100 kN/m².

Stage 1: Calculate design load
(a) Characteristic dead load, G_k, per metre length:

rc slab	$= 24 \times 0.4 \times 2.43$	$= 23.328$
spreader beams	$= 24 \times 0.225 \times 0.43$	$= 2.322$
ow wall	$= 0.2343 \times 20 \times 7.275$	$= 34.091$
		$= 59.741$ kN

(b) Characteristic superimposed load, Q_k:

on rc slab $\qquad Q_k = 100 \times 2.43 = 243$

Design load on wall:
$$n_w = (1.4 \times G_k) + (1.6 \times Q_k)$$
$$= (1.4 \times 59.741) + (1.6 \times 243)$$
$$= 472.44 \text{ kN/m}$$

Stage 2: Estimate wall thickness
Unlike solid walls, the cost of diaphragm walls does not increase significantly as the thickness increases. The two half brick leaves are merely spaced further apart to increase the overall thickness of the wall and are connected together with cross-ribs. The materials required to achieve this increased thickness are an extremely small proportion of the total. Clearly, the further apart the two leaves are, the better the resistance to overall buckling and with only a minor increase in cost. Space restrictions are likely to play a more significant part in assessing the wall thickness but for this example a 440 thick wall will be considered (see Figure 10.22).

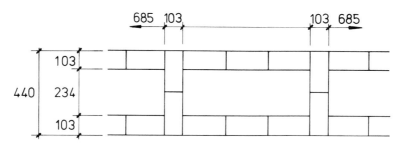

Figure 10.22 Diaphragm wall profile

Stage 3: Calculate design strength of wall
Wall properties per metre length:

Area, $A \qquad\qquad = 234.264 \times 10^3 \text{ mm}^2$

Moment of inertia, $I \ = 4579.37 \times 10^6 \text{ mm}^4$

Radius of gyration, $r \ = 139.81 \text{ mm}$

Equivalent solid wall:

From $r = \sqrt{(I/A)}$, as Example 11, $t = 484.32 \text{ mm} =$ equivalent solid thickness.

It is considered that, until a SR based on radius of gyration is introduced into BS 5628, the equivalent solid thickness of the diaphragm wall should not exceed the actual overall thickness.

Hence $t = 440 \text{ mm}$.

(a) Determine β:
To eliminate eccentricity of load in the wall from the slab, the bearing detail shown in Figure 10.23 will be used. Local stability of the leaves and ribs should be checked however, with this relatively stocky section, this will not be critical. The capacity reduction factor, β, will therefore be based on the overall section.

The continuity reinforcement tied into the rc spreader beam can be considered to provide an enhanced support condition and the effective height, h_{ef} will be taken as $0.75 \times 7.5 = 5.625 \text{ m}$.

Effective thickness of equivalent solid wall $\quad=$ actual thickness

i.e. $\qquad\qquad\qquad\qquad\qquad t_{ef} = 440$

Therefore $\qquad\qquad\qquad\qquad \text{SR} = \dfrac{5625}{440}$

$$= 12.8$$

Eccentricity of load = 0, therefore, from BS 5628, Table 7, $\beta = 0.914$ (see Table 5.15).

(b) Partial safety factor, $\gamma_m = 2.5$ as before.

(c) Characteristic strength required f_k:

$$f_k = \frac{n_w \gamma_m}{\beta \times \text{area}}$$

$$= \frac{472.44 \times 2.5 \times 10^3}{0.914 \times 234.264 \times 10^3}$$

$$= 5.52 \text{ N/mm}^2$$

Stage 4: By interpolation from BS 5628, Table 2(a) (see Table 5.4)
Use bricks with a compressive strength of 22.4 N/mm² set in a designation (iv) mortar.

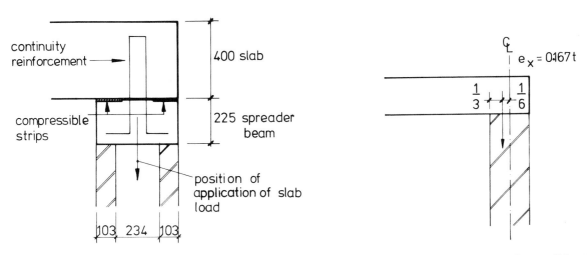

Figure 10.23 Bearing detail - diaphragm wall   **Figure 10.24** Bearing detail – solid wall

EXAMPLE 13: Comparison with solid walls
The diaphragm wall designed in Example 12 will, for purposes of comparison, now be designed as a solid wall.

Stage 1
Design load on wall as Example 12,
$n_w = 472.44$ kN/m

Stage 2
For comparison purposes a 215 thick solid wall will be designed, as this has virtually the same quantity of materials as the diaphragm wall. A 327 thick wall would obviously be better suited to minimise the buckling tendency, but requires considerably more material.

Stage 3: Calculate design strength of wall
(a) Determine β:

$$\text{Slenderness ratio} = \frac{0.75 \times 7.5}{215}$$

$$= 26$$

Eccentricity of load = 0.167t, therefore, by interpolation from BS 5628, Table 7, $\beta = 0.333$ (see Table 5.15 and Figure 10.24).

(b) Partial safety factor, $\gamma_m = 2.5$.

(c) Characteristic strength required, f_k:

$$f_k = \frac{n_w \gamma_m}{\beta \times 0.215}$$

$$= \frac{472.44 \times 2.5 \times 10^3}{0.333 \times 0.215 \times 10^6}$$

$$= 16.50 \text{ N/mm}^2$$

Stage 4: From BS 5628, Table 2(a) (see Table 5.4)
Use bricks with a compressive strength of 57.14 N/mm² set in a designation (i) mortar.

The difference in the strength requirements for virtually the same quantity of materials shows the value of the diaphragm wall for such an application. The additional workmanship must, of course, be set against the material saving for the diaphragm wall in order to relate the comparative economics of the two solutions. In both solutions it has been assumed that the stability of the platform is provided by other elements not considered in this design.

10.9 CONCENTRATED LOADS

Concentrated loads, such as occur at beam bearings and the like, are analysed using increased characteristic compressive strengths, f_k, from those shown in Table 2(a) of BS 5628 (see Table 5.4). The amount by which f_k is allowed to increase is governed by the type of bearing and three bearing types are illustrated in BS 5628, Figure 4, for guidance. Two main considerations dictate the bearing type stress increase, being:

(a) The location of the concentrated load relative to the end of the wall in which the load's capacity to disperse in both directions is considered.
(b) The length of bearing of the beam onto the wall in which the possibility of spalling due to insufficient bearing length is considered. The eccentricity of load produced in this bearing type should be considered separately in the assessment of the capacity reduction factor β for the wall as a whole.

EXAMPLE 14: Design of wall with beam bearing
Consider Example 9 and investigate the effect of the rc beam bearing onto the external gable wall. The beam bearing detail is shown in Figure 10.25 and the loadings, etc., are to be taken as for Example 1.

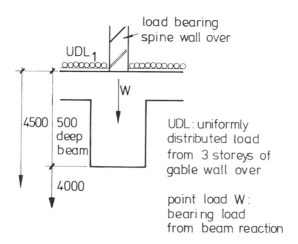

Figure 10.25

UDL = uniformly distributed load from three storeys of gable wall over
    = bearing load from beam reaction.

Stage 1: Calculate design loads on wall
UDL₁ (characteristic dead load G_k from inner leaf only not influenced by load from beam reaction).

Half brick wall $= 0.1025 \times 18 \times 8.4 \qquad = 15.50$ kN/m
Design load $\quad = 1.4 \times G_k = 1.4 \times 15.5 = 21.70$ kN/m

Point load W:

From Example 1, the design load per metre length of the spine wall $= 271.5$ kN/m. The first brick column is spaced 3.0 m away from the gable wall and, therefore, the beam reaction at the bearing onto the gable wall

$$= 271.50 \times \frac{3.0}{2}$$

$$= 407.25 \text{ kN}$$

Stage 2
The wall thickness will be assumed to be already established as a 255 thick cavity comprising two half brick leaves with a 50 mm cavity.

Stage 3: Design strength required
Consideration of beam bearings requires two design checks:

(a) The local effect immediately beneath the bearing area.
(b) The overall effect on the wall taking account of whatever other loads are already in the wall.

The recommended procedure is to design the latter condition initially to establish the minimum brick and mortar strengths required, and then to proceed to check the local condition, including a concrete spreader or padstone beneath the beam bearing if found necessary. The introduction of such a spreader is likely to be more economical than increasing the brick and mortar strength of the whole wall. Figure 10.26 shows the spread effect of the bearing load.

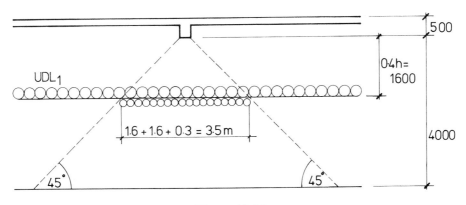

Figure 10.26

UDL_1 from wall over:
UDL_2 is spread of load at 45° through brickwork and is additive to UDL_1 at $0.4h$ for slenderness considerations.

Design load at $0.4h$ level: $\qquad \text{UDL}_1 = \text{as calculated} = \quad 21.70$

$$\text{UDL}_2 = \frac{407.25}{3.5} \qquad = 116.36$$

Total design load $\qquad\qquad\qquad\qquad\qquad\qquad\quad = \overline{138.06}$ kN/m

(a) Determine β:

$$\text{Slenderness ratio} = \frac{h_{\text{ef}}}{t_{\text{ef}}}$$

$$= \frac{0.75 \times 4.0}{\tfrac{2}{3}(102.5 + 102.5)}$$

$$= 22$$

Eccentricity of load: the load from the wall over (UDL₁), can be taken as applied on the centre line of the inner leaf in the lower storey. The beam will be assumed to be sensibly rigid with minimum rotation at the bearing. The application of this load will therefore also be on the centre line of the inner leaf, $e_x = 0$.

Therefore, from BS 5628, Table 7, $\beta = 0.62$ (see Table 5.15).

(b) Partial safety factor $\gamma_m = 2.5$, as before.

(c) Calculate characteristic strength required, f_k:

$$f_k = \frac{n_w \gamma_m}{\beta \times t \times 1.15}$$

$$= \frac{138.06 \times 2.5}{0.62 \times 102.5 \times 1.15}$$

$$= 4.72 \text{ N/mm}^2$$

Stage 4: By interpolation select brick/mortar strengths for wall from BS 5628, Table 2(a) (see Table 5.4)
Use bricks with a compressive strength of 17 N/mm² set in a designation (iv) mortar.

Now design the beam bearing using this strength of brick and mortar, introducing a spreader, if required, beneath the beam. The calculation will therefore assess the minimum width of bearing area required. The beam or spreader will take full bearing onto the wall width and is a considerable distance from the ends of the wall. By inspection, the detail can be classed as a bearing type 2 from BS 5628, Figure 4 (see Figure 5.48). The local design strength can be taken as:

$$\frac{1.5 f_k}{\gamma_m} \quad \text{and is equal to} \quad \frac{W}{bx}$$

where
    W = bearing load
    b = width of the bearing area
    x = thickness of the bearing area.

Therefore
$$b = \frac{W \gamma_m}{1.5 f_k x}$$

$$= \frac{407.25 \times 2.5}{1.5 \times 4.72 \times 102.5}$$

$$= 1.403 \text{ m}$$

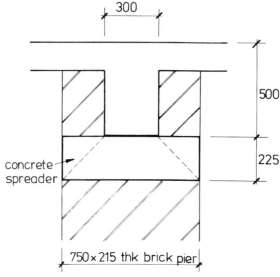

Figure 10.27

This is greater than $8x = 0.820$ m given in BS 5628, Figure 4, for bearing type 2 and is therefore not acceptable (see Figure 5.48).

The simplest solution is to introduce a brick pier for the bearing width required.

Hence, try a 215 mm thick × 750 mm long pier. The introduction of the pier will obviously strengthen the wall, so far as the initial part of the design is concerned, however, the same brick and mortar strength will be checked (see Figure 10.27):

width of bearing, $b \quad = \dfrac{407.25 \times 2.5}{1.5 \times 4.72 \times 15}$

area required, $A \quad = 0.668$ m $(8 \times 0.215 = 1.72 > 0.668$ m$)$

Use 215 × 750 long brick pier with 215 × 750 × 225 deep concrete padstone beneath beam bearing.

CHAPTER 11

DESIGN OF MASONRY ELEMENTS (2): COMBINED BENDING AND AXIAL LOADING

Chapter 10 related the basis of design, set out in Chapter 5, to specific problems of axially loaded masonry elements. This chapter will progress to the more common design conditions of combined bending and axial loading. The bending moments (BM) applied to these elements could be the result of lateral loading or eccentric loads, or a combination of both, and Chapter 10 has already introduced bending in the form of eccentric vertical loads. The recommendations of BS 5628, Part 1 are applied in the following examples to specific design problems of element design, and the guidance given in Chapter 6 will be followed.

11.1 METHOD OF DESIGN

The design method is based upon trial and error. Again, experience and familiarity will help the accuracy of initial estimates.

Some important questions which must be considered at the early stages are:

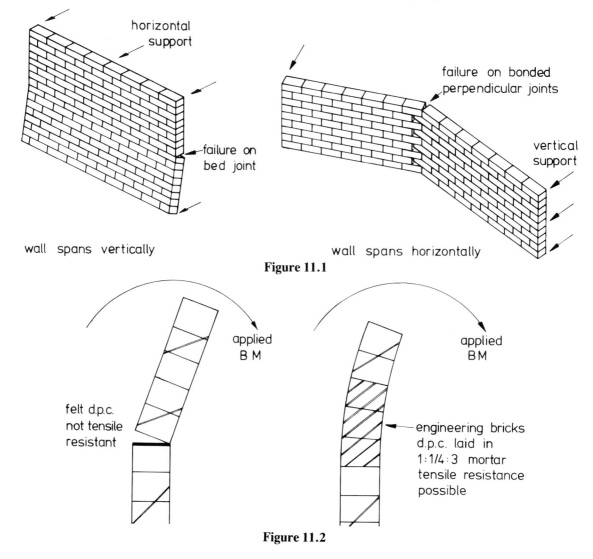

horizontal support

failure on bed joint

wall spans vertically

failure on bonded perpendicular joints

vertical support

wall spans horizontally

Figure 11.1

felt d.p.c. not tensile resistant

applied B M

applied B M

engineering bricks d.p.c. laid in 1:1/4:3 mortar tensile resistance possible

Figure 11.2

(a) In which direction, related to the bed joints, is the bending occurring (see Figure 11.1)?

(b) Is it reasonable for the particular design being considered to allow flexural tensile stresses to develop (see Figure 11.2)?

(c) Is the ratio of bending moment to axial load, i.e. M/W, high or low (where M = bending moment and W = axial load) (see Figure 11.3)?

(f) What restraint is already provided to the element, and what additional restraint could economically be achieved (see Figure 11.6)?

For example, referring back to Chapter 6, it will be noted that the flexural tensile resistance when bending is applied normal to the perpendicular joints is greater than when bending is applied normal to the bed joints. In addition, it will be noted that most dpc membranes cannot be relied

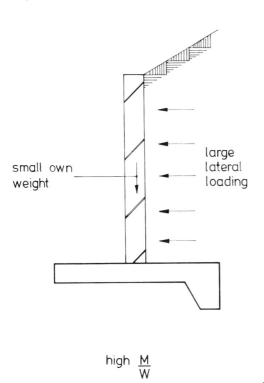

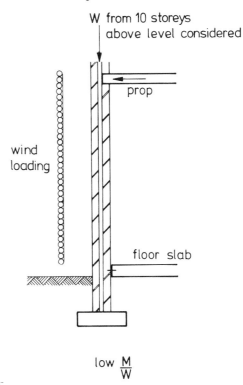

Figure 11.3

(d) Is a cracked section likely (see Figure 11.8), and is a cracked section permissible for the element being designed (see Figure 11.4)?

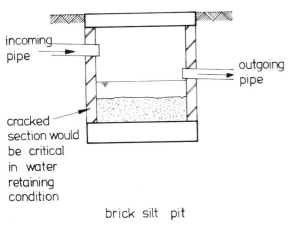

Figure 11.4

(e) Is the total wall section likely to act as a homogeneous mass, or is shear slip likely to occur within the section (see Figure 11.5)?

upon to resist tensile stresses, and that care is needed when deciding on whether or not to rely upon tensile resistance for a particular loading condition.

The ratio of bending moment to axial load, M/W, is also particularly important since, where the ratio is low, normal slender elements of solid or cavity construction would usually be suitable. However, when the bending moment to load ratio, M/W, is high, then the material can be used more economically by improving its lever arm by using a diaphragm, T or other suitable section with a high Z/A ratio, i.e. section modulus over cross-sectional area (see Chapter 9). Alternatively, post-tensioned or reinforced masonry can be used where large bending moment to load ratios are to be resisted (see Figure 11.7).

Post-tensioning or reinforcement give additional resistance moment due to the greater resistance to tensile stresses afforded by these methods.

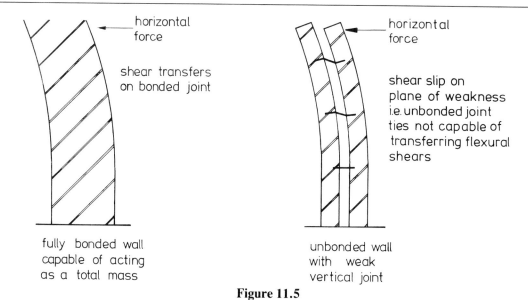

horizontal force

shear transfers on bonded joint

fully bonded wall capable of acting as a total mass

horizontal force

shear slip on plane of weakness i.e. unbonded joint ties not capable of transferring flexural shears

unbonded wall with weak vertical joint

Figure 11.5

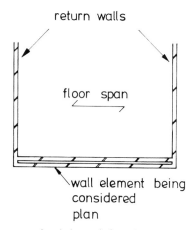

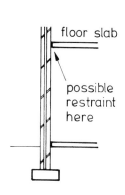

panel wall with openings

return walls

floor span

wall element being considered

plan

restraint exists at return walls bonded into wall element being designed

floor slab

possible restraint here

section

additional restraint can be achieved by tying wall into floor

Figure 11.6

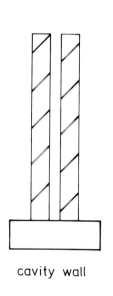

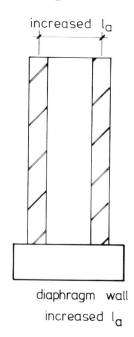

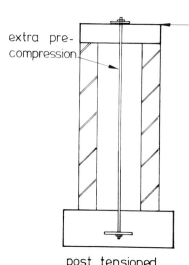

cavity wall

increased l_a

diaphragm wall

increased l_a

extra pre-compression

post tensioned diaphragm wall

increased l_a
increased load

Figure 11.7

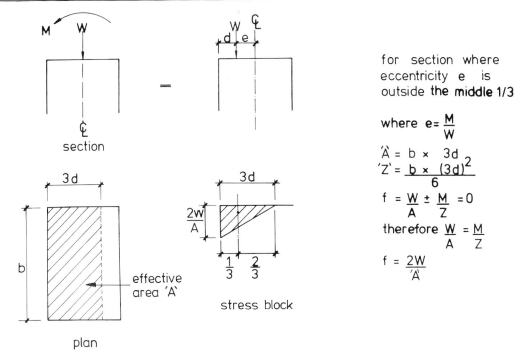

for section where
eccentricity e is
outside the middle 1/3

where $e = \dfrac{M}{W}$

$\text{'A'} = b \times 3d$

$\text{'Z'} = \dfrac{b \times (3d)^2}{6}$

$f = \dfrac{W}{A} \pm \dfrac{M}{Z} = 0$

therefore $\dfrac{W}{A} = \dfrac{M}{Z}$

$f = \dfrac{2W}{\text{'A'}}$

Figure 11.8 Consider section and stress block at working load assuming stress within elastic range

Where no flexural tensile stresses can be relied upon, the section can often be designed on the basis of a cracked section using the effective areas and section modulus of the uncracked portion of the masonry. This design could be carried out by checking the serviceability limit state under working load and ultimate limit state under ultimate load, or the design could be checked on working load only, using a suitable safety factor against overturning and flexural compressive failure (see Figure 11.8).

Care must be taken, however, when designing such walls since, in some elements, a cracked section could be undesirable from the point of view of serviceability limit state. Before calculating the section properties of a section, it must be decided whether or not the section will behave as one mass when bending is applied. For example, consider a cavity wall with butterfly ties subjected to bending. The wall when subjected to bending would tend to distort the ties (see Figure 11.9).

The two leaves deflect approximately equally being linked together by the wall ties but not fixed rigidly enough to prevent vertical shear slip from distorting the ties. The two leaves (1) and (2) shown in Figure 11.9, would tend to rotate about points O_1 and O_2 respectively. The section modulus of such a section would, therefore, be approximately equal to the sum of the separate section moduli of each leaf, and not the section modulus of the total section about its own neutral axis. In fact, BS 5628 reduces the effective cross-section to even less (see Chapter 6).

Alternatively, a wall constructed with a similar quantity of bricks, but built in a bonded form such as a totally solid wall or a diaphragm wall, would have a much increased section modulus due to the shear resistance across the bonded joints connecting the two leaves (see Chapter 13). Consideration of the possible restraints which can be provided to the element can have a large effect on the capacity of the element to resist bending. For example, a long gable wall of a single-storey building, with no restraint at roof level and no connecting walls, would require to be designed as a cantilever; whereas, a similar gable with a suitable restraint at roof level could be designed as a propped and tied cantilever (see Figure 11.10).

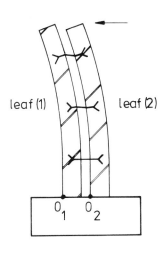

leaf (1)   leaf (2)

O_1 O_2

Figure 11.9

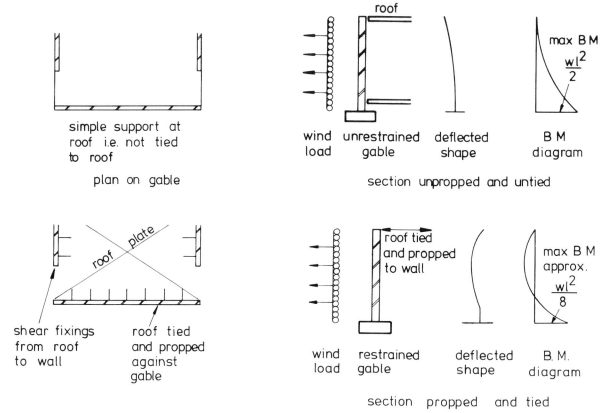

Figure 11.10

The result of such a difference in restraint is to reduce the maximum bending moment in the wall, and to increase the permissible flexural compressive stresses due to the improved effec-tive height of the wall. In the following examples these points will be highlighted in the elements designed.

EXAMPLE 1: Effects of varying the wall section

An illustration of the effects of varying the wall sections to resist bending and axial load where the ratio M/W is large.

As previously stated, the choice of wall elements will depend very much on the ratio of bending moment to axial load M/W. For example, consider a condition where the axial load is from the own weight of the wall only and a large bending moment has to be resisted. Assume the own weight of the wall to be equal to W per 102 mm of thickness per m length at a level being considered and calculate the resistance moments of various sections.

First consider a normal 254 mm cavity wall with 102 mm thick leaves (see Figure 11.11).

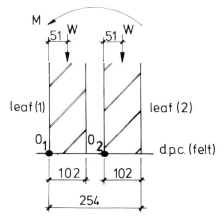

Figure 11.11

The section being considered is at dpc level, and the dpc can be assumed to have zero resistance to tensile forces at right angles to the bed joint. Normal cavity ties also can be assumed to provide very little shear resistance between the leaves and, therefore, it can be assumed that no vertical shear is transferred across the cavity.

The wall's resistance to bending at this location is therefore provided by the weight of the wall acting at its lever arm about the point of rotation. The point of rotation would occur at the compression side of the section and, for the purposes of this example, it can be assumed to be approximately at the edge of each leaf, i.e. leaf (1) rotates about point O_1 and leaf (2) rotates about point O_2.

The wall's resistance to overturning based upon zero tensile resistance at the dpc level and zero shear resistance of the cavity ties would be:

$$(W \times 51) + (W \times 51) = \text{resistance moment}$$

i.e.
$$\text{resistance moment} = 102W \text{ mm}$$

i.e.
$$\text{allowable applied BM} = \frac{102W \text{ mm}}{\text{safety factor}}$$

Now consider a condition where this resistance moment is insufficient for the design bending moment, and assume that a 102 mm bonded thickness is added to the inner leaf (see Figure 11.12).

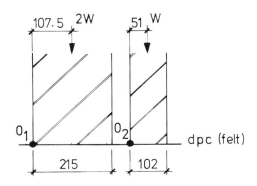

own weight of inner leaf = 2W
own weight of outer leaf = W

Figure 11.12

Based upon the assumptions previously mentioned, the resistance to overturning of this section would be:

$$\text{resistance moment} = (2W \times 107.5) + (W \times 51)$$
$$= 266W \text{ mm}$$

giving an allowable applied moment

$$= \frac{266W \text{ mm}}{\text{safety factor}}$$

This means that, by adding 50% more masonry to the wall, the resistance moment has increased 160%.

Consider now the various methods of increasing the wall's resistance moment. It can be achieved by:

(a) increasing the lever arm of the load
(b) increasing the vertical load
(c) increasing both the lever arm and the load
(d) using a dpc capable of resisting tensile stresses.

The most economical solution, in terms of the amount of masonry, would be to increase the lever arm with little or no increase in the cross-sectional area of the wall. One method of achieving this is to open up the cavity of the wall, and to introduce cross-ribs which make the two leaves interact, i.e. using a diaphragm wall (see Figure 11.13).

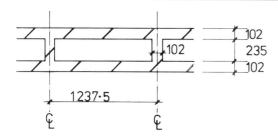

Figure 11.13

Assume that a diaphragm, as shown in Figure 11.13 is to be used, and that ribs are suitably spaced and bonded to transfer the shear forces across the cavity, and that the spacing is adequate to prevent buckling of the leaves, etc. The conditions shown in cross-section in Figure 11.14 are therefore, applicable.

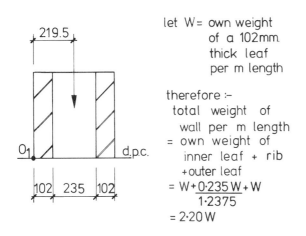

let W = own weight of a 102mm thick leaf per m length

therefore :–
total weight of wall per m length
= own weight of inner leaf + rib + outer leaf
$$= \frac{W + 0.235\,W + W}{1.2375}$$
$$= 2.20\,W$$

Figure 11.14

That is, assuming the own weight of each leaf of 102 mm thick is W per m run and that 10% extra brickwork is added to the original cavity wall in the form of cross-ribs. The resistance moment to overturning of this section can be assumed to rotate about the point O_1 indicated in Figure 11.14 since the cross-ribs provide shear resistance. Assuming, as before, zero tensile resistance and the point of rotation at the edge of the wall (note these two assumptions would have to be verified in an actual element design):

$$\text{resistance moment} = 2.2W \times 219.5$$
$$= 482.9W \text{ mm}$$
$$\text{allowable applied BM} = \frac{482.9W \text{ mm}}{\text{safety factor}}$$

This means that, by adding 10% more masonry, the diaphragm solution achieves $100 \times 482.9/102$ = 473% of the resistance moment of the original wall. Other properties of these three walls are shown in Table 11.1 for comparison purposes.

Table 11.1

Section	A m^2	Z $(\times 10^{-3})$m^3	$\dfrac{Z}{A}$ ratio
Normal cavity	0.204	3.468	0.017
Thickened leaf cavity	0.317	9.438	0.030
Diaphragm wall	0.225	27.730	0.123

Now consider the effect of post-tensioning, assuming that the compressive stresses in the diaphragms are less than the allowable, as is often the case since tensile stresses usually govern the resistance moment (see Figure 11.15).

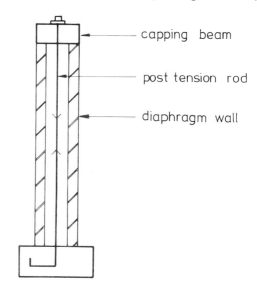

Figure 11.15

Therefore, additional compressive force can be applied at the centre line of the wall (assuming bending critical in both directions) to give maximum lever arm for the design conditions.

Post-tensioned masonry elements are dealt with in Chapter 15. However, the advantage of adding a further load by precompressing the masonry using this method can be seen since any increase in the load W increases the stability moment by the same proportion, for further details refer to Chapter 15.

In the preceding examples, gravitational weight and post-tensioning have been used in resisting bending and the advantages of varying the section have been indicated. In many cases, however, the wall section being considered in resisting bending moments will be located where the masonry will have some resistance to tensile stresses and therefore, instead of merely a gravitational condition existing, there will also be tensile stresses which can be developed. Here, the stress condition should be checked against the allowable and since, in general, the development of critical tensile stresses will occur when the compressive stresses are well within the allowable, it is likely that elastic conditions will be applicable in the compressive zone.

The stress condition can, therefore, be checked on the basis of:

$$f = \frac{P}{A} \pm \frac{M}{Z}$$

where

    M = applied bending moment
    P = applied axial load
    A = cross-sectional area
    Z = section modulus.

From Table 3.1, the advantage of increased section modulus for the diaphragm shape can be clearly seen along with the need for a high Z/A ratio for economic use of material in resisting large bending moments.

BS 5628 provides three alternative methods of calculating the design moment of resistance of walls subject to axial and lateral loading:

(a) treating the masonry section as an arch (clause 36.4.4)
(b) the 'effective eccentricity' method (clause 36.8)

(c) employing the formula $(f_{kx}/\gamma_m + g_d)Z$ (clause 36.5.3).

The authors consider the arch method as the most unreliable as it is often too dependent upon variable factors of workmanship, which are usually outside the control of the designer.

The eccentricity method is more applicable to instances of high axial load to bending moment ratio and is limited in its application to eccentricities of up to $0.3t$ which is the extent of Table 7 in BS 5628 for the calculation of β (see Table 5.15).

The use of the formula $(f_{kx}/\gamma_m + g_d)Z$ is limited, according to the wording of the code to the design of free-standing walls. The authors consider that, as this formula is based upon simple recognised structural theories, there is no reason why it should not be applied to other forms of construction, provided due consideration is given to the attendant flexural compressive stresses. In its presented form, the formula is related strictly to the flexural tensile stresses. Several examples of its use, in situations other than free-standing walls, are given in the pages which follow, and suggested methods of dealing with the attendant flexural compressive stresses are included.

The first problem which faces the designer is which formula/design method to apply to a particular problem. The answer to this can only come from the designer's own familiarity with the use and limitations of each method. Once again, there is no substitute for experience.

EXAMPLE 2: 215 mm thick solid brick wall

Design the solid brick retaining wall of the coal fuel store shown in Figure 11.16. It can be assumed that the cover slab is capable of acting as a tie to resist the reaction at this point from the wall. The fuel store is of such a length that horizontal spanning of the wall can be ignored.

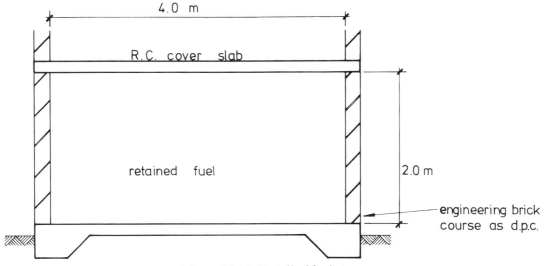

Figure 11.16 Detail of fuel store

The characteristic loadings to be used in the design will be taken as:

(i) cover slab
dead $= 3.60 \text{ kN/m}^2$
superimposed $= 2.50 \text{ kN/m}^2$

(ii) superstructure above cover slab
dead $= 80.0 \text{ kN/m run of wall}$
superimposed $= 25.0 \text{ kN/m run of wall}$

For simplicity of analysis, it can be assumed that the fuel can be filled to the full height of the wall and that the pressure at the base of the wall from the fuel, derived from Rankines formula,

$$\text{density} \times h \left(\frac{1 - \sin \theta}{1 + \sin \theta} \right)$$

is 10.5 kN/m^2. The loading from the structure above can be assumed to be applied on the centre line of the thickness of the retaining wall.

The axial loading from the structure above the fuel store cover slab will enable the retaining wall to be designed as fixed both top and bottom and the loading and bending moment diagrams are shown in Figure 11.17.

Design pressure at base of wall $= \gamma_f \times 10.5 = 1.6 \times 10.5 = 16.8 \text{ kN/m}^2$

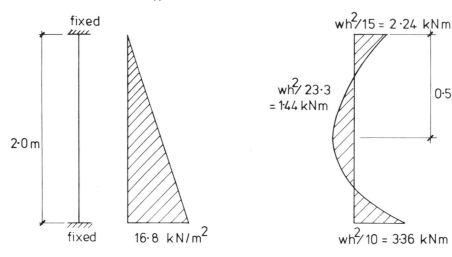

(a) loading diagram   (b) design bending moment diagram

Figure 11.17

Design BM at top of wall $= \dfrac{wh^2}{15} = \dfrac{16.8 \times 2^2}{15 \times 2} = 2.24 \text{ kN·m}$

Design BM at $0.55h$ from top $= \dfrac{wh^2}{23.3} = \dfrac{16.8 \times 2^2}{23.3 \times 2} = 1.44 \text{ kN·m}$

Design BM at base of wall $= \dfrac{wh^2}{10} = \dfrac{16.8 \times 2^2}{10 \times 2} = 3.36 \text{ kN·m}$

Characteristic vertical loads in retaining wall:
dead loads, G_k:

superstructure over $=$ as given $= 80.00$

cover slab $= 3.6 \times \dfrac{4}{2}$ $= 7.20$

ow (own weight) brickwork $= 20 \times 0.215 \times 2$ $= \dfrac{8.60}{}$

$= 95.80$ kN/m

superimposed loads, Q_k:

superstructure over $=$ as given $= 25.0$

cover slab $= 2.5 \times \dfrac{4}{2}$ $= \dfrac{5.0}{}$

$= 30.0$ kN/m

Minimum design vertical load $= G_k \times \gamma_f = 95.8 \times 0.9$ $= 86.22$ kN/m

Maximum design load $= (G_k \times \gamma_f) + (Q_k \times \gamma_f)$

$= (95.8 \times 1.4) + (30 \times 1.6) = 182.12$ kN/m

Due to the high axial load to bending moment ratio, the wall will be designed using the effective eccentricity method which takes account of the flexural compressive stresses. The amount of axial load to bending moment will ensure that the full width of retaining wall is subject to compressive stresses and no tensile stresses will develop.

Calculate eccentricity due to slenderness of wall:

$$e = t \left[\frac{1}{2400} \times \left(\frac{h_{ef}}{t_{ef}} \right)^2 - 0.015 \right]$$

$$= 215 \left[\frac{1}{2400} \times \left(\frac{0.75 \times 2}{0.215} \right)^2 - 0.015 \right]$$

$$= 1.135 \text{ mm}$$

The slenderness eccentricity diagram is shown in Figure 11.18. This eccentricity diagram should be superimposed onto the eccentricity diagram resulting from the design bending moment diagram.

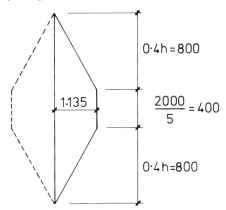

Figure 11.18 Slenderness eccentricity diagram

Calculate eccentricities from design BM diagram:

e at top of wall, min. $= \dfrac{M_A}{n_w} = \dfrac{2.24}{86.22} = 26$ mm

e at $0.55h$ level, min. $= \dfrac{M_A}{n_w} = \dfrac{1.44}{86.22} = 16.7$ mm

e at base of wall, min. $= \dfrac{M_A}{n_w} = \dfrac{3.36}{86.22} = 39$ mm

e at top of wall, max. $= \dfrac{M_A}{n_w} = \dfrac{2.24}{182.12} = 12.3$ mm

e at 0.55h level, max. $= \dfrac{M_A}{n_w} = \dfrac{1.44}{182.12} = 7.9$ mm

e at base of wall, max. $= \dfrac{M_A}{n_w} = \dfrac{3.36}{182.12} = 18.4$ mm

Eccentricity diagrams derived from BM diagram are shown in Figure 11.19.

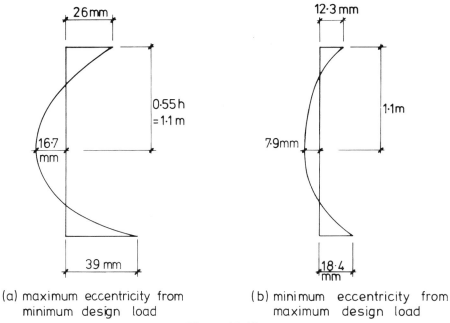

(a) maximum eccentricity from minimum design load

(b) minimum eccentricity from maximum design load

Figure 11.19

Superimpose eccentricity diagrams as in Figure 11.20.

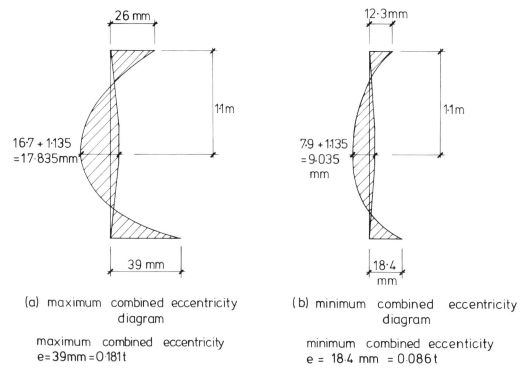

(a) maximum combined eccentricity diagram

maximum combined eccentricity
e = 39mm = 0.181t

(b) minimum combined eccentricity diagram

minimum combined eccenticity
e = 18.4 mm = 0.086t

Figure 11.20

Now design the wall as an axially loaded wall combining the minimum axial load of 86.22 kN with the maximum eccentricity, for β calculation, of 0.181t and the maximum axial load of 182.12 kN with the minimum eccentricity, for β calculation, of 0.086t.

$$\text{Slenderness ratio for both loading conditions} = \frac{0.75 \times 2000}{215} = 7$$

Consider minimum load/maximum eccentricity:

From BS 5628, Table 7, with SR = 7 and $e_x = 0.181t$, $\beta = 0.7$ (see Table 5.15).

Then, from $n_w = \dfrac{\beta t f_k}{\gamma_m}$

Required characteristic compressive strength of masonry, f_k:

$$
\begin{aligned}
f_k &= \frac{\text{min. design load} \times \gamma_m}{\beta \times t \times 1 \text{ metre length}} \\
&= \frac{86.22 \times 10^3 \times 2.5}{0.7 \times 215 \times 1000} \\
&= 1.43 \text{ N/mm}^2
\end{aligned}
$$

Now consider maximum load/minimum eccentricity:

From BS 5628, Table 7, with SR = 7 and $e_x = 0.086t$, $\beta = 0.897$ (see Table 5.15).

Hence, required characteristic compressive strength of masonry, f_k:

$$
\begin{aligned}
f_k &= \frac{\text{max. design load} \times \gamma_m}{\beta \times t \times 1 \text{ metre length}} \\
&= \frac{182.12 \times 10^3 \times 2.5}{0.897 \times 215 \times 1000} \\
&= 2.36 \text{ N/mm}^2
\end{aligned}
$$

Select, from BS 5628, Table 2(a), for a required minimum characteristic compressive strength, masonry constructed of bricks with a minimum crushing strength of 35 N/mm^2 set in a designation (i) mortar (see Table 5.4). Note that this specification provides a characteristic compressive strength far in excess of that designed. However, for the practical considerations of abrasion from the fuel and durability in use, the specification quoted is considered to be the minimum acceptable.

EXAMPLE 3: Free-standing walls

The boundary garden wall shown in Figure 11.21 is to be constructed in solid clay brickwork. The wall

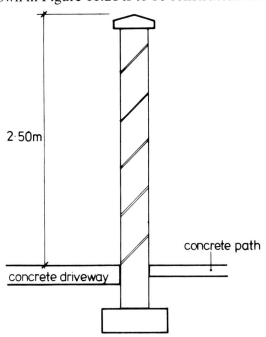

Figure 11.21 Typical section through boundary wall

is of considerable length with straight movement joints placed at 10 m centres. Close quality control of materials and workmanship can be expected, and the client is particularly interested to know the effect of introducing a felt dpc at ground level. The characteristic wind loading on the wall, W_k, may be taken as 0.6 kN/m^2 for the purpose of this example and density of the masonry will be assumed to be 20 kN/m^3.

The critical loading condition for which the wall must be designed is not that of axial loading but rather that of lateral loading due to wind pressures. The wall will act as a vertical cantilever and the loading and bending moment diagrams are shown in Figure 11.22.

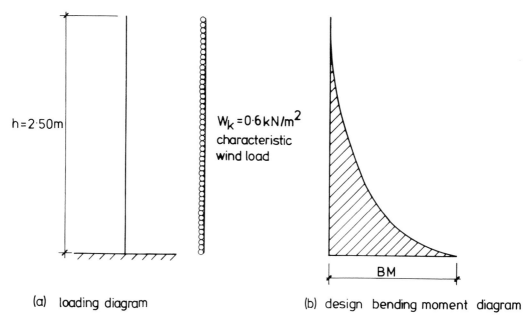

$h=2\cdot50\,\text{m}$

$W_k = 0\cdot6\,\text{kN/m}^2$
characteristic
wind load

BM

(a) loading diagram

(b) design bending moment diagram

Figure 11.22

Design bending moment at base of wall:

$$\text{BM} = W_k \times \gamma_f \times \frac{h^2}{2}$$

where

W_k = characteristic wind load
γ_f = partial safety factor for loads
h = clear height of wall above horizontal support.

Therefore,

$$\text{BM} = \frac{0.6 \times 1.4 \times 2.5^2}{2}$$

$$= 2.625 \text{ kN·m}$$

The resistance to this moment is provided by either:

(a) the tensile resistance of the wall at its base, or
(b) the gravitational stability of the wall where a felt dpc at ground level has eliminated its tensile resistance.

Case (a) will be considered first and the client's interest in the effect of introducing the felt dpc at ground level will be investigated afterwards.

Case (a): Wall capable of resisting tensile stresses at ground level (note, this can be achieved with certain engineering bricks coursed in at ground level as discussed in previous chapters). The loading condition for this example is that of low axial load to comparatively high bending moment. In this case, there is little doubt that the formula for the design moment of resistance should be:

$$\text{design MR} = \left(\frac{f_{kx}}{\gamma_m} + g_d \right) Z$$

where

f_{kx} = characteristic flexural tensile strength at the critical section
γ_m = partial safety factor for materials
g_d = design vertical load per unit area (axial compressive stress)
Z = elastic section modulus of wall.

In this particular example, the flexural compressive stress is not critical and will not be checked. However, the inexperienced designer should check the compressive stresses in all such examples in order to establish when such checks need to be carried out (this will be dealt with in greater detail in a later example). Once again, the design process is one of trial and error and a section must be selected and checked for adequacy.

Try 330 mm thick solid wall (fully cross-bonded) constructed of bricks with a characteristic compressive strength of 35 N/mm² and a water absorption of greater than 12% set in a designation (iii) mortar.

Then, for these materials:

f_{kx} = 0.30 N/mm², from BS 5628, Table 3

γ_m = 2.5 special/special

$$g_d = \frac{\gamma_f \times \text{density} \times \text{thickness} \times \text{height}}{\text{thickness}}$$

$$= \frac{0.9 \times 20 \times 0.33 \times 2.5}{0.33 \times 10^3} = 0.045 \text{ N/mm}^2$$

$$Z = \frac{bt^2}{6} = \frac{1000 \times 330^2}{6} = 18.15 \times 10^6 \text{ mm}^3$$

Therefore

$$\text{design moment of resistance} = \left(\frac{f_{kx}}{\gamma_m} + g_d \right) Z$$

$$= \left(\frac{0.3}{2.5} + 0.045 \right) \times 18.15 \times 10^6$$

$$= 2.995 \text{ kN·m}$$

Hence, the wall section is adequate provided the flexural tensile resistance required at ground level can be relied upon.

Case (b): Check the effect of introducing a felt dpc at ground level – thus the wall should be designed as a gravity structure. The design bending moment remains unaltered.

Try 440 mm thick solid wall constructed of the same materials as for Case (a). From design experience, it is known that a 330 mm thick solid wall is theoretically inadequate although, in practice, many such walls are constructed but can blow over in exceptionally high gales. The design moment of resistance of the wall is derived from the gravitational stability of its base and will be termed 'base stability moment'. The calculation of the 'base stability moment' MR_s will be based upon the rectangular stress block shown in Figure 11.23 which is derived from BS 5628, Appendix B, where:

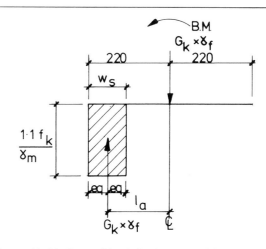

BM = design bending moment
G_k = characteristic dead loading
γ_f = partial safety factor for loads
f_k = characteristic compressive strength of masonry
γ_m = partial safety factor for materials
l_a = lever arm
w_s = width of stressed area.

Figure 11.23 Stress block for base stability moment

The essence of this design approach is that the minimum width of wall is fully stressed to create the maximum lever arm about which the dead weight of the wall rotates to achieve the maximum gravitational stability moment (MR_s) for the materials considered, hence:

$$MR_s = \gamma_f \times G_k \times l_a \text{ per m length of wall}$$
$$\gamma_f \times G_k = 0.9 \times 20 \times 0.44 \times 2.5$$
$$= 19.80 \text{ kN}$$

In order to calculate the lever arm, the minimum width of stressed area, w_s, must first be calculated. This is calculated from a simple stress = load/area consideration, hence (see Figure 11.23):

$$\frac{1.1 f_k}{\gamma_m} = \frac{\gamma_f G_k}{w_s \times 1 \text{ m length of wall}}$$

Therefore

$$w_s = \frac{19.8 \times 10^3 \times 2.5}{1.1 \times 8.5 \times 1000}$$

$$= 5.3 \text{ mm}$$

Then

$$l_a = 220 - \frac{5.3}{2}$$

$$= 217.35 \text{ mm}$$

$$MR_s = \frac{19.8 \times 217.35}{10^3}$$

$$= 4.304 \text{ kN·m}$$

Thus the specified 440 mm thick wall is adequate and it is likely that a lower strength of brick and/or mortar would be adequate if checked.

EXAMPLE 4: Collar-jointed walls

Another free-standing garden wall is to be designed, similar to that of the previous example, but reduced in height to 1.5 m. The client requires that both faces of the wall should be constructed in stretcher bond. The wall will be designed both with and without ties connecting the two stretcher bond leaves together to investigate the effect of their inclusion.

As for the previous example, the wall will act as a vertical cantilever and the design bending moment

$$BM = W_k \times \gamma_f \times \frac{h^2}{2} = 0.6 \times 1.4 \times \frac{1.5^2}{2}$$

$$= 0.945 \text{ kN·m}$$

The code definition of a collar-jointed wall is: 'two parallel single-leaf walls, with a space between not exceeding 25 mm, filled solidly with mortar and so tied together as to result in common action under load'. To ensure common action under load for a collar-jointed wall with lateral loading, the ties need to be designed to resist the shear force between the two leaves which is tending to cause 'slip' between these faces as shown in Figure 11.24.

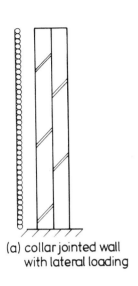
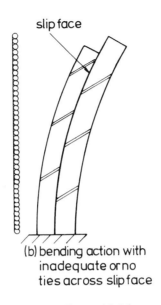
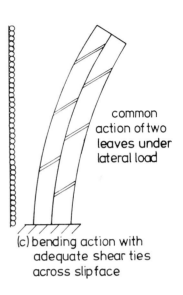

(a) collar jointed wall with lateral loading

(b) bending action with inadequate or no ties across slip face

(c) bending action with adequate shear ties across slip face

Figure 11.24

It will be assumed that the wall is capable of resisting flexural tensile stresses for its full height (i.e. no felt dpc) hence:

$f_{kx} = 0.3 \text{ N/mm}^2$ (BS 5628, Table 3)

$\gamma_m = 2.5$ special/special

$$g_d = \frac{\gamma_f \times \text{density} \times \text{thickness} \times \text{height}}{\text{area}}$$

$$= \frac{0.9 \times 20 \times 0.215 \times 1.5}{0.215 \times 1 \times 10^3} = 0.027 \text{ N/mm}^2$$

$$Z = \text{section modulus} \left(\frac{bt^2}{6}\right) = \frac{0.215^2 \times 1}{6} = 7.704 \times 10^6 \text{ mm}^3$$

Therefore

$$\text{design MR} = \left(\frac{0.3}{2.5} + 0.027\right) \times 7.704 \times 10^6$$

$$= 1.132 \text{ kN·m}$$

Thus the 215 mm thick wall specified is adequate provided adequate ties are included to ensure full interaction between the two leaves – which is what the design will now proceed to check.

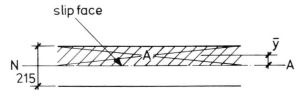

Figure 11.25 Plan on metre length of wall

Figure 11.25 shows a plan on a metre length of the wall showing the neutral axis coinciding with the slip face of the wall section.

Shear stress for which the ties must be designed occurs on the slip face between the two leaves which is the neutral axis of the full 215 thick section. Shear stress, v_h, is given by the equation:

$$v_h = \frac{VA\bar{y}}{Ib}$$

V = horizontal shear force at point of maximum rate of change in BM
A = area of shaded portion
$\bar{y}$ = distance from N/A to centroid of shaded area
I = moment of inertia of section
b = 1 metre length of wall
NA = neutral axis of full 215 mm thick wall section.

For this example:

$V = W_k \times \gamma_f \times h$ $\quad = 0.6 \times 1.4 \times 1.5 = \quad 1.26 \text{ kN}$

$A = 1000 \times 215 \times 0.5$ $\quad\quad\quad\quad\quad = 107.5 \times 10^3 \text{ mm}^2$

$\bar{y} = \dfrac{215}{4}$ $\quad\quad\quad\quad\quad\quad\quad\quad = 53.75 \text{ mm}$

$I = \dfrac{bt^3}{12}$ $\quad\quad = \dfrac{1000 \times 215^3}{12} \quad = \quad 0.828 \times 10^9 \text{ mm}^4$

Therefore, design shear stress:

$$v_h = \frac{VA\bar{y}}{Ib}$$

$$= \frac{1.26 \times 10^3 \times 107.5 \times 10^3 \times 53.75}{0.828 \times 10^9 \times 1000}$$

$$= 0.0088 \text{ N/mm}^2 = 8.8 \text{ kN/m}^2$$

This shear stress is to be resisted by double-triangle type cavity wall ties built into the bed joints, and the spacing of the ties will be calculated based on the shear forces given in BS 5628, Table 8 (see Table 6.5):

Characteristic shear force per double-triangle tie = 3.0 kN

Design shear force $= \dfrac{3.0}{\gamma_m} = \dfrac{3.0}{3.0} = 1.0 \text{ kN}$

Therefore, the number of ties required to resist the shear on the neutral axis = 8.8/1.0 = 8.8, and a minimum of 9 double-triangle ties per square metre of wall elevation should be used.

Now consider the same wall constructed in two stretcher bond leaves, but without any ties connecting them together. The design will, therefore, be based upon each leaf acting independently. The design BM remains as before = 0.945 kN·m.

$$\text{Design moment of resistance} = 2 \text{ leaves} \times \left(\frac{f_{kx}}{\gamma_m} + g_d\right) Z$$

where

$Z \text{ per leaf} = \dfrac{bt^2}{6} = \dfrac{1000 \times 102.5^2}{6}$

$\quad\quad\quad\quad = 1.751 \times 10^6 \text{ mm}^3$

Therefore

$$\text{design MR} = 2 \left(\frac{0.3}{\gamma_m} + 0.027 \right) \times 1.751 \times 10^6$$

$$= 0.515 \text{ kN·m}$$

This is less than the design bending moment of 0.945 kN·m and the section will crack. The wall should now be checked as a cracked section using a rectangular stress block to calculate the stability moment of resistance as demonstrated in an earlier example (see Figure 11.23.)

The stability moment stress block for this example is shown in Figure 11.26 in which the total design bending moment is shared equally between the two leaves, where:

$$\frac{M_A}{2} = \frac{0.945}{2} = 0.4725 \text{ kN·m}$$

$$G_k \times \gamma_f = \text{density} \times \text{thickness} \times \text{height} \times \gamma_f$$

$$= 20 \times 0.1025 \times 1.5 \times 0.9$$

$$= 2.767 \text{ kN/m per leaf}$$

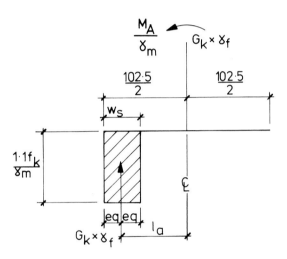

Figure 11.26 Stress block per leaf for base stability moment

Thus as demonstrated in an earlier example:

$$\frac{1.1 f_k}{\gamma_m} = \frac{G_k \times \gamma_f}{w_s \times 1 \text{ m length of wall}} \qquad \left(\text{stress} = \frac{\text{load}}{\text{area}} \right)$$

Therefore

$$w_s = \frac{G_k \times \gamma_f \times \gamma_m}{1.1 f_k \times 1 \text{ m length of wall}}$$

$$= \frac{2.767 \times 10^3 \times 2.5}{1.1 \times 8.5 \times 1000}$$

$$= 0.74 \text{ mm}$$

Then, lever arm, l_a

$$l_a = \frac{\text{Wall thickness}}{2} - \frac{w_s}{2}$$

$$= \frac{102.5}{2} - \frac{0.74}{2}$$

$$= 50.88 \text{ mm}$$

and stability moment:

$$MR_s = G_k \times \gamma_f \times l_a$$

$$= \frac{2.767 \times 50.88}{10^3}$$

$$= 0.141 \text{ kN·m}$$

The two leaves act simultaneously and the total stability moment is therefore twice that calculated above $= 0.141 \times 2 = 0.282$ kN·m. This method of construction is therefore unsuitable for the design condition, which demonstrates that it is good practice to include wall ties to make the masonry more structurally efficient.

EXAMPLE 5: Solid concrete block walls

A viewing gallery to a sports centre is to have a 200 mm thick balustrade wall constructed in solid concrete blockwork. The designer is required to check whether the wall is stable or whether any additional supports are required. A section through the wall is shown in Figure 11.27. It consists of 200 mm thick concrete blockwork constructed with blocks of 7.0 N/mm^2 crushing strength (no void, solid blocks) set in designation (iii) mortar.

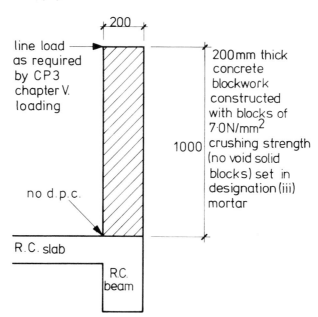

Figure 11.27 Section through balustrade wall

Characteristic superimposed
line load at top of wall, Q_k $\quad = 0.74$ kN/m (CP 3: Chapter V)
Design superimposed loading $\quad = Q_k \times \gamma_f$
$\quad = 0.74 \times 1.6 = 1.184$ kN/m
Design bending moment $\quad = 1.184 \times 1 \quad = 1.184$ kN·m (based on cantilever action of wall)
Characteristic dead loading, G_k $= $ density $\times$ height $\times$ area
$\quad = 15 \times 1 \times 0.2 = 3.0$ kN/m
Minimum design dead loading $\quad = G_k \times \gamma_f$
$\quad = 3.0 \times 0.9 \quad = 2.7$ kN/m

$$\text{Design moment of resistance} = \left(\frac{f_{kx}}{\gamma_m} \times g_d \right) Z$$

where

$f_{kx} = 0.25$ N/mm^2 (BS 5628, Table 3)

$\gamma_m = 2.5$ special/special

$$g_d = \frac{G_k \times \gamma_f}{\text{area}} = \frac{2.7}{0.2 \times 10^3}$$

$$= 0.0135 \text{ N/mm}^2$$

$$Z = \frac{200^2 \times 1000}{6} = 6.67 \times 10^6 \text{ mm}^3$$

Therefore

$$\text{design MR} = \left(\frac{0.25}{2.5} + 0.0135 \right) \times 6.67 \times 10^6$$

$$= 0.76 \text{ kN·m}$$

This is inadequate because the design bending moment has been calculated as 1·184 kN·m. Therefore, stiffening piers, post-tensioning or some other form of additional support is required. The two possibilities mentioned, stiffening piers and post-tensioning, are dealt with later.

EXAMPLE 6: External cavity wall

A three-storey building is shown in Figure 11.28. The gable wall is to be designed for the loadings given, assuming that the wall is constructed with both leaves in brickwork.

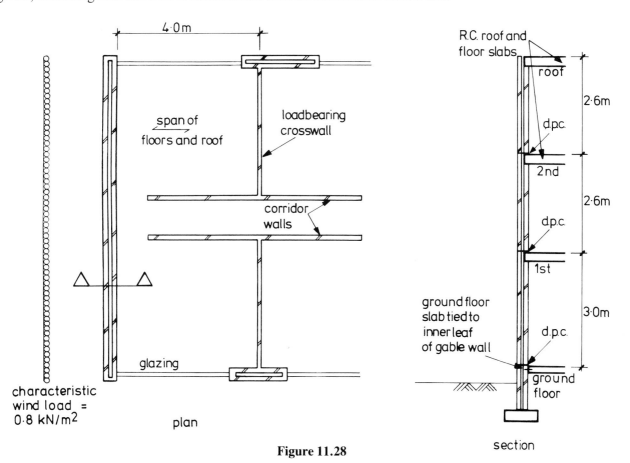

Figure 11.28

Characteristic wind loading	$= 0.8 \text{ kN/m}^2$
Roof, characteristic dead load	$= 6.5 \text{ kN/m}^2$
Roof, characteristic super load	$= 0.75 \text{ kN/m}^2$
Floors, characteristic dead load	$= 5.0 \text{ kN/m}^2$
Floors, characteristic super load	$= 1.5 \text{ kN/m}^2$
Density of masonry	$= 20 \text{ kN/m}^3$

The ground floor slab is tied to, but not supported on, the inner leaf of the gable wall to be designed.

Design wall in lowest storey:

Characteristic dead loads, G_k, (inner leaf only):

$$\text{roof} = 6.5 \times \frac{4}{2} \qquad\qquad = 13.00$$

$$2 \text{ floors} = 2 \times 5.0 \times \frac{4}{2} \qquad = 20.00$$

$$\text{ow wall} = 0.1025 \times 20 \times 8.2 = \underline{16.80}$$
$$= \overline{49.80} \ \text{kN/m}$$

Characteristic superimposed loads, Q_k, (inner leaf only):

$$\text{roof} = 0.75 \times \frac{4}{2} \qquad = 1.50$$

$$2 \text{ floors} = 2 \times 1.5 \times \frac{4}{2} = \underline{6.00}$$
$$= \overline{7.50} \ \text{kN/m}$$

Characteristic dead load, G_k, (outer leaf only):

$$\text{ow wall} = 0.1025 \times 20 \times 8.2 = 16.80 \ \text{kN/m}$$

Minimum design load (inner leaf):

$$\gamma_f \times G_k = 0.9 \times 49.80 = 44.80 \ \text{kN/m}$$

Minimum design load (outer leaf):

$$\gamma_f \times G_k = 0.9 \times 16.8 = 15.12 \ \text{kN/m}$$

Maximum design load (inner leaf):

$$(\gamma_f \times G_k) + (\gamma_f \times Q_k) = (1.4 \times 49.8) + (1.6 \times 7.5)$$
$$= 81.72 \ \text{kN/m}$$

Design wind loading:

$$\gamma_f \times W_k = 1.4 \times 0.8 = 1.12 \ \text{kN/m}^2$$

Note: The effect of wind uplift on the roof must be taken into account when calculating the minimum design load on the wall. In order to assess the moments due to wind loadings in the lowest storey of wall, the end support conditions must be considered.

The rc floor slab built in at first floor level in conjunction with the wall panel extending into the second storey above can be considered to constitute a fully fixed or continuous support condition. The lower storey height of wall extends down below the ground floor slab to the foundation. The ground floor slab, as well as the ground beneath it, will prevent inward movement of the wall. However, the earth fill against the wall externally (filling the foundation excavation) will not necessarily be adequately compacted and is considered unreliable as a support to prevent outward movement of the wall. If the span of the lower storey height of the wall is taken as between ground and first floor slabs, it will be necessary to introduce ties built into the bed courses and concreted into the ground floor slab. The ties should be capable of resisting the reaction at this point due to wind suction forces on the wall panel. Having provided ties at ground slab level and built in the slab at first floor level, it is considered reasonable to treat both supports as continuous and the bending moment diagram is shown in Figure 11.29.

Three critical sections require investigation:

(a) at the bottom of the wall panel where the inclusion of the dpc requires the section to be checked as a 'cracked section' where the moment of resistance is provided by the stability moment,
(b) at mid-height of the wall panel where the flexural tensile and compressive stresses as well as the axial compressive stresses should be checked,
(c) at the top of the wall panel where the dpc above the level considered again requires the section to be checked as a 'cracked section'. However, the moment of resistance at the upper support will be less than at the lower support, due to the reduced dead load available and, therefore, as the applied bending moment at these two levels is the same, only the upper support position needs to be checked for flexural tensile stresses.

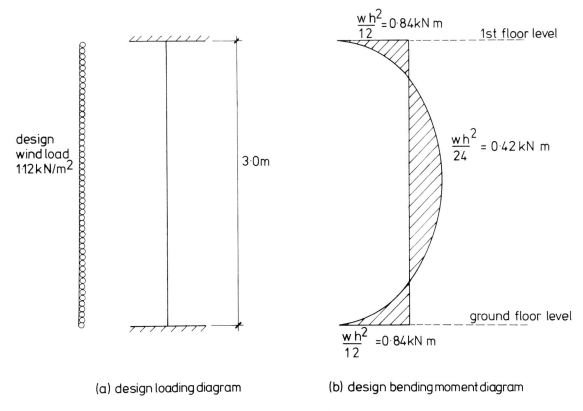

(a) design loading diagram   (b) design bending moment diagram

Figure 11.29

The design resistance moment at mid-height of the wall panel will be calculated from the formula:

$$\text{design MR} = \left(\frac{f_{kx}}{\gamma_m} + g_d\right) Z$$

rather than from the effective eccentricity. This is because the effective eccentricity method is considered more appropriate for solid, bonded walls where the effective thickness of the wall, for calculating the slenderness ratio, is the actual thickness. In cavity wall construction, the effective thickness of the wall is two-thirds the sum of the thickness of the two leaves. For this reason, the authors consider the formula $(f_{kx}/\gamma_m + g_d)Z$ a more realistic analysis of resistance moment, provided consideration is given to the flexural compressive stresses as well as the flexural tensile stresses, under design loading. The formula $(f_{kx}/\gamma_m + g_d)Z$ is based purely on flexural tensile resistance and a suggested method is given later in this example for considering the flexural compressive stresses.

The specification for the bricks and mortar for the two leaves is given as:

Outer leaf: facing bricks with a water absorption between 7 and 12% and a crushing strength of 35 N/mm² set in a designation (iii) mortar,

Inner leaf: commons with a water absorption in excess of 12% and a crushing strength of 20 N/mm² set in a designation (iii) mortar.

Check resistance moment at mid-height of wall panel:

$$\text{Resistance moment (per leaf)} = \left(\frac{f_{kx}}{\gamma_m} + g_d\right) Z$$

$$\text{Total resistance moment} = MR \text{ outer leaf} + MR \text{ inner leaf}$$

Outer leaf:

$f_{kx} = 0.4 \text{ N/mm}^2$ (BS 5628, Table 3)

$\gamma_m = 2.5$ special/special

$g_d = \dfrac{15.12}{0.1025} = 0.147 \text{ N/mm}^2$

$Z = \dfrac{1 \times 0.1025^2}{6} = 1.751 \times 10^6 \text{ mm}^3$

Inner leaf:

$f_{kx} = 0.3 \text{ N/mm}^2$ (BS 5628, Table 3)

$\gamma_m = 2.5$ special/special

$g_d = \dfrac{49.8}{0.1025} = 0.486 \text{ N/mm}^2$

$Z = \dfrac{1 \times 0.1025^2}{6} = 1.751 \times 10^6 \text{ mm}^3$

Total resistance moment:

$$= \left(\frac{0.4}{2.5} + 0.147\right) \times 1.751 \times 10^6 + \left(\frac{0.3}{2.5} + 0.486\right) \times 1.751 \times 10^6$$

$$= 0.537 + 1.061$$

$$= 1.598 \text{ kN·m}$$

The section is adequate as the resistance moment is greater than the applied bending moment of 0.42 kN·m.

It should be noted that the total resistance moment has been calculated by adding the resistance moments of the individual leaves of the cavity wall. BS 5628 permits this if wall ties which are capable of transferring the wind forces across the cavity are provided. Now check the 'cracked section' resistance moment at the upper support position.

The minimum dead loads in each leaf at this level must first be calculated as follows:

Characteristic dead loads (inner leaf), G_k:

$$\text{roof} = 6.5 \times \frac{4}{2} \qquad\qquad = 13.00$$

$$\text{floor} = 5.0 \times \frac{4}{2} \qquad\qquad = 10.00$$

$$\text{ow wall} = 0.1025 \times 20 \times 5.2 = \underline{10.66}$$

$$= \overline{33.66} \text{ kN/m}$$

Characteristic dead load (outer leaf), G_k:

$$\text{ow wall} = 0.1025 \times 20 \times 5.2 = 10.66 \text{ kN/m}$$

Minimum design load (inner leaf):

$$\gamma_f \times G_k = 0.9 \times 33.66 = 30.29 \text{ kN/m}$$

Minimum design load (outer leaf):

$$\gamma_f \times G_k = 0.9 \times 10.66 = 9.59 \text{ kN/m}$$

The design moment of resistance is now calculated from the stability moments produced by these calculated minimum design loads. The stress block producing the resistance moment for each leaf is identical to that shown in Figure 11.26, hence:

$$\frac{1.1f_k}{\gamma_m} = \frac{\text{min. design load}}{w_s \times 1 \text{ m length}}$$

therefore

$$w_s = \frac{\gamma_m \times \text{min. design load}}{1.1f_k \times 1 \text{ m length}}$$

Outer leaf:

$$w_s = \frac{2.5 \times 9.59 \times 10^3}{1.1 \times 8.5 \times 1000}$$

and

$$l_a = \frac{\text{wall thickness}}{2} - \frac{w_s}{2}$$

$$= \frac{102.5}{2} - \frac{2.56}{2} = 49.97 \text{ mm}$$

Then design resistance moment for outer leaf

$$= \frac{49.97}{10^3} \times 9.59 = 0.479 \text{ kN·m}$$

Inner leaf:

$$w_s = \frac{2.5 \times 30.29 \times 10^3}{1.1 \times 5.8 \times 1000} = 11.87 \text{ mm}$$

and

$$l_a = \frac{102.5}{2} - \frac{11.87}{2} = 45.32 \text{ mm}$$

Then design moment of resistance for inner leaf

$$= \frac{45.32}{10^3} \times 30.29$$

$$= 1.373 \text{ kN·m}$$

$$\text{Total resistance moment} = \text{MR outer leaf} + \text{MR inner leaf}$$
$$= 0.479 + 1.373$$
$$= 1.852 \text{ kN·m}$$

The section is adequate as the resistance moment is greater than the applied bending moment of 0.84 kN·m.

As described earlier, there is no need to calculate the resistance moment at the lower support as this will be larger than the 1.852 kN·m calculated above, due to the additional dead loading. However, the flexural compressive stresses at this lower level may require investigation.

The maximum axial design loading for this storey height of wall should now be calculated in accordance with the methods described in Chapter 10. The flexural compressive stresses should be calculated under dead plus superimposed plus wind loading conditions. The following design method is considered a reasonable approach – although this aspect is not covered in detail in BS 5628.

Figure 11.30 shows the stress block across a wall subject to pure axial loading and with no eccentricity of that load. The maximum compressive stress allowable in the wall section is limited by the walls tendency to buckle, hence the inclusion of the capacity reduction factor, β.

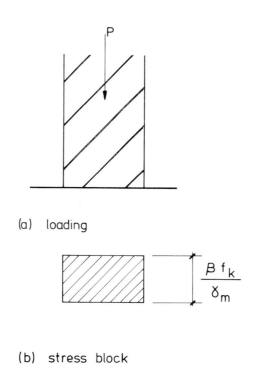

(a) loading

(b) stress block

Figure 11.30 Stress block under axial loading

Figure 11.31 shows the stress block across a wall subject to purely flexural loading conditions.

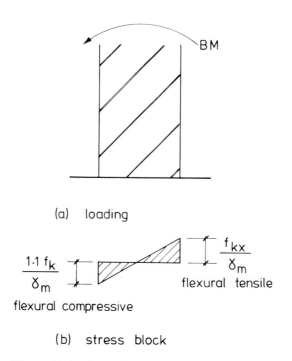

(a) loading

(b) stress block

Figure 11.31 Stress block under flexural loading

Two limiting conditions can be said to apply to the maximum allowable stresses which are given as f_{kx}/γ_m for flexural tensile stresses and $1.1f_k/\gamma_m$ for flexural compressive stresses, the latter having been adopted in Appendix B to BS 5628 for the derivation of β. The flexural tensile stresses will invariably be the limiting factor under such a loading condition. However, if the axial compressive stresses already in the wall are added to the flexural compressive stresses, this may produce a more critical design condition. Consideration must be given to the need for limiting the flexural compressive stresses due to the possibility of buckling of the section under the application of such stress. Buckling of a section, due to flexural compressive stresses, will occur perpendicular to the direction of application of the bending. Two common sections are shown in Figure 11.32.

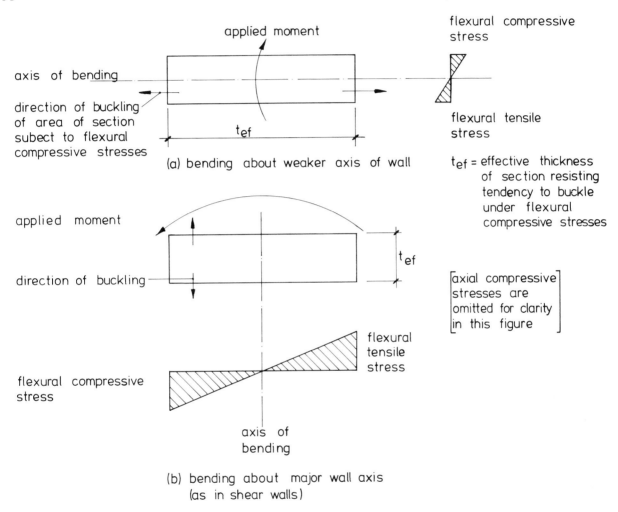

Figure 11.32

Clearly, the shear wall section shown in (b) is the more critical with respect to buckling due to flexural compressive stresses, although there is no guidance given in BS 5628 on a design method to take account of such a buckling tendency. The authors consider that the following design method provides a safe and practical solution:

(a) In the first instance, the wall should be checked for maximum axial loading only, using the design principles explained in Chapters 5 and 10. The capacity reduction factor, β, applicable to this stage of the design, should be derived from the maximum slenderness ratio. The maximum allowable stress under this loading condition is $\beta f_k/\gamma_m$.

(b) The additional compressive stress resulting from the bending due to lateral loading is then considered, and the maximum allowable combined compressive stress is $1.1f_k \times \beta/\gamma_m$, in which a 10% increase has been applied to the flexural aspect of the stress in a similar manner to Appendix B of BS 5628. The capacity reduction factor, β, should be derived from the slenderness ratio which incorporates the effective thickness appropriate to the direction of buckling tendency (i.e. perpendicular to the direction of application of the bending) as shown in Figure 11.32.

The stress blocks relevant to the two stages of the design are shown in Figure 11.33.

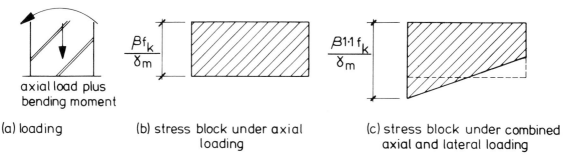

(a) loading

(b) stress block under axial loading

(c) stress block under combined axial and lateral loading

Figure 11.33

Now check the upper storey wall panel for the same gable wall. First, consider the likely action of this wall panel and the moments for which it must be designed.

At second floor level, continuity will exist to some extent as was demonstrated for the lower wall panel. At roof level, however, the amount of continuity is contentious. Whilst the inner leaf will derive some continuity from the dead loading from the roof slab, the outer leaf has no continuity and, for simplicity, this support will be treated as pin-jointed. The upper wall panel will therefore act as a propped cantilever, and the two critical sections will occur at second floor level and $3/8h$ down from roof level where the maximum wall moment occurs for a propped cantilever.

The minimum design loads must be calculated at the two critical levels.

At second floor level:

Characteristic dead loads, G_k, (inner leaf only):

$$\text{roof} = 6.5 \times \frac{4}{2} \qquad\qquad = 13.00$$

$$\text{ow wall} = 0.1025 \times 20 \times 2.6 = \underline{5.33}$$
$$= \overline{18.33} \text{ kN/m}$$

Characteristic dead loads, G_k, (outer leaf only):

$$\text{ow wall} = 0.1025 \times 20 \times 2.6 = 5.33 \text{ kN/m}$$

At $\dfrac{3}{8}h$ down from roof:

Characteristic dead loads, G_k, (inner leaf only):

$$\text{roof} = 6.5 \times \frac{4}{2} \qquad\qquad = 13.00$$

$$\text{ow wall} = 0.1025 \times 20 \times \frac{2.6 \times 3}{8} = \underline{2.00}$$
$$= \overline{15.00} \text{ kN/m}$$

Characteristic dead load, G_k, (outer leaf only):

$$\text{ow wall} = 0.1025 \times 20 \times \frac{2.6 \times 3}{8} = 2.00 \text{ kN/m}$$

Minimum design loads:

$$\text{inner leaf: 2nd floor} = 0.9 \times 18.33 = 16.50 \text{ kN/m}$$

$$\tfrac{3}{8}h \text{ level} = 0.9 \times 15.00 = 13.50 \text{ kN/m}$$

$$\text{outer leaf: 2nd floor} = 0.9 \times 5.33 = 4.8 \text{ kN/m}$$

$$\frac{3}{8}h \text{ level} = 0.9 \times 2.0 = 1.8 \text{ kN/m}$$

The loading conditions and design bending moment diagram are shown in Figure 11.34. The design wind load remains the same as for the lower wall panel.

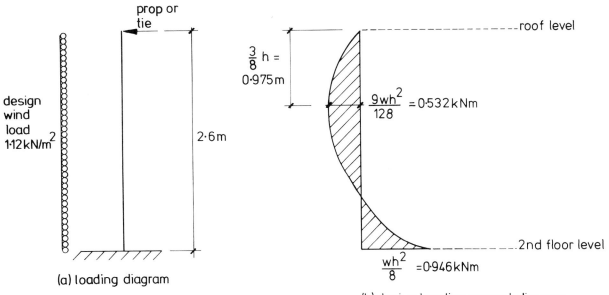

(a) loading diagram

(b) design bending moment diagram

Figure 11.34

The moment of resistance at second floor level should be designed as a 'cracked section' due to the inclusion of the dpc and, once again, the stress block producing the resistance moment for each leaf is identical to that shown in Figure 11.26, hence:

$$\frac{\beta \times 1.1 f_k}{\gamma_m} = \frac{\text{min. design load}}{w_s \times 1 \text{ m length}} \text{ (but } \beta = 1.0)$$

therefore

$$w_s = \frac{\gamma_m \times \text{min. design load}}{1.1 f_k \times 1 \text{ m length}}$$

Outer leaf:

$$w_s = \frac{2.5 \times 4.8 \times 10^3}{1.1 \times 8.5 \times 1000} = 1.28 \text{ mm}$$

and

$$l_a = \frac{\text{wall thickness}}{2} - \frac{w_s}{2}$$

$$= \frac{102.5}{2} - \frac{1.28}{2} = 50.60 \text{ mm}$$

Then design moment of resistance (for outer leaf only)

$$= 4.8 \times 0.0506$$
$$= 0.243 \text{ kN·m}$$

Inner leaf:

$$w_s = \frac{2.5 \times 16.5 \times 10^3}{1.1 \times 5.8 \times 1000} = 6.5 \text{ mm}$$

and
$$l_a = \frac{102.5}{2} - \frac{6.5}{2} = 48.0 \text{ mm}$$

Then design moment of resistance (for inner leaf only)
$$= 16.5 \times 0.048$$
$$= 0.792 \text{ kN·m}$$

$$\text{Total resistance moment} = \text{MR outer leaf} + \text{MR inner leaf}$$
$$= 0.243 + 0.792$$
$$= 1.035 \text{ kN·m}$$

This section is adequate as the resistance moment is greater than the applied bending moment of 0.946 kN·m.

Now check the resistance moment at the position of maximum design moment in the wall height (i.e. $\frac{3}{8}h$ from the top of the wall) again using the formula:

$$\text{MR} = \left(\frac{f_{kx}}{\gamma_m} + g_d \right) Z$$

$$\text{Total resistance moment} = \text{MR outer leaf} + \text{MR inner leaf}$$

Outer leaf:

$f_{kx} = 0.4 \text{ N/mm}^2$ (BS 5628, Table 3)

$\gamma_m = 2.5$ special/special

$g_d = \dfrac{1.8}{0.1025} = 0.017 \text{ N/mm}^2$

$Z = \dfrac{1 \times 0.1025^2}{6} = 1.751 \times 10^6 \text{ mm}^3$

Inner leaf:

$f_{kx} = 0.3 \text{ N/mm}^2$ (BS 5628, Table 3)

$\gamma_m = 2.5$ special/special

$g_d = \dfrac{13.50}{0.1025} = 0.132 \text{ N/mm}^2$

$Z = \dfrac{1 \times 0.1025^2}{6} = 1.751 \times 10^6 \text{ mm}^3$

Then, total resistance moment:

$$= \left(\frac{0.4}{2.5} + 0.017 \right) \times 1.751 \times 10^6 + \left(\frac{0.3}{2.5} + 0.132 \right) \times 1.751 \times 10^6$$
$$= 0.31 + 0.44$$
$$= 0.75 \text{ kN·m}$$

The section is adequate as the resistance moment is greater than the applied bending moment of 0.532 kN·m. The compressive stresses should be checked using the method suggested earlier but, by inspection, these will also be acceptable.

EXAMPLE 7: Double-leaf cavity panel wall

It is proposed to construct a steel framed industrial building with brick cladding panels up to a height of 4.0 m above ground level. The steel stanchions are to be spaced at 5.0 m centres and the steel frame has been designed to resist all the wind loading from the cladding and wall panels. Check the wall panels, under the characteristic wind load of 1.05 kN/m², for the masonry specification given below. A felt dpc is to be included at the base of the wall and the construction details of a typical wall panel are shown in Figure 11.35, γ_m can be taken as 2.5 for special/special controls.

Masonry specification:

Inner leaf: solid concrete blocks with a crushing strength of 7 N/mm² set in a designation (iii) mortar, overall density = 15 kN/m³.

Outer leaf: facing bricks with a crushing strength of 27.5 N/mm² and a water absorption of 9% set in a designation (iii) mortar, overall density = 20 kN/m³.

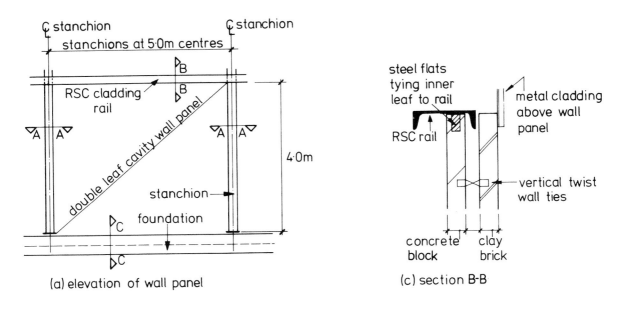

(a) elevation of wall panel   (c) section B-B

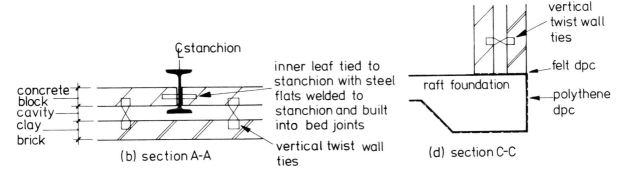

(b) section A-A   (d) section C-C

Figure 11.35

The design bending moment coefficients given in BS 5628, Table 9, will be used for this example to demonstrate their use (see Table 6.4). These coefficients are applicable only within certain dimensional proportions of the panel and this is the first check to be carried out.

Limiting dimensions in accordance with BS 5628, clause 36.3 (b) (2):

height × length = 4000 × 5000 $= 20 \times 10^6$

$$2025 \times t_{ef}^2 \quad = 2025 \times \left[\frac{2}{3}(100 + 102.5)\right]^2 = 36.9 \times 10^6$$

Therefore, height × length is less than $2025\, t_{ef}^2$ which complies with the requirements of Table 9.

Also check, $50 \times t_{ef} = 50 \times \left[\dfrac{2}{3}(100 + 102.5) \right] = 6750$

Maximum dimension of panel is 5000 which is less than $50 \times t_{ef}$ as also required for Table 9. The design can therefore be based on the bending moment coefficients from Table 9.

The table incorporates twelve different combinations of support conditions and, in order to decide which case is applicable, an assessment of the support condition at each edge of the panel to be designed must be made. The coefficients given in BS 5628, Table 9 were derived from experimental research using test panels, and this must be given due consideration when assessing the applicable support conditions. Certainly, the connection of the wall panel to the stanchions as detailed in Figure 11.35(b) can be taken to provide continuity at these vertical edges. However, the connection to the cladding rail at the top of the panel cannot be regarded as anything more than a simple support, unless the member offering support is capable of withstanding the torsional stresses which would result from a 'fixed end' condition.

The support condition at the base of the wall panel is somewhat more contentious. Whilst even with the inclusion of the dpc, the 'cracked section' at this point can certainly develop some degree of resistance moment (as has been previously demonstrated) it is considered inadvisable to treat this support as continuous. This is because the coefficients were derived from loading test panels and such tests would naturally include the benefit of the inherent resistance moment at the base of the wall panel. It would be virtually impossible to simulate a true simple support at this position for test loading purposes and the expression 'simply supported edge' given in Table 9 should be taken to include the effect of the resistance moment when applied to the base of this wall panel. The edge support conditions are therefore indicated in Figure 11.36 as well as the bending moment locations.

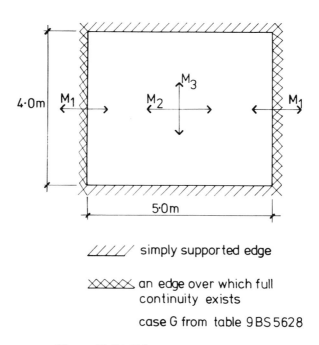

Figure 11.36 Edge support conditions

$$\text{Design wind load} = \gamma_f \times W_k$$
$$= 1.2 \times 1.05$$
$$= 1.26 \text{ kN/m}^2$$

The wall panel is constructed of a concrete block inner leaf and a clay brick outer leaf which have differing orthogonal ratios (see Chapter 6). As such, the design wind load will be shared between the two leaves in the proportions of their design moments of resistance.

Inner leaf:

$$\text{Design resistance moment} = \frac{f_{kx}}{\gamma_m} Z$$

where

f_{kx} = 0.60 N/mm^2, BS 5628, Table 3

γ_m = 2.5 special/special

Z = $\dfrac{1 \times 0.1^2}{6}$ = 1.67 $\times 10^6$ mm^3

Therefore

$$\text{design MR} = \frac{0.6}{2.5} \times 1.67 \times 10^6 = 0.4008 \text{ kN·m}$$

Outer leaf:

$$\text{Design resistance moment} = \frac{f_{kx}}{\gamma_m} Z$$

where

f_{kx} = 1.10 N/mm^2, BS 5628, Table 3

γ_m = 2.5 special/special

Z = $\dfrac{1 \times 0.1025^2}{6}$ = 1.751 $\times 10^6$ mm^3

Therefore

$$\text{design MR} = \frac{1.1}{2.5} \times 1.751 \times 10^6$$

$$= 0.7704 \text{ kN·m}$$

Therefore, the design wind load will be shared in the ratio:

$$\frac{0.4008}{0.4008 + 0.7704} = 34\% \text{ to inner leaf}$$

$$\frac{0.7704}{0.4008 + 0.7704} = 66\% \text{ to outer leaf}$$

Hence:

inner leaf design wind load = 0.34 $\times$ 1.26 = 0.428 kN/m^2
outer leaf design wind load = 0.66 $\times$ 1.26 = 0.832 kN/m^2

From BS 5628, clause 36.4.2:

$$\text{applied design horizontal bending moment} = \alpha \times W_k \times \gamma_f \times L^2$$

where

L = length of panel between stanchion supports
$W_k \gamma_f$ = proportion of design wind load
α = bending moment coefficient from BS 5628, Table 9, case G (see Table 6.4) and is dependent upon the ratio height: length (h/l) = 4000/5000 = 0.8, for this example.

$$\text{Orthogonal ratio, } \mu = \frac{f_{kx \text{ par}} + \left(\dfrac{\text{design dead load} \times \gamma_m}{\text{area}} \right)}{f_{kx \text{ perp}}}$$

in which the flexural strength in the parallel direction has been increased to allow for the stress due to the minimum design vertical load, as provided for in BS 5628, clause 36.4.2.

Now consider the outer leaf only:

$$\mu = 0.4 + \dfrac{\left(\dfrac{20 \times 0.1025 \times 0.9 \times 2.0 \times 10^3}{0.1025 \times 1.0 \times 10^6} \times 2.5\right)}{1.1}$$

$$= 0.445$$

Therefore, from BS 5628, Table 9 (G) (see Table 6.4) by interpolation for $\mu = 0.445$ and $h/L = 0.8$ BM coefficient $\alpha = 0.032$.

Now consider the inner leaf only:

$$\mu = 0.25 + \dfrac{\left(\dfrac{15 \times 0.1 \times 0.9 \times 2.0 \times 10^3}{0.1 \times 1.0 \times 10^6} \times 2.5\right)}{0.6}$$

$$= 0.53$$

Therefore, by interpolation from BS 5628, Table 9 (G) (see Table 6.4) for $\mu = 0.53$ and $h/L = 0.8$ BM coefficient, $\alpha = 0.0306$.

Design bending moments as located in Figure 11.36:

Continuity moment at stanchion support

$$M_1 = \alpha \times W_k \times \gamma_f \times L^2$$
$$M_1 \text{ outer leaf} = 0.032 \times 0.832 \times 5^2$$
$$= 0.665 \text{ kN·m}$$
$$M_1 \text{ inner leaf} = 0.0306 \times 0.428 \times 5^2$$
$$= 0.327 \text{ kN·m}$$

Moment at mid-span in horizontal plane

$$M_2 = M_1$$

Moment at approx. mid-height in vertical plane

$$M_3 = \mu \times \alpha \times W_k \times \gamma_f \times L^2$$
$$M_3 \text{ outer leaf} = 0.445 \times 0.032 \times 0.832 \times 5^2$$
$$= 0.296 \text{ kN·m}$$
$$M_3 \text{ inner leaf} = 0.53 \times 0.0306 \times 0.428 \times 5^2$$
$$= 0.173 \text{ kN·m}$$

Compare applied design moments with design moments of resistance at positions of M_1 and M_2 (negative and positive moments in horizontal span):

$$\text{Inner leaf design MR} = \dfrac{f_{kx\,perp}}{\gamma_m} Z$$

$$= \dfrac{0.6 \times 1.67 \times 10^6}{2.5}$$

$$= 0.4 \text{ kN·m}$$

The section is adequate as the design MR is greater than the applied BM of 0.327 kN·m.

$$\text{Outer leaf design MR} = \frac{1.1 \times 1.751 \times 10^6}{2.5}$$

$$= 0.77 \text{ kN·m}$$

The section is adequate as the design MR is greater than the applied BM of 0.665 kN·m. At position of M_3 (positive moment in vertical span)

$$\text{Inner leaf design MR} = \left(\frac{f_{\text{kx par}}}{\gamma_m} + g_d \right) Z$$

$$= \left(\frac{0.25}{2.5} + \frac{15 \times 0.1 \times 0.9 \times 2.0 \times 10^3}{0.1 \times 1.0 \times 10^6} \right) \times 1.67 \times 10^6$$

$$= 0.212 \text{ kN·m}$$

The section is adequate as the design MR is greater than the applied BM of 0.173 kN·m.

$$\text{Outer leaf design MR} = \left(\frac{0.4}{2.5} + \frac{20 \times 0.1025 \times 0.9 \times 2.0 \times 10^3}{0.1025 \times 1.0 \times 10^6} \right) \times 1.751 \times 10^6$$

$$= 0.343 \text{ kN·m}$$

The section is adequate as the design MR is greater than the applied BM of 0.296 kN·m.

The wall section has been checked at all critical locations of applied BM and found to be adequate for the specification and connection details given.

EXAMPLE 8: Solid concrete block crosswall acting as shear wall

Design the crosswall for the four-storey flats complex shown in Figure 11.37. The crosswalls are to be constructed in solid concrete blockwork and, acting as shear walls, they provide the sole resistance to wind forces on the glazed elevation. The floor to floor height is 2.70 m and the crosswalls are at 6.0 m centres. The following characteristic loads are to be assumed:

own weight of masonry	$= 3.0 \text{ kN/m}^2$
roof – dead load	$= 6.5 \text{ kN/m}^2$
– superimposed load	$= 1.5 \text{ kN/m}^2$
floors – dead load	$= 5.0 \text{ kN/m}^2$
– superimposed load	$= 1.5 \text{ kN/m}^2$
wind load on glazed elevation	$= 0.8 \text{ kN/m}^2$

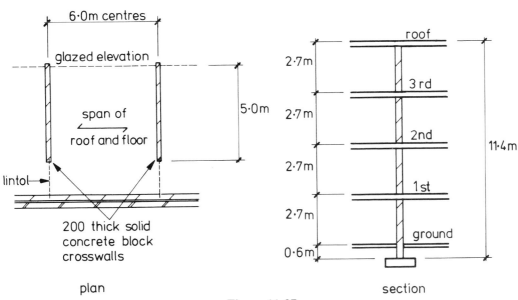

Figure 11.37

Shear walls are described in more detail in Chapter 14 dealing with multi-storey structures. Briefly, shear walls, such as those shown in Figure 11.37, act about their major axes and span as vertical cantilevers to provide stability to the structure by resisting the wind forces on the building elevation.

On the glazed elevation

$$\text{design wind loading} = W_k \times \gamma_f$$
$$= 0.8 \times 1.4 = 1.12 \text{ kN/m}^2$$

Each shear wall resists wind loading from 6.0 m of glazed elevation, thus:

$$\text{design wind load per shear wall} = 1.12 \times 6$$
$$= 6.72 \text{ kN/m}$$

The shear wall acts as a vertical cantilever and the design bending moment at the base of the wall

$$= \frac{W_k \times \gamma_f \times h^2}{2}$$
$$= \frac{6.72 \times 11.4^2}{2} = 436.67 \text{ kN·m}$$

The loading and bending moment diagrams per shear wall are shown in Figure 11.38.

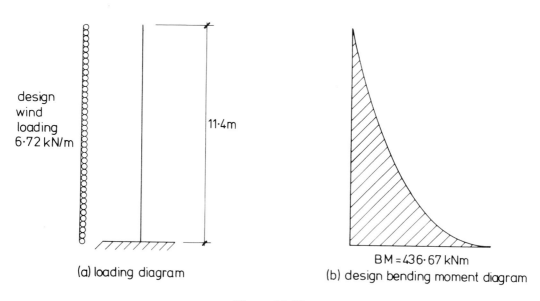

(a) loading diagram

(b) design bending moment diagram

BM = 436.67 kNm

Figure 11.38

Calculate design axial loadings on shear walls.

Characteristic dead loads, G_k:

$$\begin{array}{lll}
\text{roof} = 6.5 \times 6 & = & 39.00 \\
\text{3 floors} = 3 \times 5.0 \times 6 & = & 90.00 \\
\text{ow wall} = 11.4 \times 3 & = & 34.00 \\
& = & 163.20 \text{ kN/m}
\end{array}$$

Characteristic superimposed load, Q_k:

$$\begin{array}{lll}
\text{roof} = 1.5 \times 6 & = & 9.00 \\
\text{3 floors} = 3 \times 1.5 \times 6 & = & 27.00 \\
& = & 36.00 \text{ kN/m}
\end{array}$$

Minimum design load for dead plus wind loading combination

$$= G_k \times \gamma_f$$
$$= 163.2 \times 0.9 = 146.88 \text{ kN/m}$$

Maximum design load for dead plus superimposed plus wind loading combination

$$= (G_k \times \gamma_f) + (Q_k \times \gamma_f)$$
$$= (163.2 \times 1.4) + (36 \times 1.6) = 286.08 \text{ kN/m}$$

The design moment of resistance will be checked first against the formula

$$\left(\frac{f_{kx}}{\gamma_m} + g_d\right)Z$$

in which the limiting condition is the flexural tensile strength, f_{kx}. This check utilises the minimum design load for the calculation of g_d. However, it is essential that the flexural compressive stresses are also checked and a suggested design method is given later in this example for the second stage of the check.

Total minimum axial load on shear wall $= 5 \times 146.88 = 734.4$ kN

Shear wall properties:

area $\qquad = 0.2 \times 5 = 1000 \times 10^3 \text{ mm}^2$

section modulus, $Z = \dfrac{0.2 \times 5^2}{6} = 833.33 \times 10^6 \text{ mm}^3$

MR of shear wall $\qquad = \left(\dfrac{f_{kx}}{\gamma_m} + g_d\right) Z$

The specification for the masonry is given as solid concrete blocks with a compressive strength of 10.0 N/mm² set in a designation (iii) mortar.

γ_m can be taken as 2.5 special/special

therefore

$f_{kx} = 0.25 \text{ N/mm}^2$ (BS 5628, Table 3)

$\gamma_m = 2.5$ special/special

$g_d = \dfrac{734.4 \times 10^3}{1000 \times 10^3} = 0.7344 \text{ N/mm}^2$

$Z = 833.33 \times 10^6 \text{ mm}^3$

Hence:

$$\text{design MR} = \left(\frac{0.25}{2.5} + 0.7344\right) \times 833.33 \times 10^6$$
$$= 695.33 \text{ kN·m}$$

The section is adequate for the first stage of the design as the design MR is greater than the applied BM of 436.67 kN·m.

Now check compressive stresses under dead plus superimposed plus wind loading. The wall section needs checking at two critical levels:

(a) at foundation level where the effect of axial loading and design bending moment are greatest and,
(b) at a point 0.4h above the ground floor slab where the design bending moment will be considerably less but where the capacity reduction factor, β, reduces the vertical load resistance of the wall under combined axial and lateral loading (see Figure 11.32).

Case (a): Check stresses at foundation level

This check must be carried out in two stages as explained earlier (see Figure 11.32) considering (1) axial loading only and (2) combined axial and lateral loading.

Stage 1: Axial loading

The design stress required under axial loading is given by the equation $\beta f_k/\gamma_m$. Hence

$$\frac{\text{design load}}{\text{area}} \leqslant \frac{\beta f_k}{\gamma_m}$$

$$\frac{286.08 \times 10^3}{200 \times 1000} \leqslant \frac{\beta f_k}{\gamma_m}$$

$$1.4304 \leqslant \frac{\beta f_k}{\gamma_m}$$

At foundation level the wall is fully restrained against buckling and therefore:

$\beta = 1.0$
$\gamma_m = 2.5$ special/special
$f_k = 4.5$ N/mm^2 for the block and mortar specification given (by interpolation from BS 5628, Tables 2(b) and 2(c), for a ratio of $h/t = 1.0$) (see Tables 5.8 and 5.9). Hence

$$\frac{\beta f_k}{\gamma_m} = \frac{1.0 \times 4.5}{2.5}$$

$$= 1.80 \text{ N/mm}^2$$

(which is greater than the applied design stress 1.4304 N/mm^2).

Stage 2: Combined axial and lateral loading

The design stress required under combined axial and lateral loading is given by the equation $\beta \times 1.1 f_k/\gamma_m$.

The maximum applied design compressive stress under the loading combination dead plus super plus wind is, from a simple elastic analysis:

$$\frac{n_w}{A} + \frac{M_A}{Z}$$

where

n_w = total design vertical load under D + S + W loading combination
A = area of shear wall
M_A = applied design bending moment at foundation
Z = section modulus of shear wall.

Design vertical load, n_w:

$$n_w = 1.2 G_k + 1.2 Q_k$$

$$= (1.2 \times 163.2) + (1.2 \times 36)$$

$$= 195.84 + 43.2 \qquad = 239.04 \text{ kN/m}$$

$$A = 200 \times 5000 \times 10^{-9} \quad = 1.0 \text{ m}^2$$

$$M_A = 436.67 \text{ kN·m}$$

$$Z = \frac{200 \times 5000^2 \times 10^{-9}}{6} = 0.833 \text{ m}^3$$

Maximum applied design compressive stress:

$$= \frac{n_w}{A} + \frac{M_A}{Z}$$

$$= \frac{239.04 \times 5}{1} + \frac{436.67}{0.833}$$

$$= 1195.2 + 524.21 = 1719.41 \text{ kN/m}^2 = 1.719 \text{ N/mm}^2$$

Hence $1.719 \le \dfrac{\beta \times 1.1 f_k}{\gamma_m}$

Once again $\beta = 1.0$, $\gamma_m = 2.5$ and $f_k = 4.50 \text{ N/mm}^2$.

Therefore

$$\frac{\beta \times 1.1 f_k}{\gamma_m} = \frac{1.0 \times 1.1 \times 4.5}{2.5} = 1.98 \text{ N/mm}^2$$

(which is greater than the maximum applied design compressive stress of 1.719 N/mm^2).

Case (b): Check wall at 0.4h above ground floor slab restraint (see Figure 11.39)

Stage 1: Axial loading
Design stress required is similar to that for case (a) in which $1.4304 \le \beta f_k/\gamma_m$, where $\beta = 0.97$ (from BS 5628, Table 7) (see Table 5.15) since SR $= 0.75 \times 2.7/0.2 = 10.125$ and $e_x = 0$ to $0.05t$,

$f_k = 4.5 \text{ N/mm}^2$ and $\gamma_m = 2.5$ as before.

Hence

$$\frac{\beta f_k}{\gamma_m} = \frac{0.97 \times 4.5}{2.5}$$

$$= 1.746 \text{ N/mm}^2$$

(which is greater than the applied design stress of 1.4304 N/mm^2).

Stage 2: Combined axial and lateral load
Design stress required is $\beta \times 1.1 f_k/\gamma_m$, in which β is calculated from the slenderness ratio based upon the effective wall thickness perpendicular to the direction of application of the bending. For this example, the direction of application of the bending is parallel to the major axis of the shear wall and the effective thickness to be taken for the calculation of β is the actual thickness of 200 mm.

Hence, β is as for stage 1 = 0.97.

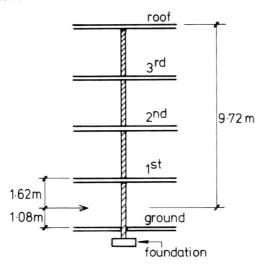

Figure 11.39 Section showing case (b) critical design moment

Hence

$$\text{design stress required} = \frac{0.97 \times 1.1 \times 4.5}{2.5}$$

$$= 1.921 \text{ N/mm}^2$$

$$\text{The maximum applied design compressive stress} = \frac{n_w}{A} + \frac{M_A}{Z}$$

where

n_w = 239.04 kN/m (as case (a))

A = 1.0 m^2

$M_A = \gamma_f W_k \times \dfrac{9.72^2}{2} \times L$ (where L is the spacing of the shear walls)

$ = 1.2 \times 0.8 \times \dfrac{9.72^2}{2} \times 6 = 272.1 \text{ kN·m}$

Z = 0.833 m^3

Therefore

$$\frac{n_w}{A} + \frac{M_A}{Z} = \frac{239.04 \times 5}{1} + \frac{272.1}{0.833}$$

$$= 1195.2 + 326.65$$

$$= 1521.85 \text{ kN/m}^2 = 1.522 \text{ N/mm}^2$$

(which is less than the design stress required of 1.921 N/mm^2).

The section is shown to be adequate for all stages of the design as the design moment of resistance is greater than the applied design bending moment of 436.67 kN·m, and the combined flexural stresses at the two critical levels are within acceptable designed limits.

EXAMPLE 9: Column design

Four rows of masonry columns supporting the main structure of an office block are to be designed and a section of the building is shown in Figure 11.40. The characteristic wind loading for the area is to be taken as 0.65 kN/m^2 and the wall, floor and roof construction is such that the columns are loaded as follows:

Column reference	A	B	C	D
Characteristic dead load (kN)	175	76	76	175
Characteristic super load (kN)	147	100	100	147

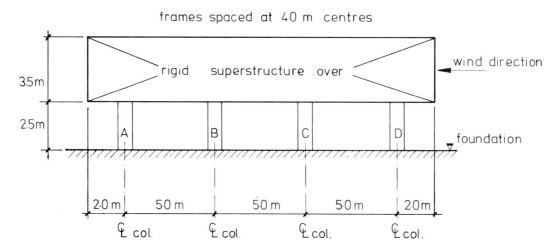

Figure 11.40 Typical cross-section

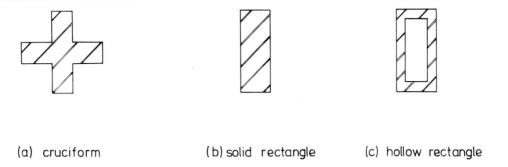

(a) cruciform   (b) solid rectangle   (c) hollow rectangle

Figure 11.41 Alternative column profile

It can be assumed that the critical loading condition occurs when the wind is in the direction of the arrow shown in Figure 11.40. The framework shown is repeated at 4.0 m centres and a felt dpc is to be incorporated at the base of each column. Three alternative cross-sections of columns are required for consideration by the client, the basic profiles of which are shown in Figure 11.41.

Minimum design loads (dead plus wind loading condition):

$$= \gamma_f \times G_k$$

Columns A and D $= 0.9 \times 175 = 157.5 \, \text{kN}$
Columns B and C $= 0.9 \times 76 = 68.4 \, \text{kN}$

Maximum design loads (dead plus super loading condition):

$$= (\gamma_f \times G_k) + (\gamma_f \times Q_k)$$

Columns A and D $= (1.4 \times 175) + (1.6 \times 147) = 480.2 \, \text{kN}$
Columns B and C $= (1.4 \times 76) + (1.6 \times 100) = 266.4 \, \text{kN}$

Design wind loading $= \gamma_f \times W_k$
$$= 1.4 \times 0.65 \times 3.5 \times 4.0$$
$$= 12.74 \, \text{kN per frame of four columns}$$

It has been assumed that the structure above the column heads will be sufficiently stiffened with crosswalls for the design wind load to be considered as acting at column head height as shown in Figure 11.42. The design wind load of 12.74 kN, acting at the mid-depth of the 3.5 m deep structure above first floor level, induces positive and negative forces into the columns and consideration is given to this later in the design. Due to the presence of the dpc at the base of the columns, the frames will be designed as fixed at the column heads and pin-jointed at their base.

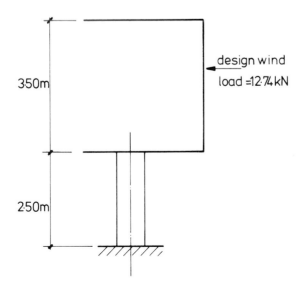

350m

design wind
load =12.74kN

2·50m

Figure 11.42 Wind loading

It has already been demonstrated that a moment can develop at a felt dpc due to the inherent stability moment. However, to simplify the analysis for this design example and at the same time, provide an additional factor of safety on the structure, the base will be treated as pin-jointed and the rigidity will be provided by the moment connection of the column heads to the structure above (see Figure 11.43). It will be assumed that the structure above can accommodate the moments at each column head. As each column is to be of identical cross-section the total design wind load will be shared equally between the four columns as shown in Figure 11.43.

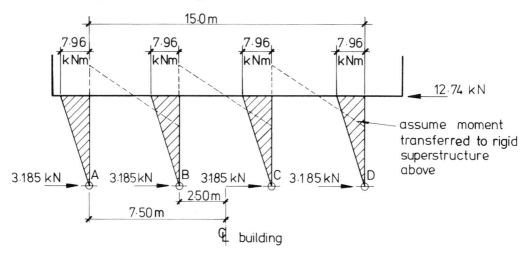

Figure 11.43 Columns bending moment diagram

It is critical to the design assumptions to ensure that the interfaces between the structure over and the tops of the four columns can transmit the calculated flexural tensile stresses. In this example, the structure over is assumed to comprise a rigid reinforced concrete slab poured directly onto the brick columns. Precast concrete beams laid 'dry' onto the brick columns could not develop the flexural tensile stresses calculated, but may generate the required moment of resistance on the basis of a 'cracked section' design.

$$\text{Design bending moment} = \frac{12.74}{4} \times 2.5$$

$$= 7.96 \text{ kN·m per column}$$

Induced compression and tension in columns A and D respectively from applied wind moment:

$$C = T = \frac{12.74 \times \left(2.5 + \dfrac{3.5}{2}\right) \times \dfrac{15}{2}}{7.5^2 + 2.5^2}$$

$$= 6.5 \text{ kN}$$

Therefore, adjusted minimum load in column D under dead plus wind load $= 157.5 - 6.5 = 151$ kN.

Worst case minimum loading is in columns B and C where the minimum design load of 68.4 kN is combined with the applied design BM of 7.96 kN.

Case (a): Cruciform columns
The proposed column profile, for trial purposes, is shown in Figure 11.44. The assessment of column trial section sizes can only be gained with experience.

Column properties:

Area $= 451.125 \times 10^3 \text{ mm}^2$
$Z = 54.13 \times 10^6 \text{ mm}^3$
$I = 26\,836 \times 10^6 \text{ mm}^4$
$r = 243.9 \text{ mm}$

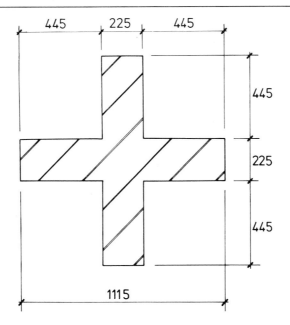

Figure 11.44 Cruciform column profile

The design resistance moment of the cruciform section based solely on the flexural tensile stresses:

$$= \left(\frac{f_{kx}}{\gamma_m} + g_d \right) Z$$

(for geometric profiles the effective eccentricity method is considered not to be a realistic analysis).

Using bricks with a crushing strength of 27.5 N/mm² and a water absorption of 9% set in a designation (iii) mortar:

$f_{kx} = 0.4 \, \text{N/mm}^2$ (BS 5628, Table 3)

$\gamma_m = 2.5$ special/special

$g_d = \dfrac{68.4 \times 10^3}{451.125 \times 10^3} = 0.152 \, \text{N/mm}^2$

$Z = 54.13 \times 10^6 \, \text{mm}^3$

Therefore

$$\text{design MR} = \left(\frac{0.4}{2.5} + 0.152 \right) \times 54.13 \times 10^6$$
$$= 16.89 \, \text{kN·m}$$

The flexural tensile resistance moment is adequate as the design MR is greater than the applied BM of 7.96 kN·m.

Check compressive stresses under applied BM and maximum design load after column has been checked for axial loading.

Check columns A and D for axial loading.

Maximum design load = 480.2 + induced compression from wind moment.
$$= 480.2 + 6.5$$
$$= 486.7 \, \text{kN (columns A and D)}$$

This load will be assumed to be applied axially with an eccentricity of 0 to 0.05t.

From the formula, radius of gyration $= \sqrt{(I/A)}$ an equivalent solid square column will be calculated in order to assess the value of β (capacity reduction factor) from BS 5628 Table 7 (see Table 5.15).

Hence

$$\text{radius of gyration} = \sqrt{\left(\frac{I}{A}\right)} = 243.9$$

Therefore

$$\sqrt{\left(\frac{bt^3/12}{bt}\right)} = 243.9$$

But, for a square section, $b = t$

therefore

$$\sqrt{\left(\frac{t^2}{12}\right)} = 243.9$$

$$t = \sqrt{(12 \times 243.9^2)} = 845 \text{ mm}$$

Hence, column section $= 845 \times 845$.

Now calculate β from BS 5628, Table 7 (see Table 5.15), for an equivalent 845 square column section:

$$\text{Slenderness ratio} = \frac{h_{ef}}{t_{ef}}$$

$$= \frac{2.5 \times 10^3}{845} = 2.96$$

Slenderness ratio based on length of outstanding leg of cruciform profile:

$$= \frac{h_{ef}}{t_{ef}}$$

$$= \frac{2 \times 330}{225} = 2.93$$

$e_x = 0$, therefore, $\beta = 1.0$ (from BS 5628, Table 7).

$$\text{Design vertical load resistance} = \frac{\beta \times \text{area} \times f_k}{\gamma_m}$$

$$= \frac{1.0 \times 451.125 \times 10^3 \times 7.1}{2.5}$$

$$= 1281 \text{ kN}$$

The section is adequate for the design axial loading as the design vertical load resistance is greater than the applied maximum design load of 486.7 kN, although the flexural compressive stresses should be checked under combined axial and lateral loading.

Check if smaller column profile will be adequate – profile shown in Figure 11.45.

Column properties:

Area $= 347.60 \times 10^3 \text{ mm}^2$
$Z = 34.94 \times 10^6 \text{ mm}^3$
$I = 13\,623 \times 10^6 \text{ mm}^4$
$r = 198 \text{ mm}$

Equivalent solid column $= 686 \text{ mm square}$.

$$\text{Design MR} = \left(\frac{f_{kx}}{\gamma_m} + g_d\right) Z$$

$$= \left(\frac{0.4}{2.5} + \frac{68.4}{347.6}\right) \times 34.94 \times 10^6$$

$$= 12.4 \text{ kN·m}$$

(The flexural tensile resistance moment is adequate as design MR is greater than applied BM.)

Now check columns A and D for axial load:

$$\text{Slenderness ratio} = \frac{2.5 \times 10^3}{686} = 3.6$$

$e_x = 0$, therefore $\beta = 1.0$ (from BS 5628, Table 7) (see Table 5.15).

$$\text{Design vertical load resistance} = \frac{\beta \times \text{area} \times f_k}{\gamma_m}$$

$$= \frac{1.0 \times 347.6 \times 10^3 \times 7.1}{2.5}$$

$$= 987.20 \text{ kN}$$

Hence, the section is adequate for the design axial loading.

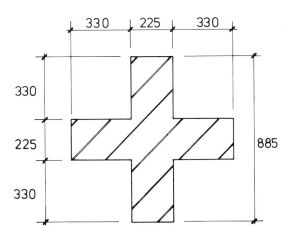

Figure 11.45 Reduced cruciform column profile

Whilst the Code of Practice does not directly cater for calculating compressive stresses under combined axial and lateral loading, it is considered that more complex geometric profiles should always be checked, as it is by no means certain that flexural tensile stress will always be the limiting factor in calculating design resistance moments. Under maximum design load and applied design bending moment, the compressive stresses at the heads of columns A and D may be calculated using the following suggested design method:

Axial compressive stress under maximum design load

$$= \frac{\text{load}}{\text{area}} \times \gamma_m$$

$$= \frac{486.7 \times 10^3}{347.6 \times 10^3} \times 2.5$$

$$= 3.5 \text{ N/mm}^2$$

Flexural compressive stress under applied BM

$$= \frac{\text{moment} \times \gamma_m}{Z}$$

$$= \frac{7.96 \times 10^6}{34.94 \times 10^6} \times 2.5$$

$$= 0.575 \text{ N/mm}^2$$

Maximum combined stress (axial plus flexural)

$$= 3.5 + 0.575$$
$$= 4.075 \text{ N/mm}^2$$

By inspection this is comfortably within the characteristic compressive stresses given in BS 5628, to which no reductions need apply as the capacity reduction factor, β, has already been calculated as 1.0.

Case (b): Solid rectangular columns
The proposed column profile, for trial purposes, is shown in Figure 11.46.

Column properties:

Area $= 292 \times 10^3 \text{ mm}^2$

$Z_{xx} = 43.1 \times 10^6 \text{ mm}^3$

$Z_{yy} = 16.1 \times 10^6 \text{ mm}^3$

Following the same sequence of design and the same brick/mortar specification as for case (a):

$$\text{Design moment of resistance} = \left(\frac{f_{kx}}{\gamma_m} + g_d \right) Z$$

$$= \left(\frac{0.4}{2.5} + \frac{68.4}{292} \right) \times 43.1 \times 10^6$$

$$= 16.99 \text{ kN·m}$$

The flexural tensile resistance moment is adequate as the design MR is greater than the applied BM of 7.96 kN·m.

Now check columns A and D for the maximum design axial load of 486.7 kN. Once again the axial load will be assumed to be applied concentrically and, therefore, $e_x = 0$.

$$\text{Slenderness ratio} = \frac{h_{ef}}{t_{ef}}$$

$$= \frac{2.5 \times 10^3}{330} = 7.6$$

$e_x = 0$, therefore, $\beta = 1.0$ from BS 5628, Table 7 (see Table 5.15)

$$\text{Design vertical load resistance} = \frac{\beta \times \text{area} \times f_k}{\gamma_m}$$

$$= \frac{1.0 \times 292 \times 10^3 \times 7.1}{2.5}$$

$$= 829 \text{ kN}$$

The axial compressive stresses are adequate as the maximum design load is only 486.7 kN and it is possible that a slightly smaller section could also be shown to be adequate.

Now check maximum flexural compressive stresses under combined axial and lateral loading:

$$\text{Axial compressive stress} = \frac{\text{load}}{\text{area}} \times \gamma_m$$

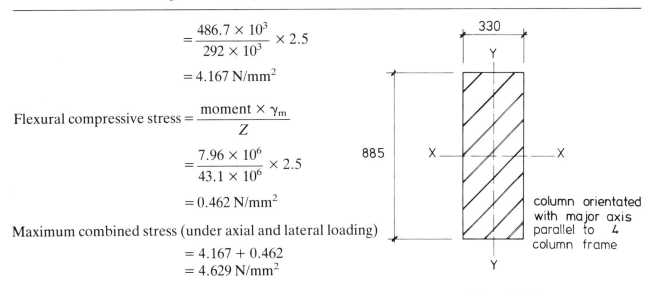

$$= \frac{486.7 \times 10^3}{292 \times 10^3} \times 2.5$$

$$= 4.167 \text{ N/mm}^2$$

Flexural compressive stress $= \dfrac{\text{moment} \times \gamma_m}{Z}$

$$= \frac{7.96 \times 10^6}{43.1 \times 10^6} \times 2.5$$

$$= 0.462 \text{ N/mm}^2$$

Maximum combined stress (under axial and lateral loading)

$$= 4.167 + 0.462$$
$$= 4.629 \text{ N/mm}^2$$

Figure 11.46

By inspection this is comfortably within the characteristic compressive stresses given in BS 5628 to which no reductions need apply as the capacity reduction factor, β, has already been calculated as 1.0. The section is therefore adequate in all respects.

The columns should also be checked for bending about the weaker axis when the wind loading is applied to the other building elevation at right angles using the same design principles.

Case (c): Hollow rectangular columns
The proposed column profile, for trial purposes, is shown in Figure 11.47 (columns orientated as column case (b)).

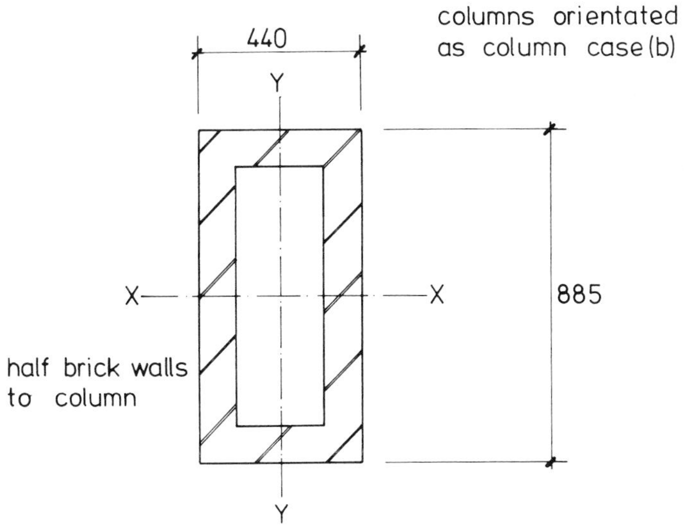

Column properties:

Area $= 229.6 \times 10^3 \text{ mm}^2$
$Z_{xx} = 43.52 \times 10^6 \text{ mm}^3$
$Z_{yy} = 25.21 \times 10^6 \text{ mm}^3$
$I_{xx} = 19\,258 \times 10^6 \text{ mm}^4$
$I_{yy} = 5547 \times 10^6 \text{ mm}^4$
$r_{xx} = 289.60 \text{ mm}$
$r_{yy} = 155.43 \text{ mm}$

Figure 11.47 Hollow rectangular column profile

From $r = \sqrt{(I/A)}$, equivalent solid column $= 1003 \times 538$ mm.

Following the same sequence of design and the same brick/mortar specification as for case (a):

$$\text{Design MR} = \left(\frac{f_{kx}}{\gamma_m} + g_d \right) Z$$

$$= \left(\frac{0.4}{2.5} + \frac{68.4}{229.6} \right) \times 43.52 \times 10^6$$

$$= 19.93 \text{ kN·m}$$

The flexural tensile resistance moment is adequate as the design MR is greater than the applied BM of 7.96 kN·m.

Now check columns A and D for the maximum design axial load of 486.7 kN:

$$\text{Slenderness ratio} = \frac{h_{\text{ef}}}{t_{\text{ef}}}$$

$$= \frac{2.5 \times 10^3}{538} = 4.6$$

As before, $e_x = 0$, therefore, $\beta = 1.0$ (from BS 5628, Table 7) (see Table 5.15).

$$\text{Design vertical load resistance} = \frac{\beta \times \text{area} \times f_k}{\gamma_m}$$

$$= \frac{1.0 \times 229.6 \times 10^3 \times 7.1}{2.5}$$

$$= 652 \text{ kN}$$

The axial compressive stresses are adequate as the maximum design load is only 486.7 kN.

Now check maximum flexural compressive stresses under combined axial and lateral loading:

$$\text{Axial compressive stress} = \frac{\text{load} \times \gamma_m}{\text{area}}$$

$$= \frac{486.7 \times 10^3}{229.6 \times 10^3} \times 2.5$$

$$= 5.299 \text{ N/mm}^2$$

$$\text{Flexural compressive stress} = \frac{\text{moment} \times \gamma_m}{Z}$$

$$= \frac{7.96 \times 10^6}{43.52 \times 10^6} \times 2.5$$

$$= 0.457 \text{ N/mm}^2$$

Maximum combined stress (under axial and lateral loading):

$$= 5.299 + 0.457$$
$$= 5.756 \text{ N/mm}^2$$

By inspection this is comfortably within the characteristic compressive stresses given in BS 5628 to which no reductions need apply as the capacity reduction factor, β, has already been calculated as 1.0. The section is therefore adequate in all respects.

The shear at the foot of each column of 3.185 kN should be checked using the manufacturers recommended values for shear forces on felt dpc's. However, it is unlikely that this will be critical. It is advisable to incorporate butterfly wall ties at the corners of such hollow column sections to reduce the risk of splitting.

EXAMPLE 10: Perforated external wall

Details are shown in Figure 11.48 of the external wall in the topmost storey of a multi-storey hotel building for which the wind loading will be assumed to be 0.9 kN/m^2.

For simplicity of analysis, it will be assumed that the roof dead loading is such that under wind loading the uplift effect on the roof is exactly cancelled out by the dead load of the roof. The inner and outer leaves of the cavity wall are to be constructed in brick with a compressive strength of 20 N/mm^2 and a water absorption of 11% set in a designation (ii) mortar. The roof structure is assumed to be capable of providing adequate support to the head of the wall panel.

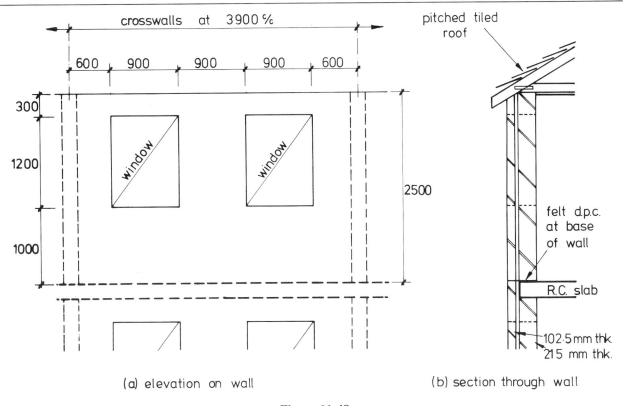

Figure 11.48

The design of more complicated elements, such as the perforated wall being considered in this design example, must include a rational judgement of the mode of action of the element, and will not necessarily conform exactly to the simple design procedures already used in the previous examples. The following suggested design method is one of a number of possible solutions.

By inspection, the area of wall between the windows is clearly the most critical for design purposes as the perforations eliminate this area of wall to span in two directions. In addition, this area of wall is also likely to be loaded with its own wind loading, plus wind loading from the windows and wall area above (and possibly below) the windows. As the central area of wall deflects under wind loading, the load supported would tend to shed towards the wall areas either side of the windows which, owing to buttressing effect of the bonded crosswalls, are considerably stiffer. Consequently, where it might, under different circumstances, be expected that half of the wind on the window areas would be supported on the brickwork each side, it is considered reasonable, in this instance, to take a proportion of this load, less than half, onto the central area of wall under consideration. The area of wind to be supported by the wall area between the windows is shown in Figure 11.49.

The loading width is calculated as:

central wall $= 900$ mm

windows $\quad = 2 \times 900 \times \dfrac{3}{8} = 675$ mm

total width $\; = 1575$ mm in which $\dfrac{3}{8}$ of the wind on the adjacent unsupported areas has been taken to the central wall area.

The section of wall below window cill level which will be assumed to be resisting the applied wind bending moment is indicated by the dotted lines on Figure 11.49. The wind on the wall areas immediately beneath the windows could equally be considered to be supported by the cill height brickwork acting as a cantilever.

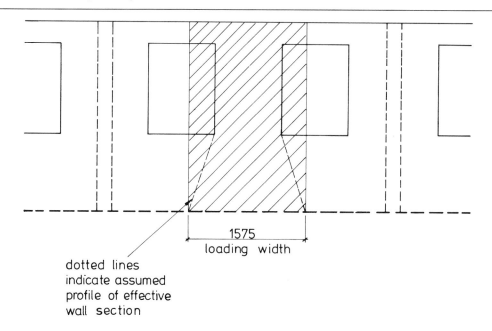

Figure 11.49 Showing area of wind loading

Wind load on central wall area $= W_k \times 1.575$
$= 0.9 \times 1.575 = 1.42$ kN/m

Design wind load $= 1.42 \times \gamma_f = 1.42 \times 1.4 = 1.988$ kN/m

The critical wall area will now be designed as a vertically spanning propped cantilever, and the loading and bending moment diagrams are shown in Figure 11.50.

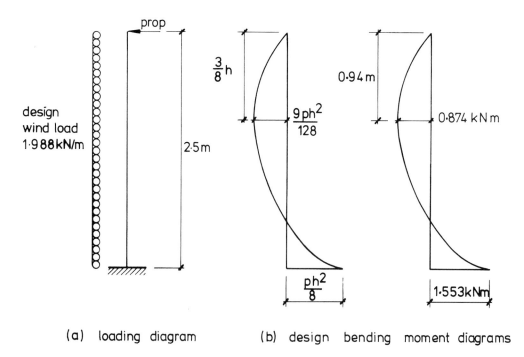

(a)   loading diagram   (b)   design   bending   moment diagrams

Figure 11.50

$$\text{Wall moment} = \frac{9 \times 1.988 \times 2.5^2}{128} = 0.874 \text{ kN·m}$$

$$\text{Base moment} = \frac{1.988 \times 2.5^2}{8} = 1.553 \text{ kN·m}$$

The design resistance moment at the $\frac{3}{8}h$ level will be based on the 900 mm wall width, whereas the resistance moment at base level will be based on the loaded width of 1575 mm as shown to be the effective section in Figure 11.49. A felt dpc has been included at the base of the wall span and the design resistance moment at this level will therefore be based on a 'cracked section' analysis.

Minimum characteristic loads at base of wall:

Outer leaf (density of masonry = 20 kN/m^3):

wall above windows   = 20 × 0.3 × 0.1025 × 1.575 = 0.970
wall between windows = 20 × 1.2 × 0.1025 × 0.9   = 2.214
wall below windows   = 20 × 1.0 × 0.1025 × 1.575 = 3.229
Total characteristic load, G_k = 6.413 kN per 1575 mm width.

Inner leaf: (215 thick) = 6.413 × $\dfrac{215}{102.5}$ = G_k = 13.452 kN per 1575 mm width.

Minimum design loads = $G_k \times \gamma_f$

Outer leaf minimum design load = $G_k \times \gamma_f$ = 6.413 × 0.9 = 5.772 kN

Inner leaf minimum design load = $G_k \times \gamma_f$ = 13.452 × 0.9 = 12.106 kN

Calculate stability resistance moments using stress block shown in Figure 11.26 (γ_m = 2.5 special/special).

$$\frac{1.1 f_k}{\gamma_m} = \frac{\text{min. design load}}{w_s \times \text{length of section}}$$

Outer leaf:

$$w_s = \frac{5.772 \times 10^3 \times 2.5}{1.1 \times 6.4 \times 1.575 \times 10^3} = 1.3 \text{ mm}$$

Then

$$l_a = \frac{\text{wall thickness}}{2} - \frac{w_s}{2}$$

$$= \frac{102.5}{2} - \frac{1.3}{2} = 50.6 \text{ mm}$$

and   design MR = design load × l_a

$$= \frac{5.772 \times 50.6}{10^3} = 0.292 \text{ kN·m}$$

Inner leaf:

$$w_s = \frac{12.106 \times 10^3 \times 2.5}{1.1 \times 6.4 \times 1.575 \times 10^3} = 2.73 \text{ mm}$$

Then

$$l_a = \frac{215}{2} - \frac{2.73}{2} = 104.77 \text{ mm}$$

and   design MR = $\dfrac{12.106 \times 104.77}{10^3}$ = 1.268 kN·m

Total design MR = design MR outer leaf + design MR inner leaf
= 0.292 + 1.268
= 1.56 kN·m

The section is adequate as the design resistance moment is greater than the applied BM of 1.553 kN·m.

Now check the 900 wide wall section at $\dfrac{3}{8} h$ level.

Outer leaf:

wall above windows	$= 20 \times 0.3 \times 0.1025 \times 1.575 = 0.970 \text{ kN}$
wall between windows	$= 20 \times (0.94 - 0.3) \times 0.1025 \times 0.9 = 1.181 \text{ kN}$
Total characteristic load, G_k	$= 2.151 \text{ kN per 900 mm width.}$
Inner leaf (215 thick)	$= 2.151 \times \dfrac{215}{102.5}$
	$= 4.512 \text{ kN per 900 mm width.}$
Minimum design loads	$= G_k \times \gamma_f$

Outer leaf minimum design load $= 2.151 \times 0.9 = 1.936 \text{ kN}$

Inner leaf minimum design load $= 4.512 \times 0.9 = 4.061 \text{ kN}$

$$\text{Design resistance moment} = \left(\frac{f_{kx}}{\gamma_m} + g_d\right) Z \text{ per leaf}$$

$$\text{Total design resistance moment} = \text{design MR outer leaf} + \text{design MR inner leaf}$$

Outer leaf:

$f_{kx} = 0.4 \text{ N/mm}^2 \text{ (BS 5628, Table 3)}$

$\gamma_m = 2.5 \text{ special/special}$

$g_d = \dfrac{1.936 \times 10^3}{900 \times 102.5} = 0.021 \text{ N/mm}^2$

$Z = \dfrac{900 \times 102.5^2}{6} = 1.576 \times 10^6 \text{ mm}^3$

Inner leaf:

$f_{kx} = 0.4 \text{ N/mm}^2 \text{ (BS 5628, Table 3)}$

$\gamma_m = 2.5 \text{ special/special}$

$g_d = \dfrac{4.061 \times 10^3}{900 \times 215} = 0.021 \text{ N/mm}^2$

$Z = \dfrac{900 \times 215^2}{6} = 6.934 \times 10^6 \text{ mm}^3$

Therefore

$$\text{design resistance moment} = \left(\frac{0.4}{2.5} + 0.021\right) \times 1.576 \times 10^6 + \left(\frac{0.4}{2.5} + 0.021\right) \times 6.934 \times 10^6$$

$$= 0.285 + 1.255$$

$$= 1.54 \text{ kN·m}$$

The section is adequate as the design MR is greater than the applied BM of 0.874 kN·m.

CHAPTER 12

DESIGN OF SINGLE-STOREY BUILDINGS

Society requires a large number and a wide range of single-storey structures – not just for factories, garages and warehouses, but also for primary schools, theatres, churches, sports halls, libraries, etc.

12.1 DESIGN CONSIDERATIONS

The design requirements can best be appreciated by considering the structural problems common to all types of walls in single-storey structures. Open-plan buildings (i.e. with no internal walls) will be discussed in particular since, structurally, these represent the worst case.

(a) Vertical loads
The vertical compressive load is rarely the critical factor in the design of masonry walls since the dead load of the roof, and its imposed load, are relatively light compared to multi-storey structures. Often, the wind suction on the roof is equal to, or greater than, its dead weight, and design cases frequently arise where the wall is subject to no vertical loading (other than its own weight) and the loading is mainly lateral.

(b) Bending stresses
The bending stresses due to wind are critical. Particularly the tensile stress due to bending, which is referred to in BS 5628 as 'flexural tension'. If the wall is treated as a free cantilever, with a uniformly distributed wind load, the maximum bending moment will be $ph^2/2$. On the other hand, if the roof is used to prop the wall, and a pinned joint is assumed at the base, the wall acts as a simply supported beam with a maximum free bending moment due to wind of $ph^2/8$ (see Figure 12.1).

(c) Roof action
To enable the roof to prop the wall, so that the wall can span as a simply supported beam, as described in (b) above, it is not sufficient to merely fix the roof beams to the top of the wall, because the wall could still act as a free cantilever (see Figure 12.2).

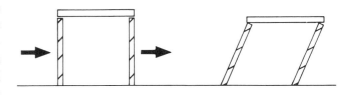

Figure 12.2

The roof must act as a plate or a wind girder so as to transfer the wind force to the gable or other transverse walls. A simple method of meeting this requirement is to fix diagonal bracing between the roof beams and purlins, this forming, in effect, a horizontal lattice (or wind) girder as

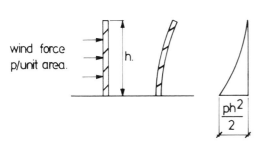

cantilever.

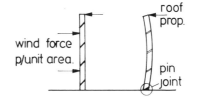

simply supported slab.

Figure 12.1

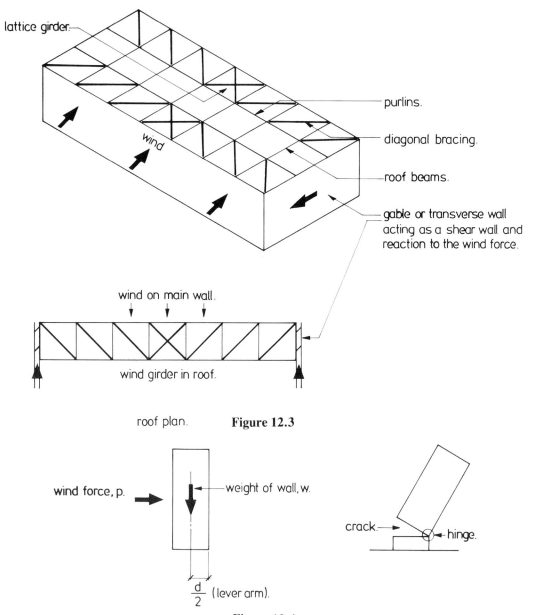

lattice girder

purlins.

diagonal bracing.

roof beams.

gable or transverse wall
acting as a shear wall and
reaction to the wind force.

wind

wind on main wall.

wind girder in roof.

roof plan.   **Figure 12.3**

wind force, p.   weight of wall, w.

crack   hinge.

$\dfrac{d}{2}$ (lever arm).

Figure 12.4

shown in Figure 12.3. It is essential, of course, that the roof is properly strapped and tied to the walls, as described in Chapter 7.

(d) Stability moment

In the free cantilever shown in Figure 12.4, the resistance of the wall to overturning, due to the force P, is its stability moment. To over-simplify for the sake of clarity, consider the stability moment of a wall = own weight of wall × its lever arm, i.e. $W \times d/2$. If the moment due to the wind is greater than the stability moment, the wall will crack at the dpc level on the windward face, and rotate at the hinge on the leeward face.

The stability moment provides some fixity at the base of the wall, and is not dependent on the structural action of the wall as a free or propped

cantilever, or partially restrained simply supported beam, etc. The stability moment is a resistance moment and, like any resistance moment in any structural element, is passive until activated by applied bending moments due to loading. The magnitude of the active stability moment is dependent on both the magnitude of the bending moment due to loading and any movement of the roof prop, if one is provided (see (e) Knife-edge condition). The magnitude of the potential stability moment depends on the wall's own weight and its lever arm. The thicker a wall, the greater is its own weight and lever arm, and, therefore, its stability moment.

In a diaphragm wall, see Chapter 13, the stability moment can be increased by increasing the depth of the void – thus increasing the lever arm.

EXAMPLE

Determine the stability moment of a 210 mm thick solid wall, a 315 mm thick solid wall and a diaphragm wall with two leaves 102.5 mm thick and a void of 210 mm. The walls are 3 m high and built of brickwork with a density of 20 kN/m³ (see Figure 12.5).

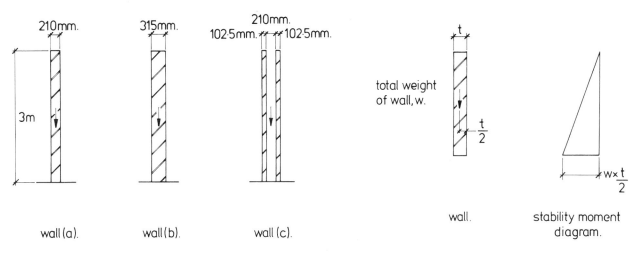

wall (a).   wall (b).   wall (c).

Figure 12.5

wall.   stability moment diagram.

Figure 12.6

Wall (a):
Weight $= 3 \times 0.210 \times 20$ $= 12.6$ kN/m run of wall
Lever arm $= \dfrac{210}{2}$ $= 105$ mm
                                                       $= 0.105$ m
Stability moment $= 12.60 \times 0.105$ m $= 1.32$ kN·m/m run

Wall (b):
Weight $= 3 \times 0.315 \times 20$ $= 18.90$ kN/m run
Lever arm $= \dfrac{315}{2}$ $= 157.5$ mm
                                                       $= 0.1575$ m
Stability moment $= 18.90 \times 0.1575$ m $= 2.977$ kN·m/m run

Wall (c):
Weight (ignoring cross-ribs) $= 3 \times (2 \times 0.1025) \times 20$ $= 12.30$ kN/m run
Lever arm $= \dfrac{210 + (2 \times 102.5)}{2}$ $= 207.5$ mm
                                                           $= 0.2075$ m
Stability moment $= 12.30 \times 0.2075$ m $= 2.55$ kN·m/m run

The stability moment is zero at the top of the wall (since there is no own weight active there) and increases uniformly, if the wall is of a constant thickness, to a maximum at the base where the full weight of the wall acts. Thus the stability moment diagram is as shown in Figure 12.6.

It should be appreciated that there is a fundamental difference between the resistance moment of steelwork or reinforced concrete and the stability moment of masonry. When the bending moment exceeds the resistance moment in steel or reinforced concrete, the resistance moment is destroyed once and for all. On the other hand, in unreinforced masonry, the stability moment is not destroyed – although it may be temporarily reduced – and on relief of the bending moment is fully restored. The example which follows may help to clarify this point (Figure 12.7).

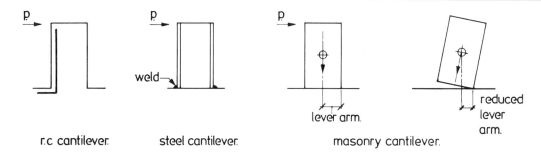

Figure 12.7

If the bending moment due to the force P exceeds the resistance moment of the reinforced concrete, the reinforcement will fail and the cantilever will have a greatly and permanently reduced moment of resistance. A similar result will occur in the steel cantilever if the tensile weld fails. In the masonry cantilever, the lever arm will decrease slightly, as the wall tilts, and thus decrease the stability moment. However, when the force P is removed, the wall will settle back into its original position and recover its full stability (resistance) moment. Since, in practice, the design force P is the maximum wind force likely to occur in the life of the building (e.g. a three second gust, once in fifty years) it is of very short and temporary duration.

Combined bending moment diagram

If the wall is propped at the top, the free bending moment due to the wind force is, as mentioned earlier, that of a simply supported beam subject to a uniformly distributed load and the maximum is $ph^2/8$. Superimposed on this is the active part of the stability moment (see Figure 12.8).

It can be shown that the maximum moment at the base due to the roof propping force, R, which is activated by wind force P/unit area, is $ph^2/8$, i.e. the propped cantilever bending moment. If this does not exceed the stability moment, the wall will act as a propped cantilever and the maximum bending moment, in the span of the wall, will be $9ph^2/128$ acting at $\frac{3}{8}h$ down the wall from the prop.

(e) Knife-edge condition

In practice, it is not possible to form a perfectly propped cantilever – just as it is difficult to form a perfectly pinned or fixed joint. Since the roof plate will be stressed, it will strain and deflect. It can be shown that deflecting the free end of the cantilever by an amount Δ will induce a moment at its support of $3EI\Delta/L^2$ (see Figure 12.9).

For a 210 mm wall, $I = \dfrac{bd^3}{12} = \dfrac{1\,000 \times 210^3}{12}$
$$= 7.718 \times 10^8 \text{ mm}^4.$$

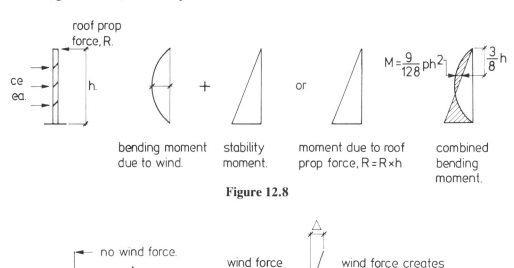

Figure 12.8

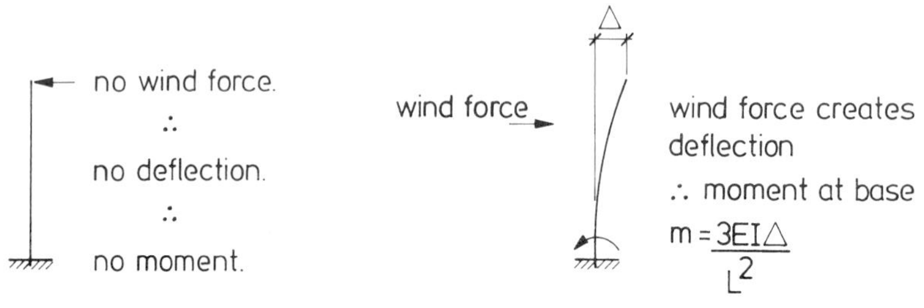

Figure 12.9

For brickwork, E_b can be taken as $= 8.68 \times 10^6$ kN/m² (medium strength brick in 1:3 mortar). For a 25 mm deflection at the top of a 3 m high wall:

$$M = \frac{3EI\Delta}{L^2}$$

$$= \frac{3 \times 8.68 \times 10^6 \times 7.718 \times 10^{-4} \times 25 \times 10^{-3}}{3 \times 3}$$

$$= 55.8 \text{ kN·m/m}$$

The stability moment for a 210 mm wall, 3 m high, is 1.32 kN·m/m (see earlier).

The moment at the base of the wall, due to deflection at the top, is far in excess of the stability moment. Thus, the wall will crack at the base or dpc, and will rotate, but not collapse (see Figure 12.10).

Theoretically, the whole weight of the wall will be concentrated on a knife-edge of zero contact area and there will be an infinitely high compressive stress at point A (see Figure 12.11).

In practice, of course, the dpc or mortar will deform and the contact area will increase from the knife-edge condition, and thus the compressive stress will decrease.

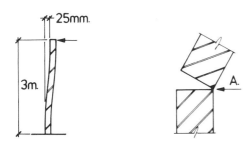

Figure 12.10   **Figure 12.11**

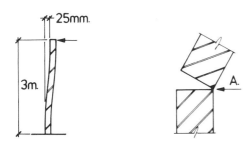

Figure 12.12

The contact area, see Figure 12.12, will now be 12 mm × 1 m per m run and the compressive stress will be

$$\frac{12.6 \times 10^3}{12 \times 10^3} = 1.05 \text{ N/mm}^2.$$

From the above, it will be seen that the moment at the base cannot exceed the stability moment – if it does, the wall cracks at the base and only the stability moment is operative.

(f) Tailing down of roof

It is not uncommon with lightweight roofs for their dead (downward) load to be less than the suction (upward) force due to wind action (see Figure 12.13).

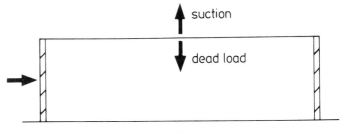

Figure 12.13

If the roof is not strapped down to the walls, or fixed to a concrete capping beam that weights it down to the walls, it will lift off. A factor of safety against uplift of 1.4 is usually adopted (see Chapter 7).

(g) Slenderness ratio

If the roof acts as a horizontal stiff plate or wind girder, and is properly fixed to the top of the wall, the wall can be considered as being adequately restrained laterally, and its effective height can be considered as being equal to its actual height. If the roof does not properly restrain the wall, the effective height should be taken as 1.5 times the actual height.

The effective thickness of walls has been dealt with in previous chapters.

The slenderness ratio must not exceed 27. So, for a 215 mm thick wall, properly restrained by the roof, the maximum height = 210 × 27 = 5.67 m. A 255 mm cavity wall (effective thickness two-thirds the sum of the thickness of the two leaves = 137 mm) must not exceed a height of 27 × 137 = 3.7 m. However, walls of this height and thickness could not withstand normal wind pressures. As stated at the beginning of this chapter, the vertical compressive load is rarely a critical design factor in practice – since the roof loading is comparatively light. It follows, therefore, that the slenderness ratio is equally rarely a critical design factor.

(h) Robustness

Consideration must be given to the robustness of

the structure (see Chapters 7 and 8). Since the main contribution to robustness in a single-storey open-plan structure comes from the interaction and connections of the walls and the roof, it is more than ever essential to ensure that the roof is properly and adequately braced, and firmly fixed to the walls.

12.2 DESIGN PROCEDURE

Usually, it is the height (vertical span) of the wall, and the bending tensile stresses (flexural tension) that will develop which govern the wall's type and thickness. The stresses at the base and at the position of the maximum bending moment in the wall's height must be checked.

For low walls, up to about 4 m depending on the wind pressure at the side, the normal solid or piered cavity wall may suffice.

For medium height walls, up to about 5 m, solid or cavity walls will need stiffening with piers or thickening.

For tall walls, possibly more than 5 m in height – and almost certainly when the height exceeds 6 m – fin or diaphragm configurations should be considered.

For tall walls, check the application of post-tensioning to produce a more economical design (see Chapter 15).

The design procedure is as follows:

1. Calculate the positive and negative wind pressures.
2. Calculate the dead, imposed and wind loading on the wall from the roof.
3. Decide on the best distribution of structural *elements* and the overall behaviour of the structure and element interaction (see Chapter 7).
4. Select a trial section (solid, cavity, piered, diaphragm, fin) and thickness. Check whether the use of post-tensioning is appropriate. Check the slenderness ratio.
5. Determine the strapping down or weighting down of the roof.
6. Check the roof plate action, i.e. provide adequate bracing and its connection to the shear walls. If there are crosswalls, or other internal transverse walls used for permanent partitions, fire breaks, etc., use should be made of them to reduce the span of the roof plate or wind girder.
7. Determine the free bending moment, $ph^2/8$, the propped cantilever span moment, i.e. $9ph^2/128$, and the stability moment at the base of the wall.
8. Calculate the position and magnitude of the maximum wind moment in the height of the wall, and the resistance moment of the wall, and compare.
9. Check stresses at the base of the wall, and at the position of the maximum span moment.
10. Revise trial section, if necessary.
11. Choose masonry unit and mortar strengths.
12. Calculate the shear stresses.
13. Check the stability of the transverse walls (shear walls) for roof plate wind reaction.

CHAPTER 13

FIN AND DIAPHRAGM WALLS IN TALL SINGLE-STOREY BUILDINGS

The authors' experience on sports halls, gymnasia, stadia, assembly halls and structures of similar form has shown that fin and diaphragm walls are well suited to tall single-storey buildings enclosing large open areas. Such buildings account for a large number of the projects constructed in Britain, and throughout the rest of the world, and their importance is particularly relevant with the present trend in this country towards providing facilities for public recreation and leisure. The vast majority of these structures have a steel or reinforced concrete framework supporting the roof loads. The framework columns are then enveloped by a cladding material, backed up by an insulating barrier and protected on the inner face by a hard lining. Frequently, the cladding, insulation and lining require a subsidiary steel framework to provide support, and both the main frame columns, and sometimes the subsidiary frame also, require fire protection. The specification for painting the structural framework depends upon its degree of exposure and accessibility, and in unfavourable conditions the costs against this item can be unexpectedly high. The resulting 'wall' thus requires up to six different materials and several sub-contractors, suppliers and trades. The framework and cladding require frequent maintenance and do not provide the durability afforded by the use of masonry for the same purpose, neither do they possess the same aesthetic qualities which are natural in masonry construction and which can be greatly enhanced by imaginative detailing.

The fin or diaphragm wall forms the structure, cladding, insulation, lining and fire barrier in one material, using one trade carried out by the main contractor. Maintenance is minimal, applied protective coatings are eliminated and durability is virtually ensured. They also have obvious applications to industrial structures where robustness to resist the hard wear of the associated operations is of prime importance. Vandal resistance is an added bonus to all projects employing fin and diaphragm wall construction.

Masonry, like all other structural materials, requires a full understanding of its strengths and weaknesses in order to employ it economically. Masonry's previously stated main weakness, low tensile strength, can be compensated for in design by providing a high Z/A ratio when bending stresses are involved. It is equally important to take full advantage of the gravitational forces involved, and the combination of these two aspects of masonry design led to the development of the diaphragm wall. An alternative solution to overcome masonry's poor tensile resistance is to provide precompression in the wall through post-tensioning rods spaced at designed centres and torqued to provide the axial loading which is usually missing from tall single-storey structures. This alternative is discussed later in this chapter and in Chapter 15.

In order to exploit both the highest Z/A ratio and gravitational resistance, the geometric distribution of the materials should be similar – that is, to place the material at its largest practical lever arm position. In arriving at the most suitable geometric profile, due consideration must be given to the shear forces involved and to the buckling tendency of the material in the compression zone of the profile.

For practical considerations, the geometric arrangement of the wall must also relate to multiples of standard brick or block dimensions.

A diaphragm wall comprises two parallel leaves of brickwork or blockwork spaced apart and joined by perpendicular cross-ribs placed at regular intervals to form box or I sections (see Figures 13.1 and 13.2).

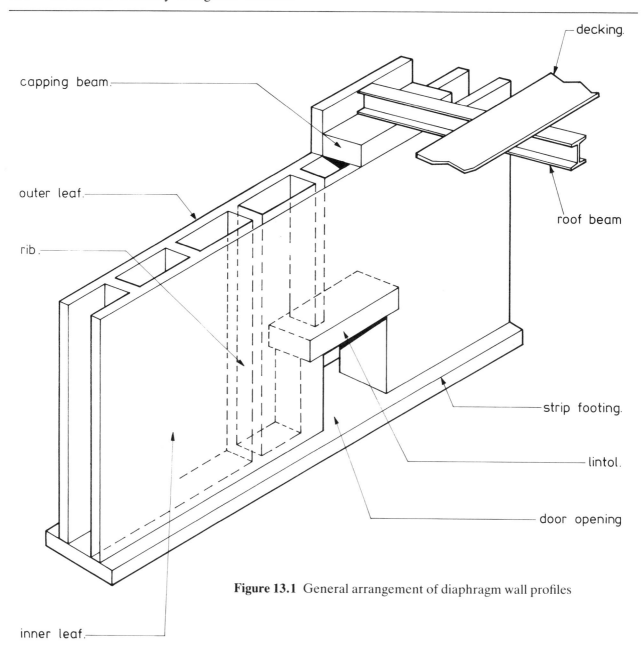

Figure 13.1 General arrangement of diaphragm wall profiles

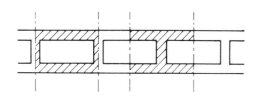

Figure 13.2 Diaphragm wall box and I section

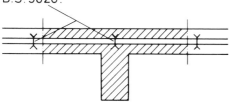

Figure 13.3 Fin wall arrangement

The two parallel leaves of the wall act as flanges in resisting the bending stresses and are stiffened by the ribs acting as webs mainly resisting shear forces. The length of the parallel leaves, which may be considered to act with the cross-ribs, is often limited by their tendency to buckle and, therefore, the section is best appraised as an I section. The length of the flange of the I section is established in a similar way to that of the T beam in reinforced concrete design, which should be familiar to many designers. The depth between the flanges is designed to meet the individual structural and other requirements of each project. Costs and space are usually minimised by designing the shallowest depths practicable.

The fin wall was developed from the diaphragm wall and its general form is shown in Figure 13.3.

The masonry T section formed by the projecting fin and the bonded leaf of the cavity wall provides the main supporting member of the structure, whilst the other leaf of the cavity wall provides either the lining or the cladding depending on whether the fins are externally or internally exposed. The whole fin plus the cavity wall is used in determining the slenderness ratio of the section, and a calculated length of the wall is considered to act with the fin as the flange of the T profile in resisting the lateral loading. It is more common to expose the projecting fins externally, as this is usually the preference of the architectural designer and greater structural economy can be achieved. However, they can be exposed internally, and the design principles involved are similar, although careful consideration must always be given to the direction of the loading, and the section available at a particular level to resist it.

13.1 COMPARISON OF FIN AND DIAPHRAGM WALLS

Having concluded that, for a particular tall single-storey project, masonry is the most suitable structural material, the next decision to be made is what form: fin, diaphragm or any other, to use for the structure. Regarding fin and diaphragm walls, each has some advantages over the other and a summary of the basic considerations is given below, from which the form most suited to the function or aesthetics of the particular project can be assessed.

Diaphragm walls
(1) Smooth, finished face both internally and externally.
(2) Better structural use of materials.
(3) Large voids available for distribution of services, etc.
(4) No cavity ties in bonded walls.
(5) Symmetrical section for simplicity of analysis.
(6) Fewer vertical plumbing lines reduces labour costs.
(7) Smaller site area required – beneficial on restricted sites.
(8) Slight cost saving.

Fin wall
(1) Less roof area is required (see Figure 13.4).
(2) Less foundation area is required (see Figure 13.5).
(3) Has greater visual impact – more scope for architectural effect.

(4) Marginally easier to post-tension when required.
(5) Less cutting of bricks for bonding can usually be achieved.

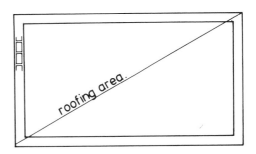

diaphragm wall

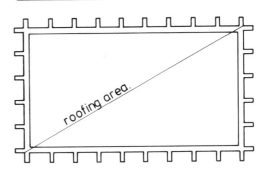

fin wall

Figure 13.4 Comparison of roofing areas

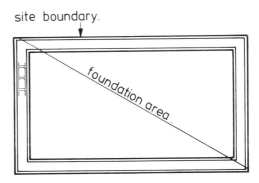

diaphragm wall.

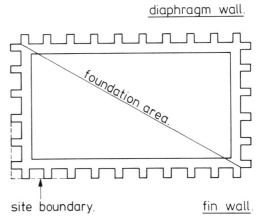

site boundary.   fin wall.

Figure 13.5 Comparison of foundation areas

Both the fin and the diaphragm walls become more economical, in comparison to other structural forms such as steel or reinforced concrete frameworks, as the height of the wall increases, and they are of little advantage on lower heights where normal cavity brickwork can often satisfy all the structural requirements. For further discussion on the application of fin and diaphragm walls see section 13.11, and Chapters 10, 11, 12 and 14.

13.2 DESIGN AND CONSTRUCTION DETAILS

Thorough consideration of the structural behaviour of the roof of the building is imperative for the maximum economy to be achieved in the overall building costs. The wall may be designed as a cantilever and the structure covered with the simplest possible roof construction. However, it has generally been found that, to obtain the greatest economy, the roof should be detailed and constructed in such a way that it can act as a horizontal plate to prop and tie the tops of the walls and to transfer the resulting horizontal reactions to the transverse walls of the building, where these reactions can then be transferred to the building foundations through the racking resistance of these shear walls (see Figure 13.26). To satisfy this design analysis, the details must provide adequately for fixing the tops of the walls to the roof plate, the roof plate must be capable of spanning between the shear walls, and the forces must be transferred from the roof plate into the shear walls.

A capping beam can be used on top of the wall to transfer the prop and tie forces into the roof plate. This has the potential advantage of being able to resist uplift forces from a lightweight roof and also of transferring the roof plate forces into the shear walls if the capping beam is continued all round the building. If, due to large roof openings or unsuitable decking, the plate action of the roof cannot be relied upon, a wind girder may be provided (see figure 13.6), in which case the capping beams can often be used as booms for this girder.

The roof decking can be constructed from a variety of materials and supported in many ways. Generally, steel universal beams, castellated beams or lattice girders have been found to be the most economical means of support, spaced at centres to suit the selected decking. They do not necessarily need to relate to the centres of ribs or fins. However, in fin wall construction, the

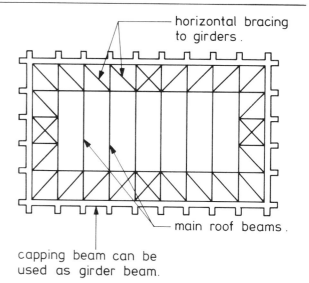

Figure 13.6 Roof girder to transmit wind forces to shear walls

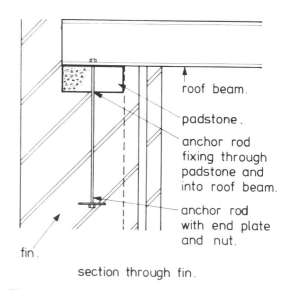

Figure 13.7 Anchoring detail for main roof beams

geometry of the building invariably leads to the roof supporting members lining up with the projecting fins. For long roof spans, a space deck can prove to be more economical, and the aesthetic value of this system combined with its economy, when applicable, makes it a popular proposition. Alternatively, timber laminated beams with solid timber decking may be used with considerable visual effect, although their economy would need to be balanced against the attractiveness of the finished product. The simplest solution in timber is, perhaps, provided by trusses with a suitably designed bracing system.

A capping beam is generally required at the top of diaphragm walls. However, for both fin and diaphragm walls where no capping beam is to be used, the main roof beam often requires strapping down to resist wind uplift forces. This can be quite easily done using rods cast into the padstone and taken down into the brickwork to a suitable level to ensure sufficient dead load, with an adequate factor of safety, to resist the uplift forces (see Figure 13.7).

When assessing the overall costs of the roof decking, it is necessary to take account of the value of its ability to act as a roof plate to resist the prop and tie forces discussed earlier. If an apparently less expensive roof decking is selected, any additional costs for strapping, bracing, etc., which would not necessarily have been required for apparently more expensive decking, must be included to arrive at the overall cost.

13.3 ARCHITECTURAL DESIGN AND DETAILING

It is generally considered that the fin wall provides greater scope for architectural expression

than the diaphragm wall. A typical simple plan layout for a fin wall building is shown in Figure 13.8 and an almost unlimited number of variations can be applied to this basic profile.

The sizes and spacing of the fins can vary, and the corner fins can be eliminated altogether as shown in Figure 13.9.

The fins themselves can be profiled on elevation, some examples of which are indicated in Figure 13.10.

The treatment at eaves level (see Figure 13.11) and the variety and mixture of the facing bricks and fin types can present unlimited and interesting visual effects.

A word of caution, however. When a mixture of bricks is to be introduced it is essential to ensure that the various bricks and/or blocks are compatible, particularly with regard to themal and moisture movements. The structural design calculations must also take account of the differing design strengths of the masonry under these circumstances. The diaphragm wall also has possibilities for architectural expression and some examples of its treatment at roof level are shown in Figure 13.12.

It is not essential that diaphragm walls should be designed with flat faces on elevation and, particularly on tall buildings, a fluted arrangement as shown in Figure 13.13 can break up a large expanse of brickwork.

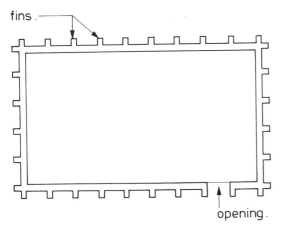

Figure 13.8 Typical simple building plan in fin wall construction

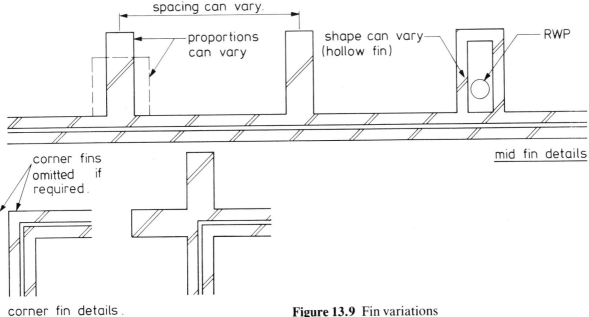

corner fin details.

Figure 13.9 Fin variations

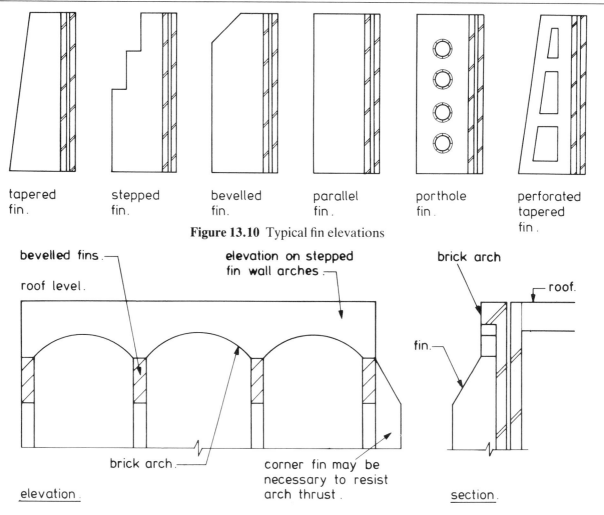

Figure 13.10 Typical fin elevations

tapered fin.   stepped fin.   bevelled fin.   parallel fin.   porthole fin.   perforated tapered fin.

Figure 13.11 Example of treatment at eaves for fin walls

The cross-ribs should, from a structural preference point of view, be bonded into the inner and outer leaves in which case they show as headers on the elevations. A different coloured brick used for the cross-ribs can create an interesting feature. It is, however, also possible to allow the cross-ribs to butt up to the inner faces of the elevational leaves, in which case the stretcher bonding would remain uninterrupted. In such a situation, designed shear ties are necessary to tie the ribs to both the inner and outer leaves to resist the shear forces involved, and it is essential to provide in the specification adequate protection for the ties to ensure that they are sufficiently durable to resist corrosion. The cost implications of bonded or unbonded cross-ribs vary from job to job, but it is unlikely to have a significant effect on the overall cost appraisal. Once again, the introduction of a different brick for the cross-ribs, or the inner leaf, would require the same check for compatibility and design strengths as was discussed earlier.

13.3.1 Services
The accommodation of building services within diaphragm wall structures presents few problems owing to the large vertical voids in the wall section. The services can be placed in service ducts incorporated into the wall profile, as shown in Figure 13.14, or can be run inside the void with access points built into the relevant leaf as required.

Openings for such access points must be checked for the possibility of local overstressing in the brickwork. Service ducts housing gas pipes placed within diaphragm wall voids should, of course, be ventilated. Careful consideration must always be given to the possibility of corrosion of services within these locations.

The accommodation of building services within fin wall structures is no different than for normal cavity walls. If necessary, however, the fin profile can be made, as shown in Figures 13.9 and 13.15, to include a void for the distribution of services and the same basic considerations apply as have been discussed for the services within diaphragm walls. Both the diaphragm wall and voided fin wall are particularly suited to the inclusion of rainwater downpipes.

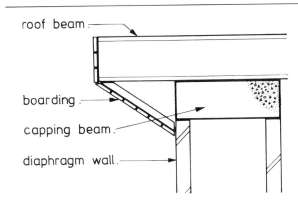

roof oversail on diaphragm.

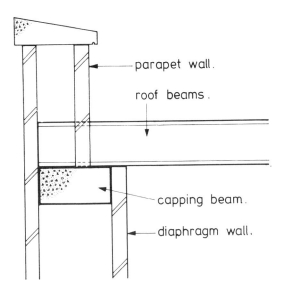

parapet upstand above diaphragm.

Figure 13.12 Detail at head of diaphragm wall

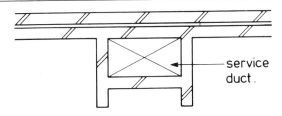

Figure13.15 Service duct detail

13.3.2 Sound and thermal insulation

Fin walls have almost the same sound and thermal insulating properties as normal cavity walls, and the same criteria for the improvement of both of these properties apply.

Diaphragm walls, however, because of the large internal void, possess better prospects for both sound and thermal insulation. Insulating boards and quilts of varying thicknesses can be quite easily fixed inside the void as shown in Figure 13.16, however, the U value of a basic diaphragm wall is estimated to be approximately 10% higher than an equivalent traditional cavity wall owing to the greater air circulation within the larger void.

13.3.3 Damp proof courses and membranes

Horizontal damp proof courses should be selected to give the necessary shear resistance to prevent sliding and should not squeeze out under vertical loading. Where flexural tensile resistance has been assumed in the structural design, particular care should be exercised in the choice and construction supervision of the damp proof course. In fin wall construction, vertical damp proof membranes separating the inner and outer

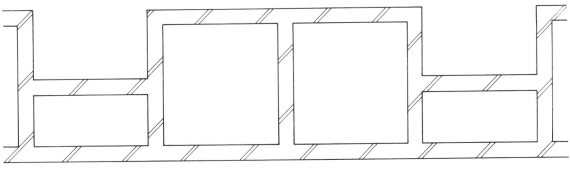

Figure 13.13 Fluted diaphragm

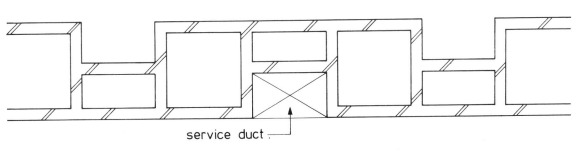

Figure 13.14 Accommodation of service duct

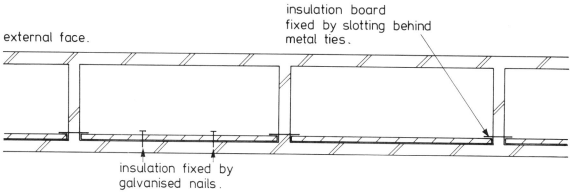

external face.

insulation board
fixed by slotting behind
metal ties.

insulation fixed by
galvanised nails.

Figure 13.16 Insulation fixing detail

leaves at door and window openings create fewer structural problems than with diaphragm walls, and can generally be quite easily accommodated. Vertical damp proof membranes are not normally necessary within diaphragm walls, except at door and window openings, provided that bricks and mortar of suitable and compatible quality are used to suit the environmental conditions. Most vertical damp proofing membranes prevent the tying of the cross-ribs to the elevational leaves and should be avoided wherever possible as this would impair the box action of the compound wall profile. If required, a bitumen based painted dpc, used in conjunction with metal shear ties, can be used in these locations. At door and window openings however, vertical dpms can be incorporated by the introduction of additional cross-ribs, as shown in Figure 13.17.

13.3.4 Cavity cleaning
There is little difference between fin walls and ordinary cavity walls with regard to the problems of cavity cleaning. With diaphragm walls, however, this problem is significantly reduced owing to the large void and, provided that normal care is excercised during construction, no elaborate methods are necessary for cleaning out the voids.

13.4 STRUCTURAL DETAILING

It is essential for any structural scheme to ensure that the assumptions made in the design process

are adequately provided for in the detailing and construction on site. This is equally true of masonry structures, and there is perhaps a good argument to suggest that masonry structures require slightly more attention to detailing than other forms of construction. Masonry structures have become accepted as the traditional type of building in which well-tested details of construction have evolved and become commonplace. This has had the effect of implanting the notion that, if it is a masonry structure, the traditional construction details will solve the structural aspects and hence an engineer's services are unwarranted. This was not an unreasonable attitude when the term 'masonry structure' automatically implied 'massive structure'. However, with the modern trend towards minimising the mass of the structure, to reduce both material and labour costs, such an attitude is likely to result in unstable construction. The services of engineers are warranted more today than at any other time in relation to masonry structures, and the wide scope for architectural design provided by fin and diaphragm walls is the most recent example of the value of their contribution. This contribution should not be limited to the provision of wall thicknesses and strengths, but should include an assessment of the economies to be achieved from the most suitable combination of all the structural elements of the building. Having advised on the most suitable combination of these elements, the most important task is to

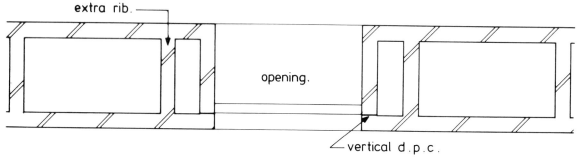

extra rib.

opening.

vertical d.p.c.

Figure 13.17 Vertical dpc at openings

ensure that they are correctly detailed and constructed in relation to each other.

13.4.1 Foundations

Generally speaking, the foundations to both fin and diaphragm walls comprise simple strip footings, as shown in Figures 13.18 and 13.19, slightly wider than normal for the diaphragm wall and with local projections to support the fins in fin wall construction. The bearing pressures involved for both wall types are invariably so low that nothing more is necessary, for a site which does not have a particular soil problem. Whatever problems are presented by the subsoil conditions, the foundation solution for fin and diaphragm walls is no more complex than for a traditional masonry structure and, in fact, the considerable stiffness provided by these geometric forms, combined with their relatively lightweight construction, has created new scope for masonry structures on difficult sites. An example of this is the sports hall of a community centre for which the authors were responsible for the structural design, and which was constructed using post-tensioned diaphragm wall construction. The foundation adopted was a cellular raft, which was necessary to cater for ground subsidence resulting from the future coal extraction beneath the site. The first of these subsidence waves from the mine workings has since traversed the site resulting in a maximum subsidence of approximately 1080 mm with a max-

imum out-of-level across the sports hall itself of approximately 130 mm. The relatively lightweight superstructure construction permitted an economical foundation design and the success of the walls is self-evident in that there is no evidence whatever of cracking or distress in the masonry due to the subsidence movements which have occurred. A structure has therefore been provided which can withstand these massive subsidence movements with virtually no attendant maintenance implications to the client.

13.4.2 Joints

Movement control joints are required in both fin and diaphragm wall construction, their requirements being no different than for simple load-bearing masonry, the recommendations for which are given in CP 121. In fin wall construction, the joints are best accommodated by introducing a double fin with the joint sandwiched between (see Figure 13.20).

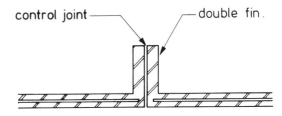

Figure 13.20 Provision of control joint in fin wall

Similarly, with diaphragm walls, a double rib can be provided as shown in Figure 13.21.

Invariably, the materials adopted for the inner leaf of fin walls or the inner face of diaphragm walls differ from those of the external faces. It is often necessary in these circumstances to provide control joints at closer centres on the inner faces, and these intermediate joints may not necessarily demand double fins or double ribs. Any joints which are introduced into the wall leaf, however, must be located with careful consideration to the design assumptions which may have been made with respect to that wall panel, i.e. the possibility of the wall panel having been designed to span horizontally between fins, etc.

13.4.3 Wall openings

Large door and window openings can create high local loading conditions from the horizontal wind loading and concentrated axial loads at the lintol supports. In both fin and diaphragm walls, these openings can easily be accommodated with an adjustment to the fin or rib sizes and/or spacing at the lintol bearings (see Figures 13.22 and 13.23).

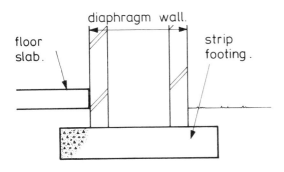

Figure 13.18 Typical diaphragm wall strip footing

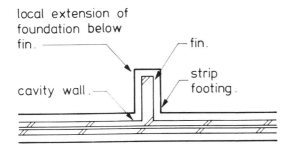

Figure 13.19 Typical fin wall strip footing

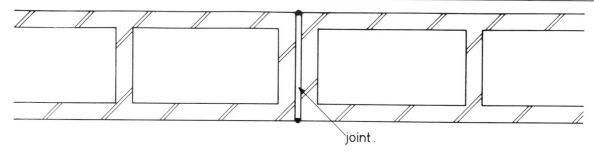

Figure 13.21 Diaphragm wall: movement joint detail

The designer must allow, in his calculations, not only for the increased vertical and horizontal loading involved at these locations, but also for the change to the geometric wall profile available to resist the increased loading.

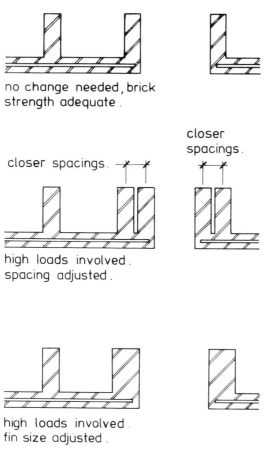

no change needed, brick strength adequate.

closer spacings.

closer spacings.

high loads involved. spacing adjusted.

high loads involved. fin size adjusted.

Figure 13.22 Typical opening details: fin walls

13.4.4 Construction of capping beam

Where appropriate, it is preferable from a structural point of view to cast the rc capping beam *in situ*. However, the beam can be precast, ideally in bay lengths, with a suitably detailed connection between to transfer the relevant forces at the joints. Precasting appears to be the more popular solution with contractors, as it eliminates the problems of protecting facing brickwork from wet concrete runs and also saves the expense of the temporary/permanent shuttering that is required for the *in situ* construction. A better quality finish can also be achieved with precasting if the capping beam is to be exposed. The capping beam is used at the seating for the roof structure as shown in Figures 13.24 and 13.25 and can be used, either alone or as part of a horizontal wind girder, to transfer the propping force at the head of the wall to the gable shear walls.

13.4.5 Temporary propping and scaffolding

Like most other walls, fin walls and diaphragm walls are in a critical state during erection prior to the roof being fixed, particularly when they have been designed as propped cantilevers. During this period, the contractor must take the normal precautions, such as temporarily propping the walls from the bricklayers' scaffolding or other means to ensure stability. Owing to the inherent stiffness of both the fin and the diaphragm wall the problem of temporary stability is

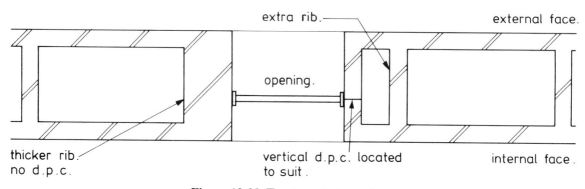

extra rib.

external face.

opening.

thicker rib. no d.p.c.

vertical d.p.c. located to suit.

internal face.

Figure 13.23 Treatment at openings

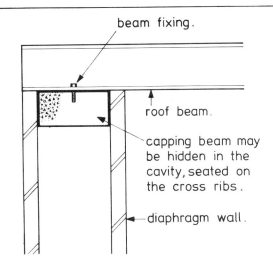

Figure 13.24 Capping beam detail: diaphragm wall

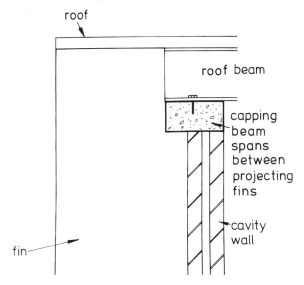

Figure 13.25 Capping beam detail: fin wall

considerably reduced from that of a simple wall, however, the height to which these walls are likely to extend must also be taken into account in assessing the propping requirements.

It is recommended that, for both fin and diaphragm wall construction, scaffolding is erected both internally and externally to ensure not only good line and plumb but also complete filling of all bed and perpendicular mortar joints. This is particularly important at the fin and rib locations, and it is considered that working over-hand from a single scaffold platform is more likely to result in poor workmanship. The double-face scaffolding arrangement should consequently provide an adequate means of temporary propping. In fin wall construction the scaffolding on the fin face of the wall must be erected in such a way as to allow the fins to be constructed at the same time as the main wall. Contractors may prefer to place the scaffold against the wall face, leaving pockets in the wall with the intention of block-bonding

the fins into the main wall at a later date. This should not be permitted, as it can not only present a hazard in its temporary state but, more important, the bonding and filling of the pockets are likely to result in an inferior quality of construction, possibly far removed from the section analysed in the calculations. It is usually essential that the fins are constructed as each course rises and the scaffolding should be arranged to cater for this requirement.

13.5 STRUCTURAL DESIGN: GENERAL

For single-storey buildings the critical design condition is rarely governed by axial compressive loading rather by lateral loading from wind forces. The limiting stresses are generally on the tensile face of the wall and it is therefore necessary to provide structural elements which are best equipped to limit these tensile stresses. Thus the development of the fin and diaphragm wall profiles in which the material of the wall is placed at a greater lever arm than conventional walls to significantly reduce the tensile stresses and in turn increase the moment of resistance of the section.

Having provided the most efficient element to reduce the flexural tensile stresses, two further considerations should be made to improve even more on this efficiency, these being:

(a) To use the roof plate as a prop to the head of the wall, transferring this propping force to the gable shear walls through the stiffness of the roof plate, or through a suitable bracing system provided for the purpose as shown in Figure 13.26. This enables the wall element to be designed as a propped cantilever, reducing the applied bending moment in the height of the wall and thus reducing the critical flexural tensile stresses.

(b) The use of post-tensioning, as shown in Figure 13.27, to increase the axial compressive stresses in the wall element and reduce the flexural tensile stresses.

Consequently, both fin and diaphragm walls, when used in tall single-storey buildings, are usually designed as propped cantilevers and the critical loading condition to consider is that of combined dead and wind loading. This takes into account the maximum wind uplift on the roof, and thus the maximum flexural tensile stresses within the masonry. The maximum compressive

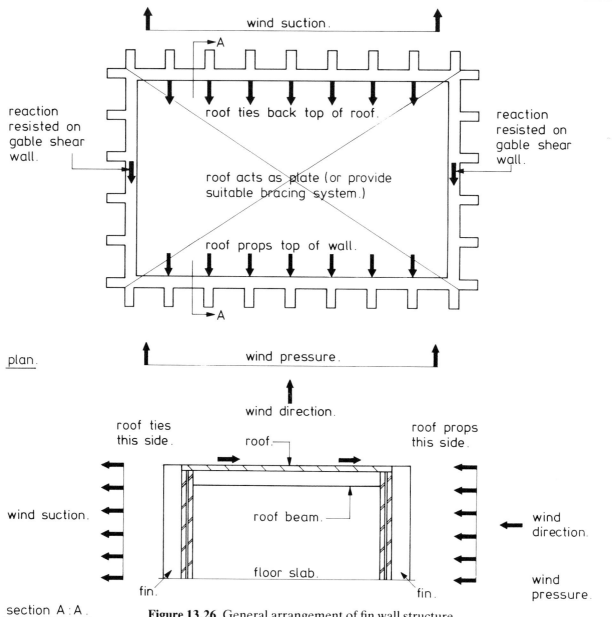

Figure 13.26 General arrangement of fin wall structure

stresses (resulting from combined dead plus superimposed plus wind loading) in a diaphragm wall are, generally, so low that the selection of a suitable brick and mortar is based almost entirely on the minimum requirements for durability and absorption. For fin walls, however, these maximum compressive stresses can become more critical, particularly when the compressive stresses at the extreme end of the fin are considered, as will be demonstrated in the worked example to follow. Hence, for fin walls, the selection of a suitable brick and mortar combination is more likely to be governed by the required compressive strengths as well as the durability and absorption criteria.

Calculations are carried out on a trial-and-error basis, by adopting a trial section and then checking the critical stresses. Guidance for the assess-

ment of trial sections for both fin and diaphragm walls is given in the worked examples which follow.

13.5.1 Design principles: Propped cantilever
Within the height of the wall, there are two locations of critical bending moments:

level A: at the base of the wall which is usually at dpc level;
level B: at a level approximately $\frac{3}{8}h$ down from the top of the wall (see Figure 13.28).

Owing to the unsymmetrical shape of the fin wall, it is essential to check the stresses at both levels and for both directions of wind loadings. However for the diaphragm wall, only the more onerous direction of wind loading need be considered, which is usually that of wind pressure.

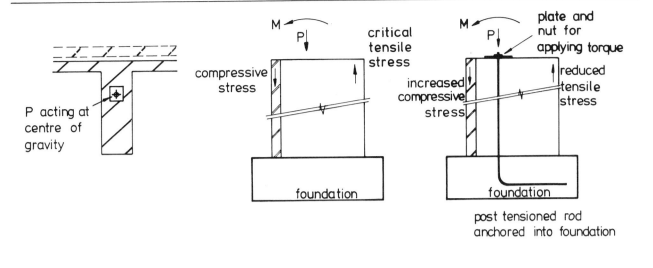

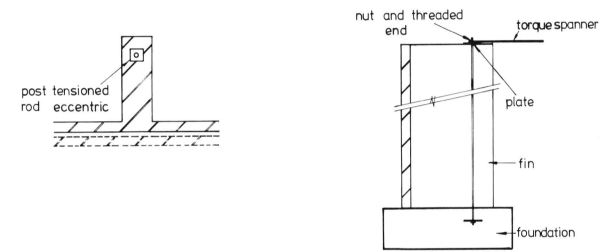

Figure 13.27 Post-tensioned fin wall details

The lateral loading will partly dictate the spacing of the fins in fin wall construction, and the spacing of the leaves and centres of the ribs in diaphragm wall construction. These aspects will be considered in greater detail in due course.

13.5.2 Calculate design loadings

It is essential to consider, at each stage of the design process for both fin and diaphragm walls, the worst combination of loading relevant to the particular check being carried out. For example,

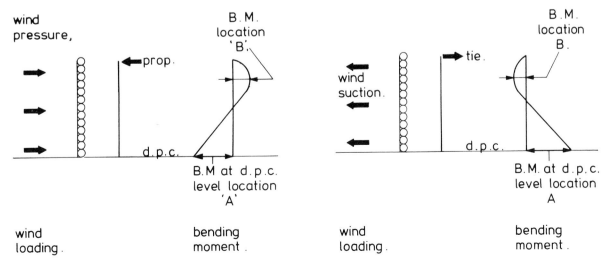

Figure 13.28 Wall action showing bending moment diagrams

if roof uplift is likely to occur, this will affect not only the flexural tensile resistance in the height of the wall but also the fixing moment of resistance for the 'cracked section' to be designed at the base of the wall. Concentrated axial loading at lintol bearings, combined with the concentration of roof uplift forces at the same location, must also be given due consideration.

13.5.3 Consider levels of critical stresses
For a uniformly distributed load on a propped cantilever of constant stiffness with no differential movement of the prop the bending moment diagram would be as shown in Figure 13.29, case (a).

However, in reality, for both fin and diaphragm walls, some deflection will occur at the head of the wall (prop location for the propped cantilever design), and the walls are not of constant stiffness throughout their height as the stability resistance moment at any particular level is related to the vertical load in the wall at that level and

this is not constant. It is, therefore, a coincidence if the resistance moment at the base is exactly equal to $ph^2/8$, which is applicable to a true propped cantilever. Thus it is usually necessary to adjust the bending moment diagram from that of a true propped cantilever, as will be explained. The upper level of critical stress does not necessarily occur at $\tfrac{3}{8}h$ from the top of the wall, but should be calculated to coincide with the point of zero shear on the adjusted bending moment diagram. The second level of critical stress to be considered will still occur at the base of the wall. This aspect of the design process is analysed in greater detail in the calculation of Design Bending Moment.

13.5.4 Design bending moments
It has been assumed that the wall acts in a similar manner to a propped cantilever, 'propped' by the action of the roof plate and 'fixed' at the base by virtue of its self-weight, the net weight of the roof structure and zero flexural tensile strength at the base. The fixed-end moment at the base due to

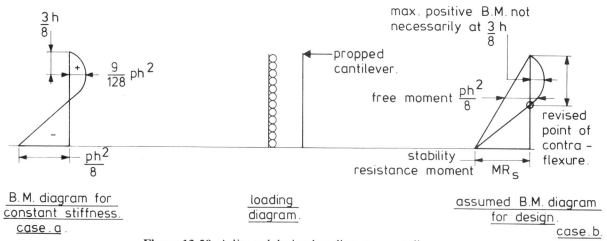

Figure 13.29 Adjusted design bending moment diagrams

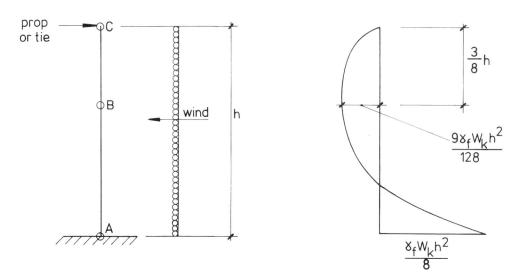

Figure 13.30 Plastic hinges in propped cantilever

the vertical loads is termed the 'stability moment'. Any wall has a stability moment of resistance (MR$_s$) throughout its height which reduces in value nearer to the top of the wall owing to the reduced self-weight. The stability moment of resistance (MR$_s$) effectively augments the design flexural strength of the wall at the higher level. However, the reason for taking zero flexural tensile strength at the base, even if a dpc capable of transferring tensile stresses is adopted, requires further explanation and involves the application of 'plastic' analysis.

The 'plastic' analysis of the wall action considers the development of 'plastic' hinges (or 'crack' hinges) and the implications of the mechanisms of failure.

Referring to Figure 13.30, three plastic hinges are necessary to produce failure of the propped cantilever shown, and these will occur at locations A, B and C. Location C, the prop, is taken to be a permanent hinge, hence, under lateral loading, the two hinges at A and B require full analysis.

As the lateral loading is applied the wall will flex, moments will develop to a maximum at A and B and the roof plate action will provide the propping force at C.

As the roof plate is unlikely to be sensibly rigid, some deflection must be considered to occur which will allow the prop at the head of the wall to move and the wall as a whole to rotate. This deflection of the roof plate will be a maximum at mid-span and zero at the gable shear wall positions (see Figure 13.44). Thus each individual fin will be subjected to slightly differing loading/rotation conditions. If, in addition to the stability moment of resistance at base level, flexural tensile resistance is also exploited to increase the resistance moment, there is a considerable danger that the rotation, due to the deflection of the roof plate prop, may eliminate this flexural tensile resistance by causing the wall to crack at base level. The effect of this additional rotation would be an instantaneous reduction in resistance moment at this level. This, in turn, would require the wall section at level B to resist the excess loading transferred to that level, and this could well exceed the resistance moment available at that level. Hence, the two plastic hinges at levels A and B would occur simultaneously, and possibly in advance of the design load having been reached. If, however, the flexural tensile resist-

ance which may be available at base level is ignored, the design bending moment diagram will utilise only the stability moment of resistance at base level, and this will remain unaffected by whatever rotation may occur.

In order to determine the required brick and mortar strengths, it is first necessary to decide the maximum forces, moments and stresses within the wall. If the applied wind moment at the base of the wall should, by coincidence, be exactly equal to the stability moment of resistance (MR$_s$), the three maxima specified above (maximum forces, moments and stresses) will be found at the base and at a level $\frac{3}{8}h$ down from the top of the wall.

If the MR$_s$ is less than the applied base wind moment of $\gamma_f W_k h^2/8$, or if significant lateral deflection of the roof prop occurs, the wall will tend to rotate and crack at the base. Provided that no tensile resistance exists at this level, the MR$_s$ will not decrease because the small rotation will cause an insignificant reduction in the lever arm of the vertical load. However, on the adjusted bending moment diagram, the level of the maximum wall moment will not now be at $\frac{3}{8}h$ down from the top and its value will exceed

$$\frac{9}{128} \gamma_f W_k h^2$$

For example: suppose the numerical value of a particular MR$_s$ is equivalent to, say $\gamma_f W_k h^2/10$ then the reactions at base and prop levels would be:

$$= \frac{\gamma_f W_k h}{2} \pm \frac{\gamma_f W_k h^2}{10h}$$

$$= 0.5\gamma_f W_k h \pm 0.1\gamma_f W_k h$$

$$= 0.6\gamma_f W_k h \text{ at base level}$$

$$= 0.4\gamma_f W_k h \text{ at prop level (see Figure 13.31)}$$

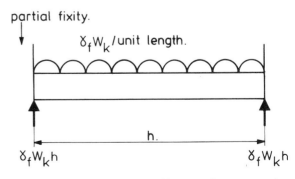

Figure 13.31 Calculation of base and prop reactions

The MR_s is inadequate to resist a true propped cantilever base moment of $\gamma_f W_k h^2/8$, and the section will crack and any additional load resistance available at the higher level will come into play. The true propped cantilever BM diagram is adjusted to allow greater share of the total load resistance to be provided by the stiffness of the wall within its height and the adjusted BM diagram for the example under consideration is shown in Figure 13.32.

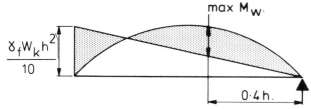

Figure 13.32 Design bending moment diagram

The applied wind moment at the level $0.4h$ down is calculated as:

$$(0.4\gamma_f W_k h \times 0.4h) - (0.4\gamma_f W_k h \times 0.2h)$$
$$= 0.08\gamma_f W_k h^2$$

which exceeds the true propped cantilever wall moment of

$$\frac{9}{128}\gamma_f W_k h^2 \ (0.07\gamma_f W_k h^2)$$

The moment of resistance provided by the wall at this level must then be checked against the calculated maximum design bending moment.

The action of the wall is, perhaps, better described as that of a member simply supported at prop level and partially fixed at base level where the partial fixity can be as high as $\gamma_f W_k h^2/8$, that of a true propped cantilever.

A rigid prop is not possible in practice (nor is a perfectly 'pinned' joint or 'fully fixed-ended' strut, etc.) but the initial assumption of a perfectly rigid prop generally provides the most onerous design condition. This is illustrated in the fin wall worked example to follow. Considering the two locations of maximum design bending moments and development of the respective moments of resistance, it is apparent that the critical design condition invariably occurs at the higher location where the resistance is dependent on the development of both flexural compressive and flexural tensile stresses in addition to the MR_s at this level. This is particularly true of the diaphragm wall. However, for fin walls, the flexural compressive stresses which occur at base level at the end of the projecting fin can become critical and, for these walls, this generalisation is less often applicable.

13.5.5 Stability moment of resistance (MR_s)

Since single-storey buildings tend to have lightweight roof construction and low superimposed roof loading, the forces and moments due to lateral wind pressure have greater effect on the stresses in the supporting masonry than they do in multi-storey buildings. Since there is little precompression, the wall's stability relies more on its own gravitational mass and resulting resistance moment. Under lateral wind pressure loading, the wall will tend to rotate at dpc level on its leeward face and crack at the same level on the windward face as indicated in Figure 13.33.

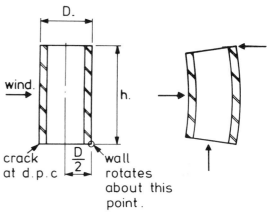

Figure 13.33 Initial rotation in diaphragm wall

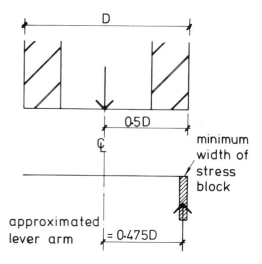

Figure 13.34 Approximate lever arm for diaphragm wall

In limit state design, the previous knife-edge concept of the point of rotation is replaced with a rectangular stressed area, in which the minimum width of masonry is stressed to the ultimate to produce the maximum lever arm for the axial load to generate the maximum stability moment of resistance, MR_s. As such, the MR_s resulting solely from the mass of the wall and ignoring any net roof loads, for each particular wall section of constant masonry density, will vary in direct proportion to its height. Unlike the fin wall, the diaphragm wall is symmetrical in its profile, and an approximation in the calculation of MR_s is warranted for the selection of a trial section. Hence, for diaphragms of normal proportions, the lever arm can be approximated to $0.475D$ as shown in Figure 13.34.

The use of this approximated lever arm for trial section purposes is illustrated in the diaphragm wall worked example to follow and in section 13.9.4.

13.5.6 Shear lag

Shear lag, the non-uniform stress distribution in the flanges of such structural members as T beams and box sections, is important in the design of thin-walled steel members subject to high bending stresses. It does not seem to be so critical in the design of normal timber box-beams, r.c. T beams etc. and most designers tend to ignore the phenomenon which tends to be allowed for in design rules.

There appears to be little or no experimental research into this phenomenon in brickwork (probably because the bulk of the research has been on solid wall sections and not box or T sections), and there seems to be no guidance in any Code or Building Regulation on this topic. Recent experiments by the first author suggest that the flange stresses at the ribs of brick diaphragm walls with extra wide rib spacings can increase by 20%. Designers may consider this an insignificant increase when it is appreciated that the true global factor of safety is probably about 8, creating a massive reserve of stress resistance. Even after tensile failure of the test diaphragm walls the shear lag stress increase had no effect on the stability of the walls – which were heavily prestressed and subject to relatively massive lateral loading.

In the authors' opinion, from experience and recent research, the phenomenon may be ignored for normal and lightly prestressed diaphragms, fins and I brick or block sections, provided the rules given for rib spacing, etc. are adhered to.

The subject will be dealt with in detail in *Advanced Structural Masonry Design* (in preparation) when the current research is completed and applied in practice.

13.5.7 Principal tensile stress

This topic, like shear lag, appears to have been practically ignored by designers, researchers and the Codes of Practice - probably because it is of little significance in solid walls. In highly stressed diaphragms, fins, etc. it can be significant, and designers should check the principal stresses in highly prestressed and highly laterally loaded section since recent research by the first author has shown that this is a failure condition. However, in normally loaded structures, dealt with in this manual (with the normal factors of safety), the principal stresses are most unlikely to be critical. On the very rare occasion that it is necessary to reduce the principal tensile stress this can be easily achieved by thickening the ribs of diaphragms or the fins, placing them at closer centres, or increasing the overall depth.

On completion of the current research and its application in practice this subject, too, will be dealt with in detail in *Advanced Structural Masonry Design*.

13.6 DESIGN SYMBOLS: FIN AND DIAPHRAGM WALLS

Certain aspects of the design processes in the two worked examples which follow will vary from the procedures given in BS 5628. As a result it has been necessary to introduce additional symbols and, in order to avoid confusion, a full list of the symbols used is now included for cross-reference.

The new symbols have been marked with *.

$*A$	cross-sectional area
$*B_r$	centre to centre of cross-ribs
b	width of section
$*b_r$	clear dimension between diaphragm cross-ribs
C_{pe}	wind, external pressure coefficient
C_{pi}	wind, internal pressure coefficient
$*D$	overall depth of diaphragm wall
$*d$	depth of cavity (void) in diaphragm wall
f_k	characteristic compressive strength of masonry
f_{kx}	characteristic flexural strength of masonry (tensile)
$*f_{ubc}$	flexural compressive stress at design load
$*f_{ubt}$	flexural tensile stress at design load
G_k	characteristic dead load
h	height of wall
h_{ef}	effective height of wall
$*I$	second moment of area
$*I_{na}$	second moment of area about neutral axis
$*K_1$	shear stress coefficient
$*K_2$	trial section stability moment coefficient

L	length
*L_f	spacing of fins, centre to centre
l_a	lever arm
*M_A	applied design bending moment
*M_b	bending moment at base (base moment)
*MR	moment of resistance
*MR_s	stability moment of resistance
*M_w	maximum bending moment in wall (wall moment)
*p_{ubc}	allowable flexural compressive stress = $1.1\beta f_k/\gamma_m$
*p_{ubt}	allowable flexural tensile stress = f_{kx}/γ_m
Q_k	characteristic superimposed load
*q	dynamic wind pressure
*r	radius of gyration
SR	slenderness ratio
*T	thickness of leaf of diaphragm wall
t	thickness of wall
t_{ef}	effective thickness of wall
*t_r	thickness of cross-rib of diaphragm wall
V	shear force
v_h	design shear stress
*W	own weight effective fin T profile per m height
W_k	characteristic wind load
*W_{k1}	design wind pressure, windward wall
*W_{k2}	design wind pressure, leeward wall
*W_{k3}	design wind pressure, uplift (on roof)

*w_s	width of stress block
*Y_1	fin dimension, neutral axis to end of fin
*Y_2	fin dimension, neutral axis to flange face
Z	section modulus
*Z_1	minimum section modulus = I_{na}/Y_1
*Z_2	maximum section modulus = I_{na}/Y_2
β	capacity reduction factor
γ_f	partial safety factor for loads
γ_m	partial safety factor for materials
*Ω	trial section coefficient ($W \times Y_2$) per m height

13.7 FIN WALL: STRUCTURAL DESIGN CONSIDERATIONS

13.7.1 Interaction between leaves

As shown in Figure 13.35, the fins are bonded to one of the leaves of a cavity wall and are considered as a T section combining the bonded leaf with the fin. The other leaf of the cavity wall is considered as a secondary member and the loading apportioned accordingly, the cavity ties being unable to transmit significant vertical shear forces but able to transmit horizontal forces across the cavity width.

It is assumed that the vertical loads applied to each leaf are taken directly on the leaf to which

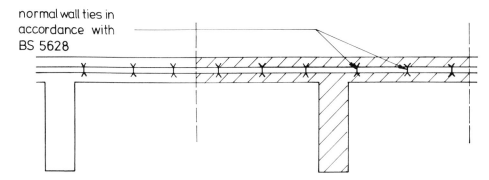

normal wall ties in accordance with BS 5628

Figure 13.35 Fin wall arrangement

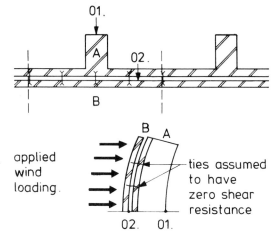

01.

A   02.

B

applied wind loading.

B A

ties assumed to have zero shear resistance

02.   01.

Figure 13.36 Assumed behaviour of fin wall

the load is applied, but that any resulting bending moments from eccentric loading and/or wind loading can be apportioned between the two members in accordance with their relative stiffnesses.

For example, in Figure 13.36 which shows in exaggerated form the assumed behaviour, the fin A and bonded leaf is considered as a T section bending about point 01. The remaining leaf, B, is considered to deflect equally, bending about point 02 and the ties deform slightly at an assumed shear resistance of zero.

13.7.2 Spacing of fins
The choice of a suitable section must take into account the cavity wall's ability to act suitably with the fin both to transfer wind forces to the overall section and to prevent buckling of the flange of the T section. This involves choosing a suitable spacing for the fin to control both these conditions, and to take into account economic spacing of the roof beams where the beams are to span onto the fins. The spacing of the fins, is, therefore, governed by the following conditions:

(a) The cavity wall acting as a continuous horizontal member subject to wind load, spanning between the fins (see Figure 13.37).

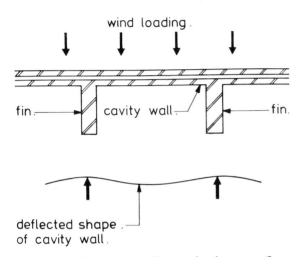

Figure 13.37 Cavity wall spanning between fins

(b) The cavity wall's ability to support vertical load without buckling. This is governed by the lesser of the effective vertical height or the effective length between fins (see Figure 13.38).

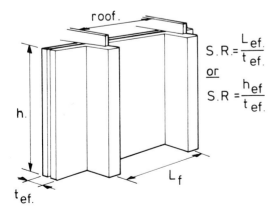

the effective length of wall is :
factor × h or factor × L_f
the factor takes into account
the restraint condition .

Figure 13.38 Slenderness ratio of wall panel

(c) The ability of the cross-section to resist the applied loading with the leaf and fin acting together to form a T beam (see Figure 13.39).

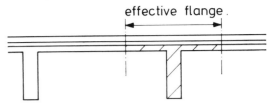

Figure 13.39 Effective length of fin flange

The effective flange of the T beam is limited to the least of:

(i) the distance between the centres of the fins;
(ii) the breadth of the fin plus twelve times the effective thickness of the bonded leaf;
(iii) one-third of the effective span of the fin.

It should be noted that clause 36.4.3 in BS 5628 embraces two of the conditions with reference to piered walls but, since it is felt that the distribution of stress into the flange is also related to the span of the fin, in a similar manner to reinforced concrete T beams, a span related limit is also necessary in accordance with item (iii).

(d) The vertical shear forces between the fin and the bonded leaf resulting from the applied bending moment on the T section (see Figure 13.40).

(e) The economic spacing of the main roof supports (where applicable).

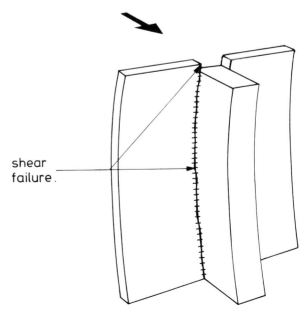

Figure 13.40 Shear failure between fin and bonded leaf

It should be noted that whilst item (c) restricts the flange width for design purposes, the actual distance between fins can be greater provided that the design of the effective flange is within the permissible design stresses and that all the other design considerations are met.

13.7.3 Size of fins
Typical fin sizes used are 0.5 m to 2 m deep at spacings of 3 to 5 m and 330 to 440 mm wide. Some typical sections and their properties are shown in the design Table 13.1.

The length and thickness of the fin is governed by the tendency of the outer edge to buckle under compressive bending stress.

The roof plate action and the stresses in the transverse walls which provide the reaction to the plate must be checked.

13.7.4 Effective section and trial section
Owing to the unsymmetrical shape of the fin wall, the geometrical properties of the effective section, when combined bending and axial forces are considered, can vary greatly under changes in loading, particularly if a 'cracked' section is being analysed. It is therefore most important, when analysing the stability moment of resistance (MR_s) to consider carefully the effective section being stressed and the effects of the 'cracked' portion on the general performance of the wall. The tensile stresses must be kept within the limit recommended in the Code of Practice but, at dpc level, the majority of membranes must be considered to have zero resistance to tensile forces.

Again, owing to the unsymmetrical section, it is not as straightforward a matter, as it is for the diaphragm wall, to provide a reasonably accurate means of assessing the trial section. A 'trial section coefficient' (Ω) has been included in the design Table 13.1 and relates to the own weight stability moment per metre height of wall when the rotation at base level occurs about the flange face. An illustration of the use of the trial section coefficient is given in the worked example to follow.

Table 13.1 Fin wall section properties

Fin reference letter	A	B	C	D	E	F	G	H
Fin size (mm)	665 × 327	665 × 440	778 × 327	778 × 440	890 × 327	890 × 440	1003 × 327	1003 × 440
Effective width of flange (m)	1.971	2.084	1.971	2.084	1.971	2.084	1.971	2.084
Neutral axis Y_1 (m)	0.455	0.435	0.524	0.500	0.589	0.563	0.654	0.626
Neutral axis Y_2 (m)	0.210	0.230	0.254	0.278	0.301	0.327	0.349	0.377
Effective area (m²)	0.3860	0.4611	0.4262	0.5152	0.4595	0.5601	0.4965	0.6098
ow of effective area per m height W(kN)	7.720	9.222	8.458	10.216	9.190	11.202	9.930	12.196
I_{na} (m⁴)	0.01567	0.01939	0.02454	0.03030	0.03590	0.04426	0.05021	0.06187
Z_1 (m³)	0.03441	0.04450	0.04684	0.06059	0.06096	0.07862	0.07677	0.09883
Z_2 (m³)	0.07462	0.08430	0.09663	0.10898	0.11928	0.13536	0.14387	0.16410
Trial section coefficient Ω (kN·m/m)	1.6212	2.1210	2.1483	2.8400	2.7662	3.6631	3.4656	4.5978

Fin reference letter	J	K	L	M	N	P	Q	R
Fin size (mm)	1115 × 327	1115 × 440	1227 × 327	1227 × 440	1339 × 327	1339 × 440	1451 × 327	1451 × 440
Effective width of flange (m)	1.971	2.084	1.971	2.084	1.971	2.084	1.971	2.084
Neutral axis Y_1 (m)	0.718	0.687	0.780	0.747	0.841	0.807	0.902	0.866
Neutral axis Y_2 (m)	0.397	0.428	0.447	0.480	0.498	0.532	0.549	0.585
Effective area (m^2)	0.5331	0.6591	0.5697	0.7084	0.6064	0.7577	0.6430	0.8070
ow of effective area per m height W(kN)	10.662	13.182	11.394	14.168	12.128	15.154	12.860	16.140
I_{na} (m^4)	0.06746	0.08312	0.08800	0.10848	0.11208	0.13826	0.13992	0.17277
Z_1 (m^3)	0.09395	0.12099	0.11282	0.14522	0.13327	0.17132	0.15513	0.19950
Z_2 (m^3)	0.16992	0.19421	0.19687	0.22600	0.22506	0.26039	0.25487	0.29530
Trial section coefficient Ω (kN·m/m)	4.2328	5.6419	5.0931	6.8006	6.0397	8.0619	7.0601	9.4419

$$Z_1 = \frac{I_{na}}{Y_1} \quad Z_2 = \frac{I_{na}}{Y_2}$$

Trial section coefficient $\Omega = WY_2$

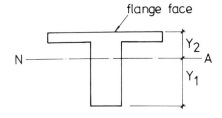

13.8 EXAMPLE 1: FIN WALL

13.8.1 Design problem

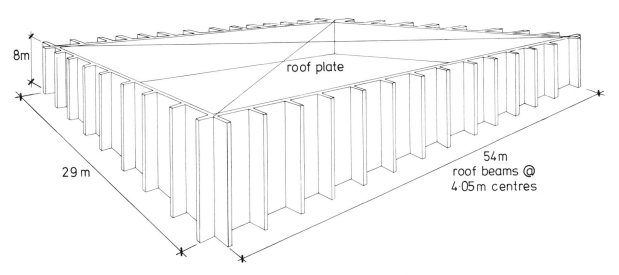

Figure 13.41 Fin wall design example

A warehouse measuring 29 m × 54 m on plan, and 8 m high, is shown in Figure 13.41. The building is to be designed in loadbearing brickwork, using fin wall construction for its main vertical structure. The fins are to project on the external face, and the wall panels between the fins are to be of 260 mm brick cavity construction. There are no internal walls within the building. The building is part of a major development where extensive testing of materials and strict supervision of workmanship will be employed.

The architect has selected particular facing bricks which are shown to have a compressive strength of 41.5 N/mm^2 and a water absorption of 8%. The bricks will be used both inside and outside the building.

13.8.2 Design approach
BS 5628 offers three options for the design of laterally loaded walls:

(a) Clause 36.4.3 in which the design moment of resistance of wall panels is given as $f_{kx}Z/\gamma_m$.
and
(b) Clause 36.8 which offers two further options:
    (i) design lateral strength equated to effective eccentricity due to lateral loads
    or
    (ii) treating the panel as an arch.

The last option can seldom be applied to single-storey buildings due to inadequate arch thrust resistance. The remaining two options take no account of flexural compressive stresses which, in the fin wall design concept, certainly require careful consideration.

For this reason, it has been considered necessary, in order to explain properly the mechanisms involved, to diverge from the BS 5628 concept of equating design loads to design strengths. The analysis considers stresses due to design loads and relates these to allowable flexural stresses in both compression and tension.

13.8.3 Characteristic loads

(a) Wind forces
The basic wind pressure on a building is calculated from a number of variables which include:

(i)   location of building, nationally;
(ii) topography of the immediate surrounding area;
(iii) height above ground to the top of the building;
(iv) building geometry.

For the appropriate conditions, the basic pressure and local pressure intensities are given in CP 3: Chapter V: Part II.

In this example, these values are assumed to have been computed as:

Dynamic wind pressure, q,	$= 0.71 \text{ kN/m}^2$
C_{pe} on windward face	$= 0.8$
C_{pe} on leeward face	$= -0.5$
C_{pi} on walls either	$= +0.2 \text{ or } -0.3$
Gross wind uplift $= C_{pe} + C_{pi}$	$= 0.60$

Therefore, characteristic wind loads are:

Pressure on windward wall
$$= W_{k1} = (C_{pe} - C_{pi})q = (0.8 + 0.3) \times 0.71$$
$$= 0.781 \text{ kN/m}^2$$

Suction on leeward wall
$$= W_{k2} = (C_{pe} - C_{pi})q = (0.5 + 0.2) \times 0.71$$
$$= 0.497 \text{ kN/m}^2$$

Gross roof uplift
$$= W_{k3} = (C_{pe} + C_{pi})q = 0.6 \times 0.71$$
$$= 0.426 \text{ kN/m}^2$$

(b) Dead and superimposed loads
(i) Characteristic superimposed load, $Q_k = 0.75 \text{ kN/m}^2$

(Assuming no access to roof, other than for cleaning or repair, in accordance with CP 3: Chapter V: Part 1.)

(ii) Characteristic dead load, G_k:

Assume: metal decking $= 0.18 \ kN/m^2$
       felt and chippings $= 0.30 \ kN/m^2$
       ow roof beams $= 0.19 \ kN/m^2$
Total G_k $= \overline{0.67 \ kN/m^2}$

13.8.4 Design loads

The critical loading condition to be considered for such a wall is usually wind + dead only, although the loading condition of dead + superimposed + wind should be checked.

Design dead load $= 0.9G_k$ or $1.4G_k$
Design wind load $= 1.4W_k$ or $0.015G_k$ whichever is the larger.

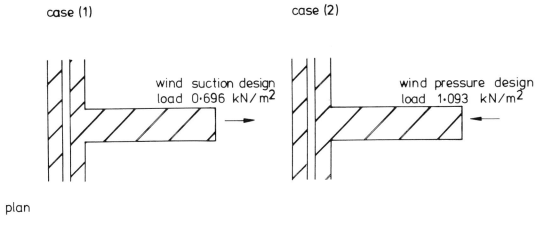

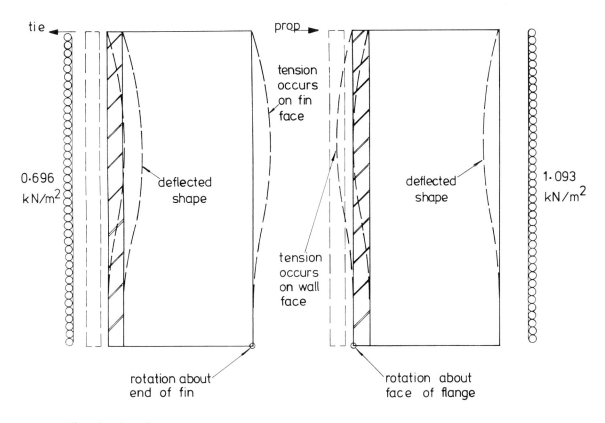

Figure 13.42

Therefore, by inspection, the most critical combination of loading will be given by:

Design dead load $= 0.9 \times 0.67 = 0.603$ kN/m^2

Design wind loads:
Pressure, from $W_{k1} = 1.4 \times 0.781$ $= 1.093$ kN/m^2
Suction, from W_{k2} $= 1.4 \times 0.497$ $= 0.696$ kN/m^2
Uplift, from W_{k3} $= 1.4 \times 0.426$ $= 0.597$ kN/m^2
Design dead-uplift $= 0.603 - 0.597 = 0.006$ kN/m^2, say equal to zero

13.8.5 Design cases (as shown in Figure 13.42)
Inner leaf offers minimal resistance and is ignored in calculations apart from assisting stiffness of flange in bending.
Note: Vertical loading from own weight of effective section only, as uplift exactly cancels out roof dead loads.

13.8.6 Deflection of roof wind girder
The wall is designed as a propped cantilever and utilises the fins bonded to the outer leaf to act as vertical T beams resisting the flexure.

The prop to the cantilever is provided by a wind girder within the roof decking system (the design of the wind girder is not covered by this book). The reactions from the roof wind girder are transferred into the transverse gable shear walls at each end of the building. Horizontal deflection of the roof wind girder, reaching a maximum at mid-span, has the effect of producing additional rotation at base level (see Figure 13.43) and this results in a less critical stress condition. However, the critical stress conditions are generally experienced in the end fins where the roof wind girder deflection is a minimum.

13.8.7 Effective flange width for T profile
The dimensional limits for the effective length of the wall permitted to act as the flange of the T profile are given in BS 5628, clause 36.4.3 (b), as 6 × thickness of wall forming the flange, measured as a projection from each face of the fin, when the flange is continuous. In this design example, as will be the general case in practice, the wall forming the flange is the outer leaf of a cavity wall, as defined in BS 5628, clause 29.1.1. It is, therefore, reasonable to take advantage of the stiffening effect of the inner leaf in resisting buckling of the outer leaf, when acting as the flange of the T profile. The effective flange length, measured from each face of the fin, is therefore calculated as 6 × effective wall thickness.

Thus, for an assumed fin width of 327 mm;

$$\text{effective wall thickness} = \frac{2}{3}\,(102.5 + 102.5)$$
$$= 137 \text{ mm}$$
$$\text{effective flange width}\ = (6 \times 137) + 327 + (6 \times 137)$$
$$= 1971 \text{ mm}$$

13.8.8 Spacing of fins
The spacing of fins has been discussed earlier – but one aspect only, the capacity of the wall panel to span between the fins, is considered here.

There is no doubt that the support provided for the wall panel at foundation level will assist in resisting the flexure due to wind forces. However, this assistance will diminish at the higher levels of the wall panel, and for this example the wall should be designed to span purely horizontally between the fins.

The wall panels are taken as continuous spans and the maximum bending moments are shown in Figure 13.44.

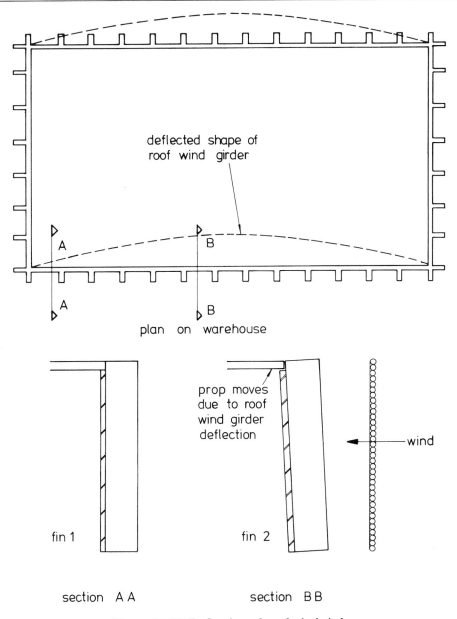

Figure 13.43 Deflection of roof wind girder

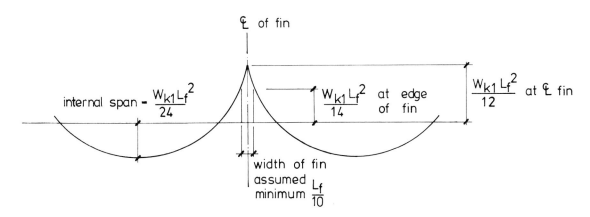

Figure 13.44 Bending moment diagram for wall panel

The maximum moment is $W_{k1}L_f^2/14$ at the edge of the fins, for an assumed fin width of $L_f/10$.

$$\text{Design bending moment} = \frac{W_{k1}L_f^2}{14} = \frac{1.093 \times L_f^2}{14} = 0.078L_f^2$$

From BS 5628, clause 36.4.3:

$$\text{design moment of resistance} = \frac{f_{kx}Z}{\gamma_m}$$

f_{kx}, for water absorption 7% to 12% set in designation (iii) mortar = 1.10 N/mm^2

Z, for two leaves = $\dfrac{2 \times 0.1025^2 \times 1.0}{6} = 0.0035$ m^3

γ_m, from BS 5628, Table 4 (see Table 5.11), special categories of manufacturing and construction control are applicable = 2.5.

Therefore

$$\text{design moment of resistance} = \frac{1.10 \times 0.0035 \times 10^9}{2.5 \times 10^6}$$

$$= 1.54 \text{ kN·m}$$

From this check maximum span of wall panel.

$$\text{Design moment} = \text{design moment of resistance}$$

i.e. $\qquad 0.078 L_f^2 = 1.54$

therefore $\qquad L_f = \sqrt{\dfrac{1.54}{0.078}} = 4.44$ m = maximum fin spacing

Therefore, 4.05 m fin spacing is acceptable (see Figure 13.45).

13.8.9 Trial section

It has been found in practice that a trial section can be reasonably obtained by providing a section which has a stability moment of resistance MR$_s$, at the level of M_b equal to $W_{k1}L_f h^2/8$ under wind pressure loading W_{k1}, i.e. when rotation at the base of the wall is about the face of the flange. For the purpose of the trial section assessment, the stability moment of resistance can be simplified to Ωh in which:

Ω = trial section coefficient from Table 13.1
h = height of fin wall.

Therefore $\qquad\qquad\qquad \dfrac{W_{k1}L_f h^2}{8} = \Omega h$

$$\frac{1.093 \times 4.05 \times 8^2}{8} = \Omega 8$$

Therefore $\qquad\qquad\qquad\qquad\qquad \Omega = 4.427$ kN·m/m height of wall.

From Table 13.1, select fin wall profile 'J'.

Note: It is important that this trial section coefficient is used only for the selection of the trial section. A thorough structural analysis must always be carried out.

—*fig 13.45*—

13.8.10 Consider propped cantilever action

With 4.050 m fin centres, design wind loads on fins are: (see Figure 13.46)

Case (i), pressure:
$$W_{k1}L_f = 1.093 \times 4.05 = 4.427 \text{ kN/m of height}$$

Case (ii), suction:
$$W_{k2}L_f = 0.696 \times 4.05 = 2.82 \text{ kN/m of height}$$

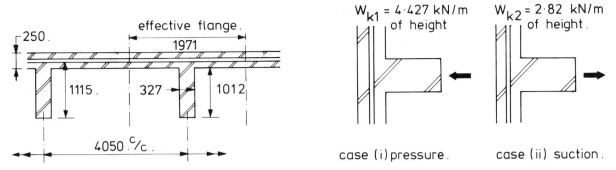

Figure 13.45 Profile of trial section – fin section 'J'   **Figure 13.46** Wind loadings

Assuming MR, is greater than $W_{k1}L_fh^2/8$ and zero deflection of the roof prop, the following BM diagrams can be drawn:

Case (i)

$$\text{Wall moment, } M_w = \frac{9W_{k1}L_fh^2}{128} = \frac{9 \times 4.427 \times 8^2}{128}$$
$$= 19.92 \text{ kN·m}$$

$$\text{Base moment, } M_b = \frac{W_{k1}L_fh^2}{8} = \frac{4.427 \times 8^2}{8}$$
$$= 35.42 \text{ kN·m}$$

Case (ii)

$$\text{Wall moment, } M_w = \frac{9W_{k2}L_fh^2}{128} = \frac{9 \times 2.82 \times 8^2}{128}$$
$$= 12.69 \text{ kN·m}$$

$$\text{Base moment, } M_b = \frac{W_{k2}L_fh^2}{8} = \frac{2.82 \times 8^2}{8}$$
$$= 22.56 \text{ kN·m}$$

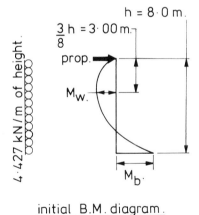

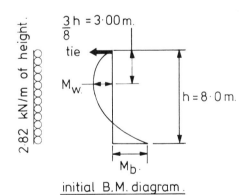

Figure 13.47 Case (i) pressure   **Figure 13.48** Case (ii) suction

The bending moment diagrams shown in Figures 13.47 and 13.48 are applicable only if it can be shown that the stability moment of resistance of the 'cracked section' MR_s at dpc level exceeds $W_kL_fh^2/8$. This should be the first check to be carried out, and if MR_s is less than $W_kL_fh^2/8$ the base moment is limited to MR_s and the BM diagram must be redrawn plotting the free moment diagram onto the fixed end moment diagram which is produced by MR_s (see Figure 13.55.)

13.8.11 Stability moment of resistance
Invariably, as is the case with this design example, there will be a damp proof course at or near to the base of the wall. Few dpcs are capable of transmitting much flexural tensile stress across the bed joint, and in this example the analysis considers the 'cracked section'.

Appendix B of BS 5628 discusses the application of a rectangular stress block under ultimate conditions, and the stability moment of resistance MR_s at the level of M_b can be assumed to be provided by the axial load in the fin section acting at a lever arm about the centroid of the rectangular stress block as shown in Figure 13.49.

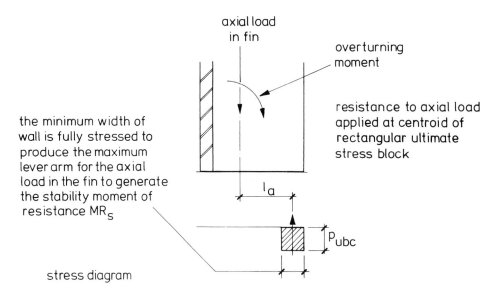

Figure 13.49 Generation of MR_s – fin wall

13.8.12 Allowable flexural compressive stresses, p_{ubc}
(taking into account slenderness, β, and material, γ_m)

Before the stability moment of resistance MR_s can be compared with the assumed base moment (M_b) of $W_k L_f h^2/8$ consideration must be given to the criteria affecting the allowable flexural compressive stresses, p_{ubc}, as this value dictates the stability moment of resistance. This is demonstrated in Figure 13.49, in which the mechanism producing the stability moment of resistance MR_s is shown.

This flexural compressive stress can become significant and must be checked, taking into account the tendency of the flange or fin to buckle at the point of application of the stress.

There is limited guidance given in BS 5628 on the effect of slenderness on the flexural compressive strength of masonry. This is because the flexural strength of masonry is assumed to be limited by the flexural tensile stresses – which is, perhaps, true of panel walls and the like, but not of the analysis of more complex geometric forms such as the fin wall.

The approach to the consideration of slenderness and flexural compressive stresses which follows is believed to provide a safe and practical design. It is expected that current research will allow more sophisticated analysis to be developed.

Identification of problem:

Case (i) pressure, showing zones of maximum values of flexural compression (Figure 13.50).

Case (ii) suction, showing zones of maximum values of flexural compression (Figure 13.51).

Considering the wind suction loading, case (ii), flexural compression is applied to the flange of the T profile at the level of M_w. The buckling stability of the flange is provided by the projecting fin and, therefore, the effective length of the flange, for slenderness considerations, can be taken as twice the outstanding length of the flange from the face of the fin. Furthermore, if the flange is properly tied to the inner leaf of the cavity wall, the effective thickness of the flange, for slenderness considerations, can be taken as ⅔ the sum of the thicknesses of the two leaves of the cavity wall.

case(i) pressure

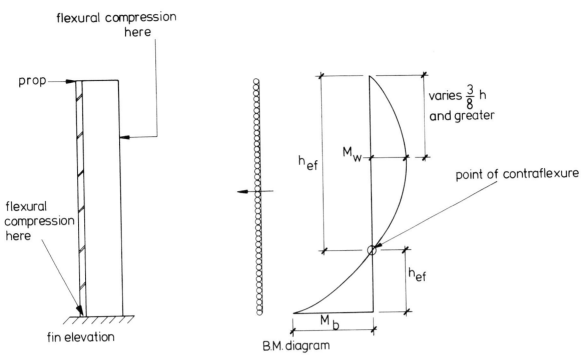

Figure 13.50 Case (i) pressure, showing zones of maximum values of flexural compression

case (ii) suction

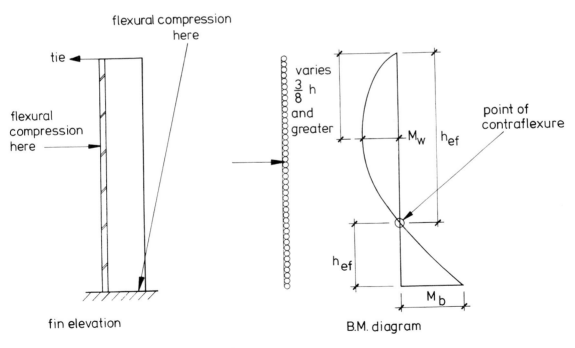

Figure 13.51 Case (ii) suction, showing zones of maximum values of flexural compression

Flexural compression is also applied to the end of the projecting fin at the level of M_b. For this design example, the foundation is assumed to comprise a reinforced concrete raft slab, as shown in Figure 13.52. The flexural compression applicable at this level is not influenced by slenderness considerations as the raft foundations can be assumed to provide full lateral stability.

Slenderness at this level would require careful consideration if the fin foundation was at a greater depth below ground level.

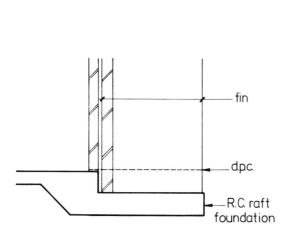

Figure 13.52 Foundation detail

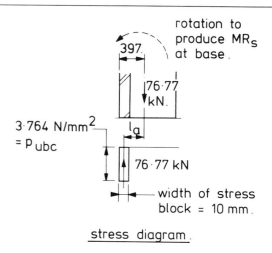

Figure 13.53 Case (i) pressure

Considering the wind pressure loading, case (i), flexural compression is applied to the end of the projecting fin at the level of M_w.

The buckling stability of the fin cannot be considered to be fully provided by the flange of the T profile, as the flange is not of comparable lateral stiffness to the fin and would tend to rotate in attempting to prevent the fin buckling. Rather, it is considered that the slenderness of the fin should relate partly to its height and, as the full height of the fin would be over-cautious, it is proposed that the height between points of contraflexure would provide adequate safety. The effective thickness of the fin for slenderness considerations is taken as the actual thickness.

The design flexural compressive stress p_{ubc} can therefore be expressed as:

$$p_{ubc} = \frac{1.1\beta f_k}{\gamma_m}$$

where
p_{ubc} = design flexural compressive stress
β = capacity reduction factor derived from slenderness ratio
f_k = characteristic compressive strength of masonry
γ_m = partial safety factor for materials.

With the lateral restraint provided by the raft foundation at M_b level, β can be taken as 1.0. Therefore, $p_{ubc} = 1.1 f_k/\gamma_m$ at M_b level (see Figure 13.52).

For this example:
f_k = 9.41 N/mm², based on 41.5 N/mm² bricks set in a designation (iii) mortar from BS 5628, Table 2(a) (see Table 5.4)
γ_m = 2.5 as previously shown.

Therefore:

$$p_{ubc} = \frac{1.1 \times 9.41}{2.5} = 4.14 \text{ N/mm}^2$$

13.8.13 Calculate MR_s and compare with M_b

(a) Consider case (i) pressure

From Table 13.1, ow = 10.662 kN/m height. Therefore:

$$\text{design axial load in fin at } M_b = 0.9 \times 10.662 \times 8$$

$$= 76.77 \text{ kN}$$

$$\text{Minimum width of stress block} = \frac{\text{axial load on fin}}{\text{flange width} \times p_{\text{ubc}}}$$

$$= \frac{76.77 \times 10^3}{1971 \times 3.764}$$

$$= 10 \text{ mm (see Figure 13.53)}$$

$$\text{Lever arm} = 397 - \frac{10}{2} = 392 \text{ mm}$$

$$\text{MR}_\text{s} = 76.77 \times 0.392 = 30.09 \text{ kN·m}$$

The stability moment of resistance is shown to be less than $M_\text{b} = W_\text{k}L_\text{f}h^2/8 = 35.42$ kN·m.

The base moment should therefore be limited to the value of stability moment, MR_s, 30.09 kN·m and the bending moment diagram adjusted accordingly.

(b) Consider case (ii) suction

From Figure 13.54, it is evident that the stability moment of resistance is provided by the flexural compressive stress at the end of the projecting fin, thus:

$$\text{minimum width of stress block} = \frac{\text{axial load on fin}}{\text{fin width} \times p_{\text{ubc}}}$$

$$= \frac{76.77 \times 10^3}{327 \times 3.764}$$

$$= 62 \text{ mm (see Figure 13.54)}$$

$$\text{Lever arm} = 718 - \frac{62}{2} = 687 \text{ mm}$$

$$\text{MR}_\text{s} = 76.77 \times 0.687 = 52.74 \text{ kN·m}$$

As this is greater than $M_\text{b} = 22.56$ kN·m (see Figure 13.48) use M_b in the design of the fin section.

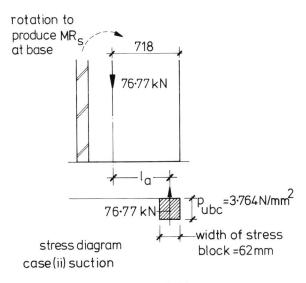

stress diagram
case (ii) suction

Figure 13.54

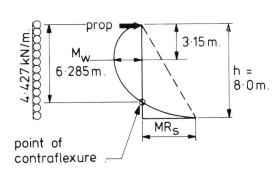

design B.M. diagram (adjusted.)

Figure 13.55 Case (i) pressure

13.8.14 Bending moment diagrams

Case (i) pressure (see Figure 13.55)
MR_s (calculated) = 30.09 kN·m.

Find M_w from zero shear

$$\text{Prop} = \left(4.427 \times \frac{8}{2}\right) - \left(\frac{30.09}{8}\right)$$

$$= 13.95 \text{ kN}$$

$$\text{Zero shear} = \frac{13.95}{4.427} = 3.15 \text{ m from top}$$

$$M_w = (13.95 \times 3.15) - \left(4.427 \times \frac{3.15^2}{2}\right)$$

$$= 43.94 - 21.96$$

$$= 21.98 \text{ kN·m}$$

Adjustment made to BM (bending moment) diagram to take account of MR_s being less than $W_{k1}L_fh^2/8$ (i.e. M_b) and therefore base moment limited to MR_s with M_w calculated by superimposing the free BM diagram onto the stability moment diagram produced by MR_s at base.

Case (ii) suction (see Figure 13.56)

$$M_w = \frac{9W_{k2}L_fh^2}{128} = \frac{9 \times 2.82 \times 8^2}{128}$$

$$= 12.69 \text{ kN·m}$$

$$M_b = \frac{W_{k2}L_fh^2}{8} = \frac{2.82 \times 8^2}{8}$$

$$= 22.56 \text{ kN·m}$$

No adjustment is necessary to BM diagram as MR_s is greater than $W_{k2}L_fh^2/8$ (i.e. M_b) and therefore maximum M_w occurs at $\frac{3}{8}h$ from top of wall.

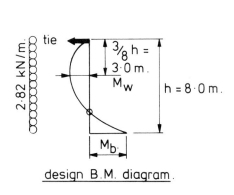

Figure 13.56 Case (ii) suction

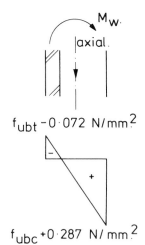

Figure 13.57 Stress diagram

13.8.15 Consider stresses at level of M_w

The stress considerations at the level of the maximum wall moment assume triangular stress distribution, using elastic analysis, but relate to ultimate stress requirements at the extreme edges of the fin or wall face, depending on the wind direction considered. For compressive stress conditions, this gives a conservative solution.

Case (i) pressure (see Figure 13.57)

Properties of effective wall section from Table 13.1, are as listed below, except that ow of effective section at level of $M_w = 0.9 \times 10.662 \times 3.15 = 30.23 \text{ kN}$.

Flexural stresses at design load:

$$\text{Flexural compressive, } f_{ubc} = + \frac{30.23 \times 10^3}{0.5331 \times 10^6} + \frac{21.98 \times 10^6}{0.094 \times 10^9}$$

$$= + 0.057 + 0.23$$

$$= + 0.287 \text{ N/mm}^2$$

$$\text{Flexural tensile, } f_{ubt} = + \frac{30.23 \times 10^3}{0.5331 \times 10^6} - \frac{21.98 \times 10^6}{0.17 \times 10^9}$$

$$= + 0.057 - 0.129$$

$$= - 0.072 \text{ N/mm}^2$$

Case (ii) suction (see Figure 13.58)
Properties of effective wall section from Table 13.1:

ow of effective section = 10.662 × 3.0
= 31.99 kN at level of M_w

Effective area = 0.5331 m²
Z minimum = 0.094 m³
Z maximum = 0.17 m³
Design axial load = ow effective section + roof dead − roof uplift
= (γ_f × 31.99) + zero
= (0.9 × 31.99) + zero
= 28.79 kN

Flexural stresses at design load:

$$\text{Flexural compressive, } f_{ubc} = + \frac{28.79 \times 10^3}{0.5331 \times 10^6} + \frac{12.69 \times 10^6}{0.17 \times 10^9}$$

$$= + 0.054 + 0.075$$

$$= + 0.129 \text{ N/mm}^2$$

$$\text{Flexural tensile, } f_{ubt} = + \frac{28.79 \times 10^3}{0.5331 \times 10^6} - \frac{12.69 \times 10^6}{0.094 \times 10^9}$$

$$= + 0.054 - 0.135$$

$$= - 0.081 \text{ N/mm}^2$$

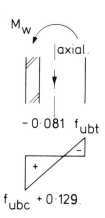

Figure 13.58 Stress diagram

13.8.16 Design flexural stress at M_w levels

(a) Design flexural tensile stress, p_{ubt}:
(taking account of materials partial safety factor, γ_m)

$$p_{ubt} = \frac{f_{kx}}{\gamma_m} \text{ (from BS 5628, clause 36.4.3)}$$

where

f_{kx} = 0.4 N/mm² for bricks with a water absorption of 7% to 12%
γ_m = 2.5 as previously shown

Therefore

$$p_{ubt} = 0.4/2.5 = 0.16 \text{ N/mm}^2$$

By comparison with the f_{ubt} values calculated and shown in Figures 13.57 and 13.58, the wall is acceptable.

(b) Design flexural compressive stresses, p_{ubc}:

$$p_{ubc} = \frac{1.1\beta f_k}{\gamma_m}$$

Calculate respective β values for case (i) and case (ii) loadings at level of M_w.

Case (ii) suction (flange in compression at M_w level)

$$\text{Slenderness ratio} = \frac{2 \times \text{flange outstanding length}}{\text{effective thickness}}$$

$$= \frac{2 \times (1971 - 327/2)}{\frac{2}{3}(102.5 + 102.5)} = 6$$

The stressed areas can be considered as axially loaded, therefore $e_x = 0$. Thus since SR = 6 from BS 5628, Table 7, $\beta = 1.0$ (see Table 5.15).

Therefore

$$p_{ubc} = \frac{1.1 \times 1.0 \times 9.41}{2.5}$$

$$= +4.14 \text{ N/mm}^2$$

Case (i) pressure (end of fin in compression at M_w level)

$$\text{Slenderness ratio} = \frac{\text{effective height between points of contraflexure}}{\text{actual thickness}}$$

$$= \frac{6285}{327} \text{ (see Figure 13.56)}$$

$$= 19.22$$

Therefore β = 0.727 from BS 5628, Table 7 (see Table 5.15).

Therefore

$$p_{ubc} = \frac{1.1 \times 0.727 \times 9.41}{2.5}$$

$$= +3.01 \text{ N/mm}^2$$

By comparison with the f_{ubc} values calculated and shown in Figures 13.57 and 13.58, the wall is acceptable.

13.8.17 Consider fins and deflected roof prop
It is evident that the deflection of the roof wind girder induces additional rotation at the level of M_b.

In this design example, the MR_s limited the moment at the base under wind pressure loading, and the additional rotation will not alter the design bending moment diagram shown in Figure 13.55. The base moment for wind suction loading, when the roof support does not deflect, is $W_{k2}L_f h^2/8$ (Figure 13.56). But, as the deflecting roof support induces further rotation at base level, the section cracks and takes full advantage of the stability moment of resistance, MR_s. The revised design bending moment diagram for this condition, when compared with Figure 13.56, is shown in Figure 13.59. The reduced wall moment value is obviously acceptable, whilst the increase in the moment at base level is also shown (Figure 13.54) to be acceptable. However, this should be fully checked if slenderness reductions are applicable at this level.

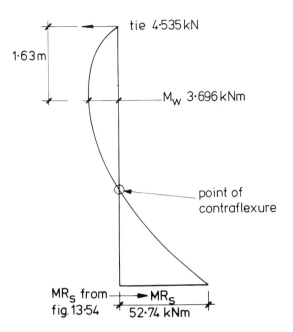

Figure 13.59 Bending moment diagram for deflected roof wind girder condition

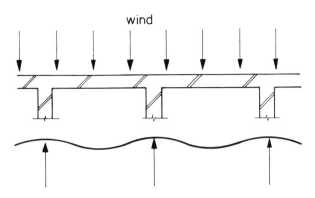

Figure 13.60 Wall spanning between ribs

(b) as a wall liable to buckling under vertical load, the effective length of the wall being determined from either the vertical height or the length measured between adjacent intersecting walls, BS 5628, clause 28.3.2, i.e. the ribs (see Figure 13.61).

SUGGESTED DESIGN PROCEDURE

After some experience, a competent designer will be able to shorten the design process considerably. A suggested design procedure for the wall is as follows:

(1) Calculate wind loadings.
(2) Calculate dead and imposed loadings.
(3) Assess critical loading conditions.
(4) Select trial section.
(5) Calculate stability moments, MR_s, at base.
(6) Calculate position of maximum wall moments.
(7) Calculate magnitude of maximum wall moment, M_w.
(8) Check compressive stress at base level.
(9) Check loadings and stresses at levels of M_w.
(10) Select brick and mortar strength required.

13.9 DIAPHRAGM WALL: STRUCTURAL DESIGN CONSIDERATIONS

13.9.1 Determination of rib centres, B_r

The centres of the ribs are governed by the following conditions:

(a) the outer leaf acting as a continuous horizontal slab, subject to wind load, supported by and spanning between the ribs (see Figure 13.60).

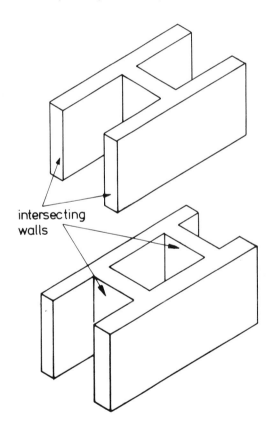

Figure 13.61 Wall restrained by ribs

(c) leaf and rib acting together to form an I section. The length of the flange of the I being restricted in a similar way to that of a concrete T beam, the requirement being that it should not exceed:
  (i) one-third of the effective span of the I beam

(ii) the distance between the centres of the ribs of the I beam
(iii) the breadth of the rib plus twelve times the thickness of the flange (see Figure 13.62).

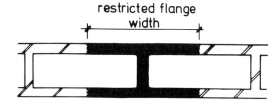

Figure 13.62 Rib and leaf acting as I beam

(d) if the ribs are spaced too widely, there will be shear failure between the ribs and the leaves, particularly where using ties (see Figure 13.63).

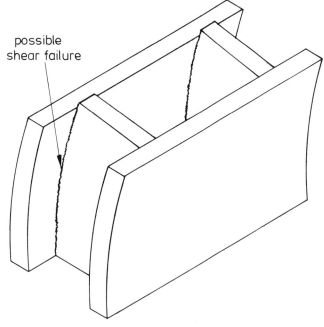

possible shear failure

Figure 13.63 Shear failure between rib and leaf

Calculating the rib centres from these conditions gives:

Case (a)

$$\gamma_f = \text{partial safety factor on loads}$$

$$M_A \text{ due to wind} = \gamma_f \frac{W_k B_r^2}{10}$$

$$Z \text{ per m height} = \frac{1 \times T^2}{6}$$

(where T is thickness of leaf)

Assume that the category of construction control is special (BS 5628, clause 27.2.2.2) and that the category of manufacturing control is special (BS

5628, clause 27.2.1.2) then, from BS 5628, Table 4, γ_m for the brickwork = 2.5 (see Table 5.11).

Assuming bricks with a water absorption greater than 12%, set in a designation (iii) mortar, from Table 3 of BS 5628 for plane of failure perpendicular to bed joints (i.e. leaf of wall spanning horizontally between ribs):

f_{kx} = characteristic flexural strength, say 0.9 N/mm²
  = 900 kN/m²

Design flexural tensile stress $\dfrac{f_{kx}}{\gamma_m} = \dfrac{900}{2.5}$ kN/m²

$$= 360 \text{ kN/m}^2$$

Example:
Characteristic wind pressure, $W_k = 0.573$ kN/m²
Thickness of leaf $\quad T = 102.5$ mm
$\quad B_r = $ spacing of ribs

Consider 1 metre height of wall

$$M_A = \gamma_f \frac{W_k B_r^2}{10} = \frac{1.4 \times 0.573 \times B_r^2}{10}$$

$$= 0.08 \times B_r^2 \text{ kN} \cdot \text{m}$$

$$Z = \frac{1 \times T^2}{6} = \frac{1 \times 0.1025^2}{6} = 1.75 \times 10^{-3} \text{ m}^3$$

$$\text{Design MR} = \frac{f_{kx} Z}{\gamma_m}$$

Therefore

$$0.08 \times B_r^2 = \frac{900}{2.5} \times 1.75 \times 10^{-3}$$

$$B_r = 2.8 \text{ m}$$

Case (b)
Maximum slenderness ratio = 27, Table 7, BS 5628 (see Table 5.15).

i.e. $\quad \dfrac{B_r}{T} = 27$

therefore

$$B_r = 27 \times 0.1025 = 2.76$$

Case (c)
The requirement is that the breadth of the flange assumed as taking compression should not exceed the least of the following:

(i) one-third of the effective span (i.e. h)
(ii) the distance between the centres of the ribs (i.e. B_r)

(iii) the breadth of the rib plus twelve times the thickness of the flange.

Then the maximum flange width is the least of $h/3$, B_r or $[(12 \times T) + t_r]$ (Figure 13.64) where h = height of the wall.

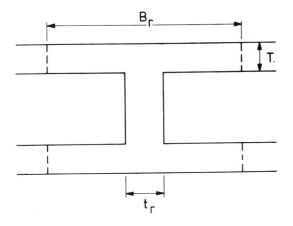

Figure 13.64 Effective length of diaphragm flange

Example:

Let h (height) = 6 m and $T = t_r = 102.5$ mm

Then

$$B_r = \frac{6}{3} \text{ or } [(12 \times 102.5) + 102.5]$$

$$= 2 \text{ m or } 1.33 \text{ m}$$

Case (d)

The shear resistance can be obtained either by bonding every alternate course of the rib into the leaf, or by using metal ties (see Figure 13.65).

From the four cases considered (a), (b), (c) and (d) the limiting dimension for the spacing of the ribs is given by case (c) as 1.33 m. Clearly, for the majority of diaphragm walls, constructed of half brick ribs and leaves throughout, this will generally be the limiting dimension for the rib spacing.

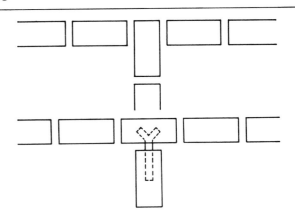

Figure 13.65 Shear ties between ribs and leaves

From experience, with wind forces around 0.6 kN/m^2 and heights of 8 m, it has been found that the rib spacings should be at about 1.0 to 1.25 m centres.

13.9.2 Depth of diaphragm wall and properties of sections

Within reason, the greater the depth of the wall, the greater is its resistance to wind forces. If the wall width becomes too large, the buckling of the cross-ribs may become critical and this would need careful consideration. Increase in depth also improves the wall's slenderness ratio, and thus its axial loadbearing capacity. From experience, with the wind forces and wall heights mentioned above, the section needs to be 0.4 to 0.7 m deep (Figure 13.66).

Breadths and depths of diaphragm walls are governed mainly by brick sizes and joint thicknesses. The engineer is free to design the diaphragm best suited to his project, and Figure 13.67 shows some typical breadths and depths, found useful in practice, based on the standard brick. Calculations for a typical section are below:

$$[\text{(bricks)} + \text{(joints)} + \text{(rib)} \quad]$$

$$B_r = [(4 \times 215) + (5 \times 10) + (2 \times \tfrac{1}{2} \times 102.5)] \times 10^{-3} = 1.0125 \text{ m}$$

$$b_r = 1.0125 - 0.1025 \qquad\qquad = 0.910 \text{ m}$$

$$D = [(2 \times 215) + 10] \times 10^{-3} \qquad\qquad = 0.440 \text{ m}$$

$$d = 0.440 - (2 \times 0.1025) \qquad\qquad = 0.235 \text{ m}$$

$$I = \frac{B_r D^3}{12} - \frac{b_r d^3}{12} \qquad\qquad = 6.21 \times 10^{-3} \text{ m}^4$$

$$Z = \frac{I}{y} = \frac{6.21 \times 10^{-3}}{0.44 \times 0.5} \qquad\qquad = 28.23 \times 10^{-3} \text{ m}^3$$

$$A = B_r D - b_r d = (1.0125 \times 0.44) - (0.91 \times 0.235) \qquad = 0.232 \text{ m}^2$$

The values per metre length of the wall are:

$I = 6.21/1.0125 = 6.13 \times 10^{-3}\ m^4$
$Z = 28.23/1.0125 = 27.88 \times 10^{-3}\ m^3$
$A = 0.232/1.0125 = 0.229\ m^2$

The section properties shown above, and others for a range of walls likely to be required, are given in Table 13.2.

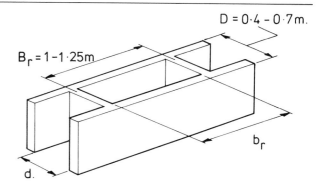

Figure 13.66 Wall profiles for typical diaphragm wall structures

Table 13.2 Diaphragm wall section properties

Section	Dimensions				Section properties per diaphragm		
	D (m)	d (m)	B_r (m)	b_r (m)	I ($10^{-3} \times m^4$)	Z ($10^{-3} \times m^3$)	A (m^2)
1	0.44	0.235	1.4625	1.36	8.91	40.49	0.324
2	0.44	0.235	1.2375	1.135	7.55	34.32	0.278
3	0.44	0.235	1.0125	0.91	6.21	28.23	0.232
4	0.5575	0.352	1.4625	1.36	16.18	58.04	0.337
5	0.5575	0.352	1.2375	1.135	13.74	49.29	0.290
6	0.5575	0.352	1.0125	0.91	11.31	40.57	0.244
7	0.665	0.46	1.4625	1.36	24.81	74.62	0.347
8	0.665	0.46	1.2375	1.135	21.12	63.52	0.301
9	0.665	0.46	1.0125	0.91	17.43	52.43	0.254
10	0.7825	0.5775	1.4625	1.36	36.56	93.45	0.359
11	0.7825	0.5775	1.2375	1.135	31.18	79.69	0.313
12	0.7825	0.5775	1.0125	0.91	25.82	66.01	0.267
13	0.89	0.685	1.4625	1.36	49.46	111.14	0.37
14	0.89	0.685	1.2375	1.135	42.4	95.3	0.324
15	0.89	0.685	1.0125	0.91	34.86	78.34	0.278

Section	Section properties per metre length			Shear stress coefficient, K_1/m^2	Trial section stability moment coefficient, K_2(kN/m) Density = 20 kN/m³
	I ($10^{-3} \times m^4$)	Z ($10^{-3} \times m^3$)	A (m^2)		
1	6.09	27.69	0.222	27.74	0.835
2	6.10	27.73	0.225	27.66	0.846
3	6.13	27.88	0.229	27.51	0.862
4	11.06	39.69	0.230	20.52	1.097
5	11.10	39.83	0.234	20.44	1.116
6	11.17	40.07	0.241	20.34	1.149
7	16.96	51.02	0.237	16.56	1.347
8	17.07	51.33	0.243	16.46	1.381
9	17.21	51.77	0.251	16.37	1.426
10	24.99	63.90	0.245	13.60	1.640
11	25.19	64.40	0.253	13.49	1.692
12	25.50	65.20	0.264	13.33	1.766
13	33.82	76.00	0.253	11.64	1.926
14	34.26	77.01	0.262	11.49	1.994
15	34.43	77.37	0.274	11.44	2.085

Note: For sections 1, 4, 7, 10, 13 the flange length slightly exceeds the limitations given in BS 5628, clause 36.4.3(b). These sections have been included since they are the closest brick sizes to the flanges recommended in the Code. If the designer is concerned at this marginal variation, he may calculate the section properties on the basis of an effective flange width of 1.33 m.

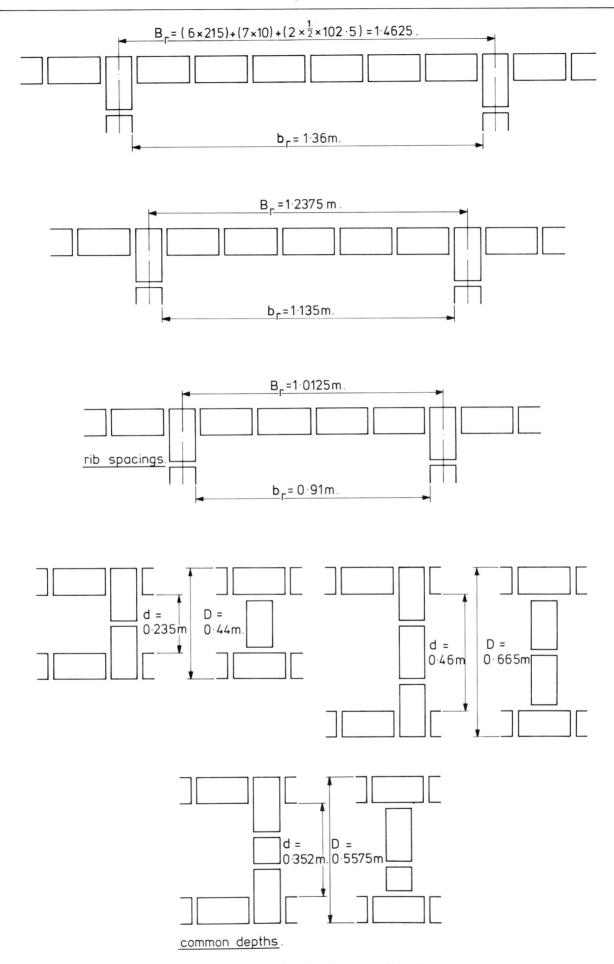

Figure 13.67 Typical diaphragm wall profiles

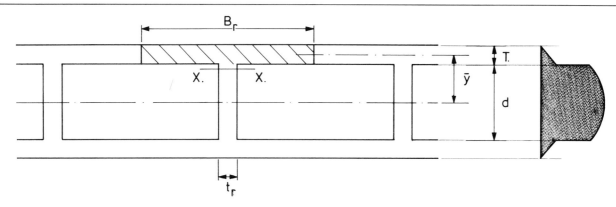

Figure 13.68 Shear stress distribution diagram

13.9.3 Shear stress coefficient, K_1
It is necessary to check the shear stress at the junction of the rib and leaves (Figure 13.68).

$$\text{Design vertical shear stress, } v_h = \frac{VA\bar{y}}{It_r}$$

where

V = design shear force

$A = B_r \times T$

$\bar{y} = \dfrac{d}{2} + \dfrac{T}{2}$

then:

Design vertical shear stress at XX,

$$v_h = \frac{V \times B_r \times T \left(d/2 + T/2\right)}{I \times t_r}$$

Generally $T = t_r = 0.1025$ m

Therefore

$$v_h = V \times \frac{B_r}{I} \left(\frac{d}{2} + \frac{0.1025}{2} \right)$$

Let

$$K_1 = \frac{B_r}{I} \left(\frac{d}{2} + \frac{0.1025}{2} \right)$$

Then

$$v_h = K_1 V$$

For section 3, Table 13.2

$$K_1 = \frac{1.0125}{6.21 \times 10^{-3}} \left(\frac{0.235}{2} + \frac{0.1025}{2} \right) \text{ per m}^2$$

$$= 27.51 \text{ per m}^2$$

Values of K_1 for other sections are given in Table 13.2.

It should be noted that the constant K_1 has been calculated on the assumption that both ribs and leaves are constructed in half brick walls – the derivation of K_1 would require adjustment for varying thicknesses of composition.

13.9.4 Trial section coefficients, K_2 and Z
Owing to the symmetrical profile of the diaphragm wall, a more direct route to a trial section has been devised and considers the two critical conditions which exist in the 'propped cantilever' action of the analysis.

Condition (i) exists at the base of the wall where the applied bending moment at this level must not exceed the stability moment of resistance of the wall.

Condition (ii) exists at approximately ⅜h down from the top of the wall where the flexural tensile stresses are a maximum and must not exceed those allowable through calculation.

Consider the two conditions

Condition (i)
The trial section analysis is simplified by assuming that the dpc at the base level cannot transfer tensile forces and that the mass contributing to the MR_s comprises only the own weight of the masonry,

$$\text{BM at base level} = \frac{\gamma_f W_k h^2}{8} \qquad (1)$$

See section 13.55 for lever arm.

MR_s at base level
$= \text{area} \times \text{height} \times \text{density} \times \gamma_f \times 0.475D$
$= 0.475(A \times h \times \gamma_f \times D \times \text{density}) \qquad (2)$

Equating (1) and (2):

$$\frac{\gamma_f W_k h^2}{8} \leq 0.475 \,(A \times h \times \gamma_f \times D \times \text{density})$$

γ_f for wind and dead loads will be taken as 1.4 and 0.9 respectively.

Hence

$$0.175 \, W_k h^2 \leq 0.4275 \,(A \times h \times D \times \text{density})$$

Now let $K_2 = 0.4275 \times A \times D \times \text{density}$

then $W_k h \leq 5.714 K_2$

and

$$h \leq \frac{5.714 K_2}{W_k} \qquad (3)$$

Condition (ii)

The trial section analysis is simplified by assuming that flexural tensile stresses control, γ_m, is taken as 2.5 and f_{kx} as 0.4 N/mm^2.

$$\text{BM at } \tfrac{3}{8}h \text{ level} = \frac{9\gamma_f W_k h^2}{128} \qquad (4)$$

$$\text{Moment of resistance} = \left(\frac{f_{kx}}{\gamma_m} + g_d\right) Z \qquad (5)$$

Equating (4) and (5)

$$\frac{9\gamma_f W_k h^2}{128} \leq \left(\frac{f_{kx}}{\gamma_m} + g_d\right) Z$$

$$\frac{9 \times 1.4 \times W_k \times h^2}{128} \leq \left(\frac{0.4 \times 10^3}{2.5} + \frac{\gamma_f \times 20 \times 3 \times h}{8}\right) Z$$

$$0.098 W_k h^2 \leq (160 + 6.75h)Z$$

$$Z = \frac{0.098 W_k h^2}{160 + 6.75h}$$

$$= \frac{W_k h^2}{1600 + 67.5h} \qquad (6)$$

Two graphs have been plotted for equations (3) and (6) and for various values of W_k, which are shown in Figures 13.69 and 13.70.

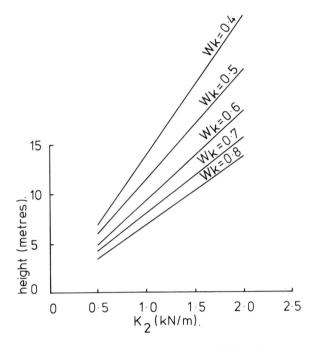

Figure 13.69 Trial section coefficient K_2

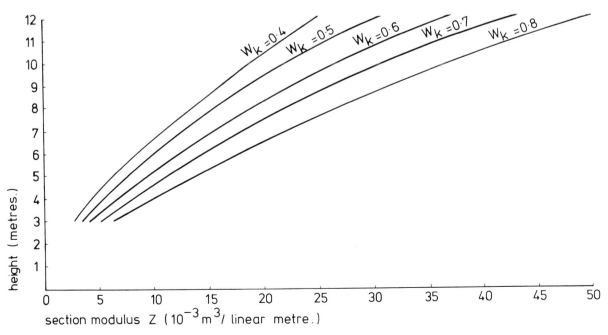

Figure 13.70 Trial section 'Z' modulus

For a known wall height and wind pressure, values of K_2 and Z may be read off the graphs and, using Table 13.2, the most suitable section can be obtained for full analysis. It should be remembered that the two trial section graphs have been drawn assuming fixed conditions for a number of variable quantities which are summarised thus:

(a) wall acts as a true propped cantilever
(b) dpc at base of wall cannot transfer tension
(c) vertical roof loads (downward or uplift) are ignored
(d) γ_m is taken to be 2.5
(e) f_{kx} is taken to be 0.4 N/mm²
(f) density of masonry is taken to be 20 kN/m³
(g) K_2 values calculated using approximated lever arm method.

The trial section graphs must be used only for the purpose of obtaining a trial section and a full analysis of the selected section must always be carried out.

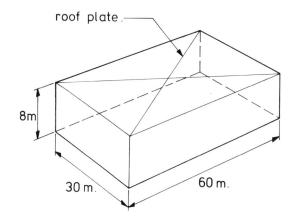

Figure 13.71 Diaphragm wall design example 2

13.10 EXAMPLE 2: DIAPHRAGM WALL

13.10.1 Design problem
A warehouse measuring 30 m × 60 m and 8 m high is shown in Figure 13.71 and is to be designed in brickwork, using diaphragm wall construction for its main vertical structure. There are no substantial internal walls within the building to provide any intermediate support. During construction extensive testing of materials and strict supervision of workmanship will be employed.

Facing bricks with a compressive strength of 41.5 N/mm² and a water absorption of 8% will be used throughout the building.

13.10.2 Characteristic and design loads
The characteristic and design loads will be taken as the same as those used for the earlier fin wall design example.

Characteristic loads
(a) Wind forces:
All as fin wall design, hence
Pressure on windward wall, $W_{k1} = 0.781$ kN/m²
Suction on leeward wall, $W_{k2} = 0.497$ kN/m²
Gross roof uplift, $W_{k3} = 0.426$ kN/m²

(b) Dead and superimposed loads:
All as fin wall design, hence
Superimposed load, $Q_k = 0.75$ kN/m²
Dead load, $G_k = 0.67$ kN/m²

Design loads
All as fin wall design, hence:

Design dead load = 0.9 × 0.67 = 0.603 kN/m²

Design wind loads:
Pressure, from W_{k1} = 1.093 kN/m²
Suction, from W_{k2} = 0.696 kN/m²
Uplift, from W_{k3} = 0.597 kN/m²

Design dead − uplift = 0.006 kN/m² (say zero).

13.10.3 Select trial section
For the wall height of 8.0 m and the characteristic wind load of 0.781 kN/m².

$K_2 = 1.16$ kN/m and $Z = 23.3 \times 10^{-3}$ m³ can be read from graphs 13.69 and 13.70 respectively. Select wall section 5 and a full analysis using this section should then be carried out.

Wall properties
$I/m = 11.10 \times 10^{-3}$ m⁴
$Z/m = 39.83 \times 10^{-3}$ m³
$A/m = 0.234$ m²
$K_1 = 20.44/$m²

The wall section is shown in Figure 13.72.

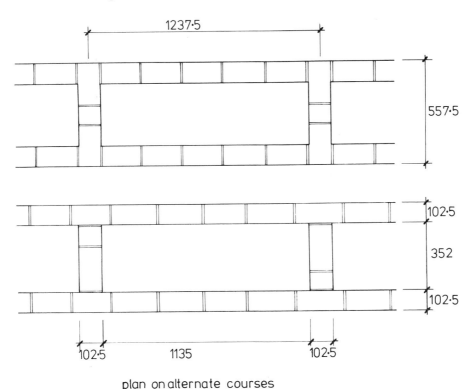

plan on alternate courses

Figure 13.72 Diaphragm wall section 5

13.10.4 Determine wind moment and MR_s at base
Considering 1 metre width of wall.

$$\text{Design wind moment at base} = 1.4W_k\ \frac{8^2}{8} = \frac{1.093 \times 8^2}{8} = 8.744\ \text{kN}\cdot\text{m}$$

The calculated stability moment of resistance at the base is found as follows:
Appendix B of BS 5628 discusses the application of a rectangular stress block under ultimate conditions, and the stability moment of resistance MR_s can be assumed to be provided by the axial load in the diaphragm section acting at a lever arm about the centroid of the rectangular stress block, as shown in Figure 13.73.

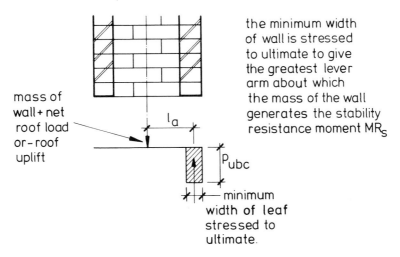

mass of wall + net roof load or – roof uplift

the minimum width of wall is stressed to ultimate to give the greatest lever arm about which the mass of the wall generates the stability resistance moment MR_s

minimum width of leaf stressed to ultimate.

Figure 13.73 Generation of MR_s for diaphragm wall

Hence:

$$\text{design dead load at base} = 0.9 \times 0.234 \times 20 \times 8$$
$$= 33.70\ \text{kN}$$

$$\text{minimum width of stress block} = \frac{\text{axial load in diaphragm}}{1\,000 \times p_{ubc}} \tag{7}$$

Now, p_{ubc} = allowable flexural compressive stress = $1.1\beta f_k/\gamma_m$, with lateral restraint provided by the foundations assuming a raft foundation and as for the fin wall design previously β can be taken as 1.0. Therefore, $p_{ubc} = 1.1 f_k/\gamma_m$ at base level.

Assuming 41.5 N/mm^2 bricks in 1:1:6 mortar then f_k from Table 2(a) in BS 5628 is 9.41 N/mm^2, and assuming $\gamma_m = 2.5$ then $p_{ubc} = 1.1 \times 9.41/2.5 = 4.14$ N/mm^2.

Substituting this value in equation (7):

$$\text{minimum width of stress block} = \frac{33.70 \times 10^3}{1\,000 \times 4.14} = 8.14 \text{ mm}$$

Therefore:

$$\text{stability moment of resistance} = 33.70 \times \left(\frac{0.5575}{2} - \frac{0.00814}{2} \right)$$

$$= 9.26 \text{ kN·m} > 8.744 \text{ kN·m} = \text{wind moment}$$

It is interesting to compare the calculated value of stability moment of resistance MR_s with the approximate lever arm method suggested earlier (see 13.5.5).

$$\begin{aligned}
\text{Approximate } MR_s &= \gamma_f \times 20 \times 0.234 \times 8 \times l_a \\
&= 33.70 l_a \\
\text{Approximate } l_a &= 0.475 D \\
&= 0.475 \times 0.5575 \\
&= 0.265 \text{ m}
\end{aligned}$$

Therefore:

$$\begin{aligned}
\text{Approximate } MR_s &= 33.70 \times 0.265 \\
&= 8.93 \text{ kN·m}
\end{aligned}$$

which is still greater than 8.744 kN·m applied BM.

Since the stability moment at the base is greater than the wind moment, the maximum span moment occurs $\frac{3}{8}h$ down from roof level.

$$M_w = \text{wind moment at } \tfrac{3}{8}h = \frac{9 \times 1.4 \times 1.093 \times 8^2}{128}$$

$$= 4.92 \text{ kN·m/m}$$

13.10.5 Consider the stress at level of M_w

The stress considerations at the level of the maximum wall moment assume triangular stress distribution, using elastic analysis, but relate to ultimate strength requirements at the extreme edges of the wall face.

$$\begin{aligned}
\text{Design axial load at } M_w \text{ level} &= 0.9 \times 0.234 \times 20 \times \tfrac{3}{8} \times 8 \\
&= 12.636 \text{ kN}
\end{aligned}$$

Flexural stresses at design load

$$\text{stress} = \frac{\text{load}}{\text{area}} \pm \frac{\text{moment}}{\text{section modulus}}$$

(i) Flexural compressive

$$f_{ubc} = \frac{12.636 \times 10^3}{0.234 \times 10^6} + \frac{4.92 \times 10^6}{39.83 \times 10^6}$$

$$= 0.054 + 0.1235$$
$$= 0.1775 \text{ N/mm}^2$$

(ii) Flexural tensile

$$f_{ubt} = 0.054 - 0.1235$$
$$= -0.07 \text{ N/mm}^2$$

These stresses must now be compared with the allowable flexural stresses at M_w level.

(a) Allowable flexural tensile stress

$$p_{ubt} = \frac{f_{kx}}{\gamma_m}$$

where

f_{kx} = 0.4 N/mm² for clay bricks with a water absorption of between 7 and 12% set in 1:1:6 mortar, taken from Table 3 of BS 5628, for the plane of failure parallel to the bed joints.
γ_m = 2.5 from Table 4 of BS 5628 for special construction control and special manufacturing control of structural units (see Table 5.11).

Therefore

$$p_{ubt} = \frac{0.4}{2.5} = 0.16 \text{ N/mm}^2$$

By comparison with calculated f_{ubt} = 0.07 N/mm² the flexural tensile stresses are acceptable.

(b) Allowable flexural compressive stresses, p_{ubc}

$$p_{ubc} = \frac{1.1\beta f_k}{\gamma_m}$$

Calculate the value of β
It is assumed that the flange of the diaphragm is liable to buckle under compressive bending loading. The effective length of the flange may be taken as 0.75 times the length of the internal hollow box and the effective thickness as the actual thickness of the flange as shown in Figure 13.74.

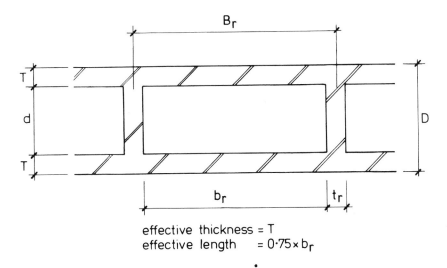

effective thickness = T
effective length   = 0.75 × b_r

Figure 13.74 Stability of diaphragm flanges

In this case slenderness ratio = 0.75 × 1.135/0.1025 = 8.30.

Further, assuming that the stressed areas are concentrically loaded (i.e. $e_x = 0$) and referring to BS 5628, Table 7 (see Table 5.15), $\beta = 0.995$ by interpolation. Therefore:

$$p_{ubc} = \frac{1.1 \times 0.995 \times 9.41}{2.5}$$

$$= 4.12 \, \text{N/mm}^2$$

By comparison with calculated $f_{ubc} = 0.1775 \, \text{N/mm}^2$ the flexural compressive stresses are acceptable.

The local buckling has been shown to be adequately analysed. The possibility of buckling of the overall section, in its height, must now be considered. This was shown to be a critical design consideration in the design of the fin wall because of the unrestrained edge of the outstanding leg of the fin from the flange. Figure 13.75 shows, that the whole of the area in compression, due to the bending, could buckle in the fin wall situation.

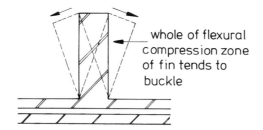

Figure 13.75 Buckling of projecting fin

The diaphragm wall, being a symmetrical section, is not prone to this mode of failure. The flanges provide adequate restraints to buckling. The possibility of the full section buckling under axial loading is outside the scope of this design example (see Chapter 10).

13.10.6 Consider diaphragm with deflected roof prop

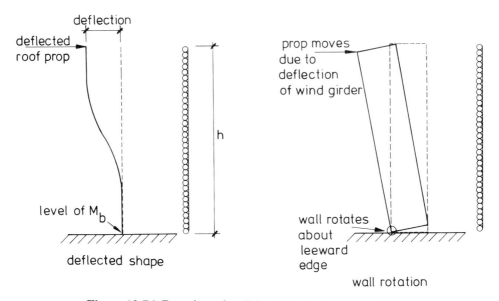

Figure 13.76 Rotation of wall for deflected prop condition

From Figure 13.76 it is evident that the deflection of the roof wind girder induces additional rotation at the level of M_b.

In this design example the stability moment of resistance $(MR_s) = 9.26 \, \text{kN·m/m}$, which is greater than the design wind moment at the base $= 8.744 \, \text{kN·m/m}$.

Thus if the roof prop should deflect, additional rotation would occur at base level causing a cracked section and thus taking advantage of the full stability moment. The revised bending moment diagram for this condition is calculated below, where it is seen that the wall moment is reduced and the wall tends toward the free vertical cantilever situation. The position of the maximum wall moment occurs at the level of zero shear.

$$\text{Prop force at the top} = \frac{1.093 \times 8}{2} - \frac{9.26}{8} = 3.214 \text{ kN/m}$$

$$\text{Point of zero shear} = \frac{3.214}{1.093} = 2.94 \text{ m from the top of the wall}$$

Therefore

$$M_w = (2.94 \times 3.214) - \frac{1.093 \times 2.94^2}{2}$$

$$= 4.73 \text{ kN·m/m}$$

This is obviously acceptable since it is less than the previously calculated $M_w = 4.92$ kN·m, which assumed the propped cantilever condition and the base moment being limited to $\gamma_f W_k h^2/8$.

The increase in the moment at the base level up to the value of MR_s has also been shown to be acceptable from the earlier calculation of MR_s, although this should be checked fully if slenderness reductions are applicable at this level. The adjusted bending moment diagram for the deflected prop condition is shown in Figure 13.77.

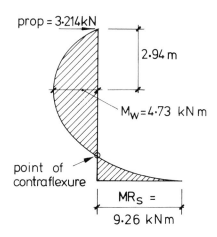

Figure 13.77 Bending moment for deflected prop condition

13.10.7 Calculate the shear stress
Reaction at base = shear force (V)

$$V = \frac{5}{8} \times 1.093 \times 8$$
$$= 5.47 \text{ kN}$$
$$v_h = K_1 V$$

Therefore

$$v_h = \frac{20.44 \times 5.47}{10^3}$$

$$= 0.112 \text{ N/mm}^2$$

Therefore:

$$\text{shear force per course} = 0.112 \times 75 \times 102.5 \times 10^{-3}$$
$$= 0.86 \text{ kN}$$

This is well within the capacity of a strip fishtail zinc-coated 3 mm × 20 mm wall tie as is shown below. Alternatively, the ribs may be fully bonded to the flange and the shear at the neutral axis checked.

Shear resistance of wall ties

Table 8 of BS 5628 (see Table 6.5) gives values of characteristic shear strengths of wall ties. For the ties specified and set in a designation (iii) mortar this value is 3.5 kN to which the partial safety factor for materials, γ_m, must be applied. γ_m for wall ties is given in BS 5628, clause 27.5 as 3.0.

Hence, design shear strength = 3.5/3 = 1.17 kN per tie.

Shear ties are required in every course in each rib at the bottom of the wall but could, if necessary, be varied in spacing throughout the height of the ribs.

13.10.8 Stability of transverse shear walls

The designer should now check the stability of, and stress in, the gable shear walls using the principles given in Chapter 11.

13.10.9 Summary

Section type (5) is acceptable using standard format 41.5 N/mm^2 crushing strength bricks with water absorption of 8% set in a designation (iii) 1:1:6 mortar. It is important for the designer to appreciate that the example sections tabulated are not the only ones which can be considered. The depth of the diaphragm and the rib spacing may be varied to suit particular requirements. If, for example, the architect wished to express the ribs externally at 2.5 m centres, this would be easily achieved as shown in Figure 13.78, or a normal intermediate rib could be added internally.

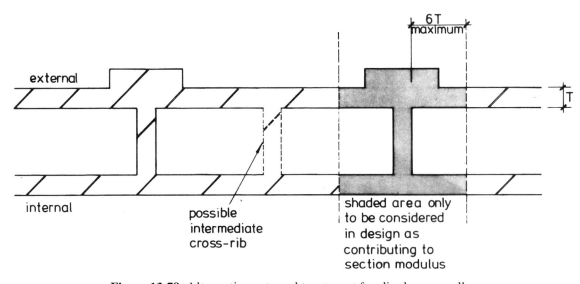

Figure 13.78 Alternative external treatment for diaphragm wall

The designer would, however, need to check the capacity of the flange to span between the ribs and give greater consideration to the design β value. In addition, for larger rib spacings, the designer should limit the length of the wall considered to be acting as the flanges of the box section to 6 × thickness of wall forming the flange in assessing the section modulus to be used in the design, see BS 5628, clause 36.4.3 (b).

13.11 OTHER APPLICATIONS

Although diaphragm and fin walls were originally developed for use in tall, single-storey, wide-span structures, they do have applications in other fields, particularly where lateral loading is more significant than axial loading.

For example, diaphragms have been used by the authors as retaining walls. On one site in particu-lar, which was covered with a large quantity of demolition rubble, the rubble was used to fill the voids in the diaphragm, and a strong and inex-pensive mass retaining wall was achieved. This wall formed part of a landscaping development scheme and is shown in Figures 13.79 and 13.80.

Diaphragms and fins are ideal forms for retaining walls and other walls which are required to resist comparatively high bending moments. They have been used in both plain and post-tensioned forms for retaining walls and in tall buildings. Diaphragm walls can also be used as sound deflectors on motorways in urban areas. Sound deflectors are presently constructed in steel frame, precast concrete or timber and it is considered that a diaphragm wall for this purpose would be less expensive, certainly more durable, and would probably possess greater aesthetic appeal.

Masonry has been a traditional choice for use in farm buildings and there is certainly scope to extend its use, by way of diaphragms and fins for storage bins for grain, potatoes, etc.

Apart from new structures, fin walls have been successfully used, by the authors, for strengthen-

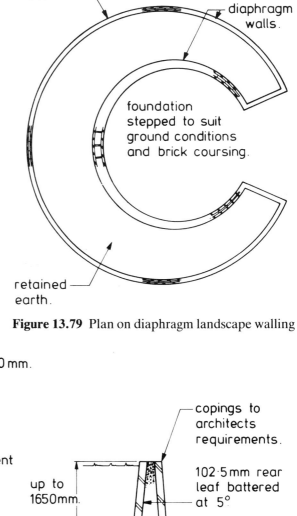

Figure 13.79 Plan on diaphragm landscape walling

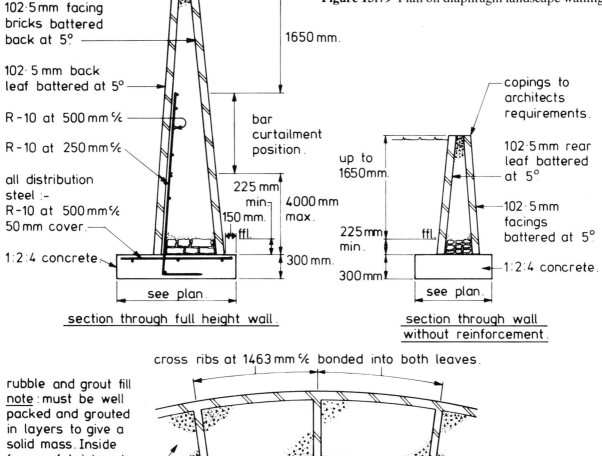

Figure 13.80 Diaphragm landscape walling

ing existing buildings. The rear wall of a grand-stand, which was showing signs of instability, was stiffened by bonding into it, at predetermined centres, a series of brick fins. The fins were designed to resist the excess of loading which the original wall was unable to support. A further application was in the use of post-tensioned fins to strengthen a retaining wall within an existing basement where a change in use of the building, which resulted in an increased lateral loading on the wall, caused it to bulge and crack. The post-tensioned brick fin proved easy to construct in an extremely confined working space with difficult access and compared with alternative schemes was shown to be the most economic solution.

The use of fin walls in conjunction with widely spaced spine walls provides a potential solution to multi-storey structures for use in open-plan office buildings, hospital ward blocks and other similar situations where the restrictions of cross-wall or cellular construction cannot be tolerated.

Whilst the discussion and calculations for the diaphragms and fins have dealt only with their effectiveness to resist lateral loading, they also both possess the correct properties to resist axial loading from tall platforms. The capacity of a simple wall to support heavy axial loading is significantly reduced by its tendency to buckle. If the load is applied at a great height, the natural compressive qualities of brickwork are not being correctly exploited if, to provide an adequate slenderness ratio, the thickness of the wall is simply increased. This approach results in a wall with low applied stresses and high material content. Both the diaphragm and fin walls will provide greatly improved slenderness ratios and, at the same time, an adequate proportion of masonry area to support the axial loading at a more efficient stress level. The slenderness ratio of a wall is often expressed as the ratio of its effective height to its effective thickness. The effective thickness of a plain solid wall is its actual thickness. The effective thickness, of the diaphragm or fin walls, assuming that the full

section is loaded in each case, may be approximately calculated from the radius of gyration giving an equivalent solid wall thickness.

Consider the diaphragm section 10 as given in Table 13.2 earlier and shown in Figure 13.81. Consider a metre length of wall:

$$\text{Radius of gyration, } r = \sqrt{\frac{I}{A}}$$

$$= \sqrt{\left(\frac{24.99 \times 10^{-3}}{0.245}\right)}$$

$$= 0.32 \text{ m} \qquad (8)$$

For a 1 m length of solid wall:

$$I = \frac{1 \times t^3}{12}$$

$$A = 1 \times t$$

$$r = \sqrt{\frac{I}{A}}$$

$$= \sqrt{\frac{t^3}{12t}} \qquad (9)$$

Equating (8) and (9):

$$0.32 = \sqrt{\frac{t^3}{12t}}$$

$$t^2 = 12 \times 0.32^2$$

$$t = 1.110 \text{ m}$$

Hence, the 782.5 mm diaphragm wall, calculated on this basis, would be equivalent, for the calculation of slenderness ratio, to a solid wall of 1 110 mm thickness. This is, however, an approximate assessment and for design purposes it is usually assumed that an effective thickness equal to the actual overall thickness should be used until a slenderness ratio based on radius of gyration is introduced into BS 5628.

A worked example of a diaphragm wall used to support high axial loading is given in Chapter 10.

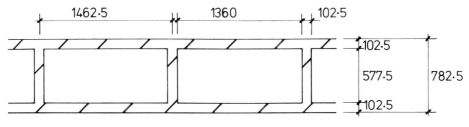

Figure 13.81 Diaphragm wall section 10

CHAPTER 14

DESIGN OF MULTI-STOREY STRUCTURES

The method commonly used in building multi-storey structures is to erect a steel or concrete frame and clad it with external walls to provide a weather-resistant and durable envelope. Internal walls are built to form partitions, acoustic or fire barriers, party walls, etc., and to enclose stair and lift wells. Thus the frame has to carry the loads from the roof and floors, and has to be strengthened to carry the weight of the walls. When, as is often the case, the walls are of brick or block, their compressive strength and structural potential are completely wasted.

In many cases, however, if forethought is given to the plan form and the structural layout, the internal and external walls can easily be designed to carry not only their own weight but also the floor and roof loads, and the cost of a structural frame of beams and columns can be saved. Depending on the type of structure, the saving can be between 8% and 10% of the total building cost. Since the number of materials, site operations and trades are reduced, and more work is carried out by the main contractor, the construction is simplified and speeded up. Savings in site construction time of more than 10% are quite common.

14.1 Structural forms

Crosswalls
Mainly used for hotel bedroom blocks, school classrooms, student hostels, town houses and other rectangular buildings with repetitive floor plans.

Cellular construction
Principally used for tall tower blocks of flats, square on plan.

Spine construction
Used where open-plan interiors are necessary in office blocks, hospital wards, warehouses and similar structures.

Column construction
An alternative to spine construction.

Before discussing the choice and design of structural forms (using plane walls: i.e. solid, cavity and piered walls, or columns, diaphragm and fin walls) it is convenient to establish briefly here, and then to consider in detail, the common factors and problems:

1. Stability under vertical loading, and from horizontal loading (mainly due to wind) on the longitudinal and lateral axes of the structure, which must be provided for.
2. External walls. Restraint of outer leaf of cavity walls. This is necessary even when the wall is non-loadbearing.
3. Provision for services. Early planning of service runs is necessary, so that openings in masonry frames can be built-in.
4. Movement joints. As with other structural materials, movement joints must be incorporated in the structure. Whilst masonry structures tend to be more resistant to damage due to movement, it is still necessary to install movement joints.
5. Vertical alignment of loadbearing walls. For simplicity, speed of construction and cost, walls should remain in the same vertical plane from foundations to roof. Where, for special reasons, the occasional wall cannot be lined up, it is not difficult to accommodate such plan changes – though it does tend to increase costs.
6. Foundations. The foundations for loadbearing masonry structures are generally simpler than those for structural frames. The loads are spread along walls founded on strip footings, so that contact pressures are low. In framed structures, loads are often concen-

trated at the column points, so that contact pressures are high.

7. Flexibility. Sometimes, over a period of time, there is a need to alter structures due to changing functional requirements. In many situations, masonry structures are more readily adaptable to change than steel or concrete frames.

8. Concrete roof slab/loadbearing wall connections. *In situ* concrete roof slabs should not be cast directly on to masonry walls. As the roof expands and contracts, due to thermal and other movements, the wall will tend to crack, particularly at the connection. A sliding joint, such as a layer of dpc, should be laid on top of the walls before casting the concrete.

9. Accidental damage. This topic is discussed in detail in Chapter 8.

10. Choice of brick, block and mortar. Whilst it is quite simple to design every wall in every storey height with a different structural masonry unit and mortar, this increases the costs, planning and supervision of the contract. On the other hand, although the use of only one brick or block laid in one class of mortar simplifies planning and supervision enormously, it may not be the most economical solution overall. For example, engineering bricks may be necessary on the lower levels of a multi-storey block; but, as these tend to be more expensive than the low strength bricks which may be adequate on the upper floors, it would be uneconomical to use them throughout the structure. Thus before making a choice, the cost implications should be carefully considered.

14.1.1 Stability

Figure 14.1 shows the main forces acting on a structure.

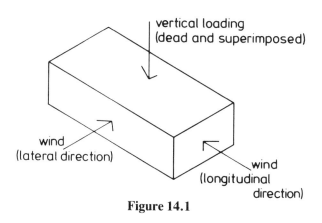

Figure 14.1

Vertical stability

It is rare for vertical instability, i.e. collapse or cracking of walls under vertical loads, to be a major problem – provided, of course, the compressive stresses in the masonry are kept within the allowable limits and the neccessary restraints to prevent buckling are provided (see Chapter 7).

Horizontal stability

The wind acts on the external walls or cladding panels which transfer the wind force to floors and roof (which can act as a horizontal plate) which, in turn, transfer the force to the transverse walls (see Figure 14.2). The wind force creates racking in the transverse walls, as shown in Figure 14.3, but walls are highly resistant to racking action. The diagonal tension or racking stresses, which could cause cracking, are either eliminated by the vertical compressive load on the wall and/or

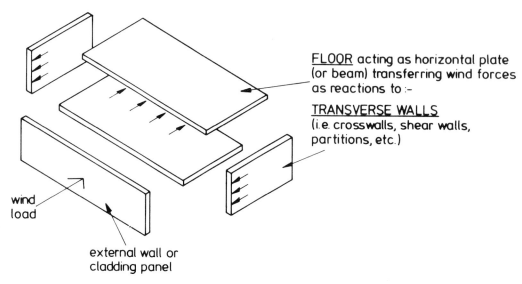

Figure 14.2

resisted by the allowable tensile stresses in the masonry. If the tensile stress should exceed the allowable limits consideration should be given to reinforcing or post-tensioning the walls.

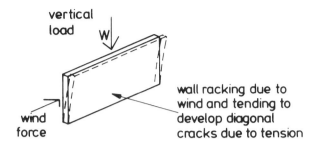

Figure 14.3

The stresses at the base of the wall are due to the combined effect of the vertical loading and the moment induced by the wind force and are determined using the normal elastic stress distribution formula (see Figure 14.4):

$$f = \frac{W}{A} \pm \frac{M}{Z}$$

There is usually little danger in multi-storey structures of a wall overturning or failing in horizontal shear.

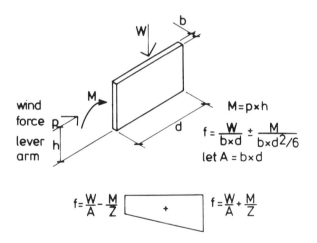

Figure 14.4

Multi-storey masonry structures tend to rely for their stability on their own weight in resisting horizontal forces due to wind. They are not capable, as can be steel or concrete frames, of being considered for design purposes as fully rigid frames. In steel or concrete structures, rigid frames tend to be necessary to resist lateral wind loading. It is not usually possible to develop as much rigidity at the junction of masonry walls and concrete floor slabs as there can be, for example, between concrete columns and beams. However, this is very rarely a difficult problem to overcome if sufficient forethought is given to the plan form and the structural layout. The use of the walls as shear walls, which replace the columns of a framed structure, can result in a very rigid design.

14.1.2 External walls
External walls can be solid (sometimes known as single-leaf masonry), cavity, piered, diaphragms or fins. It is quite common for the outer leaf of a cavity wall, or the face of a solid wall, to be in a different quality unit from the inner leaf or face. In cavity wall construction, a very frequent example is the use of a clay facing brick externally and an insulating block internally. Note that in the case of a solid wall with different bricks on the outer and inner faces, the bricks should have compatible movement characteristics.

Cavity walls are more popular than solid walls because they are more resistant to rain penetration, and have better thermal insulation properties. However, they are more expensive to build, and care and attention must be given to the choice and fixing of the wall ties. The outer leaf helps to restrain and stiffen the inner load-bearing leaf – but this action is only possible with sufficient, good, and durable ties. The external leaf should be properly and fully supported at every third storey height to prevent it bowing out, and to reduce the risk of loosening the wall ties due to differential movement of the inner and outer leaves. The only exception to this rule is in four-storey buildings, not exceeding 12 m in total height, where the restraint may be omitted at the designer's discretion.

Whilst, to some extent, both leaves carry the wind load, in addition to carrying its own weight the inner leaf supports most of the floor load. Since the outer leaf tends to carry its own weight only, the choice of facing brick or block is not so restricted by strength requirements. It has been found that, if the inner leaf is overstressed, the creep action in properly tied brickwork or blockwork tends to partly redistribute the stress to the outer leaf.

A common method of restraining the outer leaf at every third storey, as required in BS 5628, is to project the concrete slab on to it, as shown in Figure 14.5.

If the projection of the concrete slab is considered to be aesthetically undesirable, brick slips can be used to face the edge of the slab. Details of types, fixings, etc., can be found in modern textbooks on building construction. A typical detail is shown in Figure 14.6.

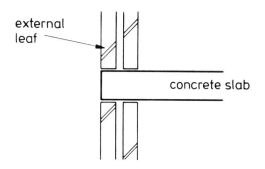

external
leaf

concrete slab

Figure 14.5

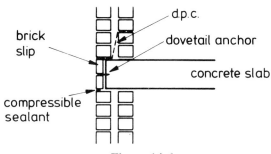

brick
slip

d.p.c.

dovetail anchor

concrete slab

compressible
sealant

Figure 14.6

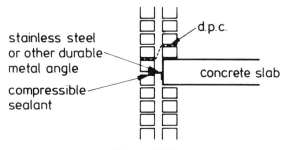

stainless steel
or other durable
metal angle

d.p.c.

concrete slab

compressible
sealant

Figure 14.7

Engineers tend to prefer bricks or blocks, rather than slips, and to anchor them (and thus restrain the outer leaf) to the slab by anchor ties or steel angles, as shown in Figure 14.7.

14.1.3 Provision for services

Inevitably, pipes for hot and cold water supply, conduits for electricity cables, ducts for air-conditioning, etc. have to pass through loadbearing masonry walls. The openings or holes for these services must always be pre-planned. Services engineers are accustomed to indiscriminate breaking out of large holes and cutting chases in relatively thick walls of traditional masonry construction when upgrading or changing the services in existing buildings. They do not always appreciate that *ad hoc* alterations cannot be permitted in modern, slender, highly stressed walls. Holes and chases should not be cut without the prior approval of the structural designer. See also Appendix 4.

Pre-formed openings can easily be arranged by leaving out bricks or blocks when building the wall. If the openings are large, or could cause overstressing or undesirable stress concentration in the surrounding masonry, reinforcement can be laid in the bed joints above the openings – and around, if necessary – to distribute the stress.

Detailed drawings of service holes and chases should be given to the contractor before the commencement of building operations. A typical builders-work drawing is shown in Figure 14.8.

Chases should be sawn out to the depth agreed by the structural designer, and must not be hacked out by hammer and chisel. Horizontal or diagonal chases are rarely permissible. Nor are vertical chases in half brick thick walls (102.5 mm thick).

Holes for vertical service runs through floor slabs form a very useful site aid in setting out and

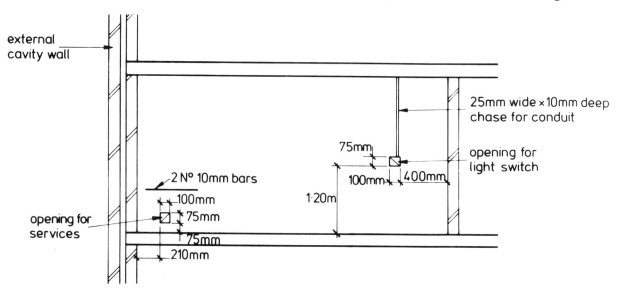

external
cavity wall

25mm wide × 10mm deep
chase for conduit

75mm

opening for
light switch

100mm   400mm

2 Nº 10mm bars

100mm

1·20m

75mm

opening for
services

75mm

210mm

Figure 14.8

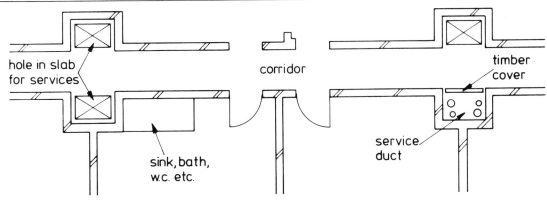

Figure 14.9

checking the vertical alignment of walls. Vertical ducts can easily be formed by making minor adjustments to the wall layouts (see Figure 14.9).

14.1.4 Movement joints
On long crosswall and spine structures, it is essential to insert movement joints to counter the effects of thermal and moisture movements. They are also advisable on structures liable to undergo excessive differential settlement. Movement joints should also be used to break up L and T plan shapes, and other similar building configurations when they are sensitive to movement. A typical method of achieving this in crosswall structures is shown in Figure 14.10.

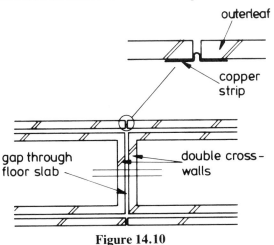

Figure 14.10

Services, finishes, etc., which have to cross the movement gap should be provided with flexible connection, as in concrete or steel-framed structures. The spacing of movement joints depends upon the masonry units used. For example, 12 m spacing is usually adequate for clay bricks, and 6 m for concrete blocks. Detailed information on movement joints is provided in Appendix 3.

14.1.5 Vertical alignment of loadbearing walls
Whilst engineers and architects have long accepted the need for column grid layouts, and are well aware of the need to line up columns (i.e. column positions should remain constant

from foundations to roof) they do not, at first, readily accept the same discipline in masonry structures – no doubt, because they have been used to placing non-loadbearing walls or partitions anywhere.

Non-loadbearing partitions can still be placed practically anywhere in a loadbearing masonry frame. But, as with steel or concrete columns, it is desirable that the loadbearing walls are lined up. They can, of course, be moved out of line – but this may mean expensive and complex beam and beam-support layouts. This factor, more than any other, has tended to militate against the use of loadbearing masonry, especially in situations where the ground floor layout differs from the upper floors. For example, in a hotel bedroom block, the ground floor may require large open spaces for restaurant, reception areas, etc. The conflicting needs of the ground floor and the upper floors can easily be reconciled by the use of podium construction (see 14.2.7).

The author's experience has shown that designers quickly adjust to the need for planning discipline, and welcome the benefits of repetition of floor layouts, windows, doors and other furniture, service runs, finishes, etc., with the accompanying savings in cost and time of erection and simplicity of construction.

Loadbearing masonry structures can accommodate a wide range of functional requirements. It is simply a question of choosing the form best suited to the function.

14.1.6 Foundations
The narrowest strip footing that can be conveniently dug by an excavator usually results in a foundation area such that the soil contact pressures are low. For example, a nine-storey hostel block with 102.5 mm crosswalls, founded on a 600 mm wide concrete strip footing, would have a contact pressure of only about 325 kN/m^2.

When the ground-bearing capacity is so low that piling is necessary, the wall itself can be treated as the compression flange of a composite reinforced concrete/masonry ground beam, with attendant savings in foundation costs (see Figure 14.11). It should be noted, however, that the use of a wall as a composite beam may limit its adaptability should it become necessary to change the structure at a later date.

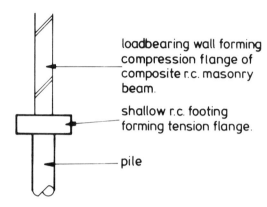

loadbearing wall forming compression flange of composite r.c. masonry beam.

shallow r.c. footing forming tension flange.

pile

Figure 14.11

Because masonry walls (particularly clay brickwork) are pliant, compared to structural steel or rc frames, they are particularly economical on sites subject to mining or other subsidence. Reinforcing the lower bed joints at each storey height results in a wall that is highly resistant to differential settlement.

14.1.7 Flexibility

Many designers think that masonry structures are inflexible – that it is difficult to alter them, once they are built. This is not so. For example, one of the author's most interesting change-of-use projects was the successful conversion of a Victorian ice-cream factory into an old people's home.

The spate of conversion, alterations and rehabilitation of masonry structures in the '70s gave masonry designers the opportunity to prove that it is often easier to alter a masonry structure than a steel or concrete structure. Admittedly, the bulk of the work was on brick structures, but the same is true – albeit, perhaps, to a somewhat lesser extent – of concrete block structures. It is often easier to demolish a masonry wall than a steel or concrete column. And it is far simpler to form an opening in a masonry wall than in a reinforced concrete wall. Generally, it is cheaper to bond in, thicken, brace or otherwise strengthen a masonry wall than a steel column. It is often easier and quicker to repair an overloaded masonry wall or arch than the equivalent in steel or concrete.

Although alterations to modern, highly stressed, loadbearing masonry structures require careful attention, it is only on rare occasions when wholesale alterations are required for a radical change of use that masonry structures become inflexible.

14.1.8 Concrete roof slab/loadbearing wall connections

Whilst it is good practice, and structurally beneficial, to cast floor slabs onto the walls, it is inadvisable to cast the roof slab directly on the top of the upper storey wall. The roof slab will tend to expand and contract with temperature variations and, if it is restrained by the slab/wall connection, either it or the wall will crack. inounde

In order to reduce this effect, the roof slab should be separated from the supporting wall. This can be done simply by laying two layers of building paper on top of the wall – although this is not considered to be acceptable good practice. A more effective separation joint can be achieved by inserting a proprietary jointing material or a layer of dpc (see Figure 14.12).

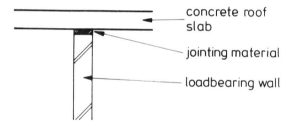

concrete roof slab

jointing material

loadbearing wall

Figure 14.12

14.1.9 Accidental damage

Although, as noted earlier, provisions against accidental damage were discussed in detail in Chapter 8, it is worth repeating here the general recommendations of BS 5628, which may be interpreted as follows:

1. The designer responsible for the overall stability of the structure should ensure that the design, details, fixings, etc., of elements or parts of the structure are compatible, whether or not the design and details were made by him or her.
2. The designer should consider the plan layout of the structure, returns at the ends of walls, interaction between intersecting walls, slabs, trusses, etc., to ensure a stable and robust design.
3. The designer should check that lateral forces acting on the whole structure are resisted by the walls in the planes parallel to those forces,

or are transferred to them by plate action of the floors, roofs, etc., or that the forces are resisted by bracing or other means.

The structure must have adequate residual stability not to collapse completely, and the Code further advises that the designer should satisfy himself or herself that '. . . . collapse of any significant portion of the structure is unlikely to occur'.

14.1.10 Choice of brick, block and mortar strengths

Generally, the bottom storey masonry will be the most highly stressed. The stress diminishes with each storey height, and the most lightly stressed storey will usually be the top one. Inevitably, within any one storey height, some walls will be more heavily stressed than others. For example in, say, a six-storey hostel block, the crosswalls may be 102.5 mm thick and the walls surrounding the staircase may, for fire protection purposes, be 215 mm thick whilst carrying only the same load as the crosswalls. Thus it follows that every storey height could be of a different strength masonry and that, within any one storey height, variations in masonry strength could be employed. However, any savings in material costs due to the widespread variation would be swallowed up by the extra costs of organising, sorting, stacking, supervising, etc.

It is generally advisable to use a maximum of only three mortar strengths: 1:¼:3 below dpc level or in extremely highly stressed work, 1:1:6 (or 1:½:4) for external and highly stressed work, and 1:2:9 for internal work (i.e. mortar designations (i), (iii) and (iv) in BS 5628, respectively).

It is difficult for administrative or supervisory staff to check, by sight, the strength of bricks, blocks, and the mix of the mortar. Reducing the cement content of the mortar produces only a minimal saving in the cost per m^2 of wall.

Every effort should be made to keep the wall of a constant thickness throughout its height. It should be kept in mind that a slender, highly stressed, wall is usually cheaper than a thick wall carrying a low stress. Brick and block strengths should generally be uniform throughout any one storey, and changes in strength should be limited to approximately every three storeys. In a recent eleven-storey contract, engineering bricks were used on the bottom three storeys, high strength bricks on the fourth to sixth floors, medium

strength on the seventh to ninth, and low strengths on the top two storeys.

Note that a top storey wall, due to its small pre-load, may have excessive flexural tensile stress due to wind forces, and may require specific brick and mortar strengths to cope with this.

14.2 CROSSWALL CONSTRUCTION

Crosswall structures are one of the simplest structural forms for multi-storey buildings, and probably have the widest application. The basic form is shown in Figure 14.13 – detailed layouts are provided later in this chapter.

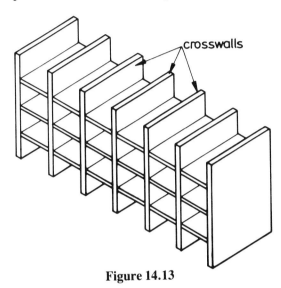

Figure 14.13

They are particularly suitable for long rectangular buildings which have repetitive compartmented floor plans. Typical examples are hotel bedroom blocks, study bedrooms in student hostels, school classroom blocks, town house developments and small four-bedded wards in hospital blocks. Crosswalls are necessary in such buildings – even if the designer uses a steel or concrete frame – for the following purposes:

(i)   Acoustic barriers. Building regulations require 102.5 mm brick, or similar, partitions between study bedrooms – it is, perhaps, regrettable that there is no such statutory requirement for hotel bedrooms. A half brick wall has an average sound reduction of 42 dB and, when plastered both sides, 50 dB.
      Similar thicknesses are required with concrete blocks.
(ii) Party walls. Building regulations require 215 mm brick, or similar, party walls between domestic units.
(iii) Fire barriers. Building regulations in many instances require 215 mm walls around stair-

cases, lift shafts, vertical service ducts, etc., in addition to fire breaks along the building. 100 mm thick clay brickwork provides 2 hours fire resistance.

14.2.1 Stability
Whilst it is a simple process to design the cross-walls to support the vertical loading, a check must be made both on them, as structural elements, and the resulting structure, to ensure that there will be no collapse (instability) or over-stressing due to horizontal loading from wind forces (see Figure 14.14).

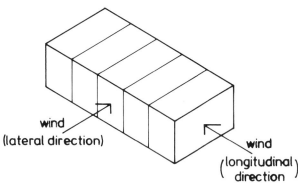

Figure 14.14

Lateral stability
Crosswalls are usually very stable under lateral loading. The stress, due solely to uniformly distributed wind loads at the base is:

$$\text{stress} = \pm \frac{\gamma_f W_k h^2/2}{bd^2/6}$$

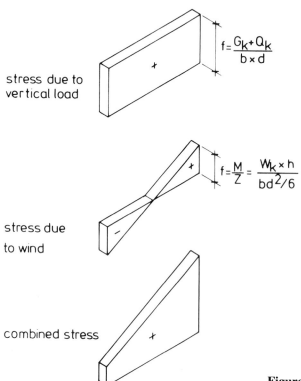

Figure 14.15

where
γ_f = partial safety factor for loads
W_k = characteristic wind load
h = height of structure
b = thickness of wall
d = depth or length of wall.

Since d is usually relatively deep, the wind stresses are minimal (see Figure 14.15).

Longitudinal stability
Unstiffened crosswall structures – i.e. crosswalls without stiffness at right angles to the plane of the wall – may not be stable under longitudinal loading from wind, and could collapse like a house of cards (see Figure 14.16).

Figure 14.16

To prevent such action, longitudinal bracing is necessary. This is usually provided (see Figure 14.17) by either:

(a) corridor walls,
(b) longitudinal external walls,
(c) stiff vertical box sections formed by the walls

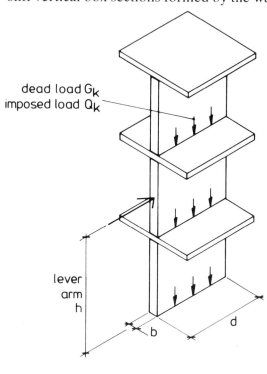

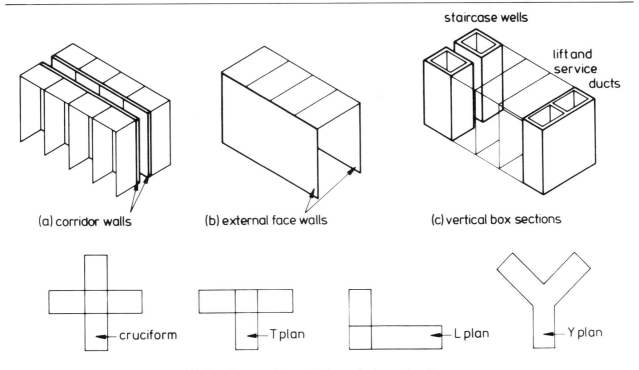

(a) corridor walls   (b) external face walls   (c) vertical box sections

cruciform   T plan   L plan   Y plan

(d) plan forms giving stiffness in two directions

Figure 14.17

to staircases, lifts and service ducts, or

(d) cruciform, T-, Y-, L-shaped block plans, or other plan forms which provide longitudinal stiffness or robustness.

14.2.2 External cladding panel walls

The external walls in Figure 14.17(b) may be subject to high lateral loads combined with only minimal vertical loads. Such masonry walls do not have a high resistance to bending perpendicular to their plane (see Figure 14.18).

The wall panels on the top storey are the most at risk because they are likely to be subject to the greatest wind pressure whilst the only compensating precompression is the vertical loading from the roof and their own weight. If a lightweight timber roof is used, there could be wind uplift forces to counteract by strapping it down to the walls. There would then be no vertical precompression in the top storey walls.

Generally, this is not a significant problem with loadbearing masonry – but it can be if the masonry is non-loadbearing and is used merely as a cladding to a steel or concrete-framed structure. It is advisable to check the stresses in such panels, following the procedure set out in Chapter 11.

14.2.3 Design for wind

The crosswalls act as shear walls under wind loading. Shear walls act as vertical, deep, stiff cantilevers in many framed structures, and resist the wind forces and moments on the structure – thus reducing the effects of wind on the frame. There is a large number of research papers on the subject, and it is discussed in detail in many good modern textbooks on the theory of structures and an example of the design of such a shear wall is given in Chapter 11.

In most loadbearing masonry crosswall structures, the stresses due to wind are insignificant compared to those due to dead and imposed loading, as the worked examples will show. Nevertheless, it may be helpful to briefly discuss the topic here.

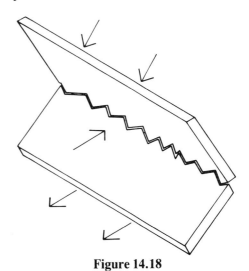

Figure 14.18

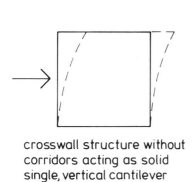

crosswall structure without
corridors acting as solid
single, vertical cantilever

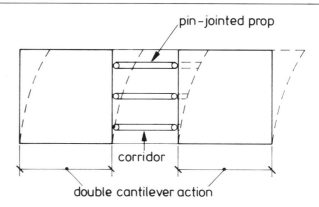

Figure 14.19

In a steel or concrete frame, the beams and columns are of relatively similar stiffness, rigidly connected, and are of the same material. However, in a loadbearing masonry structure, the walls are relatively sturdy, the floor slabs comparatively flimsy, and the structure may not act as a rigid frame. The walls, having high stiffness, act as vertical cantilevers, and the floors can be considered as acting as pin-jointed props (see Figure 14.19).

The crosswall, when broken by a corridor, acts approximately as two separate cantilevers. If both walls are of the same depth, *d,* and thickness, they share the wind force equally. When they are not of equal depth, they then share the wind force in proportion to their relative stiffness – if they deflect equally, as they are likely to do, because of the floor's action in transferring the force.

The strength of a wall is relative to its section modulus Z, $bd^2/6$, and the stiffness of a wall is relative to its second moment of area, I, $bd^3/12$.

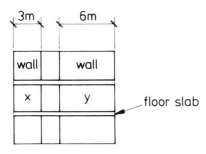

Figure 14.20

In the crosswall structure shown in Figure 14.20:

Wall x has a Z value proportional to $3^2 = 9$
Wall y has a Z value proportional to $6^2 = 36$
(Wall y is four times as strong as wall x)

Wall x has an I value proportional to $3^3 = 27$
Wall y has an I value proportional to $6^3 = 216$
(Wall y is eight times as stiff as wall x)

Since the walls are tied together by floor slabs, they are likley to deflect equally – thus wall y will carry eight times the wind force of wall x.

Walls of differing length and axes to the wind
The distribution of wind forces, particularly on tall slender crosswall structures, between walls of differing stiffnesses may need consideration. Some of the main points are illustrated below.

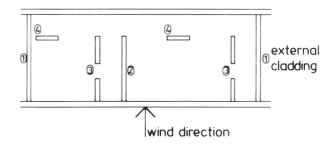

Figure 14.21

In Figure 14.21, the floor plan of a block of flats shows walls of differing length (and, therefore, stiffness) and of differing positions in relation to the wind. The main wind force would be resisted by walls 1, assisted by walls 2, with some help from walls 3, and little help from walls 4. An experienced designer would probably, at first, check only the effect of walls 1 and 2 in resisting wind, and then, if they were inadequate, consider the assistance of walls 3. He would be likely to ignore the minimal effect of walls 4 in resisting the wind forces. The use of walls 1 only, would necessitate a long span for the plate action of the roof or floors.

Walls of differing section
When external or corridor walls are bonded into crosswalls, they change the shape of a crosswall from a simple rectangle into a T, I or Z section. This can give the crosswall increased stiffness and hence increased stability.

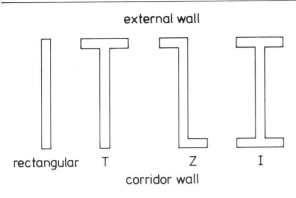

external wall

rectangular   T   Z   I

corridor wall

Figure 14.22

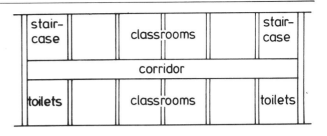

stair-case		classrooms		stair-case
corridor				
toilets		classrooms		toilets

Figure 14.24

In Figure 14.22, the I and Z sections are stiffer than the T section, which in turn, is stiffer than the rectangular section.

14.2.4 Openings in walls

Intuitively, it can be seen that, in Figure 14.23, wall (a) is stiffer than wall (b) which, in turn is stiffer than wall (c). The gable wall (d) with small, widely spaced windows, may be considered to act similarly to wall (a) if the openings are relatively small. However, if the windows are deepened, the wall approaches the condition of wall (c).

Only rarely do the calculations become very complex. However, if they do, or if the designer is in any doubt as to the stiffness of the walls or structure, he or she should either refer to one of the many computer programmes on the market, or carry out a model test. If a computer is used, the designers should satisfy themselves that the programme is suitable and well founded, and that the results of the analysis are reasonable.

14.2.5 Typical applications

School classroom blocks
These are not normally more than four storeys high, and a typical plan shape is shown in Figure 14.24.

The crosswalls usually need to be 215 mm thick to carry the load, Gable and external side walls are normally in 265 mm cavity brickwork. Corridor walls have to be at least 102.5 mm for acoustic and fire resistance. The external and corridor walls, together with the staircase, are normally more than adequate to provide longitudinal stability.

In many cases, the long floor spans are most economically formed in precast, prestressed concrete units, seated about 100 mm onto the walls. To give some continuity and resistance to the negative moments which will occur in practice (even though, in theory, the units are 'simply' supported) it is advisable to employ an rc *in situ* infill over the wall support. This will assist in preventing a floor slab collapse should a cross-wall be removed by accidental damage (see Figure 14.25). Note that it is necessary to comply with the Building Regulation covering progressive collapse from accidental damage when the building is more than five storeys high (see Chapter 8).

Where wide-span units are used to provide a fairfaced soffit, the *in situ* infill should still be provided (see Figure 14.26).

The alternative spans loaded condition, and the resulting bending moments and eccentricity of loading induced into the walls due to deflection of the floor units and rotation at the supports, are rarely critical. Nevertheless, the effect of eccen-

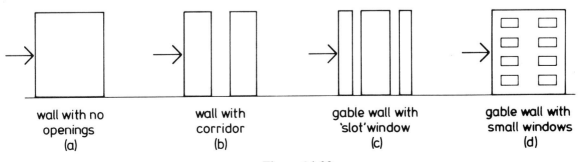

wall with no openings (a)   wall with corridor (b)   gable wall with 'slot' window (c)   gable wall with small windows (d)

Figure 14.23

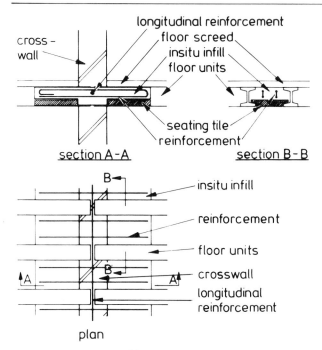

Figure 14.25

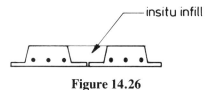

Figure 14.26

tricity on the bearing stresses should be taken into account. The reinforcement in the infill tends to reduce the effect of eccentricities and distribute the uneven stresses. (Note: Many school buildings were erected in the late '50s to early '70s using high alumina cement in the precast floor units. All these buildings had to be investigated and, as far as the authors' experience and knowledge are concerned, none of the walls showed any distress due to eccentric loading.)

Bedroom blocks

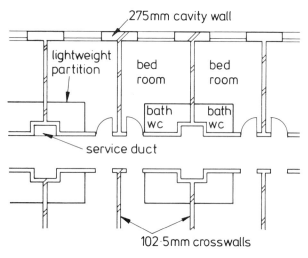

Figure 14.27

Figure 14.27 shows a typical basic floor plan of a bedroom block. Many buildings of this type are five to ten storeys high, and need to be checked for accidental damage under Building Regulation D19. Floors are usually *in situ* continuous concrete slabs. Where the external side walls and the corridor walls are loadbearing, the floor slabs may span two ways. Some minor increase in reinforcement is all that is usually necessary to cope with the accidental damage provisions.

Crosswalls usually need to be 102.5 mm thick in order to carry the load and to provide sound insulation. It is not uncommon to return the ends of the crosswalls, at their junctions with the external and corridor walls, to improve their stability.

Crosswall structures can, of course, be built much higher than ten storeys. However, as with all high-rise construction, the costs tend to increase faster than the increase in height.

Low- to medium-rise flats (up to six storeys)
A typical floor plan is shown in Figure 14.28.

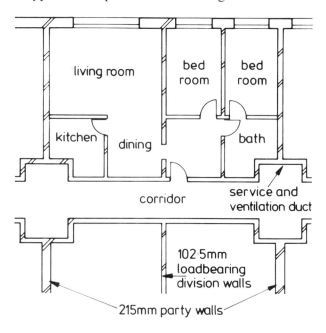

Figure 14.28

The demand for high-rise flats (which were more suited to cellular construction) has waned, and there is now more interest in medium-rise blocks. These are a hybrid form of the classroom and bedroom blocks, discussed earlier, in that they tend to comprise a mixture of 215 mm and 102.5 mm crosswalls. The party walls, spaced at about 12 m centres, need to be 215 mm thick to comply with the building regulations, and the intermediate crosswalls 102.5 mm thick for acoustic reasons. Corridor walls and external walls are generally of masonry construction, subject to the same building regulation requirements as the party walls, and are used structurally for longitudinal stability.

Floors are nearly always of *in situ* concrete construction. Timber floors could be used in low-rise construction, if fire regulations permit. If timber floors are used, increased strapping and tying is necessary, and care should be taken to ensure that the floor construction forms an efficient acoustic barrier.

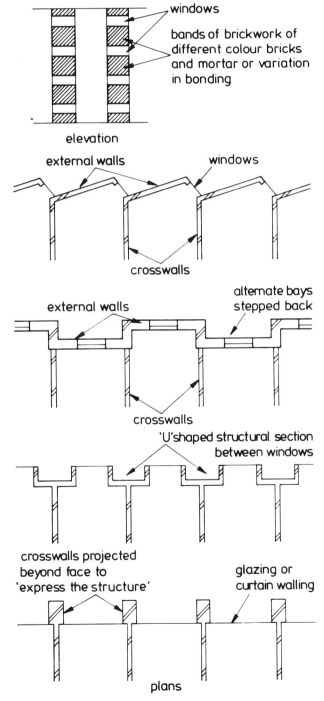

Figure 14.29

14.2.6 Elevational treatment of crosswall structures

Long side walls pierced by hole-in-the-wall windows can be visually dull. There are many ways of overcoming this – for example by using decorative brickwork and/or by modelling the elevation (see Figure 14.29).

14.2.7 Podiums

A common objection to the use of crosswall construction is that the ground floor planning requirements demand more open spaces than crosswalls permit. Typical examples are reception areas and restaurants in hotels, car parking for flats, recreation areas and shops in student hostels. But the floors above, with regular wall layouts, are ideal for crosswall construction.

Frequently, there is no need to frame the whole structure, merely because of the ground floor planning requirements. A different structural form can be used for the ground storey, and a common solution to the problem is to form a podium with steel, concrete or masonry columns supporting a concrete deck, as shown in Figure 14.30. Depending on the load from the crosswalls, the deck can be of plate or waffle slab construction, diagrid or T beam.

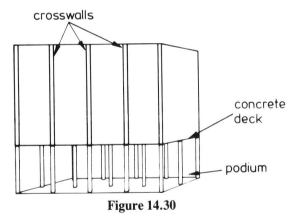

Figure 14.30

The deflection of the deck under the crosswalls should be assessed, even though clay brickwork often has an inherent flexibility that enables it to adjust to the deflection of a concrete beam. Blockwork is not so adaptable. If the deflection is of such magnitude as to cause the masonry to crack, a flexible joint should be included at the deck/wall junction and the lower courses should be reinforced, as shown in Figure 14.31. The flexible joint must not, of course, affect the loadbearing, capacity of the wall.

14.3 SPINE CONSTRUCTION

Spine construction can be used on buildings of

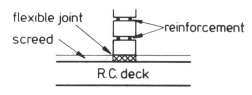

Figure 14.31

rectangular plan shape where crosswalls are either too restrictive on planning, circulation, etc., or where they cannot be lined up due to different functional requirements at each floor level. Typical examples of the first category are hospital ward blocks, storage buildings and office blocks. Buildings in the second category are much rarer – and when they do occur, it is often only because the designer will not accept the planning discipline of crosswall construction.

In many cases, the solution to these problems is to eliminate the crosswalls and to use the external side walls and the corridor walls or a spine wall as the main loadbearing elements, as shown in Figure 14.32.

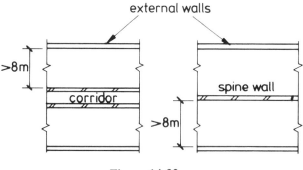

Figure 14.32

The depth from the external wall to the spine or corridor wall is usually limited to 8 m, which is about the economic limit for precast prestressed concrete floor units. If economic factors are not restrictive (which is unusual), the floor spans could be wider.

Many buildings do not really need a floor span greater than 7 or 8 m. Office workers seated at distances of more than, say, 6 m from the external windows will need permanent artificial lighting, air conditioning and expensive service runs for heating, etc. People, naturally, like to be able to see out of a window and to use it to control their physical environment. Some designers consider that 5 m should be the maximum distance of a work space from an external wall.

There are two main structural problems in the design of spine structures:

(i) the provision of lateral stability,
(ii) if the structure is over four storeys high, it must be resistant to progressive collapse following accidental damage.

14.3.1 Lateral stability

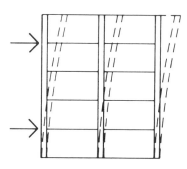

Figure 14.33

The building shown in Figure 14.33 clearly has inadequate lateral stability from wind forces, and it is necessary to introduce structural elements to resist them. This can be done externally or internally. In both cases, the floor must act as a horizontal plate, or wind girder, to transfer the wind forces to the lateral elements resisting the wind as shear walls (see Figure 14.34).

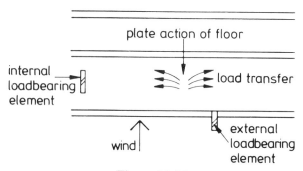

Figure 14.34

Internal loadbearing elements
These are commonly the walls around staircases, lift wells and vertical service ducts. Elements such as gable walls, fire barriers, and the occasional partition wall, can also be used, as shown in Figure 14.35.

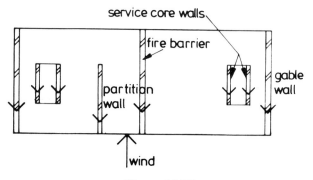

Figure 14.35

There is rarely a problem in providing lateral walls of sufficient strength to resist the wind forces, but there can be difficulty in providing the plate action of the floor if it is not of *in situ* concrete. If – perhaps for cost reasons – the floor is constructed of prestressed precast concrete beams, a reinforced concrete topping will be necessary.

External elements

Fin walls can be used to counteract the lateral wind forces. If the wind forces are appreciable, the fins may need post-tensioning (see Chapter 15). The advantage of using fin walls as vertical cantilevers resisting the wind force, is that they considerably reduce the need for the floor to act as a plate. A plan and section of a possible fin structure are shown in Figure 14.36.

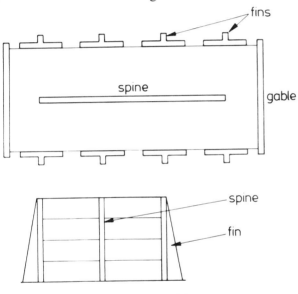

Figure 14.36

14.3.2 Accidental damage

Above four storeys, the building must be checked for accidental damage by:

(i)   a portion of the external or spine walls being removed;
(ii) a portion of the lateral internal walls being removed;
(iii) a fin being removed;
(iv) a section of the floor being removed.

A method of checking and designing against accidental damage is dealt with in Chapter 8 and in the worked example in this chapter.

14.4 CELLULAR CONSTRUCTION

Of all structural masonry forms, cellular construction is the most resistant to lateral loads and accidental damage, and was used to a limited extent for high-rise flats in the '60s. Despite the decline in high-rise construction, it is still a most valuable structural form for flats, student hostels, etc. Even below six storeys, it is still worthwhile to carry out a cost exercise to determine its economic viability (see Figure 14.37).

The technique was pioneered by the Swiss engineer, Haller, who built some spectacular high-rise blocks in Basle, Berne and Zurich. A number were built in the UK during the boom years of high-rise flat construction. However, the method did not achieve the wide popularity of the precast concrete structures – due both to government encouragement of system building, and the lack of experience of many engineers in structural masonry design. Certainly, there was no sound technical reason for the neglect of masonry construction – it was cheaper to build, more satisfactory in use, and required less maintenance than the concrete systems. Some of the system-built blocks are now causing tremendous maintenance problems due to lack of cover, leaking and inadequate joints, condensation, etc., and their vulnerability to accidental damage was tragically demonstrated in the Ronan Point disaster.

The basic 'egg crate' form (see Figure 14.38) provides stiffness in two directions at right angles, and is therefore highly resistant to wind forces. Lifts, stairs and service ducts are located at the centre of the plan, with housing units

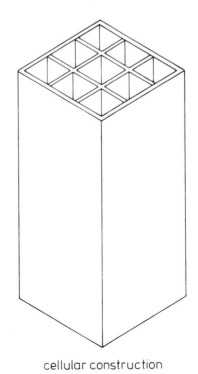

cellular construction

Figure 14.37

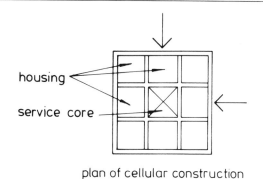

plan of cellular construction

Figure 14.38

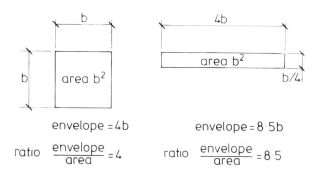

Figure 14.40

around the perimeter (see Figure 14.39). In study/bedroom blocks, the central portion can also house bathrooms, toilets and kitchens. Typical flat layouts are shown in Figure 14.39, with the main walls in heavy outline.

14.4.1 Comparison with crosswall construction

The function of a building tends to dictate its structural form. School classroom and hotel bedroom blocks are usually long rectangular buildings, and appropriate to crosswall construction. Tall, or tower, blocks of residential accommodation are often square on plan and therefore well suited to cellular construction.

In crosswall structures, it is the crosswalls that mainly carry the load, while the longitudinal walls (corridor and external side walls) serve to provide stability along the longitudinal axis. In cellular construction, all the walls carry the load and provide resistance to lateral forces on both axes.

Floors in cellular structures are nearly always of *in situ* concrete and, because of the two-way spanning possibilities and the shorter spans, tend to be more economical than in crosswall or other structural forms (see Figure 14.40).

14.4.2 Envelope (cladding) area

The cost of a building is affected by its ratio of

envelope to floor area – the smaller the ratio, the lower the cost of the envelope. Figure 14.41 shows the favourable ratio for cellular structures compared to rectangular structures.

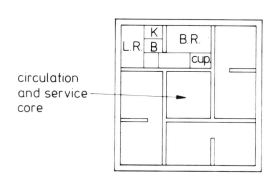

Figure 14.41

14.4.3 Robustness

Cellular construction is probably the most robust of all structural forms. If Ronan Point had been built in this form, with properly bonded intersections instead of apparently inadequately tied concrete panels, it would have been extremely unlikely to have collapsed, and the 'incident' would have been localised.

As noted earlier, the form is highly resistant to lateral loading and, because of the commonly symmetrical form, it is resistant on both axes. Since it is relatively massive – compared, for example, to steel frames with glass external cladding and lightweight internal partitions –

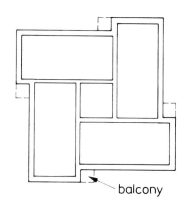

Figure 14.39

there is far less likelihood of unacceptable vibration in tower blocks due to severe wind gusting. The authors know of no case of noticeable vibration in a multi-storey cellular masonry structure – although there is evidence of such action in other materials. The inherent stiffness of the structural form makes it particularly useful in areas subject to high winds and foundation movement.

14.4.4 Flexibility
Reference has already been made to the relative ease of making alterations to masonry structures to suit changing functional requirements. Cellular structures are no exception and, because of the multiplicity of walls, they are often easier to alter than other structural forms. The designer must, of course, take the same care to ensure that the alterations do not overstress the structure.

14.4.5 Height of structure
Very tall structures can be built using cellular construction. Apart from cost, the main factor affecting the height is stability under wind loading. A simple calculation shows that a 20 m square block, 15 storeys (40 m) high has a generous factor of safety against overturning from the wind forces. A detailed check must, of course, be made to see that the structure has adequate stiffness. (See Figure 14.42.)

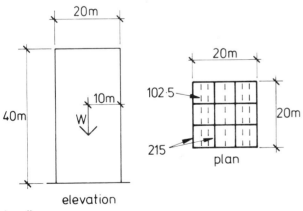

loading
walls $8(20 \times 2.4 \times 0.215 \times 20 kN/m^3) = 1651$
$+5(20 \times 2.4 \times 0.103 \times 20 kN/m^3) = 494$
floors $13(20 \times 20 \times 3 kN/m^2) = \underline{15600}$
17745

stability $M = 17745 \times 10 = 177450 kNm$
wind $M = 40 \times 20 \times 20 \times 1 kN/m^2 = 16000 kNm$

Figure 14.42

14.4.6 Masonry stresses
Since the load is shared by all the walls, the stresses tend to be lower than in other structural masonry forms – which makes cellular construc-

tion particularly suitable for high-rise buildings, and when considering lateral loads on external walls. The wall thicknesses need to meet building regulation requirements for fire, sound and thermal insulation, party walls, etc., and are often greater than the thickness required to carry the loads. Calculations show that, quite often, the most heavily stressed wall under imposed and wind loading (the external wall on the leeward face) needs only to be a normal 255 mm cavity wall.

14.4.7 Foundations
The foundations of cellular structures tend to be cheaper than those for other structural forms. The loads are spread over more walls, more uniformly, and at closer spacings. Contact pressures, therefore, are generally lower. On soils of good bearing capacity, it is not uncommon to merely thicken the ground floor slab under the internal walls.

14.5 COLUMN STRUCTURES

Masonry columns are scarcely a new structural form, and modern masonry column structures are simply a development of an old technique. For example, many medieval cathedrals are basically column structures in which the lateral thrust from the roof is resisted by flying buttresses. When, in a modern column structure, the columns are not adequate to resist the lateral thrust from the wind, the restraint can be provided by shear walls, fins, etc.

Use
Masonry column structures are mainly used for buildings whose functions require large open spaces, such as warehouses, department stores and, occasionally, open-plan offices.

The spacing of the column grid is usually a compromise between the client's preferences, cost and engineering feasibility. The closer the spacing, the cheaper the structure, but the greater the loss of spatial freedom. As so often happens with concrete structures, the client tends to want a grid spacing of 12 m, but eventually settles for about 7.5 m.

Floors
Most column structures have *in situ* concrete floors, of either plate or waffle slab construction to eliminate the inconvenience of beams which reduce headroom. The floors must act as horizontal plates to transfer the lateral wind thrusts to shear walls.

14.5.1 Advantages
It may appear odd to use masonry columns in a structure when so much *in situ* concrete is used in the floors and foundations. Masonry columns are chosen for:

(a) Speed of erection
It is quicker to build a masonry column than an *in situ* concrete column, or to fabricate, erect, plumb and fire-protect a steel column. To ensure continuity of the masonry labour force, it is usually advisable to use masonry construction for the external cladding, shear walls, etc.

(b) Economy
Masonry columns are generally cheaper than alternative materials. Admittedly, economy is rarely a major consideration, since the cost of the columns is only a minor part of the total construction costs.

(c) Durability
Corners of columns are easily chipped and damaged in warehouses, etc. The corners of concrete columns can be rounded off by using ¼ round bead fillets in the shuttering, and the fire protection cladding to steel columns can be strengthened by adding steel angles at the corners. Both methods add to costs. Radiused bricks or bull-nosed bricks are not expensive and are usually readily available.

(d) Aesthetics
Since the columns are likely to be heavily stressed, engineering or high strength facing bricks are often required. Both categories are available in a wide range of colours and textures, and are likely to be far more visually attractive than encased steelwork or plastered concrete.

14.5.2 Cross-sectional shape
Square and rectangular columns are the cheapest and simplest to build since they only involve four corners (and thus four plumbing lines), do not require specially shaped bricks, and are easily bonded. However, there are sometimes structural or aesthetic advantages in using other sections.

(a) Cruciform
These give higher lateral resistance in two directions. No guidance is given in BS 5628 on the slenderness ratio or section modulus of such sections. However, the authors suggest that the radius of gyration of the section be determined, related to a square section of equal radius of gyration, and the thickness of the square section taken as the effective thickness of the cruciform section. The section modulus can be determined from first principles, in the same way as the section modulus of cruciform sections in other materials.

(b) I sections
Similar effectiveness and methods apply to I sections. A further advantage of this section is that services can be run up the faces of the webs.

(c) Hollow square, rectangular or circular sections
These have the advantages of a high second moment of area and section modulus, but require a larger overall cross-sectional area. They are particularly useful as structural and permanent shuttering to reinforced concrete columns. Care must be exercised in grouting up each column lift to ensure that all mortar droppings are removed from the surface of the previously poured grout.

14.5.3 Size
With masonry columns, there is less need to maintain the same profile all the way up the structure – as there is with concrete (to reduce shuttering cost) or with steelwork (to simplify connections). Nevertheless, to avoid complications in standard details, fixings, etc., it is advisable to keep the number of changes of section to a minimum. This can be achieved by using high strength masonry in the lower storeys (perhaps, with the addition of reinforcement) and lower strength masonry in the upper storeys to produce an economical balance.

Typical floor to floor heights for offices are about 3–4 m, for department stores around 5 m and for warehouses 4–6 m. Grid spacings are generally from 5–7 m in both directions. Thus the size of the columns can vary enormously since it depends on their height, load (dependent on the grid spacing and the use of the structure) and the strength of the masonry. It is not difficult to reduce the cross-section of a column by adding reinforcement, if this is economically worthwhile.

Recent experience has shown that, in terms of cost and speed of construction, it is well worth considering the use of masonry columns instead of steel or concrete columns.

14.6 DESIGN PROCEDURE

The design procedure for multi-storey masonry

structures is similar to that for other structural materials and is as follows:

1. Layout. Wall positions should be chosen to suit the function of the building. This is usually the architect's responsibility, but the engineer should certainly advise in this regard, and on the types and direction of floor spans and roofs, joint locations, restraint considerations, etc. All too frequently, the two professions – architects and engineers – do not confer at an early enough stage in the creative process.
2. Wall thicknesses should be chosen to comply with:
   (a) building regulations (fire resistance, acoustic and thermal requirements, etc.),
   (b) material dimensions (215 or 102.5 mm brickwork, 100 or 140 mm blockwork),
   (c) serviceability needs,
   (d) estimated trial section for load carrying.

3. Check required thickness against trial section thickness for load carrying.
4. Preliminary appraisal of liability to accidental damage (span of floors, returns to walls, etc.).
5. Determine dead, imposed and wind loadings, and worst combination.
6. Determine stresses in masonry elements.
7. Choose masonry and mortar strengths.
8. Check for 'column action' or overstressing of areas of wall in external walls with large window or other openings, internal walls with large door openings or service access, particularly corridor walls with clerestorey lights.
9. Check upper storey (or storeys) for flexural tensile stress under minimum axial loading.
10. Check for stability.
11. Check again for accidental damage.
12. Add straps and ties, where necessary.
13. Check details (end bearing stresses, stresses around service holes, etc.).

14.7 EXAMPLE 1: HOTEL BEDROOMS, SIX FLOORS

Basic data (see Figure 14.43)

Overall height	15.0 m
Floor to floor height	2.50 m
Span of floors	3.0 m
Overall length	30.0 m
Overall width	20.5 m
Density of masonry	20 kN/m³
Density of reinforced concrete	24 kN/m³

The floor and roof construction is of reinforced *in situ* concrete slabs supported by loadbearing masonry. Precast concrete floors were not chosen in this example because *in situ* floors tend to give greater rigidity to the structure when it has to be designed to meet accidental damage loadings. Furthermore, on such relatively short spans, *in situ* concrete slabs tend to be more economical than precast, and a suitable bearing onto thin crosswalls is more easily achieved.

14.7.1 Characteristic loads

Roof

Dead loads, G_k, 125 mm rc slab = 24×0.125 = 3.0 kN/m²
lightweight screed to falls (average) allow 1.0 kN/m²
= 4.0 kN/m²

Imposed load, Q_k, (CP 3: Chapter V: Part 1) = 0.75 kN/m²

Floors

Dead loads, G_k, 125 mm rc slab = 24×0.125 = 3.0 kN/m²
partitions, allow 1.0 kN/m²
finishes and services, allow 0.3 kN/m²
allow floor screed 1.0 kN/m²
= 5.3 kN/m²

Imposed load, Q_k, (CP 3: Chapter V: Part 1) = 2.0 kN/m²

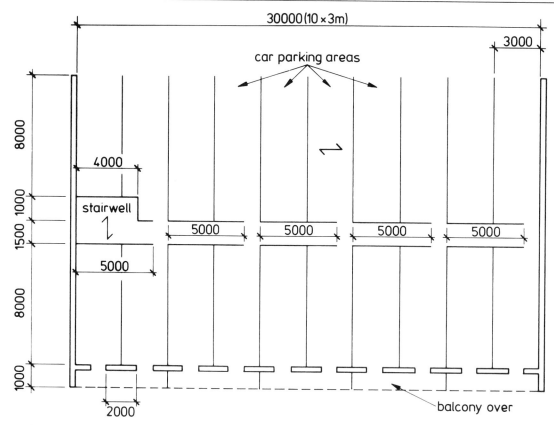

↩ denotes direction of floor spans

Ground Floor Plan

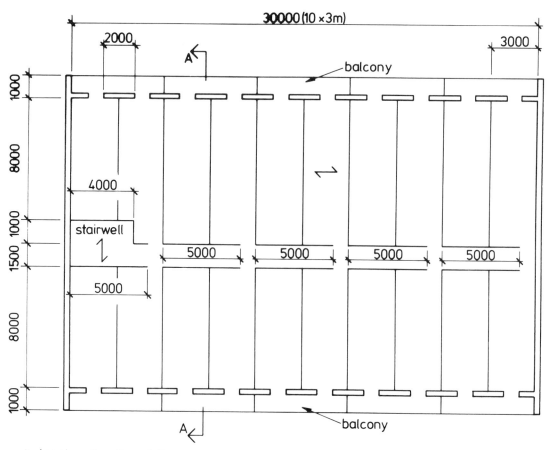

↩ denotes direction of floor span

General Floor Plan

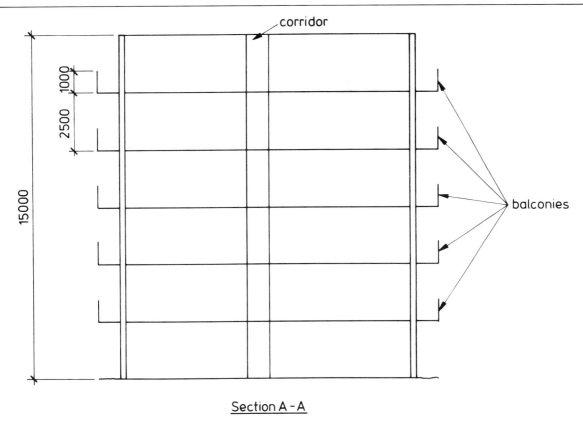

Section A-A

Figure 14.43

Wind loading
The choice of the characteristic wind load for design purposes is covered in CP 3: Chapter V: Part 2, 1972, and is outside the scope of this example. Assume that the maximum characteristic wind pressure, W_k, is 0.894 kN/m². Also assume that the roof is flat and that $C_{pi} = +0.2$ or -0.3 and $C_{pe} = +1.0$.

14.7.2 Design of internal crosswalls
Having decided on the basic design parameters and loadings, the designer can now go on to design the structure itself. It should be remembered that careful consideration must be given to the location joints within the concrete slabs and masonry walls, since these will affect the overall stability of the structure.

Loading

Table 14.1

Position	Characteristic dead load, G_k, kN/ per metre run due to floors and roof	Characteristic imposed load, Q_k, kN/per metre run	Imposed load reduction factor (%) (Table 2, CP 3: Chapter V: Part 1)
Roof	$3 \times 4 = 12$	$3 \times 0.75 = 2.25$	0
5th floor	$12 + 3 \times 5.3 = 27.9$	8.25	10
4th floor	43.8	14.25	20
3rd floor	59.7	20.25	30
2nd floor	75.6	26.25	40
1st floor	91.5	32.25	40

The self-weight of the masonry must be added to the above in obtaining the total design load. For a trial section, try a 102 mm thick wall. At position A, the characteristic load = height × thickness × density = $15.0 \times 0.102 \times 20 = 30.6$ kN/m run due to the masonry.

Assume $\gamma_f = 1.4$ for dead loads or 0.9 as appropriate, 1.6 for imposed loads (from clause 22, BS 5628).

Therefore, at A (see Table 14.1) the total design load

$$= 1.4(91.5 + 30.6) + 1.6(0.6 \times 32.25)$$
$$= 170.9 + 30.96$$
$$= 201.85 \text{ kN/m run}$$

Now assuming that the floor slab, if built into the wall, provides enhanced resistance to lateral movement, then the effective height of the wall (clause 28.3.1.1, BS 5628) may be taken as $0.75 \times 2500 = 1875$.

The effective thickness of a half brick wall is the actual thickness = 0.102 m.

Therefore, slenderness ratio $= 1.875/0.102 = 18.4$

Check maximum eccentricity at first floor level (see clause 31, BS 5628). Assume the loading given in Figure 14.44 for the calculation of the eccentricity.

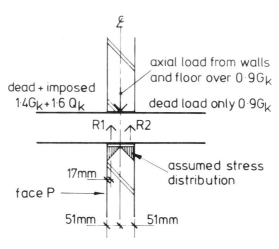

Figure 14.44

The resultants R_1 and R_2 are assumed to act at one-third of the depth of the bearing area from the loaded faces of the wall, i.e $(102/2) \times \frac{1}{3} = 17$mm in from the face.

Now
$$R_1 = (1.4 \times 5.3 \times \tfrac{3}{2}) + (1.6 \times 0.6 \times 2 \times \tfrac{3}{2}) = 11.13 + 2.88$$
$$= 14.01 \text{ kN/m run}$$
and
$$R_2 = (0.9 \times 5.3 \times 1.5)$$
$$= 7.1 \text{ kN/m run}$$

The load in the wall above the first floor level may be assumed to be axial

and
$$= 0.9(75.6 + 12.5 \times 0.102 \times 20)$$
$$= 91 \text{ kN/m run}$$

Considering a 1 m length of wall and taking moments about face P, let R be the distance to the resultant of R_1, R_2 and the axial load:

$$(91 \times 0.051) + (14.01 \times 0.017) + (7.1 \times 0.085) = (91 + 14.01 + 7.1)R$$
Therefore
$$R = 0.049$$

Therefore, eccentricity at the top of the wall $= 0.051 - 0.049 = 0.002$ m.

Now, thickness, $t = 0.102$, therefore eccentricity as a proportion of $t = 0.002/0.102 = 0.0196t$.

So, from Table 7, BS 5628 (see Table 5.15), β is unchanged for values of eccentricity up to 0.05t, and thus the resultant eccentricity even under the worst loading conditions at first floor level has no effect on the β value. Hence, β = 0.76.

14.7.3 Partial safety factor for material strength (Table 4, BS 5628 – see Table 5.11)

Case (a) Manufacturing and construction control, both special: $\gamma_m = 2.5$
Case (b) Manufacturing and construction control, both normal: $\gamma_m = 3.5$

Note: Normally, the designer would use only one value of γ_m. In this example, two cases are used for comparison purposes only.

14.7.4 Choice of brick in the two design cases, at ground floor level

Case (a)

Design strength $= \dfrac{\beta t f_k}{\gamma_m}$ per m run

Design load   =   202 kN/m run, β = 0.76, t = 102 mm, $\gamma_m = 2.5$
Design strength   ⩾   design load

Therefore

$$f_k \text{ required} = \frac{202 \times 2.5 \times 10^3}{0.76 \times 102 \times 1.15 \times 10^3}$$

$$= 5.66 \text{ N/mm}^2$$

where the factor 1.15 in the denominator is the stress increase from clause 23.1.2 for narrow brick walls.

Assuming a mortar designation (iii) (1:1:6) (from Table 2(a), BS 5628 – see Table 5.4) bricks with a compressive strength of 20 N/mm² are required.

Case (b)

$$f_k \text{ required} = \frac{202 \times 3.5 \times 10^3}{0.76 \times 102 \times 1.15 \times 10^3}$$

$$= 7.93 \text{ N/mm}^2$$

Bricks with a compressive strength of 35 N/mm² set in a mortar designation (iii) would be required. The strength of brick can be reduced at higher levels, normally at every third floor.

14.7.5 Choice of brick in the two design cases, at third floor level

Design load = 1.4(43.8 + 15.3) + 1.6(0.8 × 14.25)
          = 101 kN/m run

It should be noted that, as before, the eccentricity of loads under the worst condition is still less than 0.05t and thus β is 0.76 as previously calculated.

Case (b)

$$f_k \text{ required} = \frac{101 \times 3.5 \times 10^3}{0.76 \times 102 \times 1.15 \times 10^3}$$

$$= 3.96 \text{ N/mm}^2$$

15 N/mm² bricks set in a designation (iii) mortar would be satisfactory (manufacturing and construction control both normal) for both cases.

It may be noted that common bricks are normally at least 20 N/mm² crushing strength.

14.7.6 Design of gable cavity walls to resist lateral loads due to wind

Assume clay bricks are used for the inner and outer leaves. A critical design case occurs in the top storey of a building under lateral loading, because the compression in the wall from the dead loads is small, and the wind can cause uplift on the roof, further reducing the compressive load. Walls under high lateral loading and low compressive load (particularly when the walls are used as mere cladding to a steel or concrete frame) are more likely to fail due to flexural tensile cracking, rather than axial compressive crushing or buckling.

14.7.7 Uplift on roof

From clause 22, BS 5628, γ_f = 0.9 for dead load
                           and = 1.4 for wind load
Design dead load = 0.9 × 4 = 3.6 kN/m^2
Design wind load (uplift)   = 1.4 × 0.89 (1 + 0.2)
                            = 1.5 kN/m^2

Therefore, net dead load, contributing to compressive load in the walls,
$$= 3.6 - 1.5$$
$$= 2.10 \text{ kN/m}^2$$

14.7.8 Design of wall

There are three general methods for the design of walls under lateral loading mentioned in BS 5628, and definitive guidance is not provided as to which method should be used in a given case. The three methods are as follows:

1. Effective eccentricity method

Clause 36.8 covers the lateral strength of axially loaded walls and columns from consideration of the effective eccentricity, and using β factors.

2. Arching – horizontal or vertical

Clause 36.8 also mentions design using the formula:

$$q_{\text{lat}} = \frac{8tn}{h^2 \gamma_m}$$

3. Designing on the basis of a cracked or uncracked section

By assessing moments or using the method for panel walls in BS 5628.

By implication, methods 1 and 2 above, are normally used where the axial loads are high in relation to the lateral loads. In the top storey axial loads are low, so that there is little thrust to resist the arching effect. Also, the moment is high in relation to the axial load, so that the effective eccentricity is high, giving a large capacity reduction (i.e. the β value from BS 5628, Table 7 is very small – see Table 5.15). Thus adopt method 3.

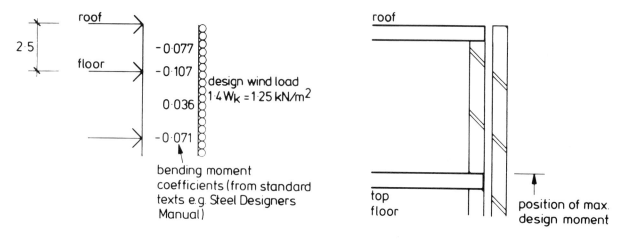

Figure 14.45   Figure 14.46

Wall ties of the vertical twist type should be provided in accordance with Table 6 of BS 5628 – see Table 6.2 (i.e. 2.5 ties per m²). Therefore, both leaves may be treated as acting together, but not as a homogeneous section.

14.7.9 Calculation of design wall moment

Assume that the wall behaves as a slab spanning continuously over supports provided by the floors (see Figure 14.45).
Design moment occurs at top floor level (see Figure 14.46)

$$= 1.25 \times (2.5)^2 \times 0.107$$
$$= 0.84 \text{ kNm/m length}$$

14.7.10 Resistance moment of wall (Figure 14.46)

Assume:

1. Clay bricks used in both leaves, with water absorption 7%, in designation (iii) mortar, i.e. $f_{kx} = 0.5$ N/mm².
2. Special manufacturing and construction control, i.e. $\gamma_m = 2.5$.
3. The total resistance moment is the sum of the resistance moments of the two individual leaves, i.e. fZ, where $f = g_d + f_{kx}/\gamma_m$.

This design method assumes that flexural tensile resistance can be developed and relied upon at this level.

Design dead load in inner leaf at position of maximum moment due to the net dead load from the roof, and a storey height of 102 mm thick brickwork

$$= (2.1 \times \tfrac{3}{2}) + (0.9 \times 2.5 \times 0.102 \times 20) = 3.15 + 4.59$$
$$= 7.74 \text{ kN/m run}$$

Therefore
$$g_d = \frac{7.74 \times 10^3}{102 \times 10^3} = 0.076 \text{ N/mm}^2$$

Design dead load outer leaf $= 4.59$ kN/m run

Therefore
$$g_d = \frac{4.59 \times 10^3}{102 \times 10^3} = 0.045 \text{ N/mm}^2$$

$$Z = \frac{1000 \times 102^2}{6} = 1.734 \times 10^6 \text{ mm}^3$$

Total resistance moment
$$= \left(\frac{0.5}{2.5} + 0.076 \right) \times 1.734 \times 10^6 + \left(\frac{0.5}{2.5} + 0.045 \right) \times 1.734 \times 10^6$$
$$= 0.48 + 0.42$$
$$= 0.9 \text{ kN·m/m}$$

The resistance moment is greater than the design moment, thus the wall is satisfactory at this level. At other levels, where the compressive dead load is higher, there is less risk of flexural tensile failure of the bed joints, and the governing factor in design is the axial compressive, rather than the flexural, strength of the wall.

14.7.11 Overall stability check

1. Stability in y direction (see Figure 14.47)
This is provided by the gable walls and by crosswalls 1 to 18, at 3 m centres. These walls are deep (8 m

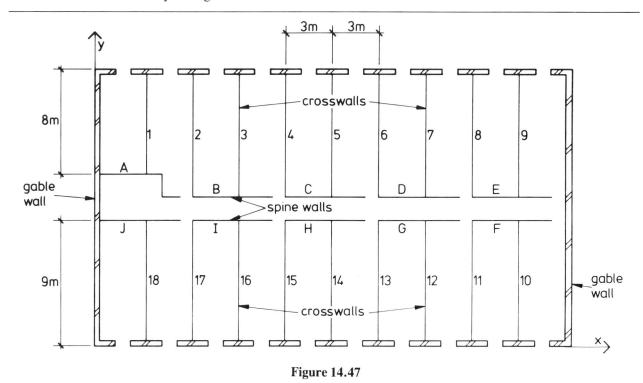

Figure 14.47

and 9 m, in the direction of the wind) and, by inspection, the building is rigid in the y direction. Overturning of the structure as a whole is not a problem for a building of this height and with these proportions. The additional stresses due to the wind on the crosswalls are very small (see Figure 14.48). For the method of calculation, see 2 (below) on the stability in x direction.

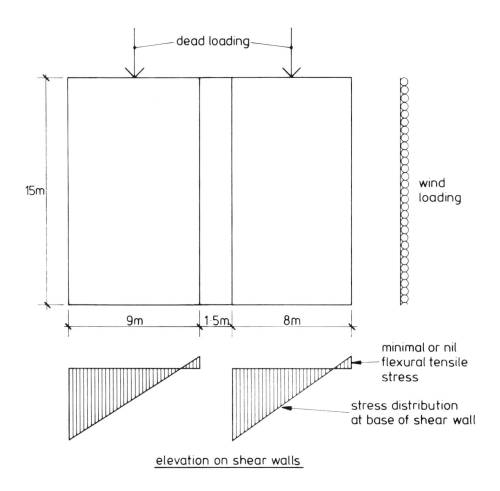

elevation on shear walls

Figure 14.48

2. *Stability in x direction*
The wind is taken by the two corridor walls and the two external walls. For simplicity in this example, it is assumed that the load is shared equally by the four walls. Since the resulting stresses are likely to be low, this approximation is not unreasonable, although a more accurate analysis would be to share the load in proportion to the relative stiffnesses of the walls providing the stability. Consider the two corridor walls each consisting of five shorter walls. The walls may be assumed to act as vertical cantilevers, each wall taking an equal proportion of the total wind load. It will be assumed that the floors are relatively flexible and are not able to transfer flexural shears, hence the assumed deflected shape is as shown in Figure 14.49.

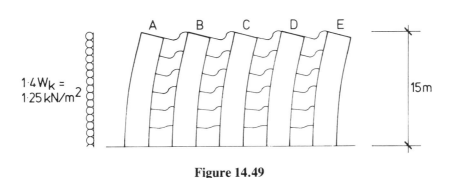

Figure 14.49

If the floors were made rigid and able to transfer the flexural shears, a much higher lateral load could be carried or, for the same load, the stress due to lateral loading would be much lower.

A critical design case for the wall is at its mid-height, between ground and first floor level. There are two cases to consider:

Case (i)
Dead + wind load acting, $\gamma_f = 0.9$ on dead load and $\gamma_f = 1.4$ on wind load, and flexural tensile stresses are considered.

$$\text{Design moment on the section} = W_k \times \gamma_f \times \frac{h^2}{2} \times \frac{\text{breadth of building}}{\text{no. of shear walls} \times \text{no. of walls}}$$

$$= 1.25 \times \frac{15^2}{2} \times \frac{20.5}{4} \times \frac{1}{5}$$

$$= 144 \text{ kN·m}$$

Design axial load acting:
due to self-weight of masonry $\quad = 0.9 \times 15 \times 0.102 \times 20 \quad = 27.5 \text{ kN/m}$
due to design dead load from
roof and five floors in 1 m wide
corridor area $\quad\quad = 0.9 (4 + 5.3) \times 5 \times 0.5 = 13.7 \text{ kN/m}$
$$= \overline{41.2 \text{ kN/m}}$$

Total design axial load over
5 m long wall $\quad\quad\quad = 5 \times 41.2 \quad\quad\quad\quad = 206 \text{ kN}$

Wall properties (Figure 14.50)

$A = 100 \times 5000 \quad = 50 \times 10^4 \text{ mm}^2$

$Z = \dfrac{100 \times 5000^2}{6} = 416.7 \times 10^6 \text{ mm}^4$

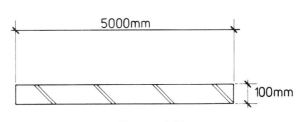

Figure 14.50

$$\text{Design moment of resistance} = \left(\frac{f_{kx}}{\gamma_m} + g_d\right) Z$$

Using 20 N/mm² clay bricks in designation (iii) mortar with water absorption less than 7% then $f_{kx} = 0.5$ N/mm².

Assume special construction and manufacturing control ($\gamma_m = 2.5$).

Therefore

$$\text{design moment of resistance} = \left(\frac{0.5}{2.5} + \frac{206 \times 10^3}{5 \times 10^5}\right) \times \frac{416.7 \times 10^6}{10^6} = 255 \text{ kN·m}$$

This is greater than the design moment = 144 kN·m

Case (ii)
The wall should also be checked for compressive buckling failure. Possible load combinations are $1.4G_k$ and $1.4W_k$ or $1.2G_k$; $1.2Q_k$ and $1.2W_k$. In this case, the former will give the worst design condition. Check at mid-span of wall.

Design stress due to dead load
$$= \frac{1.4(4 + 5.3) \times 5 \times 0.5 \times 10^3}{10^3 \times 100}$$
$$= 0.21 \text{ N/mm}^2$$

Design stress due to ow masonry
$$= \frac{1.4 \times 13.75 \times 0.10 \times 20 \times 10^3}{1000 \times 100}$$
$$= 0.39 \text{ N/mm}^2$$

Total axial compressive stress due to dead loads = 0.21 + 0.39
$$= 0.6 \text{ N/mm}^2$$

The flexural compressive stress $= M/Z$

The design bending moment $= 1.25 \times \dfrac{13.75^2 \times 20.50 \times 1}{2 \times 4 \times 5} = 121.12$ kN·m

Therefore

flexural compressive stress
$$= \frac{121.12 \times 10^6}{416.7 \times 10^6}$$
$$= 0.291 \text{ N/mm}^2$$

14.7.12 Eccentricity of loading

Design moment on wall $= 121.12$ kN·m

Design axial load on wall $= \dfrac{0.6 \times 5000 \times 100}{10^3} = 300$ kN

Eccentricity $= \dfrac{121.12}{300} = 0.404$ m, i.e. $0.07b$ (see Figure 14.51)

The slenderness ratio, SR, as previously calculated = (2500/100) × 0.75 = 18.75. Hence, β = 0.7.

Therefore

$$\text{maximum design strength} = \frac{\beta f_k \times 1.15}{\gamma_m}$$

$$= \frac{0.7 \times 5.8 \times 1.15}{2.5} = 1.85 \text{ N/mm}^2$$

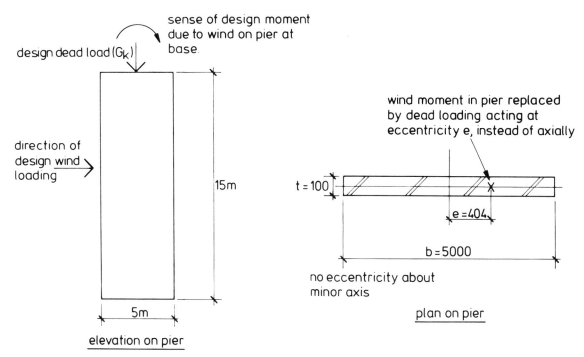

Figure 14.51

This is greater than the sum of the design dead load stresses and bending stresses = 0.6 + 0.29 = 0.89 N/mm^2.

Hence, the wall is satisfactory.

14.7.13 Accidental damage
Section five of the Code covers accidental damage, and Chapter 8 has explained its requirements in detail. The partial safety factors, γ_f, to be used are given in clause 22 (d) as follows:

$0.95G_k$ or $1.05G_k$
$0.35Q_k$ ($1.05Q_k$ where the building is used predominantly for storage)
$0.35W_k$

Values of γ_m for materials may be taken as half the values given in Table 4 of BS 5628 (see Table 5.11).

This example is for a Category 2 building, i.e. one of five storeys or more. Table 12 of the Code gives three options for detailing and designing the structure to withstand accidental damage (see Table 8.1). It is considered that Option 1 (i.e. vertical and horizontal elements, unless protected, proved removable one at a time without causing collapse) would be adopted here.

Crosswalls removed
The normal span of concrete floor is in the x direction. If, say, crosswall CH is removed, the slab can be designed to span – using increased distribution steel if necessary – in the y direction from the spine walls to an edge beam spanning from B to D. In addition the slab will also tend to hang in catenary action between walls BJ and DG (see Figure 14.52).

Gable wall removed
The Code of Practice BS 5628 states that for walls without vertical lateral supports the whole length of external walls must be considered removable whilst for similar internal walls only 2.25h need be considered as the removable length. The Building Regulations do not differentiate between internal and external walls but limit the removable length to 2.25h for all walls. It seems, to the authors, particularly harsh to consider, say in a spine wall structure of 30 m or more in length, the possibility of an incident capable of removing such a disproportionate length of external wall. Hence it is suggested

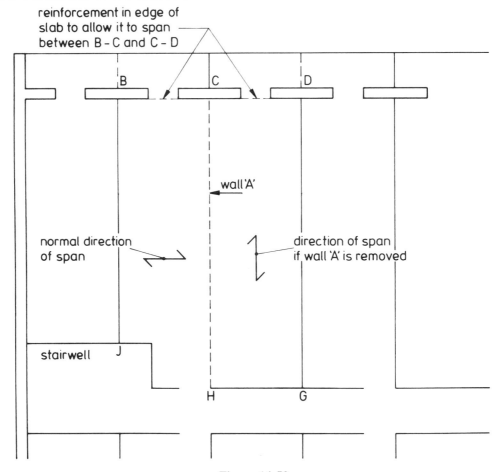

Figure 14.52

that in certain circumstances the designer should use his discretion in assessing a realistic but reasonable length for removal.

Having assessed the removable length of gable wall, consideration can now be given to the alternative means of support for the structure following its removal. If the length removed is not excessive, consideration may be given to composite action of the masonry over acting with the floor slab immediately above the removed length of wall. This, together with the arching effect of the masonry to spread the loads over to either side of the removed length of wall, may be all that is necessary with the additional reinforcement, if any, being added peripherally in the *in situ* floor slab. A more complex analysis might consider two adjacent floor slabs acting as the flanges of deep I beams with the spine walls between them acting as the webs of the same beams. These composite sections may be used to cantilever from the last crosswall and could support, at the end of the cantilever, a similar I–shaped composite beam utilising the gable wall as the web. Hence, a framework of composite beams is provided, and reinforced accordingly, to support the structure over (see Figure 14.53).

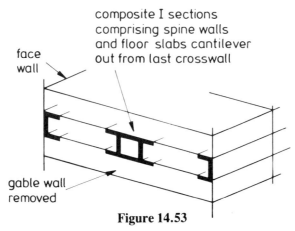

Figure 14.53

It may well be the case that, at the lower levels of a loadbearing brickwork structure, there is enough compressive dead load from above to enable the wall to withstand a lateral force of 34 kN/m² and thus it is a protected member as defined in clause 37.1.1.

14.8 EXAMPLE 2: FOUR-STOREY SCHOOL BUILDING

Design of an internal crosswall to resist vertical loading in a four-storey school-type building, shown in Figure 14.54. Overall height is 12 m.

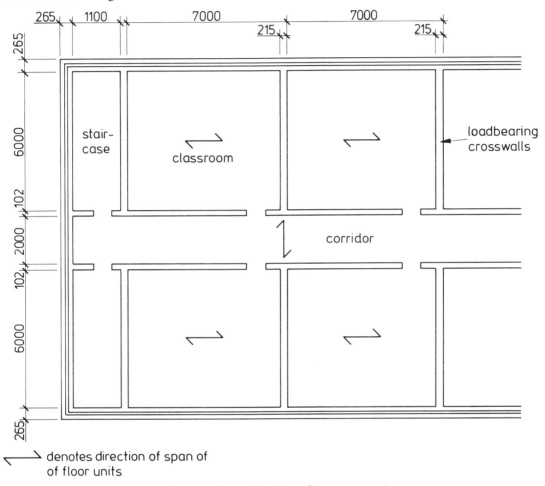

denotes direction of span of
of floor units

Plan on School Building (four storeys)

Figure 14.54

14.8.1 Characteristic loads

Roof

Dead loads, G_k, precast concrete units   = 3.6 kN/m²
            screed to falls, allow   <u>1.4 kN/m²</u>
                                      = 5.0 kN/m²

Imposed load, Q_k = 0.75 kN/m²

Floors

Dead loads, G_k, precast concrete units   = 4.8 kN/m²
            screed, finishes, allow   <u>1.2 kN/m²</u>
                                       = 6.0 kN/m²

Imposed load, Q_k = 3.0 kN/m²

14.8.2 Design of wall at ground floor level

Reduction in imposed load: four floors = 30%

Roof + Floors

Design dead load $= 1.4(5 + 6 + 6 + 6)$ $= 32.2 \text{ kN/m}^2$
Design imposed load $= 1.6(0.75 + 3 + 3 + 3) \times 0.7$ $= 10.9 \text{ kN/m}^2$
$\overline{= 43.1 \text{ kN/m}^2}$

Design load per m run $= 43.1 \times \frac{7}{2}$ $= 151 \text{ kN/m}$

For trial section try a 215 mm thick wall.

Walls

Characteristic load per m run due to masonry self weight: (brickwork density 20 kN/m^3).

$20 \times 12.0 \times 0.215$ $= 52 \text{ kN/m}$
Design load 52×1.4 $= 72 \text{ kN/m}$
Therefore, total design load, n_w $= 72 + 151 = 223 \text{ kN/m}$
Effective height $= 0.75 \times 3.0 = 2.25 \text{ m}$
Effective thickness $= 0.215 \text{ m}$
Slenderness ratio: $2.25/0.215 = 10.5$ and $e_x = 0$ to $0.5t$
Reduction factor for slenderness $\beta = 0.96$

Vertical load resistance per metre given by $\beta t f_k / \gamma_m$

Therefore, required minimum characteristic strength of masonry $f_k = \gamma_m n_w / \beta t$

Case (a)

Manufacturing and construction control both special, $\gamma_m = 2.5$.

$$f_k \text{ required} = \frac{2.5 \times 223 \times 10^3}{0.97 \times 215 \times 10^3} = 2.7 \text{ N/mm}^2$$

Assume mortar designation (iii) and standard format bricks.

Common brick (20 N/mm^2) should give ample strength ($f_k = 5.8 \text{ N/mm}^2 > 2.7 \text{ N/mm}^2$ required)

Case (b)

Manufacturing and construction control both normal, $\gamma_m = 3.5$.

$$f_k \text{ required} = \frac{3.5 \times 223 \times 10^3}{0.9 \times 215 \times 10^3} = 3.7 \text{ N/mm}^2$$

Again use a common brick (20 N/mm^2) set in designation (iii) mortar.

14.9 EXAMPLE 3: FOUR-STOREY OFFICE BLOCK

14.9.1 Column structure for four-storey office block

A simplified plan of a brick column structure is shown in Figures 14.55 and 14.56. Although the structure is mainly open-plan, a number of internal walls and partitions have been omitted for clarity. Since the building is not more than four storeys high, there is no statutory requirement to check for progressive collapse from accidental damage. The basic T-shaped plan, the staircase, lift wells and service cores, the internal partitions and the plate action of the floor slab and beams would make for a robust structure. Wind forces are transferred to the gable and internal shear walls by the plate action of the floors and roof.

The external gable walls are 255 mm cavity walls, the shear walls and walls to staircases, lifts, etc, are 215 mm solid walls, and the columns are 330 mm square in the upper storey, and 440 mm square below this.

The columns and the shear walls may be designed as follows:

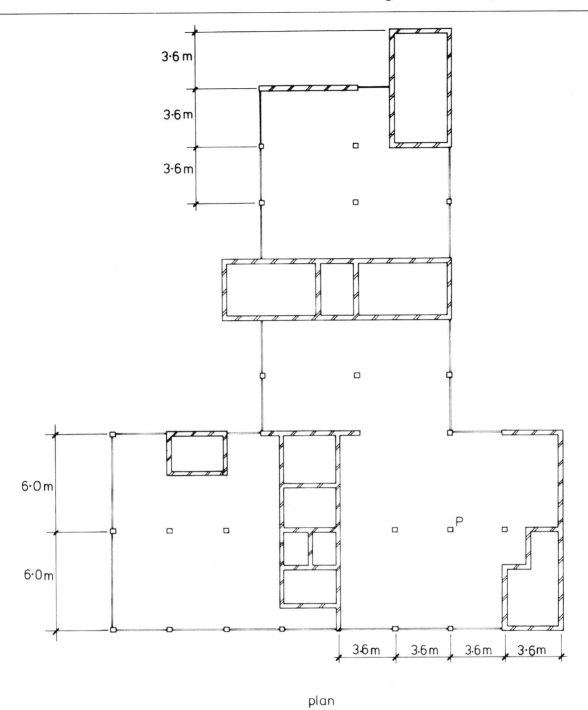

plan

Figure 14.55

14.9.2 Characteristic loads

Roof

Trussed timber roof, G_k	$= 1.25 \text{ kN/m}^2$
Imposed load, Q_k	$= 0.75 \text{ kN/m}^2$

Floor

Precast beam and pot floor with structural screed, G_k	$= 4.9 \text{ kN/m}^2$
Partitions, G_k	$= 1.0 \text{ kN/m}^2$
	$= \overline{5.9 \text{ kN/m}^2}$
Imposed load, Q_k	$= 2.5 \text{ kN/m}^2$

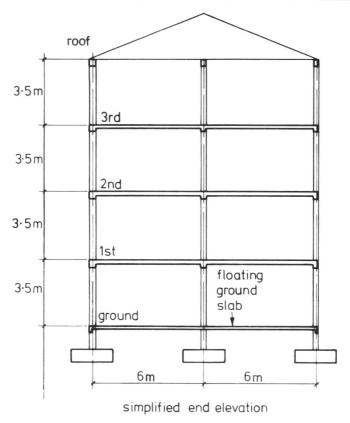

Figure 14.56

Wind loading

Assume that the maximum characteristic wind pressure (W_k) is 0.8 kN/m^2 and that the wind uplift on the roof is 1.0 kN/m^2.

Before proceeding with the design of the structure as a whole, the designer should consider carefully the location of joints, and the like, since these will affect the stability and strength of the structure. In this example, it has been assumed that the floors act as a plate to transfer the wind load to the shear walls which are designed as vertical cantilevers to resist the lateral loading. Therefore, it may be assumed that there is no wind load on the columns.

14.9.3 Design of brick columns

The first step in the design is to rationalise the number of load cases to be considered since, in general, each column will have a different load. By inspection of Figure 14.55 the column marked P will be one of the most heavily loaded, and this example will show the design of column P from ground floor to roof level. The design of other columns is performed in a similar manner. The second step is to calculate the design loadings for the selected groups of columns (see Table 14.2).

Table 14.2

Position	Characteristic dead load (kN) due to floors and roof	Characteristic imposed load (kN)	Imposed load reduction factor (Table 2, CP 3: Chapter V: Part 1)
Roof	6 × 3.6 × 1.25 = 27	6 × 3.6 × 0.75 = 16	0%
3rd floor	27 + (6 × 3.6 × 5.9) = 154	16 + (6 × 3.6 × 2.5) = 70	10%
2nd floor	282	124	20%
1st floor	409	178	30%

14.9.4 Loading on column P
The dead load of the masonry must be added to the above in obtaining the total design axial load.

Assume the characteristic density of the brickwork $= 20\,\text{kN/m}^3$

Therefore, characteristic dead load at A
due to brick column above
$$= 20 \times 0.33^2 \times 4 \times 3.5$$
$$\text{(density} \times \text{area} \times \text{height)}$$
$$= 30\,\text{kN}$$

Assuming brick columns are 330 mm $\times$ 330 mm.

Design load at ground floor (A)
For this combination of loading $\gamma_f = 1.4$ for dead loads and 1.6 for imposed loads (clause 22, BS 5628). Therefore:
$$\text{design load at A} = (1.4 \times 409) + (1.6 \times 0.7 \times 178) + (1.4 \times 30)$$
$$= 573 + 199 + 42$$
$$= 814\,\text{kN}$$

Slenderness ratio of column
The effective height of the column is the actual height between lateral supports, therefore, $h_{ef} = 3.50$ m. Assume for a trial section a 330×330 brick column constructed of standard format facing bricks with a compressive strength of 35 N/mm^2, laid in a designation (iii) mortar. The least lateral dimension is 330 mm therefore, the slenderness ratio = 3.5/0.33 = 10.6.

Characteristic compressive strength of brickwork, f_k
$f_k = 8.5$ N/mm^2 (Table 2(a), BS 5628 – see Table 5.4 – for 35 N/mm^2 compressive strength facing bricks in designation (iii) mortar).

The area of the brick column = 0.11 m^2 and clause 23.1.1 of BS 5628 states that, where the horizontal cross-sectional area of a column is less than 0.2 m^2, the characteristic compressive strength should be multiplied by the factor $(0.70 + 1.5A)$ where A is the horizontally loaded cross-sectional area of the column. In this case the area reduction factor = $0.7 + (1.5 \times 0.11) = 0.865$.

Calculation of β value for design
First determine the eccentricity of the loading (see clause 31, BS 5628) (see Figure 14.57).

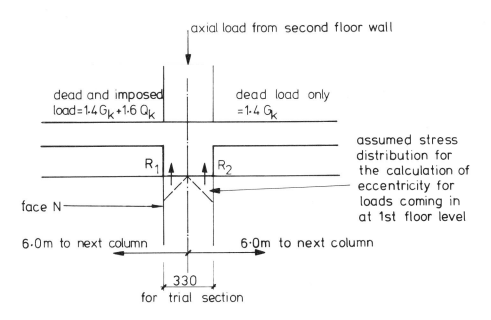

Figure 14.57

The resultants R_1 and R_2 are assumed to act at one-third of the depth of the bearing area from the loaded face of the column, i.e. $(330/2) \times \frac{1}{3} = 55$ mm in from the face.

Now
$$R_1 = \left(1.4 \times 5.9 \times \frac{3.6}{2} \times 6\right) + \left(1.6 \times 2.5 \times \frac{3.6}{2} \times 6\right)$$
$$= 89.21 + 43.2$$
$$= 132.41 \text{ kN}$$

and
$$R_2 = 89.21 \text{ kN}$$

The load in the column just above the first floor may be assumed to be axial and
$$= (1.4 \times 282) + (1.6 \times 0.8 \times 124) + (1.4 \times 20 \times 0.33^2 \times 3 \times 3.5)$$
$$= 395 + 159 + 32$$
$$= 586 \text{ kN}.$$

Taking moments about face N, let R be the distance to the resultant of R_1 and R_2 and the axial load.

$$\left(586 \times \frac{0.33}{2}\right) + (132.41 \times 0.055) + (89.21 \times 0.275) = (586 + 132.41 + 89.21)R$$

Therefore
$$R = 0.159 \text{ m from face N}$$

Therefore, eccentricity at the top of the column $= 0.33/2 - 0.159 = 6.3 \times 10^{-3}$ m $= 6.3$ mm $= 0.02t$.

Hence, for slenderness ratio of 10.6 and eccentricities up to $0.05t$ (Table 7, BS 5628 – see Table 5.15) by interpolation: $\beta = 0.958$.

Design vertical load resistance of column

The design vertical load resistance $= \dfrac{\beta t f_k}{\gamma_m} \times$ area reduction factor.

Now assume the manufacturing and construction controls are both special, therefore, $\gamma_m = 2.5$ (Table 4, BS 5628 – see Table 5.11)

$$\text{Design vertical load resistance} = \frac{0.958 \times 330 \times 330 \times 8.5 \times 0.865}{2.5}$$
$$= 307 \text{ kN}$$

This is less than the required vertical load resistance of 814 kN. Hence, column section designed is inadequate.

Increase trial column size
Try a 440 mm square brick column constructed of standard format bricks with a compressive strength of 50 N/mm² laid in designation (iii) mortar.

f_k $= 10.6$ N/mm² (Table 2(a), BS 5628 – see Table 5.4)
Area reduction factor $= 0.7 + (1.5 \times 0.44^2) = 0.99$
Design load $= 814$ kN + extra load due to increased size of column
Extra column load $= (0.44^2 - 0.33^2) \times 4 \times 3.5 \times 20$
                            $= 23.7$ kN
Design load $= 814 + 23.7$
                            $= 837.7$ kN

By comparison with 330 square column, β will be 1.0 for 440 square column.
Vertical load resistance of 440 square column

$$= \frac{1 \times 440^2 \times 10.6 \times 0.99}{2.5} = 812 \text{ kN}$$

This is just less than the design load of 814 kN. The minimum required characteristic strength of the masonry is:

$$= \frac{2.5 \times 837.7 \times 10^3}{0.99 \times 440^2}$$

$$= 10.92 \text{ N/mm}^2$$

From Figure 1 of BS 5628, the required compressive strength of bricks laid in designation (iii) mortar = 58 N/mm², in a 440 × 440 column. If it is not possible to obtain facing bricks of this strength, then the use of engineering bricks should be considered, or the column could be reinforced.

Design of column between 1st and 2nd floor levels
Assume a 440 × 440 column

Design vertical load = $(1.4 \times 282) + (1.6 \times 0.8 \times 124) + (3 \times 3.5 \times 20 \times 0.44^2 \times 1.4)$
(see Figure 14.58)
$$= 395 + 158 + 57$$
$$= 610 \text{ kN}$$

The eccentricity of loading in the column should be recalculated at each level but will be assumed to be zero for this case also.

Minimum required compressive strength of masonry

$$f_k = \frac{2.5 \times 610 \times 10^3}{0.99 \times 440^2}$$

$$= 7.95 \text{ N/mm}^2$$

From Figure 1, BS 5628, the required compressive strength of bricks is 33 N/mm² in designation (iii) mortar. Facing bricks can be obtained with this strength for use in a 440 × 440 brick column.

The design of the brick columns at higher levels is performed in a similar manner, and the section may be reduced, or the brick strength varied, as required, to suit both architectural and economic considerations.

Simplified design for lateral loading due to wind
By inspection of the plan, the structure appears to be rigid and robust with many crosswalls, and the stresses induced in the structure due to lateral loading are small. Consider wind loading on the south elevation. The main elements resisting the loading are shown in Figure 14.58.

As in the previous shear wall design example, the structure will be assumed to behave as a vertical cantilever under lateral loading. Hence, the deflected shape is as shown in Figure 14.59.

The total design bending moments should be shared between the shear walls in proportion to their stiffnesses, and the designs of the individual walls should then proceed using the principles demonstrated in example number 8 in Chapter 11. A careful check should be made at all floor levels, particularly in the uppermost storey, as the dead loading available to eliminate flexural tensile stresses is considerably reduced.

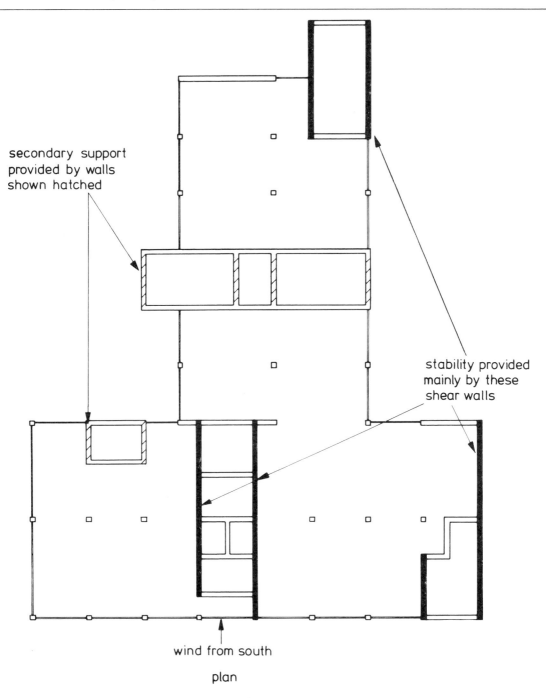

secondary support
provided by walls
shown hatched

stability provided
mainly by these
shear walls

wind from south

plan

Figure 14.58

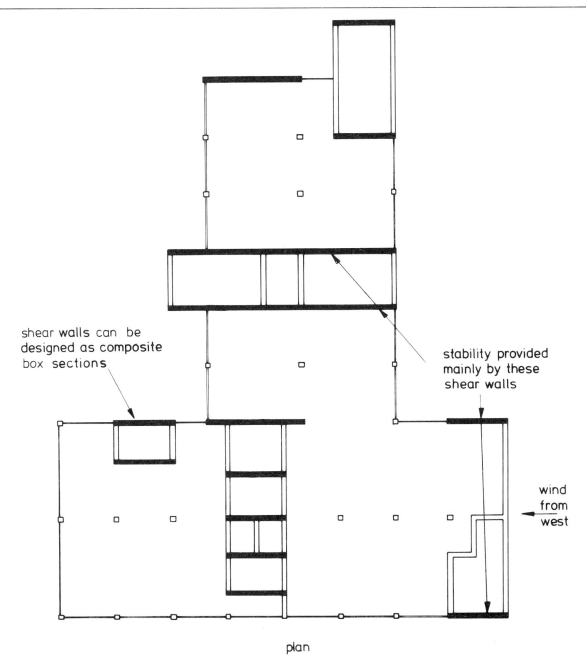

plan

Figure 14.59

CHAPTER 15

REINFORCED AND POST-TENSIONED MASONRY

In the preceding chapters, discussion has been largely confined to the basic principles and assumptions underlying the design of plain structural masonry – ordinary bricks or blocks and mortar construction. However, in that it is strong in compression and relatively weak in tension, the structural applications of plain masonry tend to be restricted to walls, columns, arches and other elements carrying mainly compressive loads. When plain structural masonry elements are subjected to lateral loading, from wind or retained earth and other causes, they need thickening or a change of geometric shape to resist the resulting tensile stresses. In short, the material's relatively low tensile strength tends to govern the design. As a result, it's high compressive strength is often partly wasted.

As most engineers know, concrete – which is also strong in compression but weak in tension – is commonly either reinforced with steel to carry the tensile stresses, or prestressed to eliminate them. Similar principles can be applied to the design of structural masonry (see Figure 15.1) with corresponding gains in the extension of its field of application.

Such an obvious concept is not new. In fact, the reinforcing of brickwork long preceded the reinforcing of concrete. About 1820, Sir Marc Brunel reinforced the brick shafts of the Wapping – Rotherhithe tunnel. At the turn of the present century, Sir Alexander Brebner used reinforced brickwork in India, and his example was followed in the '20s and '30s in Japan and other countries subject to earthquakes. Since the last war, there has been an increasing application of the technique in the USA.

Nevertheless, and despite these historical precedents, the development of reinforced and prestressed masonry has lagged far behind reinforced and prestressed concrete. This is hardly surprising. There has been relatively little research on the subject, there are hardly any technical papers, Codes of Practice, design guides, etc., to assist the engineer, and practically no engineering student receives any instruction in the subject during his studies.

This chapter, it is hoped, will help to alleviate the situation. There is no doubt, in the authors' minds, that industry and society as a whole are missing out on a valuable and worthwhile technique of construction, in that reinforced and post-tensioned masonry maintain all the advantages set out in Chapter 2, including speed, simplicity and economy, see Figure 15.2.

15.1 GENERAL

15.1.1 Design theory

Being based on limit state principles, the design philosophy for reinforced and prestressed masonry is exactly the same as for plain masonry. Reference is made to CP 110: The Structural Use of Concrete, which is also based on limit state principles. This is not because there is any direct relationship between masonry and concrete (see 15.1.2), but because of the progress that has been made in the research and development of reinforced and prestressed concrete and the comparative neglect of masonry.

The assessment of loadings and member forces is made in exactly the same way as for plain masonry. Since no specific recommendations exist for reinforced and prestressed masonry, the analysis of sections is generally based on the methods given in BS 5628 for plain masonry, together with those for reinforced and prestressed concrete in CP 110.

15.1.2 Comparison with concrete

Because masonry is analogous to concrete, some engineers tend to consider them as almost iden-

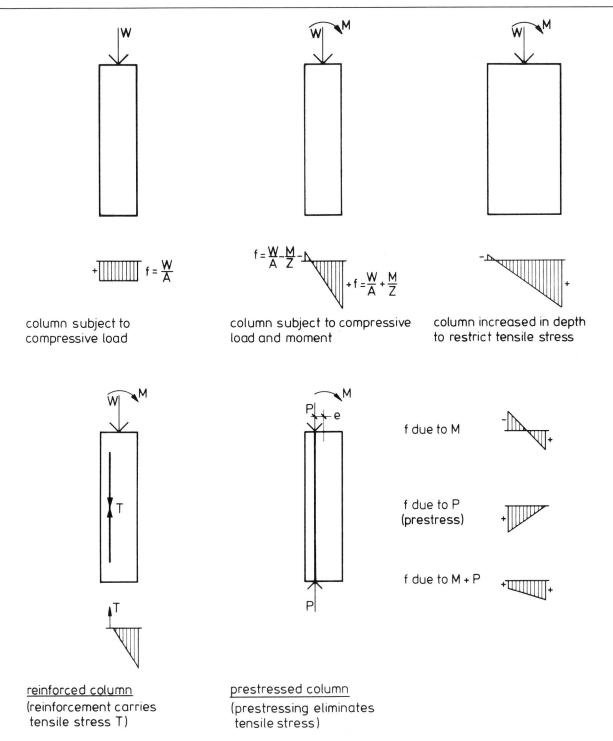

column subject to
compressive load

column subject to compressive
load and moment

column increased in depth
to restrict tensile stress

f due to M

f due to P
(prestress)

f due to M + P

<u>reinforced column</u>
(reinforcement carries
tensile stress T)

<u>prestressed column</u>
(prestressing eliminates
tensile stress)

Figure 15.1

tical materials in design terms. They are not –
and the analogy can be pushed too far.

Unlike concrete, masonry – brickwork particularly – is not homogeneous or isotropic. Concrete
shrinks as it matures and brickwork expands, and
this affects bond strength, creep losses, etc.
Cracking on the tensile face of reinforced concrete members will be spread along the face, and
the cracks are likely to be minute. Cracking on
the tensile face of a reinforced masonry member

will be concentrated at the mortar joints, and the
cracks may well be larger.

Whilst the bulk of concrete is reinforced, only
some is prestressed. It would certainly appear
likely that, with further experience, the reverse
situation will occur in masonry. Reinforcing concrete is generally simpler than prestressing it.
Quite the opposite applies to brickwork. Prestressed concrete usually calls for high stresses
needing sophisticated stressing equipment, high

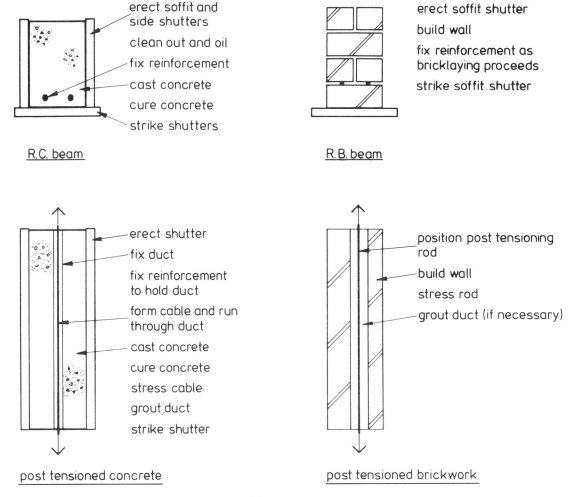

R.C. beam

erect soffit and
side shutters

clean out and oil

fix reinforcement

cast concrete

cure concrete

strike shutters

R.B. beam

erect soffit shutter

build wall

fix reinforcement as
bricklaying proceeds

strike soffit shutter

post tensioned concrete

erect shutter

fix duct

fix reinforcement
to hold duct

form cable and run
through duct

cast concrete

cure concrete

stress cable

grout duct

strike shutter

post tensioned brickwork

position post tensioning
rod

build wall

stress rod

grout duct (if necessary)

Figure 15.2

strength materials, complicated duct installation, high tensile steel tendons and careful site supervision. On the other hand, post-tensioned brickwork is concerned with relatively low stresses, requiring an almost rudimentary technique using everyday materials and methods.

Although it is true that masonry and concrete are not identical, nevertheless they are sufficiently alike to enable some similar design concepts of reinforcing and prestressing to apply. On the other hand, they are sufficiently different as to require different design methods, detailing and construction. The designer *must* be aware of these differences, and must not blindly apply the methods and techniques of *in situ* or precast concrete to masonry.

15.1.3 Applications
Reinforced masonry has been used to enable walls to act as beams, lintols, cantilevers, etc., with the reinforcement in the bed joints (i.e. horizontal reinforcement – see Figure 15.3).

It was, and still is, used for vertical members subject to lateral loading, such as retaining walls. Its most common, economical and simplest use is in grouted cavity construction (see Figure 15.4).

Another use, and one likely to continue, is to

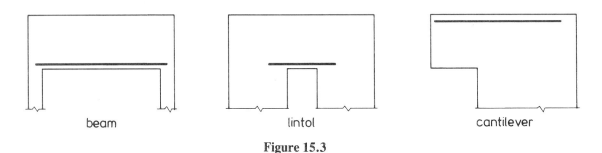

beam   lintol   cantilever

Figure 15.3

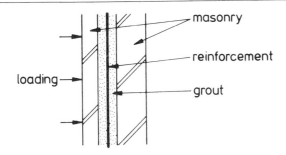

Figure 15.4

enhance the loadbearing capacity of brick columns when there are restrictions on their cross-sectional area. In the limited amount of prefabricated brickwork built in recent years, reinforcement has often been added to cope with the erection and handling stresses. In addition to being a useful and economical alternative to concrete, brickwork can also be effectively used in association with it. There are occasions when it can act compositely with concrete, e.g. in retaining and balcony walls (see Figure 15.5).

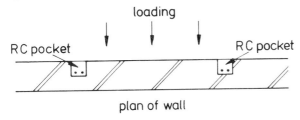

Figure 15.5

Some engineers have appreciated the fact that a relatively thin reinforced concrete footing to a wall can result in a composite beam, where the masonry wall forms the compression flange and the rc footing the tensile flange (see Figure 15.6). This has led to worthwhile savings in foundation costs, particularly on soils subject to significant differential settlement.

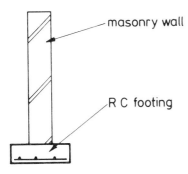

Figure 15.6

Brickwork has been increasingly used in recent years as a veneer to rc cladding panels, where it can act as a permanent shutter and provide an attractive mask to the often unacceptable face of concrete. Too often, however, the compressive strength of the brickwork has been neglected, resulting in a less economical design than was

possible. The use of an unstressed brickwork veneer is considered by some to be structurally 'dishonest' and makes inefficient use of the material.

15.1.4 Prestressing

Of the two techniques of prestressing, i.e. pretensioning and post-tensioning, the former method has been most widely used in concrete, particularly in precast units. In structural masonry, however, post-tensioning, has so far been found to be the most successful, and certainly the simplest method. Briefly, the procedure is to anchor one end of a high tensile steel bar and build the masonry around it – leaving a space for grouting up, if felt necessary. On completion of the bricklaying, a plate is placed over the end of the rod – which is threaded – to act as a load-dispersal anchorage. A nut is then screwed on, and tightened up to the required post-tensioning stress with a torque spanner (see Figure 15.7).

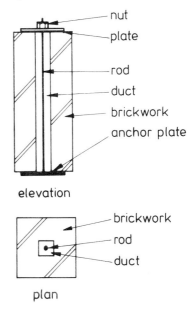

Figure 15.7

Prestressed concrete sections are more structurally efficient if the section has a high section modulus/cross-sectional area ratio (Z/A), as in an I or T beam, and the same principle applies to prestressed masonry. The post-tensioning of diaphragm walls (I sections) and fin walls (T sections) shows great potential for tall walls subject to high lateral loading and low axial loading.

15.1.5 Methods of reinforcing walls

Masonry can be constructed with reinforcement incorporated in pockets in the face of the wall, in vertical holes inside the wall and in the voids of cavity construction (see Figure 15.8).

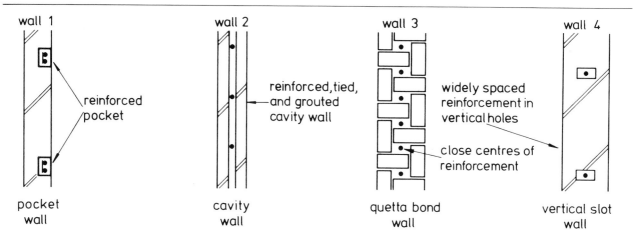

Figure 15.8

In the case of walls 2, 3 and 4, shuttering is not necessary and the void can be filled as the work proceeds. As far as simplicity is concerned, walls 2 and 4 are generally much easier and quicker to construct than walls 1 and 3 – wall 2 has the simplest bond. From the construction point of view, the quetta bond wall is usually expensive and slow.

Methods of filling around the reinforcement must be considered (see Figure 15.9). If grouting is used, then difficulties of keeping the void clear of mortar droppings and projections, of keeping the brickwork adequately tied to resist hydrostatic pressure, and of preventing air being trapped below the grout, all become very real problems. Generally, the simplest, most successful and economical method is to fill the void with a suitable quality mortar as the work proceeds, and to make the bricklayers aware of the need for complete filling of the void to ensure adequate bond and protection against corrosion.

The designer must be very aware of the problems at the design detail stage, and should make adequate adjustments to suit the methods and likely quality of construction. If grouting is chosen for a particular situation, then lifts with vent holes, to prevent air locks and to monitor

the filling, can help considerably – as can clear communications to site operatives, and good supervision.

To summarise:

(a) complicated details should be avoided wherever possible;
(b) shuttering should be reduced to a minimum;
(c) hydrostatic head should be reduced to a minimum;
(d) bonding should be kept simple;
(e) the majority of the work should be kept under direct control of the bricklayer.

The designer should make allowances for the effects the above points may have on the expected quality. For example, if the bond of the mortar to rods is expected to be reduced, the stresses used in design should be reduced and lap lengths increased accordingly. If corrosion could be a problem due to the expected quality of workmanship, non-ferrous metal or galvanised steel should be used and, again, the stresses adjusted to suit these materials and any possible loss of bond. See also 15.1.8.

15.1.6 Composite construction

As mentioned earlier, this is another form of

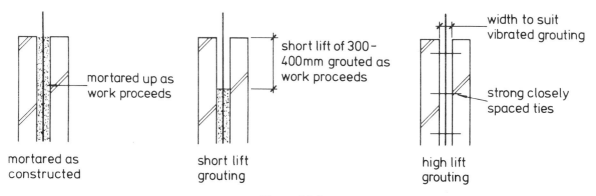

Figure 15.9

reinforced masonry. Generally, a shallow reinforced concrete beam is designed for two basic conditions. First, as a member to support the temporary condition of the first lifts of wet masonry. Second, and after the masonry has cured and acts with the beam, to support compositely the loads likely to be applied. This method of construction can prove to be very simple and economical to construct, particularly at foundations and floor slab levels where reinforced concrete is frequently already being used.

The principles of design for composite constructions are similar to those for reinforced masonry, and careful attention should be paid to the points summarised in 15.1.5. Again, the designer should keep the details simple. He should also consider carefully the differential movements of the two materials, and the temporary construction loading (see Figure 15.10), and the final loading including suitable allowance for openings, damp proof courses, etc., which may affect the composite action.

15.1.7 Economics

Even using the results of detailed cost surveys, it is notoriously difficult to quantify cost savings. However, the authors' experience suggests that structural masonry can show 10% savings in building costs and, by reinforcing or post-tensioning, these savings can often be further increased.

Savings in construction time tend to be between 10% and 30%, depending on the type of structure. Reinforcing and post-tensioning add little to the time-saving implications of strucutral masonry – their main value lies in widening its scope, increasing its range of application and in making it an even better and more economical alternative to other structural materials.

15.1.8 Corrosion of reinforcement and prestressing steel

Since clay bricks, calcium silicate bricks and concrete cellular blocks are porous, understandable concern has been expressed about the possibility of corrosion in reinforcing steel and post-tensioning rods. There are a number of buildings, in various countries, where careful detail, good workmanship and proper supervision have shown that the problem can be solved. In reinforced masonry walls and columns, the rods should have a minimum cover of 100 mm, and the bed and perpend joints should be completely and properly filled with dense and durable mortar. When reinforcement is placed in the cavity of a cavity wall, or the core of a hollow column or fin, the void should be fully grouted up with a well-designed cement grout. In post-tensioned diaphragm and fin walls, a 'belt-and-braces' approach can be adopted by using a larger rod than necessary, to allow for the loss of cross-sectional area due to corrosion, and by reducing the risk of corrosion by coating the rod with a bitumen paint and wrapping it in a proprietary waterproof tape. Alternatively, stainless steel rods can be used. Although these are currently about four times the cost of high tensile steel rods, the extra cost per m^2 of the completed wall is not usually significant.

15.2 CHOICE OF SYSTEM

The authors have designed many structures using post-tensioned and/or reinforced masonry, and the reasons for choosing either method in any particular situation have varied – as, indeed, have the economic conditions existing at the time of construction. For each scheme, the choice is essentially a matter for the designer's judgement. There are no hard and fast rules. Nevertheless, the aim of all design should be to achieve simple, safe and economical details and, bearing this in

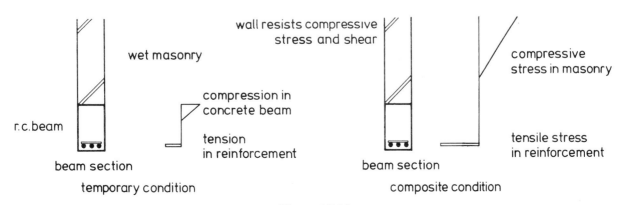

Figure 15.10

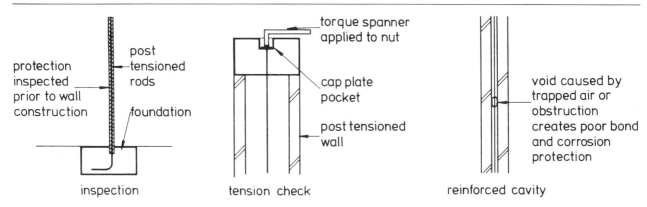

inspection   tension check   reinforced cavity

Figure 15.11   **Figure 15.12**

mind, several general points emerge which should help to guide the reader towards a satisfactory decision.

Vertically post-tensioned masonry is generally simpler than vertically reinforced masonry from both the construction and supervision points of view, and particularly when a cavity is to be maintained.

Protection for post-tensioning rods can be more easily provided and relied upon. Rods can be galvanised, painted, wrapped, etc., above the foundation anchorage, without any worries over loss of bond stress at these levels, and with the simplest of supervision. Rods can easily be inspected for protection, and can be checked for tension simply by applying a torque spanner to the nut after post-tensioning (see Figure 15.11).

Reinforcement, on the other hand, relies upon bond which, in turn, relies upon adequate compaction of the grout or mortar. Reinforcing steel also requires adequate cover to prevent corrosion (this is absolutely vital to walls exposed to driving rain). These requirements can be very difficult to achieve, supervise and check (Figure 15.12).

For horizontal members, however, post-tensioned sections generally become more difficult, and details less simple, whilst reinforced masonry becomes more reliable from the point of view of achieving a satisfactory bond and a properly grouted cross-section (see Figure 15.13).

Again, the need to give adequate protection to the reinforcement should be stressed. This cannot be over-emphasised and, where cover to bars is minimised in attempts to support masonry which is forming, say, the soffit shutter for the grout, the bars should either be made of non-ferrous metal, or be protected by galvanising or by other means.

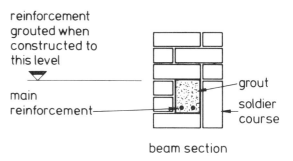

beam section

Figure 15.13

All too often, there is a tendency for engineers to introduce secondary reinforcement into locations which later result in problems of bursting, due to corrosion, when the omission of such reinforcement would have produced a suitable and reliable detail. This is very apparent in many existing reinforced concrete buildings, and must be avoided in the next generation of structures

Provided that the above points are watched, reinforced masonry is often suitable for horizontal members. Particularly, of course, when a masonry finish is required.

Thus to summarise the situation – post-tensioning is generally better for vertical conditions, and reinforcement for horizontal work. Occasions do arise, however, where the balance of advantage is only marginal. For example, both reinforced and post-tensioned masonry can prove to be very economical for balcony walls and retaining walls (see Figure 15.14).

Fundamentally, masonry is a 'walling' material and it is likely therefore, that greater use will be made of post-tensioned masonry than reinforced. Since cracking, due to bending (possibly causing durability problems), is more likely to occur in reinforced masonry than post-tensioned masonry, designers may tend to opt for the latter.

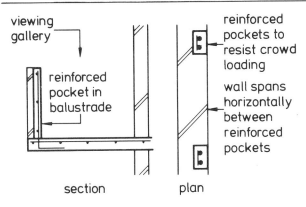

Figure 15.14

The principal use of both techniques is in members resisting large bending moments. However, if we consider large compressive loads, post-tensioning would tend to reduce the wall's ability to resist such loads, whereas reinforcement would improve the wall's strength in compression (see Figure 15.15).

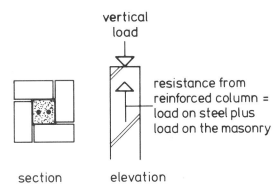

Figure 15.15

Reinforcement and post-tensioning can be exploited in the construction of precast masonry panels, which require added tensile strength to resist handling stresses.

Similar conditions arise in areas subject to mining subsidence, in that the most critical condition for the masonry is the effect of the tensile stresses which can be produced in direct form and in bending (see Figure 15.16).

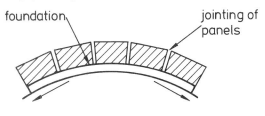

tensile stresses in ground
from subsidence

Figure 15.16

The stresses can be reduced, and the masonry's resistance improved, by jointing the walls in short lengths and post-tensioning the resulting panels to increase the compression on the bed joints – thereby increasing the joint's resistance to applied tensions. Horizontally, the bonding of the masonry provides a greater resistance than the weaker bed joints and, generally, the panel size is restricted to maintain the stresses within acceptable limits (see Figure 15.17).

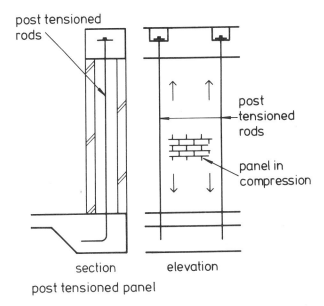

post tensioned panel

Figure 15.17

15.3 DESIGN OF REINFORCED BRICKWORK

15.3.1 Partial factors of safety

Loadings
Partial factors of safety on ultimate loadings and details of the various load combinations to be considered are all as for plain masonry. Details of the various values are given in Chapter 5. Partial factors of safety for serviceability limit state are given in Table 15.1.

Table 15.1 Partial factors of safety – serviceability limit state

	γ_m
Dead and imposed	1.0
Dead and wind	1.0
Dead, imposed and wind	$\begin{cases}1.0 \times G_k\\0.8 \times Q_k\\0.8 \times W_k\end{cases}$

Materials

The partial factors of safety on materials remain as previously, reference Chapters 5 and 6 however, certain additional factors are required in respect of the additional materials. The partial factor of safety for steel reinforcement is termed γ_{ms}, and γ_{mb} is the partial safety factor for the bond between mortar or grout and steel. The appropriate values for these partial factors of safety are given in Table 15.2. Values for γ_m the partial factor of safety on strength of materials are given in Table 5.11 and for γ_{mv} for shear strength in section 6.11.1.

Table 15.2 Partial factor of safety on materials – ultimate and accidental damage limit states

	Limit state	
	Ultimate	Accidental damage
Steel γ_m	1.15	1.0
Compressive strength of masonry	See Chapter 5	50% of ultimate values
Bond between mortar/grout and steel	1.5	1.0

15.3.2 Strength of materials

In order to analyse reinforced masonry the characteristic strengths of the various materials used must be determined. The properties of both the masonry and the reinforcement must be determined, in addition to the properties of the combined materials.

Properties of masonry

Characteristic compressive strength (f_k):

The characteristic compressive strength of the masonry is determined in exactly the same way as previously for plain masonry, details are given in Chapter 5 as is the capacity reduction factor.

Properties of reinforcement

Characteristic tensile strength (f_y):

The characteristic tensile strengths of reinforcement for various types of steel are given in the appropriate British Standards. Some values are given in Table 15.3.

Table 15.3 Characteristic tensile strength of reinforcement

Designation	Nominal Sizes (mm)	Characteristic tensile strength f_y (N/mm^2)
Hot rolled mild steel BS 4449	All	250
Hot rolled high yield steel BS 4449	All	410
Cold worked high yield BS 4461	16 Over 16	460 425
Hard drawn steel wire	12	485

Characteristic compressive strength

The design compressive strength of steel is taken as less than the design tensile strength. The design tensile strength being given by the expression:

$$\frac{f_y}{\gamma_{ms}}$$

The design compressive strength being given by the expression:

$$\frac{f_y}{\gamma_{ms} + \dfrac{f_y}{2000}}$$

There is thus an additional partial factor of safety on the design compressive strength. For mild steel and high yield bars to BS 4449 the design strength for the ultimate limit state becomes:

mild steel

$$\frac{f_y}{1.15 + \dfrac{250}{2000}} = 0.78f_y = \frac{0.9f_y}{\gamma_{ms}}$$

high yield

$$\frac{f_y}{1.15 + \dfrac{410}{2000}} = 0.74f_y = \frac{0.85f_y}{\gamma_{ms}}$$

For cold worked bars to BS 4461

$$= 0.72f_y = \frac{0.83f_y}{\gamma_{ms}}$$

Hence, as a simplification the characteristic compressive strength is taken as $0.8f_y$.

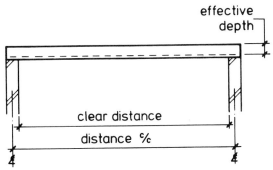

simple supports:effective span lesser of clear distance +effective depth or distance ℄

Figure 15.18

15.3.3 Design for bending: reinforced masonry

The design loading on a particular structural element is determined from the combination of the characteristic loadings from the relevant Codes of Practice and the partial factors of safety appropriate to the case being considered.

In the case of elements subject to bending, e.g. beams, retaining walls, etc., the following points should be used in assessing the design loads as illustrated in Figures 15.18 and 15.19.

Effective span

The effective span of a simply supported member should be taken as the lesser of the distance between the centres of bearing or the clear distance between supports plus the effective depth.

The effective span of a continuous member should be taken as the distance between centres of supports.

The effective span of a cantilever should be taken as its length to the face of the support plus half its effective depth except where it forms the end of a continuous beam where the length to the centre of the support should be used.

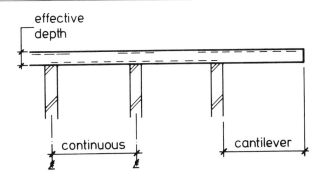

Figure 15.19

15.3.4 Lateral stability of beams

To ensure lateral stability a simply supported or continuous beam should be so proportioned that the clear distance between lateral restraints does not exceed $60b_c$ or $250b_c \times 2/d$, whichever is the lesser, where b_c = breadth of compression face and d = effective depth. For walls the slenderness ratios should be limited to the values given in Table 15.4.

Table 15.4 Limiting slenderness ratios for walls

End condition	Ratio
Simply supported	35
Continuous	45
Cantilever with up to 0.5% reinforcement*	18

*The percentage of reinforcement should be based on the gross cross-sectional area of the brickwork. For higher percentages of reinforcement special consideration should be given to deflection.

15.3.5 Design formula for bending: moments of resistance for reinforced masonry

The calculation of the moment of resistance for a rectangular section is based on the assumption that at ultimate load the compression can be approximated to a semi-parabola extending from the neutral axis for a distance corresponding to

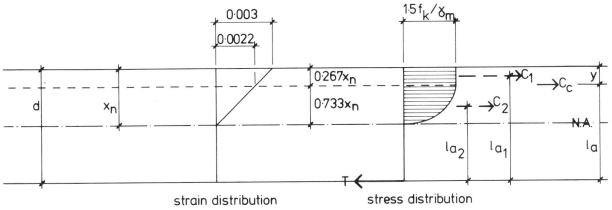

strain distribution   stress distribution

Figure 15.20

the transition strain and rectangular thereafter. (This is a similar assumption to that made in the design of reinforced concrete.)

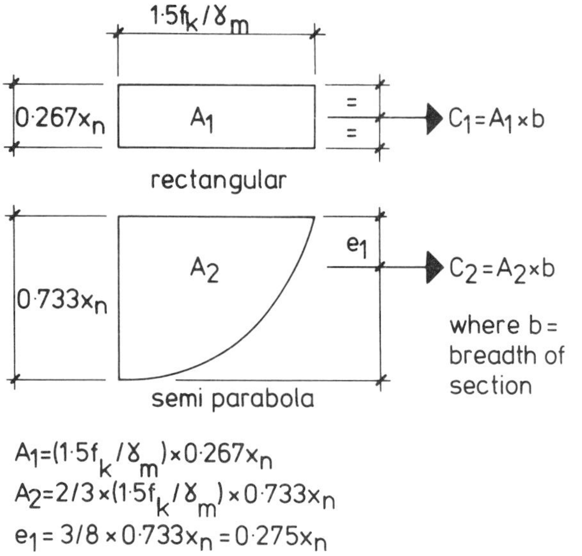
1·5fk/γm
0·267xn
A1
C1=A1×b
rectangular
e1
A2
C2=A2×b
0·733xn
where b= breadth of section
semi parabola
A1=(1·5fk/γm)×0·267xn
A2=2/3×(1·5fk/γm)×0·733xn
e1=3/8×0·733xn=0·275xn

Figure 15.21

The flexural compressive resistance is thus made up from the resistance of the rectangular section, A_1, and the parabolic section, A_2 (see Figure 15.21).

The flexural compressive stress is related to the characteristic direct compressive stress, f_k, modified by the relevant partial factor of safety and a factor relating flexural compressive stress to direct compressive stress. This factor has been taken as 1.5 which appears reasonable although further research is required on this subject.

With reference to Figures 15.20 and 15.21 for equilibrium total compressive force = total tensile force:

$$C_c = C_1 + C_2 = T$$

i.e.

$$A_1 b + A_2 b = T = \frac{A_s f_y}{\gamma_{ms}}$$

Therefore:

$$\frac{1.5 f_k}{\gamma_m} b\, x_n \left(0.267 + \frac{2}{3} \times 0.733\right) = \frac{A_s f_y}{\gamma_{ms}}$$

Therefore, depth to neutral axis

$$x_n = \left(\frac{A_s f_y}{\gamma_{ms}} \times \frac{\gamma_m}{1.5 f_k}\right)\left(\frac{1}{0.76b}\right)$$

Let

$$K_a = \frac{f_y}{\gamma_{ms}} \times \frac{\gamma_m}{1.5 f_k}$$

Hence

$$x_n = \frac{K_a A_s}{0.76b} \qquad (1)$$

Values are given in Table 15.5 and graphically in Figures 15.22–15.29.

Table 15.5 $K_a = \dfrac{f_y}{\gamma_{ms}} \times \dfrac{\gamma_m}{1.5 f_k}$

f_y	γ_{ms}	γ_m	f_k	K_a
250	1.15	2.5	4	90.6
			5	72.5
			6	60.4
			7	51.8
			8	45.3
			9	40.3
			10	36.2
			11	32.9
250	1.15	2.8	4	101.4
			5	81.1
			6	67.6
			7	57.9
			8	50.7
			9	45.1
			10	40.6
			11	36.9
250	1.15	3.1	4	112.3
			5	89.9
			6	74.9
			7	64.2
			8	56.2
			9	49.9
			10	44.9
			11	40.8
250	1.15	3.5	4	126.8
			5	101.4
			6	84.5
			7	72.5
			8	63.4
			9	56.4
			10	50.7
			11	46.1
410	1.15	2.5	4	148.6
			5	118.8
			6	99.0
			7	84.9
			8	74.3
			9	66.0
			10	59.4
			11	54.0
410	1.15	2.8	4	166.4
			5	133.1
			6	110.9
			7	95.1
			8	83.2
			9	73.9
			10	66.6
			11	60.5
410	1.15	3.1	4	184.2
			5	147.4
			6	122.8
			7	105.3
			8	92.1
			9	81.9
			10	73.7
			11	67.0

f_y	γ_{ms}	γ_m	f_k	K_a
410	1.15	3.5	4	208.0
			5	166.4
			6	138.6
			7	118.8
			8	104.0
			9	92.4
			10	83.2
			11	75.6

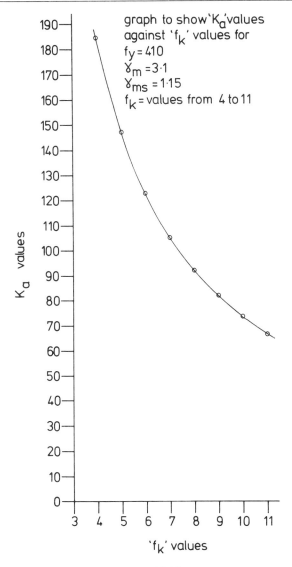

Figure 15.23

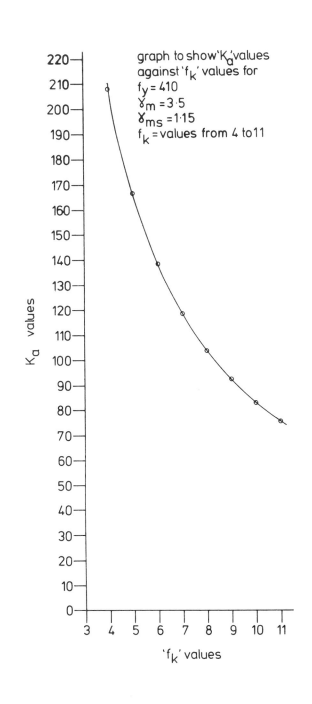

Figure 15.22

The lever arm, l_a, is obtained as follows:
Taking moments about the top of the section in Figure 15.20.

$$C_c y = C_1(0.267x_n) + C_2(0.267 + 0.275)x_n$$

$$= C_1(0.267x_n) + C_2(0.542x_n)$$

$$= \frac{1.5f_k}{\gamma_m} \times 0.267^2(x_n^2 b)$$

$$+ \frac{2}{3} \times \frac{1.5f_k}{\gamma_m} \times 0.733 \times 0.542(x_n^2 b)$$

$$= \frac{1.5f_k}{\gamma_m}(x_n^2 b)\left(0.267^2 + \frac{2}{3} \times 0.733 \times 0.542\right)$$

$$= \frac{1.5f_k}{\gamma_m}(x_n^2 b) \times 0.336$$

But $C_c = T$

Therefore $\quad y = \frac{1}{T} \times \frac{1.5f_k}{\gamma_m}(x_n^2 b) \times 0.336$

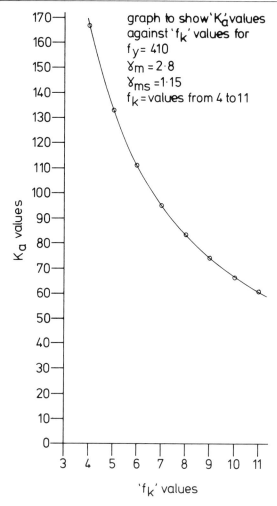

graph to show 'K$_a$' values against 'f$_k$' values for
f$_y$ = 410
γ_m = 2·8
γ_{ms} = 1·15
f$_k$ = values from 4 to 11

Figure 15.24

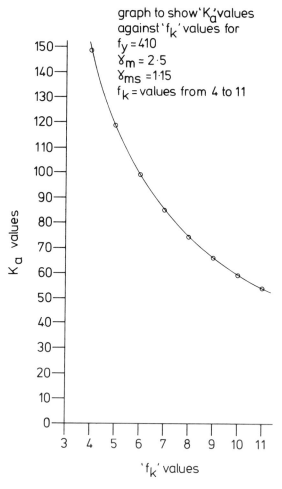

graph to show 'K$_a$' values against 'f$_k$' values for
f$_y$ = 410
γ_m = 2·5
γ_{ms} = 1·15
f$_k$ = values from 4 to 11

Figure 15.25

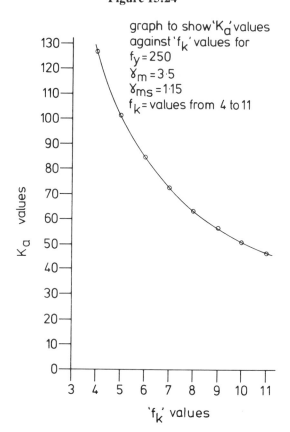

graph to show 'K$_a$' values against 'f$_k$' values for
f$_y$ = 250
γ_m = 3·5
γ_{ms} = 1·15
f$_k$ = values from 4 to 11

Figure 15.26

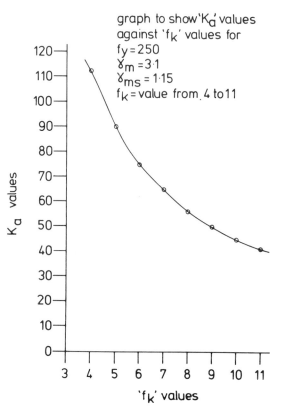

graph to show 'K$_a$' values against 'f$_k$' values for
f$_y$ = 250
γ_m = 3·1
γ_{ms} = 1·15
f$_k$ = value from 4 to 11

Figure 15.27

i.e.

$$y = \frac{\gamma_{ms}}{A_s f_y} \times \frac{1.5 f_k}{\gamma_m} (x_n^2 b) \times 0.336$$

$$= \frac{1}{K_a A_s} (x_n^2 b) \times 0.336$$

where

$$R = \frac{1.5 f_k}{\gamma_m} \times 0.303$$

Values of R may be obtained from Table 15.6 for various f_k and γ_m values or from Figures 15.30–15.33.

Substitute for x_n from equation (1)

$$y = \frac{1}{K_a A_s} \left(\frac{K_a A_s}{0.76 b} \right)^2 b \times 0.336$$

$$= \frac{0.58 K_a A_s}{b}$$

Now $l_a = d - y$

$$= d - \left(\frac{0.58 K_a A_s}{b} \right)$$

$$= d \left[1 - \left(\frac{0.58 K_a A_s}{bd} \right) \right] \qquad (2)$$

i.e.

$$= d \left[1 - \left(\frac{0.58}{bd} \right) \left(\frac{A_s f_y}{\gamma_{ms}} \right) \left(\frac{\gamma_m}{1.5 f_k} \right) \right]$$

The moment of compressive resistance:

$$M_{rc} = C_c l_a$$

$$= C_1 l_{a1} + C_2 l_{a2}$$

$$= \frac{1.5 f_k}{\gamma_m} (0.267 x_n b) \left[d - \left(\frac{0.267 x_n}{2} \right) \right] + \frac{2}{3}$$

$$\times \frac{1.5 f_k}{\gamma_m} (0.733 x_n b) [d - (0.275 + 0.267) x_n]$$

$$= \frac{1.5 f_k}{\gamma_m} \times b (0.267 x_n d - 0.036 x_n^2$$

$$+ 0.489 x_n d - 0.265 x_n^2)$$

$$= \frac{1.5 f_k}{\gamma_m} \times b (0.756 x_n d - 0.301 x_n^2)$$

but x_n is limited to $\leq d/2$ (i.e. NA cannot be greater than half depth)

Therefore maximum M_{rc} will be when $x_n = d/2$.

i.e.

$$M_{rc} = \frac{1.5 f_k}{\gamma_m} \times b \left[0.756 \times \frac{d}{2} \times d \right.$$

$$\left. - 0.301 \times \left(\frac{d}{2} \right)^2 \right]$$

$$= \frac{1.5 f_k}{\gamma_m} \times b \times 0.303 d^2$$

$$= \left(\frac{1.5 f_k}{\gamma_m} \times 0.303 \right) bd^2$$

Then $M_{rc} = R b d^2 \qquad (3)$

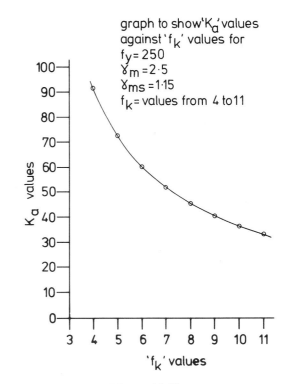

graph to show 'K_a' values against 'f_k' values for
$f_y = 250$
$\gamma_m = 2.5$
$\gamma_{ms} = 1.15$
f_k = values from 4 to 11

Figure 15.28

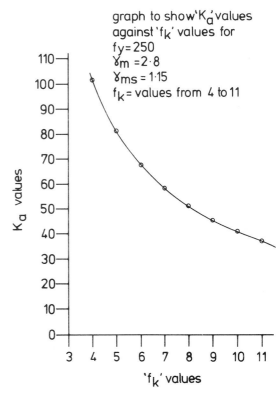

graph to show 'K_a' values against 'f_k' values for
$f_y = 250$
$\gamma_m = 2.8$
$\gamma_{ms} = 1.15$
f_k = values from 4 to 11

Figure 15.29

Table 15.6 Values of $R = \dfrac{1.5 f_k}{\gamma_m} \times 0.303$

γ_m	f_k	$R(\text{N/mm}^2)$	γ_m	f_k	$R(\text{N/mm}^2)$
2.5	4	0.70	3.1	4	0.59
	5	0.90		5	0.73
	6	1.10		6	0.88
	7	1.27		7	1.03
	8	1.45		8	1.17
	9	1.64		9	1.32
	10	1.82		10	1.47
	11	2.00		11	1.61
2.8	4	0.65	3.5	4	0.52
	5	0.81		5	0.65
	6	0.97		6	0.78
	7	1.14		7	0.91
	8	1.30		8	1.04
	9	1.51		9	1.17
	10	1.62		10	1.30
	11	1.79		11	1.43

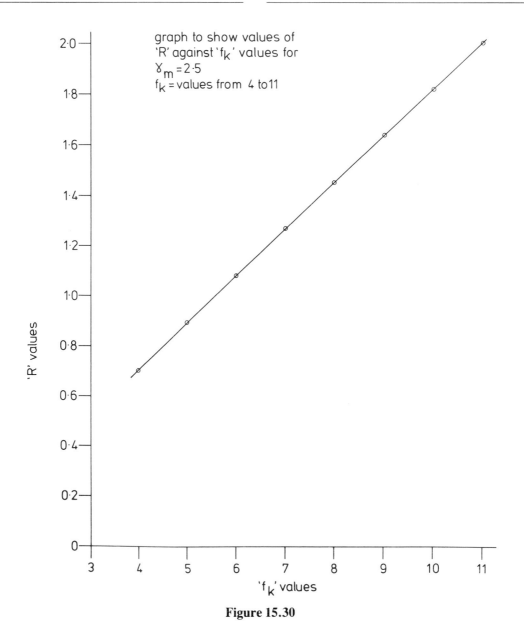

graph to show values of
'R' against 'f$_k$' values for
$\gamma_m = 2.5$
f_k = values from 4 to 11

Figure 15.30

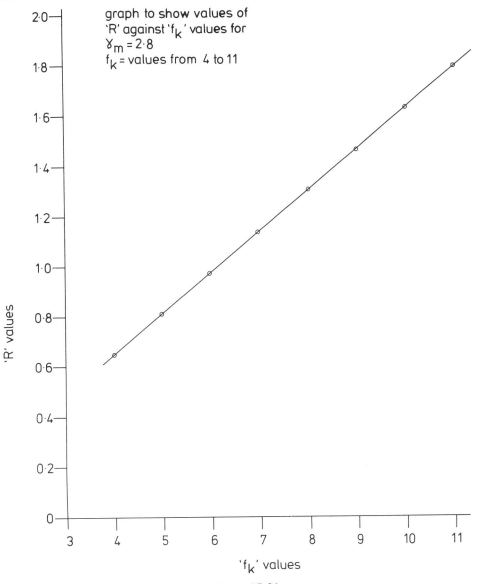

Figure 15.31

The moment of tensile resistance

$$M_{rs} = Tl_a$$

$$= \frac{A_s f_y}{\gamma_{ms}} l_a \qquad (4)$$

Where the lever arm l_a may be obtained from equation (2): l_a is limited to less than or equal to $0.95d$.

The applied design moment due to loading must thus, be less than or equal to the lesser of the compressive or tensile design moments of resistance.

$$M_A \leq \text{least of } M_{rc} \text{ and } M_{rs}$$

15.3.6 Design formula: shear stress
The behaviour of reinforced masonry members to flexural shear forces is more complex than when subjected to bending moments. The flexural shear resistance is influenced by many factors including the span to depth ratio of the member, the type of loading, etc. The design for shear resistance is thus based more on past experience than exact analysis and the theory for the estimation of shear stress is as follows (see Figure 15.34).

Consider the equilibrium of two planes AB and CD taken perpendicular to the horizontal axis of the member. The tensile force in the reinforcement = T_0 and T_1 respectively. The change in the tensile force over length, $T_0 - T_1$, is equal to the horizontal shearing force on any plane below the neutral axis.

The horizontal shearing force = $v_h \times \delta L \times b$ where v_h = shear stress, therefore

$$v_h \times \delta L \times b = T_0 - T_1$$

but $\qquad T_0 = M_0/l_a$, and $T_1 = M_1/l_a$,
where l_a = lever arm

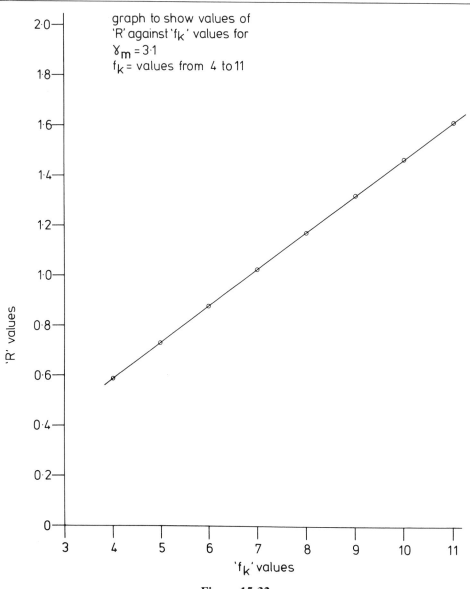

graph to show values of
'R' against 'f$_k$' values for
$\gamma_m = 3.1$
f$_k$ = values from 4 to 11

'R' values

'f$_k$' values

Figure 15.32

i.e.
$$v_h = \left(\frac{M_0 - M_1}{\delta L}\right) \times \frac{1}{l_a b}$$

but $(M_0 - M_1)/\delta L$ is the rate of change of bending moment which is equal to the shear force, V.

Therefore
$$v_h = \frac{V}{l_a b} \qquad (5)$$

This value of shear stress is then related to the design shear stress of the section, i.e. the characteristic value divided by the relevant partial factor of safety. If the actual shear stress exceeds the design shear stress, reinforcement in the form of links or bent bars must be provided to resist the excess stress. Even with additional shear reinforcement, the maximum value of v_h should be limited to $2.0/\gamma_{mv}$ N/mm^2.

Recent Codes of Practice have tended to modify the expression for shear by replacing the lever arm, l_a, by the effective depth and adjusting the characteristic shear stress accordingly. Thus expression (5) becomes:

$$v_h = \frac{V}{bd} \qquad (6)$$

The characteristic flexural shear strength, f_{vb}, appears to be independent of the amount of tensile reinforcement and very dependent on the ratio of the shear span to the effective depth. The characteristic value has been modified, as noted previously, and is termed f_{vb} to distinguish it from the value f_v from Chapter 6. The shear span, a_v, is defined as the distance from concentrated load to the support (Figure 15.35).

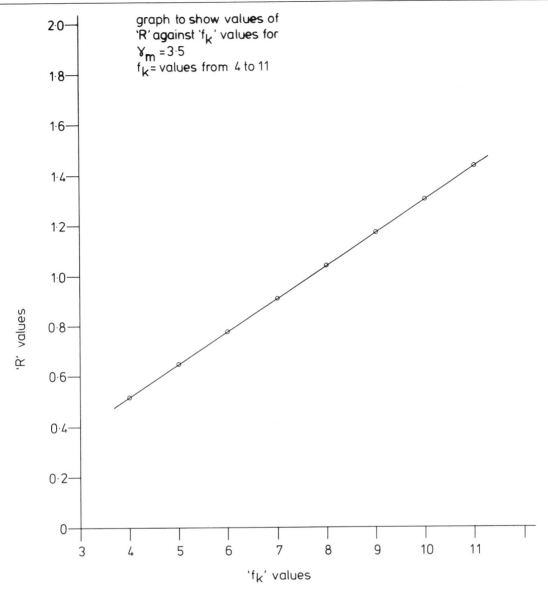

Figure 15.33

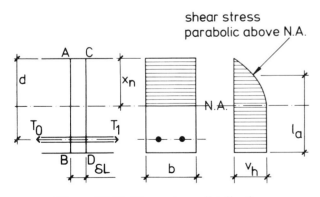

Figure 15.34 Shear stress distribution

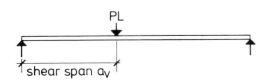

simply supported beam

Figure 15.35

For bricks having a characteristic compressive strength of not less than 7 N/mm² built with mortar designations (i), (ii) or (iii), f_{vb} may be taken as 0.35 N/mm².

However, where the point of application of loading is close to the support there may be some justification for considering a higher value. Further research work however, is required.

Thus if $v_h = V/bd \leqslant f_{vb}/\gamma_{mv}$ no shear reinforcement required.

If $v_h > f_{vb}/\gamma_{mv}$ but $\leqslant 2.0/\gamma_{mv}$ N/mm² shear reinforcement required. In no case must $v_h > 2.0/\gamma_{mv}$ N/mm².

15.3.7 Shear reinforcement

Shear reinforcement in the form of links is designed on the basis of the truss analogy (as explained in many textbooks on reinforced con-

364 Structural Masonry Designers' Manual

crete). The vertical tie force to be provided by the reinforcement must be greater than, or equal to, the shear force across the section less the resistance provided by the member itself. The spacing of links when required, should not exceed 0.75d to ensure that links will pass through any potential failure plane (see Figure 15.36).

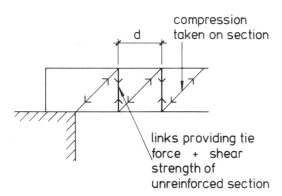

Figure 15.36

i.e. tensile force, $T \geq V$

The tensile force provided by the reinforcement

$$= A_{sv} \frac{f_{yv}}{\gamma_{ms}}$$

i.e. area of bars × design stress where f_{yv} is the characteristic strength of the shear reinforcement.

Therefore $\quad A_{sv} \frac{f_{yv}}{\gamma_{ms}} + \frac{f_{vb}}{\gamma_{mv}} \times bd \geq V$

but $V = v_h bd$

therefore $\quad A_{sv} \frac{f_{yv}}{\gamma_{ms}} + \frac{f_{vb}}{\gamma_{mv}} \times bd \geq v_h bd$ (7)

Links will be provided at centres less than the effective depth, therefore, several ties will become effective and the contribution of the reinforcement will be increased according to the ratio of the link spacing to the effective depth. Therefore, number of ties or links = d/S_v, where S_v = spacing of links.

Therefore equation (7) becomes:

$$\frac{d}{S_v} \times A_{sv} \times \frac{f_{yv}}{\gamma_{ms}} + \frac{f_{vb}}{\gamma_{mv}} \times bd \geq v_h bd$$

Hence $\quad \dfrac{A_{sv}}{S_v} \geq \dfrac{bd\left(v_h - \dfrac{f_{vb}}{\gamma_{mv}}\right)}{d\dfrac{f_{yv}}{\gamma_{ms}}}$

$$\geq \frac{b\left(v_h - \dfrac{f_{vb}}{\gamma_{mv}}\right)}{f_{yv}} \gamma_{ms} \quad (8)$$

15.3.8 Design formula: local bond

It is essential that the reinforcement provided in a flexural member is able to act with the mortar/grout/concrete. Referring to Figure 15.34, it has previously been established that the change in tensile force in the reinforcement between AB and CD is given by $T_0 - T_1$ and that this may be expressed as $(M_0 - M_1)/l_a$.

This change in tensile force is balanced by the shearing stresses in the surrounding mortar/grout. The forces must be transmitted between steel and mortar or grout via the adhesion between the two. That is to say, the strength of the adhesive bond between the two must be greater than or equal to the change in force. The strength of the adhesion may be defined as the product of the bond stress and the contact area between steel and mortar grout,

i.e. $(f_{bs}/\gamma_{mb}) \times \delta L \times \pi$ Dia

where

f_{bs} is the characteristic local bond strength

γ_{mb} is the partial factor of safety (see Table 15.2)

δL is the length of bar being considered (see Figure 15.34)

Dia   is the diameter of the bar.

π Dia may be replaced by Σu where Σu is the sum of the perimeters of the bars providing the tensile reinforcement, i.e. strength of adhesive bond is $(f_{bs}/\gamma_{mb}) \times \delta L \times \Sigma u$.

Thus the total design adhesion must be greater than or equal to the design force to be transferred.

i.e. $\quad \dfrac{f_{bs}}{\gamma_{mb}} \times \delta L \times \Sigma u \geq \dfrac{M_0 - M_1}{l_a}$

Rearranging this becomes:

$$\frac{f_{bs}}{\gamma_{mb}} \geq \frac{M_0 - M_1}{\delta L \times \Sigma u \times l_a}$$

But, as before, $(M_0 - M_1)/\delta L$ is equal to the rate of change in bending which in turn is equal to the shear force V, as explained earlier.

Thus $\quad \dfrac{f_{bs}}{\gamma_{mb}} \geq \dfrac{V}{\Sigma u \times l_a}$ (9)

As before, with the shear stress, this is generally expressed as:

$$\frac{f_{bs}}{\gamma_{mb}} \geqslant \frac{V}{\Sigma u \times d} \qquad (10)$$

The characteristic local bond stress, f_{bs}, between mortar or grout and steel may be taken as:

2.1 N/mm^2 for plain bars
2.8 N/mm^2 for deformed bars.

The characteristic anchorage bond strength, f_b, between mortar or grout and steel may be taken as:

1.5 N/mm^2 for plain bars
2.0 N/mm^2 for deformed bars.
The detailing of reinforcement should generally follow the guidance given in CP 110, but adjusted as necessary to take account of the differences in materials and construction.

15.3.9 Design for axial loading
The design loadings are determined in the normal manner. When determining the design strength, the effective height or length of an element should be as defined in BS 5628.

Effective thickness
The effective thickness should be as defined in BS 5628, clause 28.3.1. For grouted cavity walls the effective thickness should be taken as the actual thickness.

Design formula for axial loading
The design vertical axial strength of a reinforced element is simply the combination of the vertical axial strength of the masonry in compression, as determined for unreinforced masonry, plus the vertical axial strength in compression of the reinforcement. The capacity reduction factor is applied to account for slenderness effects as with plain masonry.

Design vertical axial strength = β (compressive strength of masonry + compressive strength of reinforcement), where β = capacity reduction factor based on the slenderness ratio as determined from BS 5628, clause 28.

$$\text{Compressive strength of masonry} = f_k \times \frac{A}{\gamma_m}$$

where
f_k = characteristic compressive strength of masonry
A = cross-sectional area
γ_m = partial safety factor for materials

$$\begin{matrix}\text{Compressive strength} \\ \text{of reinforcement}\end{matrix} = 0.8 f_y \times \frac{A_t}{\gamma_{ms}}$$

where
f_y = characteristic tensile strength of steel
A_t = total area of reinforcement
γ_{ms} = partial safety factor for steel reinforcement

Thus design vertical axial strength,

$$N_d = \beta \left(\frac{f_k A}{\gamma_m} + \frac{0.8 f_y A_t}{\gamma_{ms}} \right) \qquad (11)$$

See section 15.7, Example 4.

15.4 EXAMPLE 1: DESIGN OF REINFORCED BRICK BEAM

Reinforced brickwork beams are required to span 3.0 m, spaced at 3.0 m centres supporting a concrete floor as shown in Figure 15.37.

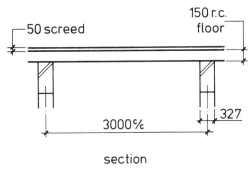

Figure 15.37

Masonry stresses
Characteristic compressive strength of masonry, f_k, using standard format bricks in designation (iii) (1:1:6 cement–lime–sand) mortar = 5.8 N/mm^2

Characteristic shear strength, f_{vb}, $= 0.35$ N/mm^2

Steel strength
Characteristic tensile strength using:

hot-rolled high yield $= 410$ N/mm^2
mild steel $\qquad = 250$ N/mm^2

Partial factors of safety
Loads: dead + imposed load, combination

Design dead load $\quad = 0.9G_k$ or $1.4G_k$
Design imposed load $= 1.6Q_k$

Materials:

For brickwork in compression and shear, assume special manufacturing and construction control, γ_m $= 2.5$

Bond between mortar and steel:

bricks, $\gamma_{mb} = 1.5$
steel, $\gamma_{ms} \ = 1.15$ $\Big\}$ ultimate limit state

Density of brickwork $= 20$ kN/m^3

Loading
Characteristic dead load:

150 mm rc floor	$24 \times 0.15 =$	3.6 kN/m^3
50 mm screed	$24 \times 0.05 =$	1.2 kN/m^3
Partition allowance	$=$	1.0 kN/m^3
	$=$	5.8 kN/m^3

Design dead load	$1.4 \times 5.8 =$	8.1 kN/m^2
Characteristic imposed load	$=$	1.5 kN/m^2
Design imposed load	$1.6 \times 1.5 =$	2.4 kN/m^2
Total design load	$8.1 + 2.4 =$	10.5 kN/m^2

Design UDL on beam	$10.5 \times 3 \ =$	31.5 kN/m
Design dead load due to self-weight of beam (estimated)	$=$	8.5 kN/m
	$=$	40.0 kN/m

Beam is simply supported

Design bending moment $\qquad = \dfrac{40 \times 3.02^2}{8} = 45$ kN·m

Maximum design shear force, V, $= \dfrac{40 \times 3}{2} \quad = 60$ kN

Effective depth, d, required to provide a moment of resistance equal to this value: $M_{rc} = Rbd^2$. Therefore

$$d = \sqrt{\left(\frac{M_{rc}}{Rb}\right)}$$

for $\gamma_m = 2.5$ and $f_k = 5.8$ N/mm^2. The value of $R = 1.0$ (Table 15.6), therefore

$$d = \sqrt{\frac{45 \times 10^6}{Rb}}$$

Take $b = 327$ mm which is the width of the supports.

$$d \text{ required} = \sqrt{\frac{45 \times 10^6}{1.0 \times 327}}$$

$$= 371 \text{ mm}$$

Depth required to resist shear without shear reinforcement

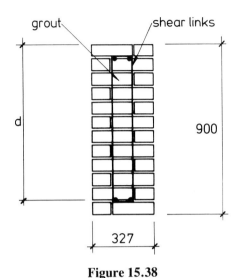

Figure 15.38

Design shear strength, $v_h = \dfrac{0.35}{2.5} = 0.14 \text{ N/mm}^2$

$$d \text{ required} = \frac{V}{v_h b}$$

$$= \frac{60 \times 10^3}{0.14 \times 327}$$

$$= 1311 \text{ mm}$$

Therefore provide shear reinforcement and try 327 wide × 900 deep overall beam, which is a multiple of the height of one brick course and gives an effective depth of 790 mm, i.e. allowing one course of brickwork plus 25 mm cover plus half bar diameter (see Figure 15.38).

Ultimate compressive moment of resistance of beam

$$\begin{aligned}
M_{rc} &= Rbd^2 \\
&= 1.0 \times 327 \times 790^2 \\
&= 240 \text{ kN·m}
\end{aligned}$$

(Design bending moment = 45 kN·m which is less than the moment of resistance.)

Moment of tensile resistance

$$M_{rs} = \frac{A_s f_y l_a}{\gamma_{ms}}$$

Therefore, area of reinforcement required:

$$A_s = \frac{M_A \gamma_{ms}}{f_y l_a}$$

Estimating the lever arm l_a as $0.90d$,

$$A_s = \frac{45 \times 10^6 \times 1.15}{410 \times 0.90 \times 790}$$

$$= 154 \text{ mm}^2$$

Use two Y12 bars (226 mm^2).

Check value of lever arm
The actual value of the lever arm is now calculated to ensure that the value assumed before in the calculation of A_s was reasonable.

$$l_a = d \left[1 - \left(\frac{A_s f_y}{\gamma_{ms}} \right) \left(\frac{\gamma_m}{1.15 f_k} \right) \left(\frac{0.58}{bd} \right) \right]$$

$$= d \left[1 - \frac{226 \times 410 \times 2.5 \times 0.58}{1.15 \times 1.5 \times 5.8 \times 327 \times 790} \right]$$

$$= 0.95d$$

This is greater than the assumed value and is therefore satisfactory.

Check percentage of reinforcement in grouted void

There is a maximum practical percentage of reinforcement which can be accommodated in the grout void which may be assumed to be 4% of the grouted area.

$$\text{Reinforcement percentage} = \frac{226}{100 \times 150}$$

$$= 1.5\% \text{ of the grouted area which is acceptable}$$

Shear links

Flexural shear strength, f_{vb}	$= 0.35 \text{ N/mm}^2$
Design shear load, V	$= 60 \text{ kN}$

Design shear stress, v_h $= \dfrac{V}{bd} = \dfrac{60 \times 10^3}{327 \times 790}$

$$= 0.23 \text{ N/mm}^2$$

Design shear strength $= \dfrac{f_{vb}}{\gamma_{mv}} = \dfrac{0.35}{2.5}$

$$= 0.14 \text{ N/mm}^2$$

Thus design shear strength is less than design shear stress. Links should normally be provided unless the design shear strength is greater than the applied design shear stress.

Positioning of links

According to the type of bonding, continuous vertical mortar joints occur at intervals. At these positions 8 mm diameter bars could be accommodated (if necessary corners of the adjacent bricks could be removed to provide more space). Figure 15.39 shows that continuous joints suitable for accommodating links occur at 225 mm centres. Links would not be required where the shear stress was less than the shear strength. Where links are provided but not calculated to provide a specific shear value, they are termed nominal links.

Design of links

$$\frac{A_{sv}}{S_v} \geqslant \frac{b \left(v_h - \dfrac{f_{vb}}{\gamma_{mv}} \right) \gamma_{ms}}{f_{yv}}$$

i.e.

$$\frac{A_{sv}}{S_v} \geqslant \frac{327(0.23 - 0.14)1.15}{250}$$

$$\frac{A_{sv}}{S_v} \geqslant 0.135$$

This is provided by R8 links at 225 centres.

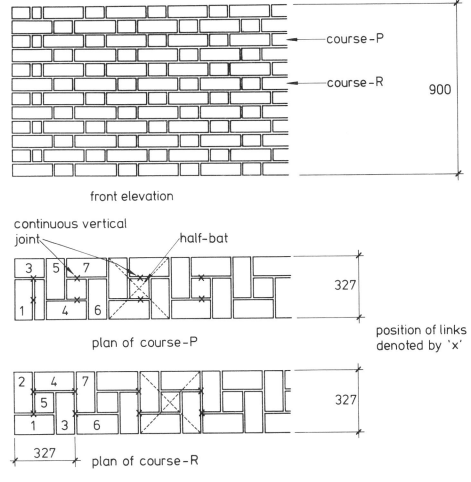

Figure 15.39 Position of shear links in double Flemish bond brickwork beam 1½ bricks thick

Local bond

Characteristic local bond strength, f_{bs}		$= 2.1 \text{ N/mm}^2$
Design local bond strength	$= \dfrac{2.1}{1.5}$	$= 1.4 \text{ N/mm}^2$

Local bond stress

$$= \frac{V}{\Sigma u \times d}$$

$$= \frac{60 \times 10^3}{2\pi \times 12 \times 790} = 1.0 \text{ N/mm}^2$$

Design local bond stress $<$ design local bond strength.

15.5 EXAMPLE 2: ALTERNATIVE DESIGN FOR REINFORCED BRICK BEAM

For use when the depth of the beam is to be restricted. Beam depth is arrived at as follows: consider what shear reinforcement it is possible to get in the beam; calculate minimum depth based on shear resistance thus obtained. This gives a smaller depth than the first method. The first part of the calculation is as before.

Depth required to resist shear

According to the type of bonding, continuous vertical mortar joints occur at intervals. At these positions 8 mm diameter links could be accommodated in the joint (if necessary the corners of the adjacent bricks could be removed to provide more space). Figure 15.39 shows that, for a beam built in double Flemish bond, suitable locations for links occur at 225 mm centres. Assuming then, that R8 shear links at 225 mm centres will be used, a suitable depth of beam can be calculated.

Minimum depth required, using R8 shear links at 225 centres:

$$\frac{A_{sv}}{S_v} \geq \frac{b\left(v_h - \dfrac{f_{vb}}{\gamma_{mv}}\right)\gamma_{ms}}{f_{yv}}$$

A_{sv} $= 2 \times 50.3$ (R8 link)
S_v $= 225$ mm
b $= 327$ mm
f_{vb} assume this approach yields a depth less than 750 mm; therefore the lower value of f_{vb} must be used, i.e. 0.35 N/mm^2
γ_{mv} $= 2.5; f_{vb}/\gamma_{mv} = 0.14$ N/mm^2
γ_{ms} $= 1.15$
f_{yv} $= 250$ N/mm^2 for mild steel.

Therefore

$$\frac{50.3 \times 2}{225} \geq \frac{327(v_h - 0.14)1.15}{250}$$

$$\frac{50.3 \times 2 \times 250}{225 \times 327 \times 1.15} + 0.14 \geq v_h$$

$$v_h \leq 0.44 \text{ N/mm}^2$$

$v_h = V/bd$ where $V = 60$ kN and $b = 327$ mm, therefore

$$d = \frac{60 \times 10^3}{327 \times 0.44}$$

$$= 417 \text{ mm}$$

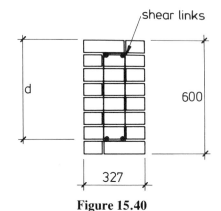

Figure 15.40

Minimum effective depth $= 417$ mm

$\dfrac{\text{bar diameter}}{2}$, say, $= 10$

cover $= 25$

1 course of brickwork below $= 75$
Overall depth of beam $= 527$ mm

The nearest size greater than this figure which is also a multiple of the height of one brick course is 600 mm (giving an effective depth of 490 mm) see Figure 15.40. Use 327 wide × 600 overall deep beam.

It can be seen that the designer has great scope to design a suitable beam for many different visual circumstances and within certain limits can vary the design to achieve a desired depth.

Ultimate compressive moment of resistance of beam

$$M_{rc} = Rbd^2$$
$$= 1.0 \times 327 \times 490^2$$
$$= 79 \text{ kN·m}$$

(Design bending moment $= 45$ kN·m which is less than the moment of resistance.)

Ultimate moment of tensile resistance:

$$M_{rs} = \frac{A_s f_y l_a}{\gamma_{ms}}$$

Area of reinforcement required:

$$A_s = \frac{M_A \gamma_{ms}}{f_y l_a}$$

Estimating l_a as $0.80d$

$$A_s = \frac{45 \times 10^6 \times 1.15}{410 \times 0.80 \times 490}$$

$$= 322 \text{ mm}^2$$

Use 2 Y16 bars (provides 402 mm^2).

Check value of lever arm

The actual value of the lever arm is now calculated to ensure that the value assumed above in the calculation of A_s was reasonable:

$$l_a = d \left[1 - \left(\frac{A_s f_y}{\gamma_{ms}} \right) \left(\frac{\gamma_m}{1.15 f_k} \right) \left(\frac{0.58}{bd} \right) \right]$$

$$= d \left[1 - \frac{402 \times 410 \times 2.5 \times 0.58}{1.15 \times 1.5 \times 5.8 \times 327 \times 490} \right]$$

$$= 0.85d$$

This is greater than the assumed value and is therefore satisfactory.

Local bond

Characteristic local bond strength, f_{bs}		$= 2.1 \text{ N/mm}^2$
Design local bond strength	$= \dfrac{2.1}{1.5}$	$= 1.4 \text{ N/mm}^2$
Design local bond stress	$= \dfrac{V}{\Sigma u \times d}$	
	$= \dfrac{60 \times 10^3}{2\pi \times 16 \times 490} = 1.22 \text{ N/mm}^2$	

Design local bond stress $<$ design local bond strength.

15.6 EXAMPLE 3: REINFORCED BRICK RETAINING WALL

The retaining wall shown in Figure 15.41 is to be constructed in two leaves of brickwork, the outer leaf being thicker than the inner, with a 100 mm cavity to take the reinforcement, which is to be grouted after construction. Design height $= 1.8$ m.

The loading on the wall will include a particularly severe surcharge as indicated in the design calculations.

In accordance with CP 121; Part 1: 1973, Table 4.1, special quality bricks in designation (i) mortar are to be used. To comply with the requirement of special quality, use bricks having a water absorption not greater than 7%, and crushing strength, say, 27.5 N/mm^2. The wall retains dry sand to a depth of 1.8 m. Assume manufacturing and construction control both special, therefore $\gamma_m = 2.5$.

Characteristic compressive strength of bricks of 27.5 N/mm^2 crushing strength in designation (i) mortar, f_k $= 9.2 \text{ N/mm}^2$

Characteristic flexural strength of bricks having a water absorption less than 7% in designation (i) mortar, f_{kx} $= 0.7 \text{ N/mm}^2$

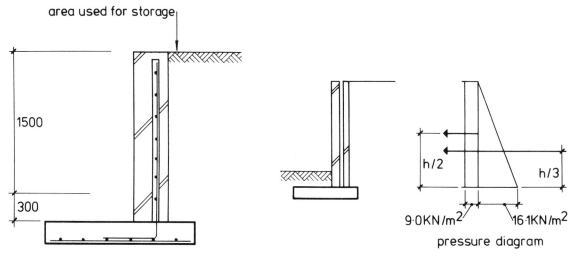

area used for storage

1500

300

Figure 15.41

h/2

h/3

9·0KN/m²   16·1KN/m²

pressure diagram

Figure 15.42

Design load

Retained material, fine dry sand:

Angle of internal friction $= 30°$

Density $= 17 \, kN/m^3$

$q_1 = k_1 \times \text{density} \times h$, where $k_1 = \dfrac{1 - \sin 30°}{1 + \sin 30°} = 0.33$

Allow 1 m surcharge to allow for surcharge loading at high level.

Characteristic horizontal pressure due to retained material behind wall:

at top $= 0.33 \times 17 \times 1 \quad = \quad 5.6 \, kN/m^2$
at base, $q_1 = 0.33 \times 17 \times 2.8 = 15.7 \, kN/m^2$

Retained material should be considered as imposed load.

Design load $= Q_k \quad \times \gamma_f$
Design load (top) $= \quad 5.6 \times 1.6 = \quad 9.0 \, kN/m^2$
          (base) $= 15.7 \times 1.6 = 25.1 \, kN/m^2$

See Figure 15.42.

Overturning moment

$\dfrac{9.0 \times 1.8^2}{2} \qquad = 14.6 \ kN{\cdot}m$

$16.1 \times \dfrac{1.8}{2} \times \dfrac{1.8}{3} = \dfrac{8.7}{}$
                       $= \overline{23.3} \ kN{\cdot}m$

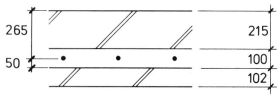

265

50

215

100

102

Figure 15.43

Try the following section

Note that the width of the grouted cavity shown here (100 mm) is the minimum practical width to ensure sufficient cover to the reinforcement. The use of galvanised reinforcement should also be considered depending on the required life of the wall.

Effective depth $d = 215 + \dfrac{100}{2} = 265$ mm (see Figure 15.43).

$$M_{rc} = Rbd^2$$

For $f_k = 9.2\ \text{N/mm}^2$ and $\gamma_m = 2.5$, $R = 1.65$ (see Figure 15.30), therefore

$$M_{rc} = 1.65 \times 1000 \times 265^2 = 116\ \text{kN·m}$$

$$A_s = \frac{M_A \gamma_{ms}}{f_y l_a}$$

$f_y = 410\ \text{N/mm}^2$
$l_a = 0.85d$

Therefore

$$A_s = \frac{23.3 \times 10^6 \times 1.15}{410 \times 0.85 \times 265}$$

$$= 290\ \text{mm}^2/\text{m}$$

Use Y12s at 300 centres.

This reinforcement is vertical reinforcement and some longitudinal bars should be provided to tie these together for ease of construction. Therefore, provide Y10 bars longitudinally at 600 mm vertical centres.

Check shear

$$V = 9.0 \times 1.8 + 16.1 \times \frac{1.8}{2}$$

$$= 16.2 + 14.5$$

$$= 30.7\ \text{kN}$$

$f_{vb} = 0.35\ \text{N/mm}^2$

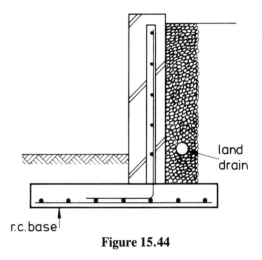

Design shear strength $= \dfrac{f_{vb}}{\gamma_{mv}} = \dfrac{0.35}{2.5} \qquad = 0.14\ \text{N/mm}^2$

Shear stress $\quad = v_h \quad = \dfrac{30.7 \times 10^3}{1000 \times 265} = 0.116\ \text{N/mm}^2$

r.c. base

Figure 15.44

$$< 0.14\ \text{N/mm}^2$$

Check local bond

Characteristic local bond strength, f_{bs} $\qquad = 2.8\ \text{N/mm}^2$

$\gamma_{mb} \qquad = 1.5$

Design local bond strength $\qquad = \dfrac{f_{bs}}{\gamma_{mb}} \qquad = 1.87\ \text{N/mm}^2$

Design local bond stress $\qquad = \dfrac{V}{\Sigma u \times d} = \dfrac{30.7 \times 10^3}{12\pi \times \dfrac{1000}{300} \times 265}$

$$= 0.92\ \text{N/mm}^2 < 1.87\ \text{N/mm}^2$$

The main vertical reinforcement must, of course, be bonded into the concrete base. A concrete base should now be designed, its size being such as to give a suitable factor of safety against overturning and sliding (see Figure 15.44). This is beyond the scope of this book.

15.7 EXAMPLE 4: COLUMN DESIGN

Calculate the area of reinforcement required for the column shown in Figure 15.45 subject to the loading shown. See section 15.3.9.

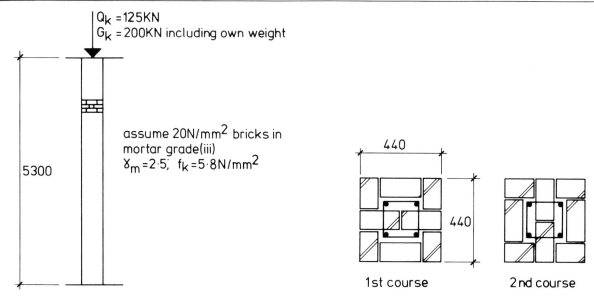

Figure 15.45

Slenderness ratio $\dfrac{h_{ef}}{t_{ef}} = \dfrac{5300}{440} = 12$, therefore $\beta = 0.93$.

Design vertical axial strength

$$N_d = \beta \left(f_k \frac{A}{\gamma_m} + 0.8f_y \frac{A_t}{\gamma_{ms}} \right)$$

Design vertical axial strength, $N_d \geqslant$ applied design load $(1.4G_k + 1.6Q_k) = (1.4 \times 200) + (1.6 \times 125)$ = 480 kN.

Therefore

$$480 \text{ kN} \leqslant \beta \left(f_k \frac{A}{\gamma_m} + 0.8f_y \frac{A_t}{\gamma_{ms}} \right)$$

If A_t required is the area of reinforcement required

$$A_t \text{ required} \geqslant \left(\frac{480}{\beta} \times 10^3 - f_k \frac{A}{\gamma_m} \right) \frac{\gamma_{ms}}{0.8f_y}.$$

Substituting for γ_m etc. and assuming mild steel reinforcement:

$$A_t \text{ required} \geqslant \left(\frac{480 \times 10^3}{0.93} - \frac{5.8 \times 440^2}{2.5} \right) \times \frac{1.15}{0.8 \times 250}$$

$$\geqslant 385 \text{ mm}^2$$

Adopt four No. R12 diameter bars which give an area of 452 mm².

Links should be provided in accordance with the recommendations of CP 110.

It should be noted that the provision of four R12 bars plus links increases the load carrying capacity of the column by only 17.5% over and above a solid masonry column.

15.7.1 Members subject to bending and axial loading

Sections may be analysed on a similar basis to that given above to determine the ultimate resistance to the applied moment and axial force. The analysis may be undertaken for the rectangular/semi-parabolic stress block similar to that shown previously in Figure 15.20 or to simplify the analysis an equivalent rectangular stress block may be assumed as shown in Figure 15.46. A member subject to both axial loading and a moment may fail either by crushing of the masonry – termed compressive failure – or by crushing of the masonry after the reinforcement has yielded – termed tensile failure. The transition condition between compressive failure and tensile failure is termed a balance failure

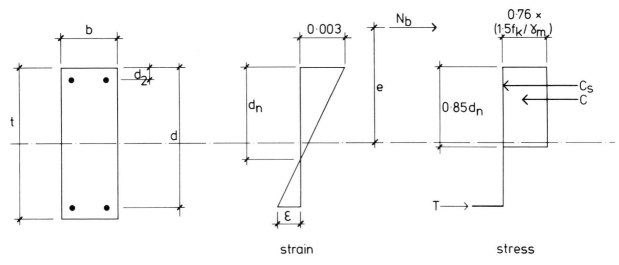

the simplified stress block is rectangular and has an assumed depth of $0.85d_n$

Figure 15.46 Symmetrically reinforced rectangular column subject to an eccentric load

and occurs when the masonry crushes just as the reinforcement is yielding. This is the condition shown in Figure 15.46 with symmetrical reinforcement.

Equating the axial forces:

$$N_b = C + C_s - T$$

$$= 0.76 \left(1.5 \frac{f_k}{\gamma_m} \right) b(0.85d_n) + A_{sc} \left(\frac{0.8f_y}{\gamma_{ms}} \right) - A_s \frac{f_y}{\gamma_{ms}}$$

From the strain diagram:

$$\frac{d_n}{0.003} = \frac{d - d_n}{\epsilon}$$

where ϵ is the strain in the reinforcement. This strain may be obtained as follows in terms of the stress and Young's modulus:

$$\frac{\text{stress}}{\text{strain}} = \text{Young's modulus}$$

$$\text{strain} = \frac{\text{stress}}{\text{Young's modulus}}$$

$$= \left(\frac{f_y}{\gamma_m} \right) \frac{1}{E}$$

With $f_y = 250 \text{ N/mm}^2$: mild steel
and $E = 21 \times 10^4 \text{ N/mm}^2$

Hence

$$\epsilon = \left(\frac{250}{\gamma_m} \right) \frac{1}{21 \times 10^4}$$

$$= \frac{1.19 \times 10^{-3}}{\gamma_m}$$

With $f_y = 410 \text{ N/mm}^2$: high yield steel

$$\epsilon = \left(\frac{410}{\gamma_m} \right) \frac{1}{21 \times 10^4}$$

$$= \frac{1.95 \times 10^{-3}}{\gamma_m}$$

Substituting these values in the equation of strains it can be shown that for:

mild steel $\qquad d_n = \left(\dfrac{\gamma_m}{0.4 + \gamma_m}\right) d = S_1 d$

high yield steel $\quad d_n = \left(\dfrac{\gamma_m}{0.65 + \gamma_m}\right) d = S_2 d$

Substituting this in the equation for axial forces:

$$N_b = 0.76b\left(\frac{1.5f_k}{\gamma_m}\right)\left(0.85 S_n d\right) + A_{sc}\left(\frac{0.8f_y}{\gamma_{ms}}\right) - A_s\left(\frac{f_y}{\gamma_{ms}}\right)$$

This section is symmetrically reinforced and thus $A_s = A_{sc} = A_t/2$ where A_t is the total area of reinforcement.

Therefore, substituting $A_t/bt = r$, i.e. $A_t = rbt$

$$\frac{N_d}{f_k bt} = \frac{0.97 S_n}{\gamma_m}\left(\frac{d}{t}\right) - \frac{0.1f_y}{\gamma_{ms}}\left(\frac{r}{f_k}\right) \tag{12}$$

This is in a non-dimensional form basically of the form

$$\frac{N_d}{f_k bt} = X_1\left(\frac{d}{t}\right) - X_2\left(\frac{r}{f_k}\right)$$

where X_1 and X_2 are constants.

Taking moments about the tensile steel

$$N_b\left[e + \left(\frac{d - d_2}{2}\right)\right] = 0.76b\left(\frac{1.5f_k}{\gamma_m}\right)(0.85 S_n d)\left(d - \frac{S_n d}{2}\right) + A_{sc}\left(\frac{0.8f_y}{\gamma_{ms}}\right)(d - d_2)$$

If $M_{rb} = N_b e$ is substituted in this equation, it becomes:

$$M_{rb} + N_b\left(\frac{d - d_2}{2}\right) = b\left(\frac{0.97f_k}{\gamma_m}\right)(S_n d)\left(d - \frac{S_n d}{2}\right) + A_{sc}\left(\frac{0.8f_y}{\gamma_{ms}}\right)(d - d_2)$$

Substituting for N_b as obtained previously:

$$M_{rb} + b\left(\frac{0.97f_k}{\gamma_m}\right)(S_n d)\left(\frac{d - d_2}{2}\right) = b\left(\frac{0.97f_k}{\gamma_m}\right)(S_n d)\left(d - \frac{S_n d}{2}\right) + A_{sc}\left(\frac{0.8f_y}{\gamma_{ms}}\right)(d - d_2)$$

i.e. $\qquad\qquad \dfrac{M_{rb}}{f_k bt^2} = \dfrac{0.97 S_n}{2\gamma_m}\left(\dfrac{d}{t^2}\right)(d - S_n d + d_2) + \dfrac{0.4 f_y r\,(d - d_2)}{\gamma_{ms}(f_k t)}$

But $d_2 = t - d$, therefore:

$$\frac{M_{rb}}{f_k bt^2} = \frac{0.49 S_n}{\gamma_m}\left(\frac{d}{t^2}\right)(t - S_n d) + \frac{0.4 f_y r\,(2d - t)}{\gamma_{ms}(f_k t)}$$

$$\frac{M_{rb}}{f_k bt^2} = \frac{0.49 Sn}{\gamma_m}\left(\frac{d}{t}\right) - \frac{0.49 S_n^2}{\gamma_m}\left(\frac{d}{t}\right)^2 + \frac{0.8 f_y r\,(d/t - 0.5)}{\gamma_{ms} f_k} \tag{13}$$

Thus for a given value of d/t, there is for each value of r/f_k a value of $N_b/f_k bt$ and a corresponding value of $M_{rb}/f_k bt^2$. These values are then plotted graphically for various values of r/f_k, d/t being kept at a particular value for each chart.

The maximum load a column can carry is when the load acts at the centroid and this may be obtained from equation (11):

$$N_0 = \beta\left(\frac{f_k A}{\gamma_m} + \frac{0.8 f_y A_t}{\gamma_{ms}}\right)$$

This may be re-written as follows:

$$\frac{N_0}{f_k\,bt} = \beta \left[\frac{f_k(bt)}{\gamma_m\,(f_k bt)} + \frac{0.8f_y\,(rbt)}{\gamma_{ms}\,(f_k bt)}\right]$$

i.e.

$$\frac{N_0}{f_k\,bt} = \frac{\beta}{\gamma_m} + \frac{0.8f_y\beta r}{\gamma_{ms}f_k} \tag{14}$$

If a graph of $N_d/f_k bt$ is plotted against $M_{rb}/f_k bt^2$ for a constant d/t, for each value of r/f_k a series of points may be obtained relating to the balance condition. M_{rb} and N_b, another set of points, may be obtained for the load applied at the centroid (see Figure 15.47).

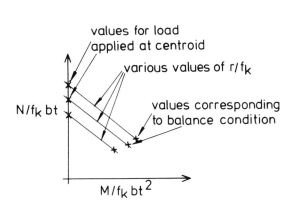

Figure 15.47

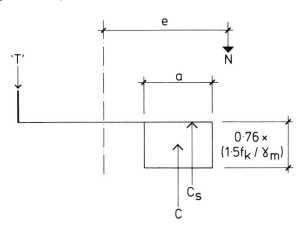

Figure 15.48 The simplified stress block for a column failing in tension

If a linear relation between the points relating to the balance and centroid load is assumed, a series of curves may be drawn for compressive failure. When considering tensile failure, the conditions are as shown in Figure 15.48. In this case, it is not necessary to consider the strains as the stress in the steel is that at the yield point.

Equating forces as before:

$$N_d = C + C_s' - T$$

$$= 0.76\left(\frac{1.5f_k}{\gamma_m}\right)ab + A_{sc}\left(\frac{0.8f_y}{\gamma_{ms}}\right) - A_s\left(\frac{f_y}{\gamma_{ms}}\right)$$

Which may be rewritten:

$$\frac{N_d}{f_k\,bt} = \frac{1.14}{\gamma_m}\left(\frac{a}{t}\right) - \frac{0.1f_y}{\gamma_{ms}}\left(\frac{r}{f_k}\right) \tag{15}$$

where a is the depth of the stress block.

Taking moments about the tensile steel as before:

$$N_d\left(e + \frac{d - d_2}{2}\right) = 0.76\left(\frac{1.5f_k}{\gamma_m}\right)(ab)\left(d - \frac{a}{2}\right) + A_{sc}\left(\frac{0.8f_y}{\gamma_{ms}}\right)(d - d_2)$$

i.e.

$$M_{rb} + N_d\left(\frac{d - d_2}{2}\right) = \left(\frac{1.14f_k}{\gamma_m}\right)(ab)\left(d - \frac{a}{2}\right) + A_{sc}\left(\frac{0.8f_y}{\gamma_{ms}}\right)(d - d_2)$$

or

$$\frac{M_{rb}}{f_k\,bt^2} + \frac{N_d}{f_k\,bt}\left(\frac{d}{t} - 0.5\right) = \frac{1.14a}{\gamma_m t}\left(\frac{d}{t} - \frac{a}{2t}\right) + \left(\frac{0.8f_y}{\gamma_{ms}}\right)\left(\frac{r}{f_k}\right)\left(\frac{d}{t} - 0.5\right) \tag{16}$$

For design purposes a in equation (15) would be substituted into equation (16) and N_d could then be found. Alternatively, a design chart may be constructed. For a given value of r/f_k and an assumed value of $N_d/f_k bt$, the value of a/t in equation (15) may be found and this then substituted in equation (16). The value of $M_{rb}/f_k bt^2$ is thus obtained and may be plotted against $N_d/f_k bt$ to obtain a series of curves to complete those in Figure 15.47; a typical graph is given in Figure 15.49.

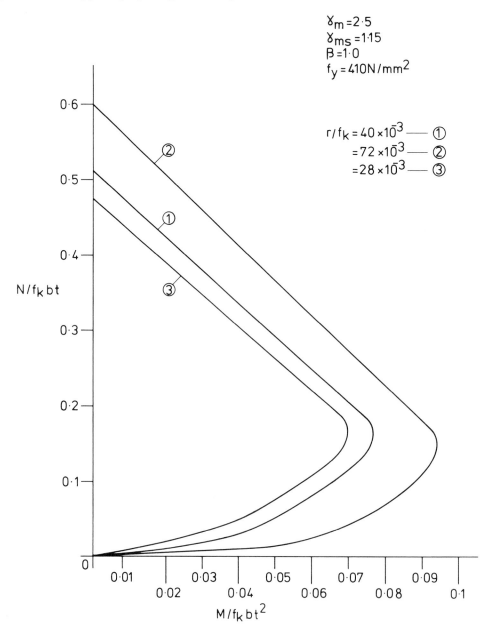

Figure 15.49 Column design chart

The basic procedure for column design is thus to calculate the applied design loads, determine whether the column will fail in tension or compression, and determine the appropriate value of 'r' from the relevant equation. This is an involved process and may be more easily accomplished by the use of design charts as exemplified in Figure 15.49. Generally, it can be seen whether the bending moment or axial loading is dominant, and hence, whether failure is more likely to be tensile than compressive. To obtain approximate sizes where tensile failure is likely, the section may be designed preliminarily as a purely flexural member, allowing for a slight increase in the area of reinforcement to accommodate the axial load. Similarly, if compressive failure is likely, the section may be designed preliminarily as a member subject to direct compression, only, with no bending. Having thus obtained approximate sizes, the design may be checked against the full equations.

Where the slenderness ratio of the member exceeds 12, an additional applied moment must be considered which is induced by the lateral deflection.

$$\text{Additional applied moment} = \frac{N h_{cf}^2}{2000b} \tag{17}$$

This value is based on research and depends on various factors including the stiffness of the member, etc, where:

N = applied design vertical load
h_{ef} = effective height or length about the appropriate axis
b = the dimension of the column in the plane of bending.

As an alternative to the full column design procedure, short columns, that is – on the basis of reinforced concrete columns – those with a slenderness ratio of 12 or less, may be designed as members subject only to bending under an additional moment of:

$$\left(\frac{N}{2}\right) b \tag{18}$$

when the resultant eccentricity, e, is not less than $t/2 - d_2$, see 15.8, Example 5.

Slender columns may be assumed to be satisfactory if the following expression is satisfied:

$$\frac{M_A}{M_d} + \frac{N}{N_d} \leq 1 \tag{19}$$

where
M_A = applied moment due to design loads
M_d = design moment of resistance, see equations (3) and (4)
N = applied vertical load due to design loads
N_d = design vertical axial strength (equation (11))

See 15.9, Example 6.

15.8 EXAMPLE 5: DESIGN OF A REINFORCED BRICK COLUMN (1)

Calculate the size and area of reinforcement required for the column shown in Figure 15.50.

By inspection, it can be seen that the bending moment predominates and thus, tensile failure will govern the design. Therefore, size member on the basis of a purely flexural member.

The moment of resistance of a rectangular section in compression is given by the formula given in equation (3): $M_{rc} = Rbd^2$. Thus assuming values of f_k and γ_m, a value for R may be obtained from Table 15.6 and assuming a column width, the effective depth required can be obtained as follows:

$$d \text{ required} = \sqrt{\left(\frac{M_{rc}}{Rb}\right)}$$

but $M_{rc} = 25$ kN·m, assume $b = 440$ mm to suit brick dimensions and $f_k = 5.8$ N/mm^2, $\gamma_m = 2.5$ therefore $R = 1.0$ from Figure 15.30.

$$d \text{ required} = \sqrt{\left(\frac{25 \times 10^6}{1.0 \times 400}\right)}$$
$$= 238 \text{ mm}$$

With 440 deep column section, as shown in Figure 15.50:

$$\text{Effective depth} = \text{actual depth} - \text{brickwork} - \text{cover} - \left(\frac{\text{bar diameter}}{2}\right)$$

$$d = 304 \text{ mm}$$

Adopt 440 square section (see Figure 15.51).

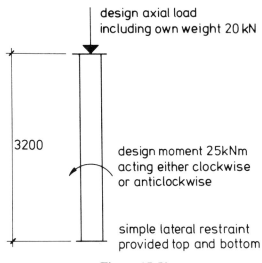

Figure 15.50

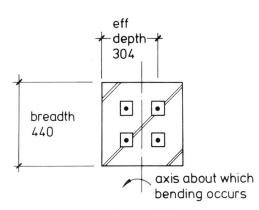

Figure 15.51

Having thus established the column section the slenderness ratio may be determined.

$$\text{Slenderness ratio} = \frac{h_{ef}}{t_{ef}} = \frac{3200}{440} = 7.3$$

i.e. < 12

Thus the member may be designed as a purely flexural member to resist a design bending moment equal to: $M + Nb/2$, i.e. $25 + (20 \times 0.44/2) = 29.4 \text{ kN·m}$

Check compression:

$$M_{rc} = Rbd^2$$

$$= \frac{1.0 \times 440 \times 304^2}{10^6}$$

$$= 40.6 \text{ kN·m}$$

Area of tensile reinforcement required obtained from equation (4):

$$M_A = \left(\frac{A_s f_y}{\gamma_{ms}}\right) l_a$$

$$A_s = \frac{M_A \gamma_{ms}}{f_y l_a}$$

The lever arm, l_a, may be obtained from equation (2):

$$l_a = d\left[1 - \left(A_s \frac{0.58 K_a}{bd}\right)\right]$$

To obtain an approximate area of reinforcement assume $l_a = 0.8d$ and then check back in equation (2) using high yield steel

$$f_y = 410 \text{ N/mm}^2$$
$$\gamma_{ms} = 1.15$$

Therefore

$$A_s = \frac{29.4 \times 10^6 \times 1.15}{410 \times 0.8 \times 304}$$

$$= 339 \text{ mm}^2$$

Try two Y16 bars.

$$A_s = 402 \text{ mm}^2$$

$$\text{Actual } l_a = d \left(1 - \frac{402 \times 100 \times 0.58}{440 \times 304} \right)$$

From Figure 15.26, $K_a = 100$

i.e. $\qquad\qquad\qquad\qquad\qquad l_a = 0.82d$

This value is approximately as assumed, and thus the area of reinforcement will be used in each face as the moment is reversible. Adopt 440 square section, four Y16 bars plus links in accordance with CP 110 recommendations.

15.9 EXAMPLE 6: DESIGN OF A REINFORCED BRICK COLUMN (2)

A reinforced brick column, 552.5×327.5 supporting vertical loading and moment about the major axis only, due to an unsymmetrical arrangement of beams framing into the column. The design loadings have already been calculated and are as shown in Figure 15.52.

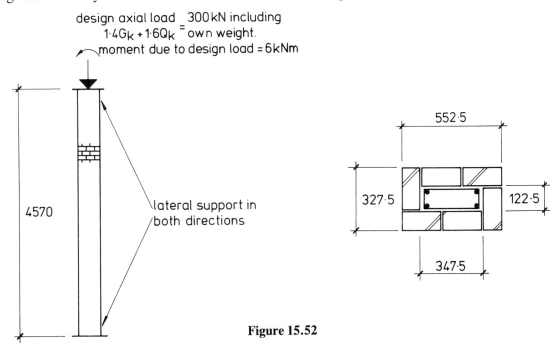

Figure 15.52

Slenderness ratio

$$\text{Major axis } = \frac{4570}{552.5} = 8.3$$

$$\text{Minor axis } = \frac{4570}{327.5} = 13.9$$

The column is thus a slender column, having a slenderness ratio in excess of 12, and must be designed for an additional moment induced by the vertical load due to lateral deflection. The column is therefore, designed so that the sum of the ratios of applied to resistance moments and applied vertical loading to vertical strength is less than or equal to unity.

Applied design moment

The applied design moment is equal to the moment due to the design loads plus the additional moment due to lateral deflection. The additional moment may be taken as: $Nh_{ef}^2/2000b$, where b is the dimension of the column in plane of bending (see equation (17)).

But

$\quad N \quad = 300 \text{ kN}$
$\quad h_{ef} = 4570$
$\quad b \quad = 552.5$

Therefore, additional moment $= \dfrac{300 \times 4.57^2}{2000 \times 0.5525}$

$\qquad\qquad\qquad\qquad\qquad\quad = 5.7 \text{ kN·m}$

Applied design moment $\qquad = 6 + 5.7$

$\qquad\qquad\qquad\qquad\qquad\quad = 11.7 \text{ kN·m}$

Design moment of resistance

Assume $f_k \quad = 5.8 \text{ N/mm}^2$ (20 N/mm^2 bricks in mortar designation (iii))

$\qquad \gamma_m \quad = 2.5$

$\qquad \gamma_{ms} \ = 1.15$

Try four Y12 bars $f_y = 410 \text{ N/mm}^2$

Effective depth $= 552.5 - 102.5 - 30 - 6 = 414 \text{ mm}$

Moment of resistance in compression:

$$M_{rc} = \frac{1.5 \times 5.8 \times 0.303 \times 327.5 \times 414^2}{2.5 \times 10^6}$$

$$= 59 \text{ kN·m}$$

Moment of resistance in tension:

$$l_a = d \left(1 - \frac{226 \times 410 \times 2.5 \times 0.58}{1.15 \times 1.5 \times 5.8 \times 327.5 \times 414} \right)$$

$$= 0.9d$$

$$M_{rs} = \frac{226 \times 410 \times 0.9 \times 414}{1.15 \times 10^6}$$

$$= 30 \text{ kN·m which becomes the design moment of resistance}$$

Ratio of applied moment to the design moment of resistance:

$$\frac{M_A}{M_d} = \frac{11.7}{30} = 0.39$$

Applied design vertical loading: $Q = 300 \text{ kN}$

Design vertical strength:

SR = 14, therefore, $\beta = 0.89$.

Thus $\qquad\qquad\qquad\qquad N = 0.89 \left(\dfrac{5.8 \times 327 \times 552.5}{2.5 \times 10^3} + \dfrac{0.8 \times 410 \times 452}{1.15 \times 10^3} \right)$

$$= 488 \text{ kN}$$

Ratio of applied load to strength

$$\frac{N}{N_d} = \frac{300}{488} = 0.61$$

Thus $\qquad\qquad\qquad\qquad \dfrac{M_A}{M_d} + \dfrac{N}{N_d} = 0.39 + 0.61 = 1$

which is satisfactory.

15.10 DESIGN OF POST-TENSIONED BRICKWORK

15.10.1 General
In order to introduce the idea of post-tensioned masonry in a little more detail, and before considering specific aspects of design and design examples, it may be of value to first briefly consider its use in a particular application – post-tensioned fin walls.

As discussed earlier, because of masonry's low resistance to tensile stresses, it is often advantageous when high bending moments and/or uplift forces have to be resisted in locations of low gravitational loads to incorporate a material which is good in resisting tensile stresses. This can either be in the form of reinforcement to resist the tensile stress, or post-tensioned rods to create additional compressive stress in the masonry to cancel out the excessive flexural tensile stress. In many cases, it is more economical to post-tension than to reinforce and, in the case of fin walls, and other applications, experience to date favours the use of post-tensioning.

Simplified design philosophy
Consider a section of a brick fin wall of cross-sectional area, A, and section modulus, Z, resisting a compressive load, P, and a bending moment, M (see Figure 15.53).

For the normal stress condition, the stress in the extreme fibres $= P/A \pm M/Z$ for an uncracked section. For a normal section under combined bending and compression, in a typical fin wall situation, it would be the allowable flexural tensile stress which would limit the resistance moment. Therefore, it is $P/A - M/Z$ which is critical, and not $P/A + M/Z$. It follows that increasing P/A improves the critical condition by reducing the flexural tensile stress when the bending moment, M, is applied. Thus where

possible, it is logical to make use of the remaining allowable compressive stress by increasing P by post-tensioning with an applied force, T, so that when the bending moment is applied the resulting flexural tensile stress is below the allowable limit. A check must also be made for the combination of $P/A + M/Z$, plus the post-tensioning stress in the brickwork to check the maximum combined compressive stress against the allowable value.

Location of post-tensioning rods
The post-tensioning rod should be located within the critical section so that M is also reduced by the eccentricity of the rod (see Figure 15.54), but regard must be given to the effect of this eccentricity at the other critical section for the reverse applied bending moment induced by the eccentric post-tensioning force.

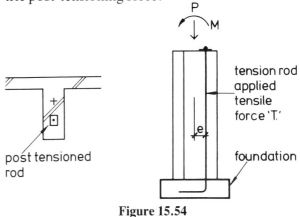

Figure 15.54

The resulting stress at the tensile face of the section shown in Figure 15.54 would now be

$$\frac{P}{A} - \frac{M}{Z} + \frac{Te}{Z}$$

giving a much reduced flexural stress. The effect of the post-tensioning is, therefore, to reduce the section size required to resist the combined bending and direct load condition.

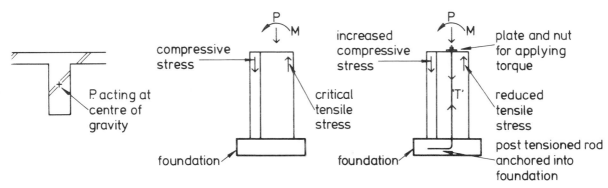

Figure 15.53

It should be noted that, depending on the location of the rods, there may be a change in the strain due to bending, which may affect the post-tensioning force. As with prestressed concrete design to CP 110, pending further research, this may be ignored.

Method and sequence of construction

The foundations are cast with the post-tensioning rods anchored in. The masonry wall is then constructed, leaving suitable voids around the rods. At the top of the wall, the threaded upper ends of the rods are allowed to project. When the masonry has cured, a plate is placed over the upper end of each rod on the top of the wall. Nuts are screwed onto the threads and tightened down to a predetermined torque using a calibrated torque spanner. The threads must be kept clear of foreign matter and be lubricated to give a calculable force in the rod when the torque is applied.

On completion of the post tensioning, voids can, if required, be filled with a slurry grout, in which case they should be constructed with vent holes up the height of the wall. The amount of grout should be predetermined by calculation so as to provide a check that the voids have been completely filled. However, in most cases, the voids are left ungrouted – the rods being protected above the foundation anchorage by a tape and paint product such as 'Denso' (see Figure 15.55).

The method of corrosion protection should be chosen to suit the exposure conditions and the stress condition in the rods. It should be noted, however, that where the anchorage of a rod relies upon bond stress, any protective coating that is likely to slip under load must not be used within the designed length of anchorage (see Figure 15.56).

At the location of beam bearings, it is often convenient to use the post-tensioning rods as an anchorage for the roof beams (see Figure 15.57).

High walls in post-tensioned construction

In high walls, post-tensioning rods should be

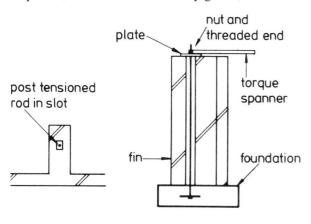

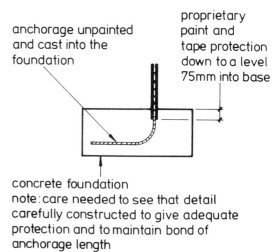

Figure 15.55

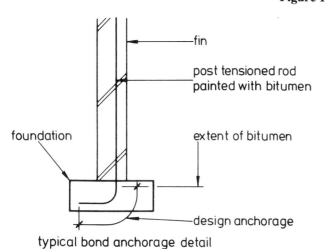

typical bond anchorage detail

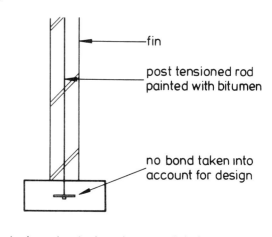

typical mechanical anchorage detail

Figure 15.56

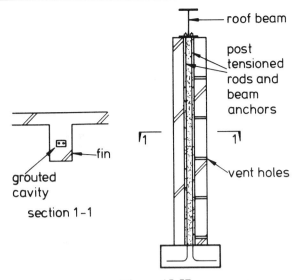

Figure 15.57

jointed in order to restrict the projecting length to a manageable amount. The length should be determined from consideration of the bar diameter, the possibility of temporary support from scaffolding, etc., and the likely wind exposure during construction. The joints must be of sufficient strength to resist the forces involved, should be simple to construct, and adequately protected against corrosion. Bottle connectors or welded joints are suitable.

On very high or multi-storey buildings, where temporary stability of the post-tensioned wall becomes more critical if the total post-tensioning is left until completion of the brickwork, a phased sequence of post-tensioning can be operated as the work progresses. For example, at pre-set levels within the height of the wall, the rods can be post-tensioned as the brickwork reaches each level, leaving a projection of

threaded rod onto which an extension sleeve can be fixed to continue the rod to the next stage. Alternatively, a number of rods can be used, relating the necessary height of each rod to the design condition and curtailing it at a suitable level. This will result in rods which curtail at various heights up the wall. Post-tensioning of the rods as the wall reaches the curtailment level will give added stability to the wall as the work progresses to the next height. The addition of loading on a previously post-tensioned section may reduce the post-tensioning in that section by changing the strain in the rod. This effect should be considered when using this technique (see Figure 15.58).

15.10.2 Post-tensioned masonry: design for flexure

In reinforced masonry in bending, the lack of tensile strength is overcome by providing reinforcement to resist the tensile stresses. Another solution to this problem is to provide sufficient precompression within the member so that the net tensile stresses, when flexural stresses are combined with this compressive stress, are very small or zero. For masonry it is considered that, under the application of working loads, the combined flexural tensile stress should be limited to zero.

Post-tensioning, as applied to masonry, is basically the addition of compressive axial loading, which may or may not be applied eccentrically, to reduce the flexural tensile stresses set up in the member due to bending. By applying the axial loading eccentrically, for example in a member subject to lateral loading from one direction

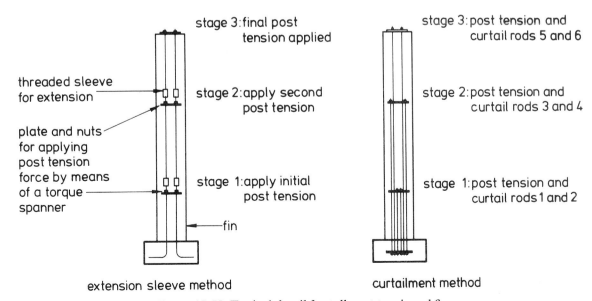

Figure 15.58 Typical detail for tall post-tensioned fins

only, flexural tensile and compressive stresses may be induced in reverse to those set up under the application of the working lateral load. The combination of axial loading with the applied lateral loading produces an increase in the total compressive stress which the masonry is more able to accommodate and a reduction in the tensile stresses (see Figures 15.59 and 15.60).

The applied force and its eccentricity may be varied to produce design stresses within the allowable limits of the tensile and compressive strengths of the masonry.

The simple basic idea of post-tensioning is made complicated, as far as the design is concerned, by the natural properties of the materials involved which may cause changes in the applied force after its application. The initial applied force may be reduced by losses or increased by gains which

may occur due to any or all of the following phenomena:

(a) moisture movements in the masonry,
(b) relaxation of the post-tensioning steel,
(c) elastic deformation of the masonry,
(d) friction,
(e) natural growth of clay brickwork,
(f) creep of brickwork and blockwork,
(g) thermal movement.

These are the main factors influencing losses or gains in the post-tensioning force as applied, in its simplest form, to masonry.

The stresses must therefore, be checked for the conditions appertaining when the post-tensioning force is initially applied and then again after all or most of the losses have occurred. If gains are expected, which is generally less likely, allowance for this should be made in assessing the

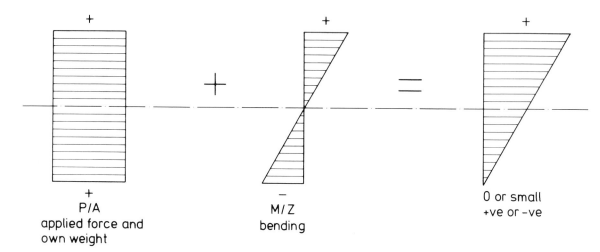

$$\frac{+}{P/A}$$
applied force and own weight

$$\frac{-}{M/Z}$$
bending

0 or small
+ve or -ve

Figure 15.59

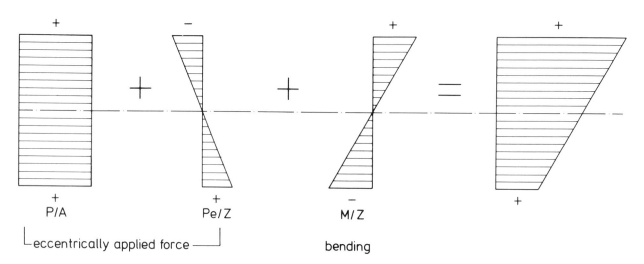

$$\frac{+}{P/A}$$

$$\frac{+}{Pe/Z}$$

eccentrically applied force

$$\frac{-}{M/Z}$$
bending

$$+$$

P = applied post tension force and own weight plus vertical load.
A = area of section
M = applied bending moment
Z = section modulus

Figure 15.60

stresses applicable at any particular design stage and the strength assessments, which are described in the following pages, must be adjusted accordingly.

As the flexural tensile stresses are eliminated (or reduced to nominal acceptable amounts) owing to the application of the load to be supported, an elastic analysis is appropriate. The design is therefore, based on design flexural compressive stresses related to design strengths using the appropriate partial safety factors and capacity reduction factors where applicable.

15.10.3 Design strengths
The basic characteristic compressive strength of masonry, f_k, may be increased to allow for:

(a) the permanence of the strength requirement,
(b) flexural compressive stresses as opposed to axial compressive stresses.

The partial loss of post-tensioning force, which commences immediately after its application, may be taken to constitute a temporary force prior to the losses and a permanent force after losses. The magnitude of the total losses allowed for in the design is approximately 20% which will be discussed later in this chapter and it is therefore considered reasonable to increase the characteristic compressive strength of the masonry by 20% to accommodate the short term duration of this loading condition.

The percentage increase on the characteristic compressive strength, to take account of its application as a flexural compressive strength, should vary depending on the ratio of axial to flexural compressive stress being applied. If the applied stress is wholly axial f_k should not be increased at all. If the applied stress is wholly flexural, it is considered reasonable to increase f_k to $1.5f_k$ as is used in reinforced masonry design. A sliding scale of increase factors should be devised from research on the basis of the principles described and possibly incorporating a unity factor relating actual and allowable stresses for both axial and flexural loading. In the absence of such a scale it is proposed that a global figure of $1.25f_k$ should be adopted for general situations, with adjustments made to this figure, at the discretion of the designer, when extreme cases are considered.

Care should be exercised in assessing the stress conditions at all critical stages of loading for each different type of element to be analysed. For example, consider the following two totally differing elements described briefly as:

Case (1) Masonry beam, spanning horizontally and supporting predominantly dead loads.
Case (2) Masonry wall panel spanning vertically supporting equal wind forces in either direction.

Case (1) element would most likely be post-tensioned eccentrically to induce the reverse stress diagram to that of the working loads. Hence, the situation prior to losses would be that of a temporary strength requirement limited by the flexural compressive stress on one side of the stress block and zero flexural tensile stress on the other side. It is considered reasonable to limit the design flexural compressive strength in this situation to 1.20 ($1.25f_{ki}/\gamma_m$) in which both increase allowances have been exploited and f_{ki} is the characteristic compressive strength of the masonry at the age at which the post-tensioning is applied.

For the same element, the next loading stage to be analysed is after all losses are assumed to have taken place and the applied load added (including removal of props from beneath the beam). This is now the permanent working state for this element and the reversed stress diagram should now limit the design flexural compressive stress to $1.25f_k/\gamma_m$ in which only the flexural aspect of the increase is exploited and the masonry is assumed to have gained its full characteristic compressive strength value of f_k. The stress diagrams for these loading situations on case (1) element are shown in Figure 15.61.

The application of the capacity reduction factor, β, to these strength limitations is applicable in certain circumstances and is discussed in the worked examples which follow.

Consider now case (2) element in which the post-tensioning force would be most effective if applied concentrically to have an equal effect on the flexural tensile stresses from each direction of wind loading. The situation prior to losses would be that of a temporary strength requirement for axial compressive stresses only and shows as a rectangular stress block in which the design compressive strength is limited to $1.2\beta f_{ki}/\gamma_m$. After losses and under the application of the wind loading the situation could be considered as that of a more permanent strength requirement

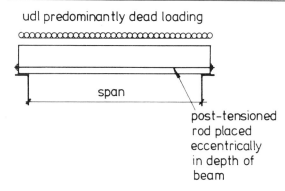

Elevation of post-tensioned masonry beam

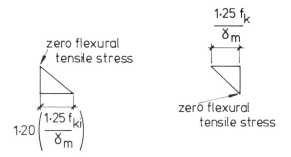

(a) initial stresses before losses and before application of loading

(b) permanent stresses after losses and after application of permanent load

Figure 15.61

limited by the flexural compressive stress on one side of the stress block and zero flexural tensile stress on the other side. In this instance for solid walls the design flexural compressive stress may be limited to $1.25f_k/\gamma_m$, and the capacity reduction factor, β, is applied inversely to the actual stress resulting from the axial load only as it is considered inappropriate to apply it to the flexural compressive stress in solid walls.

Hence, the design equation is written as:

$$f_{ubc} + \frac{f_{uac}}{\beta} \leqslant \frac{1.25f_k}{\gamma_m}$$

where

f_{ubc} = design flexural compressive stress

f_{uac} = design axial compressive stress

β = capacity reduction factor

f_k = characteristic compressive strength

γ_m = partial safety factor on materials.

For geometric profiles, such as the fin or diaphragm wall, where under certain conditions the whole of the stressed area can buckle under the combined flexural and axial compressive loading the factor should be applicable in the normal way and the design equation in this situation may be written as:

$$\text{actual combined stress} \leqslant \frac{1.25\beta f_k}{\gamma_m}$$

Whilst the application of the wind loading is described as a more permanent strength requirement, it could equally be argued that wind loading is of a particularly temporary nature and thus, a temporary strength allowance could be considered. The compressive strength requirement may, however, be dominated by the post-tensioning force which is now a more permanent load and it is therefore, recommended that the temporary nature of the wind load is ignored so far as stress increases are concerned. In addition, its temporary nature is catered for in the allocation of partial safety factors (γ_f) for loads for differing load combinations. With this element, a third and more critical loading situation requires analysis, this being the more common permanent working state of the wall, in which the post-tensioning has been effected, losses have occurred but no wind loading is applied. This loading situation is similar to that which occurred before losses but without the increased stress benefit of the anticipated losses. Hence, the design compressive stress is limited to $\beta f_k/\gamma_m$ and shows as a rectangular stress block. The stress diagrams for each of these three loading situations are shown in Figure 15.62.

15.10.4 Steel stresses
The maximum initial stress in the post-tensioning rods should be limited to 70% of the normally applicable design strength of f_y/γ_{ms} in which γ_{ms} has been given in Table 15.2 as 1.15. The purpose of this additional partial safety factor is to limit the stress relaxation in the steel which has a significant effect on the loss of post-tensioning force to be expected.

15.10.5 Asymmetrical sections
Particular care is required when analysing asymmetrical sections such as the fin wall as the section has two Z (section modulus) values. The appropriate Z value must be carefully selected for each level considered, for both directions of lateral loading and for both flexural tensile and flexural compressive stresses.

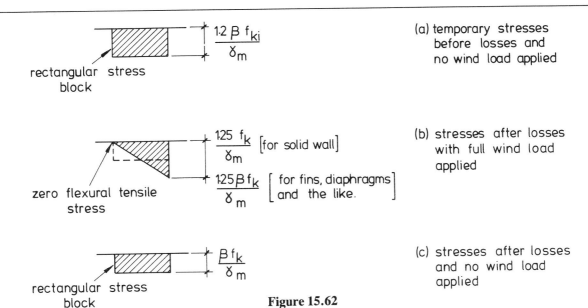

(a) temporary stresses
before losses and
no wind load applied

$1.2 \beta \dfrac{f_{ki}}{\gamma_m}$

rectangular stress
block

$1.25 \dfrac{f_k}{\gamma_m}$ [for solid wall]

$1.25 \beta \dfrac{f_k}{\gamma_m}$ $\begin{bmatrix} \text{for fins, diaphragms} \\ \text{and the like.} \end{bmatrix}$

zero flexural tensile
stress

(b) stresses after losses
with full wind load
applied

$\beta \dfrac{f_k}{\gamma_m}$

rectangular stress
block

(c) stresses after losses
and no wind load
applied

Figure 15.62

15.10.6 Losses of post-tensioning force

For most applications of post-tensioning to masonry as presented here, it is considered that 20% losses in the post-tensioning force due to all of the various factors may be assumed. A more accurate figure could only be determined by a programme of research into the various effects. The need for a more accurate figure is debatable, bearing in mind the quality control and practicalities of construction, and that the 20% figure has proved satisfactory over a considerable period. In particularly specialised applications with very strict quality control, it may be considered worthwhile to determine a more accurate figure. However, more research is required with regard to gains of post-tensioning force in clay brickwork due to the long term expansion noted earlier, and little is understood about this phenomenon in relation to post tensioning.

Assessment of losses

(a) Creep and moisture movements
When loading is applied to masonry, there is an instantaneous shortening due to contraction from compressive forces which is termed the elastic deformation, and then a further deformation which is time-dependent and known as creep. Much research has been undertaken into creep in concrete. However, little has been done with regard to masonry. The mechanism of creep in masonry appears to be similar to that of concrete, although there are important differences.

Moisture causes various reversible and irreversible movements in masonry clay. Bricks undergo a comparatively small and reversible expansion and contraction due to wetting and drying. A

much larger and irreversible expansion occurs immediately after firing. Most of this expansion occurs within the first seven days for the majority of units, continuing at a decreasing rate for about six months, when a limiting value is reached. In addition to these two types of movement, there appears to be, in some clay bricks, a much longer term permanent expansion movement, possibly related to moisture movements within the bricks. The longer term expansion may be greater than the contraction due to creep, thus clay brickwork can expand or 'grow' after the initial elastic compression. Thus in the long term, there may be no net losses in the post-tensioning force due to creep and moisture movements but, in some cases, a possible effective gain. This increase may cause problems of cracking and local crushing around bearing plates, etc., and should be considered in the design.

The effect of creep should also be considered, as the two effects cannot reasonably be assumed to counteract one another. Figures for moisture and thermal movements in brickwork are given in various BDA publications as well as in Appendix 3 and additional information on the behaviour of the particular type of brick or block should be obtained from the manufacturer.

Concrete brickwork and blockwork do not generally 'grow' with age but rather are more likely to shrink fairly rapidly. The total shrinkage may not however, have taken place when the units are built into a post-tensioned wall and the shrinkage which may yet be to follow will constitute losses in the post-tensioning force, values of anticipated total shrinkage of concrete units are given and discussed in Appendix 3.

(b) Relaxation of post-tensioning steel

The amount of relaxation or creep of steel depends on the quality of the steel, and also the stress level within the steel. Detailed information on this subject is available from specialist manufacturers of various proprietary bars. Figures for the maximum relaxation after 1000 hours duration are given in various British Standards, and these are based on a 70% stress level. However, bars larger than those required for the designed loads are often provided in post-tensioned brickwork to allow for possible corrosion. It is thus less likely that bars will be stressed to their maximum stress levels. The losses due to relaxation may be assumed to decrease from about 8% at the 70% stress level down to 0% at the 50% stress level. An appropriate figure for the losses should be assessed from the above for any particular application, depending on the stress level within the bars.

(c) Elastic deformation of masonry

The instantaneous elastic deformation of the masonry will, for all practical purposes, be taken up during the stressing of the bars and no compensating calculations are necessary for this in the assessment of losses in post-tensioning force but would be necessary for the similar situation in prestressed work. The amount of this elastic deformation may be determined from an elastic method of analysis with a modular ratio based on Young's modulus for masonry given by the following assumption that:

$$E_b = 900 f_k$$

where

E_b = Young's modulus for masonry
f_k = characteristic compressive stress.

(d) Friction

Frictional losses occur at the stressing stage and will be concerned mainly with tightening down of the bearing plates which transfer the stress from the bars into the masonry. It is recommended that a torque wrench be used when tightening down the nuts. The manufacturers of such wrenches are generally able to advise on the reduction for any friction losses, and it is usually unnecessary to make a specific allowance in the calculation of post-tensioning forces. It should be noted, however, that the information provided by the torque wrench manufacturer will relate to certain specific conditions prevailing when the torque is applied. Such conditions are likely to include the type of thread and the lubrication quality (i.e. lightly oiled) of the thread on both the rod and the nut, as well as the surface condition of the spreader plate against which the nut is turned. It is vital that these conditions are provided and, during the construction of the wall, adequate protection of the threads must be maintained to ensure this.

15.10.7 Bearing stresses

The post-tension force in the bars is applied to the masonry, generally, by means of steel plates bearing on the top of the wall. The compressive force, applied locally at these points, then disperses into the remainder of the element. The relatively high local bearing stresses should thus be checked to ensure that the loading is applied over a sufficiently large area to avoid local overstressing. This may be calculated by the methods given in Chapter 10, which relates to BS 5628, clause 34, for the ultimate limit state.

15.10.8 Deflection

In general, post-tensioned masonry should comply with the slenderness ratio requirements for plain masonry, and this should limit deflections to acceptable values. Deflections may, however, need to be considered in more detail in the case of cantilevered walls, and also where the post-tensioning force is applied eccentrically. In such cases, an elastic analysis should be undertaken to determine the deflection at working load to be compared with acceptable values based on particular requirements for each application.

15.10.9 Partial safety factor on post-tensioning force

The partial safety factor for loads, γ_f, is applicable, to some degree, to the post-tensioning force. The values given in BS 5628 for γ_f have been determined statistically for dead, superimposed and wind loadings. No research has been carried out to determine, statistically, the appropriate γ_f values for the post-tensioning operation and an assessment of the likely variable factors must be made by the designer in order to arrive at a reasonable combination of extreme values. In the absence of any other values it is proposed that the values already determined for dead loadings should be used for application to post-tensioning force calculations.

15.11 EXAMPLE 7: HIGH CAVITY WALL WITH WIND LOADING

The 255 thick cavity wall shown in Figure 15.63 is to be constructed in two leaves of brickwork comprising bricks with a minimum compressive strength of 20 N/mm^2 and a water absorption of 9% set in a designation (iii) mortar. The partial safety factor for material will be taken as 2.5; the characteristic wind load will be taken as 0.9 kN/m^2 and the characteristic wind uplift will be taken as 1.0 kN/m^2. There are no internal walls offering support to the external wall. The density of the masonry will be taken as 19.0 kN/m^3 and the characteristic superimposed roof load will be taken as 0.75 kN/m^2. The plain cavity wall can be shown, by calculation, using the design principles given in Chapter 11, to be unstable, hence, post-tensioning will be introduced to provide the additional strength required.

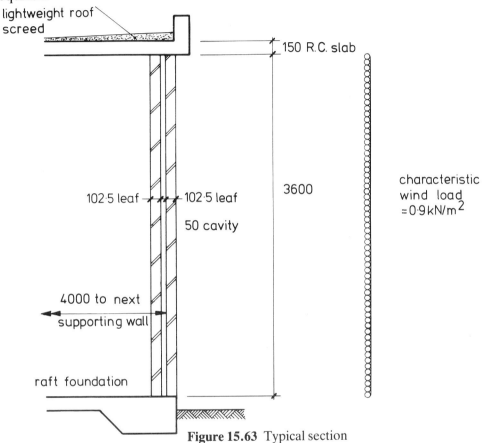

Figure 15.63 Typical section

Stresses should be checked for the following conditions:

(a) After losses (or gains)
These may occur in the post-tensioning force due to the factors discussed in 15.10.6:

flexural tensile stresses: with the wall subject to dead plus wind loading only (this condition also determines the magnitude of the post-tensioning force required).

flexural compressive and axial compressive stresses: with the wall subject to the various combinations of dead plus superimposed, dead plus wind and dead plus superimposed plus wind loadings.

(b) Before losses
axial compressive stresses: with the wall subject to dead plus superimposed loads only. In this situation, where the wind loading may be applied as both suction and pressure, the post-tensioning force will be applied concentrically on the wall. It will be assumed that the post-tensioning force will not be applied until the masonry has achieved its characteristic strength. If the force should be applied before this strength is achieved and it is not recommended that the post-tensioning force is applied before fourteen days, allowance must be made in the calculations for the strength available for a particular loading situation and in the assessment of any additional losses which may be expected as a result.

For each of the above loading situations, the dead loading should be considered as comprising both the dead loading of the wall and roof slab together with the applied post-tensioning force.

The most economical design for the wall would utilise the roof slab as a prop to the head of the wall and the upper anchorage of the post-tensioning rod will be made through a spreader plate placed on top of the roof slab as shown in Figure 15.64.

This detail would have the added advantage of ensuring that the post-tensioning force was shared equally between the two leaves of the wall and the stiffness of the slab may be assumed to place the force into each leaf with zero eccentricity. High yield steel rods will be used for the post-tensioning.

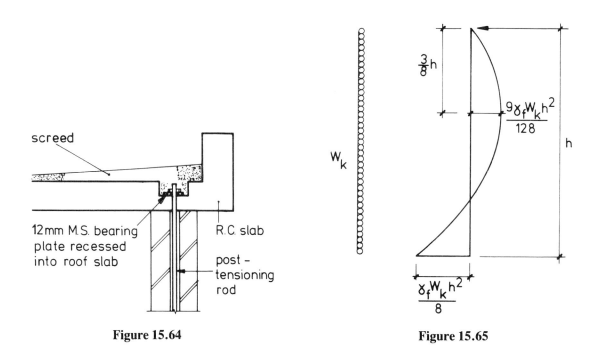

Figure 15.64   **Figure 15.65**

15.11.1 Capacity reduction factor, β

The wall is, for the major part of its life, subject to axial loading (dead plus post-tensioning plus superimposed) only, and only intermittently is it subject also to the lateral wind loadings which generally dictate the most onerous design cases. Hence, the effect of buckling under axial loading must be considered and is based on the capacity reduction factor, β, as derived from the slenderness ratio of the wall and the eccentricity of the loading.

$$\text{Slenderness ratio} = \frac{\text{effective height}}{\text{effective thickness}}$$

$$= \frac{0.75 \times 3600}{\tfrac{2}{3}(102.5 + 102.5)}$$

$$SR = 19.9$$

Eccentricity of load, $e_x = 0$ to $0.05t$ as previously stated, although consideration should be given to the possibility of eccentricity of the dead and superimposed roof loads which may result from rotation of the slab at the bearing due to its deflection at mid-span. The net eccentricity for this example will be assumed to lie within 0 and $0.05t$ for each leaf. Hence, capacity reduction factor, $β = 0.70$ (from BS 5628, Table 7 – see Table 5.15).

For solid and cavity walls the capacity reduction factor is applicable only to the axial compressive stresses whereas for geometric wall profiles, such as the fin or diaphragm wall, the capacity reduction factor is applicable to both axial compressive and flexural compressive stresses (as discussed in 15.10.3) as was the case for the design of plain fin and diaphragm walls covered in Chapter 13.

15.11.2 Characteristic strengths

Masonry
Characteristic compressive strength, f_k = 5.8 N/mm^2
Characteristic flexural compressive strength, f_{kx} = 1.25 × 5.8 = 7.25 N/mm^2
Characteristic flexural tensile strength = limited to zero

Steel
Characteristic tensile strength, f_y = 410 N/mm^2

15.11.3 Design strengths (after losses)

Design compressive strength (at base)
$$= \frac{1.15 f_k}{\gamma_m} \text{ (1.15 narrow wall factor)}$$

$$= 1.15 \times \frac{5.8}{2.5} \qquad = 2.68 \text{ N/mm}^2$$

Design compressive strength (in wall height)
$$= \frac{1.15 \beta f_k}{\gamma_m}$$

$$= 1.15 \times \frac{0.70 \times 5.8}{2.5} = 1.87 \text{ N/mm}^2$$

Design flexural compressive strength
$$= \frac{1.15 f_{kx}}{\gamma_m}$$

$$= 1.15 \times \frac{7.25}{2.5} \qquad = 3.34 \text{ N/mm}^2$$

Design flexural tensile strength = limited to zero.

15.11.4 Section modulus of wall
For a cavity wall, tied with vertical twist type wall ties in accordance with BS 5628, the section modulus of the wall may be taken as the sum of the section moduli of the individual leaves.

Hence
$$Z = \frac{2 \times 1000 \times 102^2}{6} = 3.47 \times 10^6 \text{ mm}^3$$

15.11.5 Design method
The design method adopted for this example is to calculate the theoretical flexural tensile stress which would be likely to develop as a result of the lateral loading and in the absence of any post-tensioning force. An axial compressive stress is then applied by way of the post-tensioning force and in addition to the existing axial stress from the minimum vertical loading, to eliminate the theoretical flexural tensile stress previously calculated. Checks are then carried out to establish that the wall specified can support the applied vertical and lateral loading as well as the post-tensioning force for the various design conditions prevailing.

15.11.6 Calculation of required post-tensioning force
The wall will be considered to act as a propped cantilever in that the post-tensioning force will be calculated to ensure that the wall section at its base does not 'crack' (tensile stresses are limited to zero) and the bending moment diagram for the wall is as shown in Figure 15.65.

For the loading condition dead plus wind only, the applicable partial safety factors for dead and wind loads are 0.9 and 1.4 respectively.

Hence, the design bending moments are calculated as:

at base level, $M_b = \dfrac{\gamma_f W_k h^2}{8}$

$$= \dfrac{1.4 \times 0.9 \times 3.6^2}{8} \qquad = 2.0 \text{ kN·m}$$

at $\dfrac{3}{8} h$ level, $M_w = \dfrac{9\gamma_f W_k h^2}{128}$

$$= \dfrac{9 \times 1.4 \times 0.9 \times 3.6^2}{128} = 1.15 \text{ kN·m}$$

Characteristic dead load, G_k

roof slab	$= 0.15 \times 24 \times 2$	$=$	7.20
roof finishes		$=$	5.00
ow wall	$= 2 \times 0.102 \times 19 \times 3.6$	$=$	13.95
		$=$	26.15 kN/m

G_k at $\dfrac{3}{8} h$ level $= 7.2 + 5.0 + \left(13.95 \times \dfrac{3}{8} \right) = 17.43$ kN/m

Characteristic wind uplift

$$= 1.0 \times \dfrac{4}{2} = 2.0 \text{ kN/m}$$

Net design dead load

$$(\gamma_f \times G_k) - (\gamma_f \times \text{wind uplift}) = (0.9 \times 26.15) - (1.4 \times 2.0) = 20.74 \text{ kN/m}$$

The design dead load will be considered to be shared equally by the two leaves of the cavity wall hence, the axial compressive stress due to the design dead load:

$$g_d = \dfrac{20.74 \times 10^3}{2 \times 102 \times 1000} = 0.10 \text{ N/mm}^2$$

Theoretical flexural tensile stress (in the absence of the application of the post-tension force) equals:

$$g_d - \dfrac{M_A}{Z}$$

where
M_A = design bending moment per leaf
Z = section modulus per leaf
g_d = axial compressive stress from dead, wind, uplift.

$$\text{Theoretical flexural tensile stress} = 0.10 - \dfrac{2.0 \times 10^6 \times 0.5}{3.47 \times 10^6 \times 0.5}$$

$$= -0.48 \text{ N/mm}^2$$

Hence, a design post-tensioning stress of $+0.48 \text{ N/mm}^2$ should be provided to eliminate this theoretical flexural tensile stress and these processes are shown in Figure 15.66.

In order to achieve a minimum design post-tensioning stress of 0.48 N/mm^2, a characteristic post-tensioning stress of $0.48/\gamma_f$ should be provided by the post-tensioning force in the rods; where γ_f would be taken as 0.9 as discussed in 15.10.9. Hence, characteristic post-tensioning stress $= 0.48/0.9 = 0.53 \text{ N/mm}^2$.

The spacing of the post-tensioning rods, the force required per rod, the torque required to produce that force and the local bearing stresses will be considered later in this example. The next stage of the design process is to consider the effect of these compressive stresses on the wall section provided to ensure that stability is maintained.

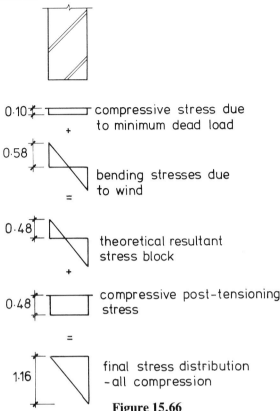

Figure 15.66

15.11.7 Consider compressive stresses: after losses

Owing to the variations in the factors for partial safety on loading, it is not immediately obvious which combination of loading, (dead plus superimposed, dead plus wind, dead plus superimposed plus wind) may produce the most onerous design condition. The three loading cases stated above should therefore be considered individually taking account of the capacity reduction factor, β, where applicable.

Case (a) dead plus superimposed

$$\text{Design axial stress, } f_{uac} = \left[\frac{(\gamma_f G_k) + (\gamma_f Q_k)}{\text{area}} \right] + \text{post-tensioning stress}$$

where
  G_k = characteristic dead load = 26.15 kN/m
  Q_k = characteristic superimposed load
  γ_f = partial safety factors for loads; for this loading combination $\gamma_f = 1.4$ and 1.6 for G_k and Q_k respectively.

Hence $f_{uac} = \dfrac{(1.4 \times G_k) + (1.6 \times Q_k)}{2 \times 102 \times 1000} + \text{post-tensioning stress}$

$= \dfrac{(1.4 \times 26.15) + (1.6 \times 0.75 \times 2)}{2 \times 102 \times 1000} + (\gamma_f \times \text{characteristic post-tensioning stress})$

$= 0.191 + (1.4 \times 0.53)$

$= 0.933 \text{ N/mm}^2$

in which the partial safety factor for loads of 1.4 has been applied to the characteristic post-tensioning stress of 0.53 N/mm² calculated earlier to give the design post-tensioning stress for this situation.

$$\text{Design strength in span of wall} = \frac{1.15\beta f_k}{\gamma_m}$$

$$= \frac{1.15 \times 0.70 \times 5.8}{2.5} = 1.87 \text{ N/mm}^2$$

As the design strength of 1.87 N/mm^2 exceeds the design axial stress of 0.933 N/mm^2 the wall is acceptable for this loading condition.

Case (b) dead plus wind

Design axial stress, f_{uac}, at base level

$$= \left[\frac{\text{design dead load} - \text{design wind uplift}}{\text{area}}\right] + \text{post-tensioning stress}$$

$$= \left[\frac{(\gamma_f G_k) - (\gamma_f \times \text{wind uplift})}{\text{area}}\right] + (\gamma_f \times 0.53)$$

$$= \left[\frac{(1.4 \times 26.15) - (1.4 \times 2.0)}{2 \times 102 \times 1000}\right] + (1.4 \times 0.53)$$

$$= 0.166 + 0.742$$

$$= 0.908 \text{ N/mm}^2$$

Design axial stress, f_{uac}, at $\frac{3}{8}h$ level

$$= \left[\frac{(1.4 \times 17.43) - (1.4 \times 2.0)}{2 \times 102 \times 1000}\right] + (1.4 \times 0.53)$$

$$= 0.106 + 0.742$$

$$= 0.848 \text{ N/mm}^2$$

$$\frac{f_{uac}}{\beta} = \frac{0.848}{0.70} = 1.211 \text{ N/mm}^2$$

The lateral wind loading produces the design bending moments shown in Figure 15.65. Design flexural stress, $f_{ubc} = \pm M_A/Z$ (compressive and tensile).

Combined design stress =

(i) at base level $= f_{uac} + f_{ubc}$

(ii) at $\frac{3}{8}h$ level $= \left(\dfrac{f_{uac}}{\beta}\right) + f_{ubc}$

(i) at base level

$$f_{ubc} \qquad = \pm \frac{2.0 \times 10^6}{3.47 \times 10^6} \qquad = \pm 0.6 \text{ N/mm}^2$$

Combined design stress $= 0.908 \pm 0.6$

$$= +1.508 \text{ N/mm}^2 \text{ or } +0.308 \text{ N/mm}^2$$

(ii) at $\frac{3}{8}h$ level

$$f_{ubc} \qquad = \pm \frac{1.15 \times 10^6}{3.47 \times 10^6} \qquad = \pm 0.33 \text{ N/mm}^2$$

Combined design stress $= \dfrac{0.848}{0.7} \pm 0.33$

$$= +1.541 \text{ N/mm}^2 \text{ or } +0.881 \text{ N/mm}^2$$

The stress diagrams for this loading situation at the top critical levels considered are shown in Figure 15.67.

$$\text{Design strength of wall} = \frac{1.15 \times 1.25 f_k}{\gamma_m}$$

$$= \frac{1.15 \times 1.25 \times 5.8}{2.5}$$

$$= 3.34 \text{ N/mm}^2$$

The design strength exceeds the combined design stresses hence, the wall is acceptable for this loading condition.

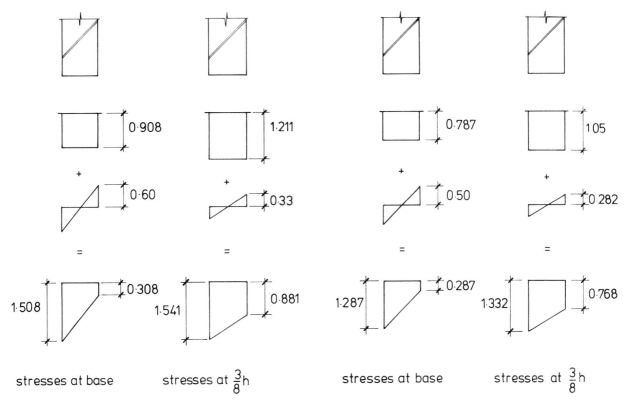

| stresses at base | stresses at $\frac{3}{8}$h | stresses at base | stresses at $\frac{3}{8}$h |

Figure 15.67   **Figure 15.68**

Case (c) dead plus superimposed plus wind
Design axial stress, f_{uac} at base level

$$= \left[\frac{(\gamma_f G_k) - (\gamma_f \times \text{wind uplift}) + (\gamma_f Q_k)}{\text{area}} \right] + \text{post-tensioning stress}$$

$$= \left[\frac{(1.2 \times 26.15) - (1.2 \times 2.0) + (1.2 \times 1.5)}{2 \times 102 \times 1000} \right] + (\gamma_f \times 0.53)$$

$$= 0.151 + (1.2 \times 0.53)$$

$$= 0.787 \text{ N/mm}^2$$

Design axial stress, f_{uac}, at $\frac{3}{8}h$ level

$$= \left[\frac{(1.2 \times 17.43) - (1.2 \times 2.0) + (1.2 \times 1.5)}{2 \times 102 \times 1000} \right] + (1.2 \times 0.53)$$

$$= 0.099 + 0.636$$

$$= 0.735 \text{ N/mm}^2$$

$$\frac{f_{uac}}{\beta} = \frac{0.735}{0.70} = 1.05 \text{ N/mm}^2$$

Design M_b at base $\quad = \frac{\gamma_f W_k h^2}{8} = \frac{1.2 \times 0.9 \times 3.6^2}{8} = 1.75 \text{ kN·m}$

Design M_w at $\frac{3}{8}h$ level $= \frac{9\gamma_f W_k h^2}{128} = \frac{9 \times 1.2 \times 0.9 \times 3.6^2}{128} = 0.98 \text{ kN·m}$

(i) at base level

$f_{ubc} \qquad\qquad = \pm \frac{1.75 \times 10^6}{3.47 \times 10^6} \qquad\qquad = \pm 0.5 \text{ N/mm}^2$

Combined design stress $= 0.787 \pm 0.5$

$$= +1.287 \text{ N/mm}^2 \text{ or } +0.287 \text{ N/mm}^2$$

(ii) at $\frac{3}{8}h$ level

$f_{ubc} \qquad\qquad = \pm \frac{0.98 \times 10^6}{3.47 \times 10^6} \qquad\qquad = \pm 0.282 \text{ N/mm}^2$

Combined design stress $= \frac{0.735}{0.70} \pm 0.282$

$$= +1.332 \text{ N/mm}^2 \text{ or } +0.768 \text{ N/mm}^2$$

The stress diagrams for this loading situation at the top critical levels considered are shown in Figure 15.68.

The design strength of 3.34 N/mm² calculated earlier again exceeds the combined design stresses and the wall is acceptable for this loading condition.

15.11.8 Consider compressive stresses: before losses
Dead plus superimposed load only will be considered.

$$\text{Design axial stress}, f_{uac} = \left[\frac{(1.4G_k) + (1.6Q_k)}{\text{area}} \right] + \text{post-tensioning stress before losses}$$

The design post-tensioning stress will be increased by 20% in anticipation of that amount of loss – hence:

$$f_{uac} = \left[\frac{(1.4 \times 26.15) + (1.6 \times 1.5)}{2 \times 102 \times 1000} \right] + \frac{1.4 \times 0.53}{0.8}$$

$$= 0.191 + 0.927$$

$$= 1.118 \text{ N/mm}^2$$

$$\text{Design strength of wall before losses} = \frac{1.20 \beta (1.15 f_{ki})}{\gamma_m}$$

where f_{ki} is the characteristic compressive strength of the masonry at the age at which the post-tensioning is applied.

In this example it will be assumed to have achieved its full characteristic strength, i.e. f_k (based upon the 28 day test).

Hence

$$\text{design strength before losses} = \frac{1.20\,\beta\,(1.15 f_{ki})}{\gamma_m}$$

$$= \frac{1.20 \times 0.70 \times 1.15 \times 5.8}{2.5}$$

$$= 2.24 \text{ N/mm}^2$$

The design strength before losses exceeds the design axial stress hence, the wall is acceptable for this loading condition.

15.11.9 Design of post-tensioning rods
The characteristic post-tensioning stress required, to be provided in the brickwork by the post-tensioning rods, is 0.53 N/mm^2.

This is the equivalent of $0.53 \times 2 \times 102 \times 1000 = 108$ kN per m run of wall. To allow for 20% losses in the post-tensioning force the initial equivalent load $= 108/0.8 = 135$ kN per m run.

In order to limit relaxation of the steel and hence minimise the losses, the stress in the post-tensioning rods is limited to

$$\frac{0.7 f_y}{\gamma_{ms}} = \frac{0.7 \times 410}{1.15} = 250 \text{ N/mm}^2$$

$$\text{steel area required per m} = \frac{135 \times 10^3}{250} = 540 \text{ mm}^2 \text{ per m run}$$

High yield bars of 25 mm diameter placed at 900 mm centres (which provide an area of 546 mm^2 per m) will be used.

Torque to provide rod tension
Considerable variation exists in the recommendations, given by manufacturers and suppliers of torqueing equipment, for the calculation of torque requirements.

The amount of tension induced by a particular torque is dependent on numerous factors, the two most significant being the pitch and type of the thread and the coefficients of friction between the contact surfaces of nuts, bolts, spreader plates, etc. This latter aspect is largely dependent on the type and quality of the original finish to these components and the degree of lubrication and general protection during the construction of the works prior to the post-tensioning.

The calculation provided below is based on a general engineering formula derived from test research and utilises lightly oiled, metric threads with self-finish nuts and bolts and a hardened washer between the nut and the spreader plate.

$$\text{Torque required} = \frac{\text{bolt tension} \times \text{bolt diameter}}{5}$$

$$\text{Bolt tension (in kgf)} = \frac{135 \times 10^3}{9.81} \text{ per m run of wall}$$

$$= \frac{135 \times 0.9 \times 10^3}{9.81} \text{ per rod at 900 mm centres}$$

$$= 12\,385 \text{ kgf per rod}$$

$$\text{Torque required} = \frac{12\,385 \times 0.025}{5}$$

$$= 61.9 \text{ kgf·m}$$

Spreader plate design

$$\text{Maximum design force per rod} = 135 \times \gamma_f$$
$$= 135 \times 1.4$$
$$= 189 \text{ kN}$$

$$\text{Design compressive strength of wall} = \frac{1.5 \times 1.15 \times 5.8}{2.5}$$

$$= 4.0 \text{ N/mm}^2$$

in which a 1.5 strength increase factor has been incorporated to take account of the local bearing condition of the spreader plate on the brickwork.

$$\text{Area of spread required} = \frac{189 \times 10^3}{4.0}$$

$$= 47\,250 \text{ mm}^2$$

$$\text{Length of spread along two leaves} = \frac{47\,250}{2 \times 102}$$

$$= 232 \text{ mm}$$

Allowing for a 45° dispersion of load from the spreader plate through the rc slab as shown in Figure 15.69.

A 150 mm long × 100 mm wide × 12 mm thick spreader plate would be suitable.

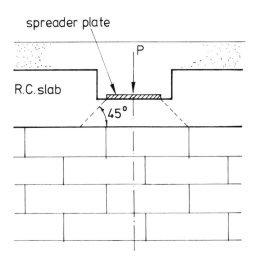

Figure 15.69

15.12 EXAMPLE 8: POST-TENSIONED FIN WALL

A warehouse measuring 27 m × 46 m on plan and 10 m high, as shown in Figure 15.70 is to be constructed in loadbearing brickwork using fin wall construction for its main vertical structure. The planning requirements are for the fins to be as small as possible, hence post-tensioning of the fins is proposed to provide for this requirement. The wall panels between fins will be of normal 255 cavity construction and there are no internal walls within the building. The roof construction and detailing will be assumed to provide an adequate prop to the head of the wall and will similarly ensure stability of the structure by transferring this propping force to the gable shear walls. The type of brick to be used throughout will have a compressive strength of 30 N/mm² and a water absorption of 10%. The partial safety factor for materials may be assumed as 2.5. The fin profile to be used is reference 'E' from Table 13.1 and the various dimensions and properties may be obtained from that table and Figure 15.70.

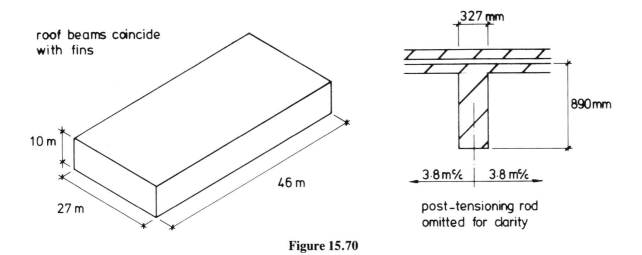

Figure 15.70

15.12.1 Design procedure

The design procedure is similar to that used for the previous example but consideration must be given to the fin profile being asymmetrical. The post-tensioning force was positioned concentrically in the previous example because the section is symmetrical and the wind loading (suction and pressure) in each direction could be assumed to be of a similar magnitude. For the asymmetrical fin wall profile, with similar wind loading, the stresses at the extreme edges of the fin and flange would differ considerably owing to the two values of section modulus Z_1 and Z_2. Hence, the post-tensioning force is applied eccentrically, in this situation, to make the maximum use of the eliminating compressive stress at each of the extreme edges.

Figures 15.71 and 15.72 show the theoretical stresses which may occur in the wall, in the absence of the post-tensioning force, for wind suction and wind pressure loadings respectively. For each direction of wind loading, a theoretical flexural tensile stress may be calculated, for the extreme edges of either fin or flange and at either base level or at maximum span moment level, as applicable.

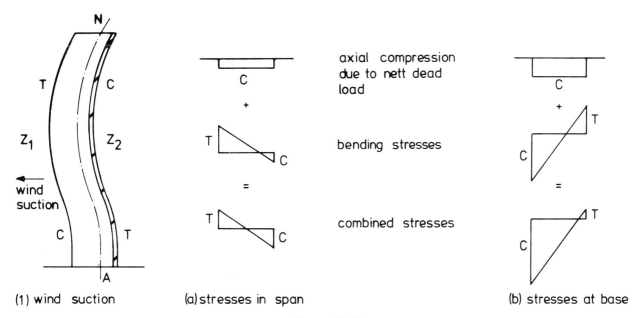

Figure 15.71

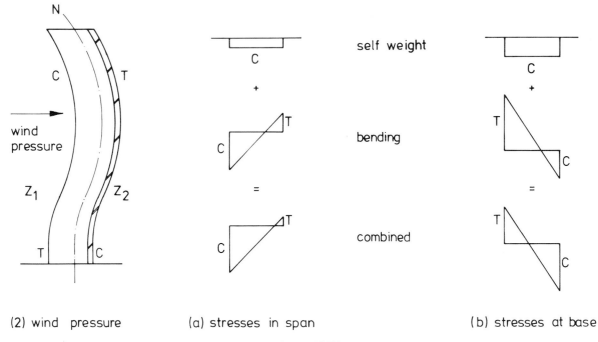

(2) wind pressure   (a) stresses in span   (b) stresses at base

Figure 15.72

Having established the maximum theoretical flexural tensile stress at each of the extreme edges, an eccentric post-tensioning force may be calculated to eliminate these stresses. The basic theory of this process is indicated in Figure 15.73 in which f_{t1} and f_{t2} are the maximum theoretical flexural tensile stresses at the extreme edges of the fin and flange respectively.

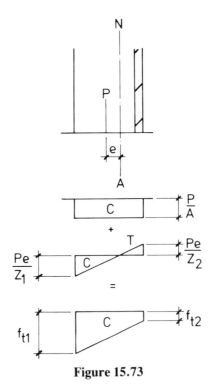

Figure 15.73

In Figure 15.73:

P = design post-tensioning force
A = area of effective fin section (from Table 13.41)
e = eccentricity of P about neutral axis
Z_1 = minimum section modulus (from Table 13.41)
Z_2 = maximum section modulus (from Table 13.41)

The design should then proceed to check that the compressive (combined axial and flexural) stresses which result are acceptable in the same manner as for the previous example.

15.12.2 Design post-tensioning force and eccentricity

Figure 15.73 represents the design post-tensioning force as P and its eccentricity, about the neutral axis (N.A.) of the section, as e. The stresses at each of the extreme edges may be expressed as:

$$f_{t1} = \frac{P}{A} + \frac{Pe}{Z_1} \tag{20}$$

$$\text{and} \quad f_{t2} = \frac{P}{A} - \frac{Pe}{Z_2} \tag{21}$$

A pair of simultaneous equations may be written from equations (20) and (21) to solve for P and e as follows:

Multiply (20) throughout by Z_1 and (21) throughout by Z_2 giving:

$$f_{t1}Z_1 = \frac{PZ_1}{A} + Pe \tag{22}$$

$$f_{t2}Z_2 = \frac{PZ_2}{A} - Pe \tag{23}$$

adding (22) to (23) gives:

$$(f_{t1}Z_1) + (f_{t2}Z_2) = \frac{P}{A}(Z_1 + Z_2)$$

Hence

$$P = \frac{[(f_{t1}Z_1) + (f_{t2}Z_2)]A}{Z_1 + Z_2} \tag{24}$$

The value of P calculated in equation (24) can now be substituted into equation (21) to calculate the value of e.

Equation (21) transposed gives:

$$e = \left(\frac{P}{A} - f_{t2}\right)\frac{Z_2}{P}$$

$$= \left(\frac{1}{A} - \frac{f_{t2}}{P}\right)Z_2 \tag{25}$$

15.12.3 Characteristic strengths

For the materials given in Example 8 the following properties may be obtained from BS 5628.

Masonry

Characteristic compressive strength, f_k	$= 7.6\,\text{N/mm}^2$
Characteristic flexural tensile strength (failure plane parallel to bed joints)	$= 0.4\,\text{N/mm}^2$
Characteristic flextural tensile strength (failure plane perpendicular to bed joints)	$= 1.1\,\text{N/mm}^2$
Characteristic flexural compressive strength ($1.25f_k$)	$= 9.5\,\text{N/mm}^2$

Steel

Characteristic tensile strength, $f_y = 410\,\text{N/mm}^2$

15.12.4 Loadings

The following data will be assumed to have been provided for the calculation of the design loadings:

(a) Wind loads

Dynamic wind pressure, q $\qquad = 0.74\,\text{kN/m}^2$
Characteristic wind loads:

suction on leeward face, W_{k1} $= 0.81$ kN/m^2
pressure on windward face, W_{k2} $= 0.56$ kN/m^2
gross wind uplift on roof, W_{k3} $= 0.39$ kN/m^2

(b) Dead and superimposed loads
Characteristic superimposed load, $Q_k = 0.75$ kN/m^2
Characteristic dead load, G_k $= 0.60$ kN/m^2

The wall panel is assumed to have been checked as being adequate to span horizontally between fins spaced at 3.80 m centres.

For the loading combination, dead plus wind, the partial safety factors for loads may be taken as 0.9 and 1.4 or 1.4 and 1.4 for dead and wind loads respectively.

Design wind loads per fin

Case (i) suction
Design wind load $= 3.8 \times 1.4 \times 0.56$
                     $= 2.98$ kN/m height of fin wall

Case (ii) pressure
Design wind load $= 3.8 \times 1.4 \times 0.81$
                     $= 4.31$ kN/m height of fin wall

Case (iii) uplift
Design wind load $= 1.4 \times 0.39$
                     $= 0.55$ kN/m^2
Dead load – roof uplift $= (0.9 \times 0.6) - (1.4 \times 0.39)$
                             $= 0.54 - 0.55$
            say $=$ zero

Hence, only the own weight of the masonry can be considered in providing resistance to the flexural tensile stresses.

15.12.5 Design bending moments
The wall will be designed, taking advantage of the prop provided by the roof plate, as a propped cantilever. Unlike the plain fin wall, there should not normally be any requirement to adjust the bending moment diagram from the propped cantilever proportions as the post-tensioning force can be made sufficiently large to avoid this. The design bending moment diagram is as shown in Figure 15.74. Hence, the design bending moments are:

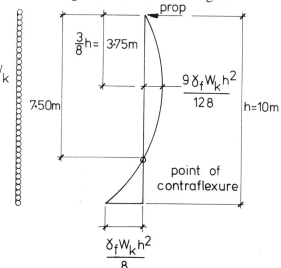

Case (i) suction
(a) At $\frac{3}{8}h$ level, $M_w = \dfrac{9 \times 2.98 \times 10^2}{128} = 21$ kN·m

(b) At base level, $M_b = \dfrac{2.98 \times 10^2}{8} = 37.3$ kN·m

Case (ii) pressure
(a) At $\frac{3}{8}h$ level, $M_w = \dfrac{9 \times 4.31 \times 10^2}{128} = 30.3$ kN·m

(b) At base level, $M_b = \dfrac{4.31 \times 10^2}{8} = 53.9$ kN·m

Figure 15.74 Bending moment diagram

15.12.6 Theoretical flexural tensile stresses

Case (i) suction

(a) At $\frac{3}{8}h$ level:

axial compressive stress due to ow $= \dfrac{0.9 \times 19 \times 3.75}{1000} = +0.064 \text{ N/mm}^2$

theoretical flexural stresses, $\pm \dfrac{M_w}{Z_1} = \pm \dfrac{21 \times 10^6}{0.061 \times 10^9} = \pm 0.344 \text{ N/mm}^2$

theoretical flexural tensile stress $\qquad\qquad = -0.280 \text{ N/mm}^2$

(b) At base level:

axial compressive stress due to ow $= \dfrac{0.9 \times 19 \times 10}{1000} = +0.171 \text{ N/mm}^2$

theoretical flexural stresses, $\pm \dfrac{M_b}{Z_2} = \dfrac{37.3 \times 10^6}{0.12 \times 10^9} = \pm 0.310 \text{ N/mm}^2$

theoretical flexural tensile stress $\qquad\qquad = -0.139 \text{ N/mm}^2$

Case (ii) pressure

(a) At $\frac{3}{8}h$ level:

axial compressive stress due to ow $=$ as case (i) $\qquad = +0.064 \text{ N/mm}^2$

theoretical flexural stresses, $\pm \dfrac{M_w}{Z_2} = \pm \dfrac{30.3 \times 10^6}{0.12 \times 10^9} = \pm 0.253 \text{ N/mm}^2$

theoretical flexural tensile stress $\qquad\qquad = -0.189 \text{ N/mm}^2$

(b) At base level:

axial compressive stress due to ow $=$ as case (i) $\qquad = +0.171 \text{ N/mm}^2$

theoretical flexural stresses, $\pm \dfrac{M_b}{Z_1} = \pm \dfrac{53.9 \times 10^6}{0.061 \times 10^9} = \pm 0.884 \text{ N/mm}^2$

theoretical flexural tensile stress $\qquad\qquad = -0.713 \text{ N/mm}^2$

By inspection of these theoretical flexural tensile stresses, cases (ii) (b) and (ii) (a) give the most onerous values of f_{t1} and f_{t2} respectively for the calculation of the required post-tensioning force and its eccentricity.

i.e. $f_{t1} = 0.713 \text{ N/mm}^2$
and $f_{t2} = 0.189 \text{ N/mm}^2$

15.12.7 Calculations of *P* and *e*

From 15.12.2, equation (24)

$$P = \frac{[(f_{t1}Z_1) + (f_{t2}Z_2)]A}{Z_1 + Z_2}$$

From 15.12.2, equation (25)

$$e = \left(\frac{1}{A} - \frac{f_{t2}}{P}\right)Z_2$$

where

$f_{t1} = 0.713 \text{ N/mm}^2$

$f_{t2} = 0.189 \text{ N/mm}^2$

$A = 0.46 \times 10^6 \text{ mm}^2$ (from Table 13.1)

$Z_1 = 0.061 \times 10^9 \text{ mm}^3$ (from Table 13.1)

$Z_2 = 0.12 \times 10^9 \text{ mm}^3$ (from Table 13.1)

Hence

$P = 168 \text{ kN}$ and $e = 124 \text{ mm}$

15.12.8 Spread of post-tensioning force

The area, A, used in the above equations for the calculations of P and e is the effective area of the effective fin section which is based on a flange width of 1971 mm. The spacing of the fins for this example is 3800 mm leaving a central length of wall, 1829 mm long, which was not taken into account in calculating P and e. The effective fin section is shown in Figure 15.75.

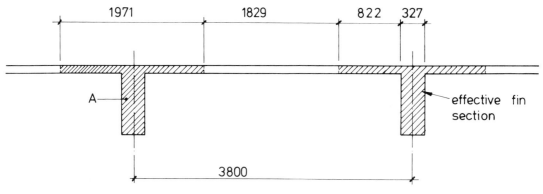

Figure 15.75

It may be argued that the effect of the post-tensioning force will spread into this central, 1829 mm long, area and therefore, account of it should be included in the calculation of P and e. However, whatever effect P may have on this central area should be compensated for by a larger effective section giving higher section moduli, Z_1 and Z_2. In the absence of any research work to investigate this phenomenon it is considered that a reasonable and safe solution would be provided by considering the post-tensioning force to be effective over the area of the effective section only.

15.12.9 Characteristic post-tensioning force P_k

In order to achieve a minimum design post-tensioning force of 168 kN, a characteristic post-tensioning force of $168/\gamma_f$ should be provided in the rods where γ_f would be taken as 0.9 as discussed in 15.10.9.

Hence, characteristic post-tensioning force $P_k = 168/0.9 = 187 \text{ kN}$.

This force is now used to check the design compressive stresses in the wall and to establish the size of the post-tensioning rods.

15.12.10 Capacity reduction factors, β

At the base of the wall it will be assumed that a raft foundation has been provided which is able to provide full buckling restraint to both the fin and the flange depending upon the particular direction of wind loading. It may be noted that a more robust raft foundation may be necessary for the post-tensioned fin wall than for the same section of plain fin wall.

In wall span

The calculation of the relevant capacity reduction factors follows the same design basis as was used for the plain fin wall design example given in Chapter 13.

Case (i) suction

(maximum combined compressive stress in flange:)

$$\text{Slenderness ratio, SR} = \frac{2 \times \text{outstanding length of flange}}{\text{effective cavity wall thickness}}$$

$$= \frac{2 \times 822}{\frac{2}{3}(2 \times 102.5)}$$

$$= 12$$

The eccentricity of the compressive stress in the flange of the fin wall may be taken to be 0 to $0.05t$, hence $\beta = 0.93$.

Case (ii) pressure
(maximum combined compressive stress at end of fin:)

$$\text{Slenderness ratio, SR} = \frac{\text{distance between points of contraflexure}}{\text{actual thickness of fin}}$$

$$= \frac{7500}{327}$$

$$= 23$$

The eccentricity of this compressive stress may again be taken as 0 to $0.05t$, hence, $\beta = 0.58$.

15.12.11 Check combined compressive stresses

The critical loading condition will be either case (a) dead plus wind (where the partial factors of safety for loads are $1.4G_k$ and $1.4W_k$) or case (b) dead plus superimposed plus wind (where the partial safety factors for loads are $1.2G_k$, $1.2Q_k$ and $1.2W_k$).

Characteristic vertical loads in fin

Case (a) dead plus wind
Roof wind uplift cancels out roof dead load hence, only own weight of brickwork applicable for this loading case.

G_k at $\frac{3}{8}h$ level $= 9.19 \times 3.75$ $= 34.46$ kN

G_k at base level $= 9.19 \times 10.0$ $= 91.90$ kN

Case (b) dead plus superimposed plus wind

Q_k $= 0.75 \times 3.8 \times \dfrac{27}{2} = 38.48$ kN

G_k at $\frac{3}{8}h$ level $=$ as case (a) $= 34.46$ kN

G_k at base level $=$ as case (a) $= 91.90$ kN

Characteristic wind loads per fin

Case (i) suction $= 3.8 \times 0.56 = 2.128$ kN/m of height
Case (ii) pressure $= 3.8 \times 0.81 = 3.078$ kN/m of height

Design bending moments

Case (a) dead plus wind $(1.4W_k)$

Case (i) suction:

$$\text{at } \frac{3}{8}h \text{ level, } M_w = \frac{9\gamma_f W_k h^2}{128} = \frac{9 \times 1.4 \times 2.128 \times 10^2}{128} = 20.95 \text{ kN·m}$$

$$\text{at base level, } M_b = \frac{\gamma_f W_k h^2}{8} = \frac{1.4 \times 2.128 \times 10^2}{8} = 37.24 \text{ kN·m}$$

Case (ii) pressure:

$$\text{at } \frac{3}{8}h \text{ level, } M_w = \frac{9\gamma_f W_k h^2}{128} = \frac{9 \times 1.4 \times 3.078 \times 10^2}{128} = 30.30 \text{ kN·m}$$

$$\text{at base level, } M_b = \frac{\gamma_f W_k h^2}{8} = \frac{1.4 \times 3.078 \times 10^2}{8} = 53.86 \text{ kN·m}$$

Case (b) dead plus superimposed plus wind $(1.2W_k)$

Case (i) suction:

$$\text{at } \frac{3}{8}h \text{ level, } M_w = 20.95 \times \frac{1.2}{1.4} = 17.95 \text{ kN·m}$$

$$\text{at base level, } M_b = 37.24 \times \frac{1.2}{1.4} = 31.91 \text{ kN·m}$$

Case (ii) pressure:

$$\text{at } \frac{3}{8}h \text{ level, } M_w = 30.30 \times \frac{1.2}{1.4} = 25.97 \text{ kN·m}$$

$$\text{at base level, } M_b = 53.86 \times \frac{1.2}{1.4} = 46.16 \text{ kN·m}$$

Combined compressive stresses for case (i) suction

Case (a) dead plus wind loading – at base level

Design axial stress due to own weight, G_k

$$= +\frac{\gamma_f G_k}{A} = +\frac{1.4 \times 91.9 \times 10^3}{0.46 \times 10^6} = +0.280 \text{ N/mm}^2$$

Design axial stress due to post-tension force, P_k

$$= +\frac{\gamma_f P_k}{A} = +\frac{1.4 \times 187 \times 10^3}{0.46 \times 10^6} = +0.598 \text{ N/mm}^2$$

Design flexural stresses due to post-tensioning force, P_k

$$= +\frac{\gamma_f P_k e}{Z_1} = +\frac{1.4 \times 187 \times 10^3 \times 124}{0.061 \times 10^9} = +0.532 \text{ N/mm}^2$$

or

$$= -\frac{\gamma_f P_k e}{Z_2} = -\frac{1.4 \times 187 \times 10^3 \times 124}{0.12 \times 10^9} = -0.270 \text{ N/mm}^2$$

Design flexural stresses due to applied moment

$$= +\frac{M_b}{Z_1} = +\frac{37.24 \times 10^6}{0.061 \times 10^9} = +0.610 \text{ N/mm}^2$$

or

$$= -\frac{M_b}{Z_2} = -\frac{37.24 \times 10^6}{0.12 \times 10^9} = -0.310 \text{ N/mm}^2$$

Hence:

Maximum combined compressive stress $= +2.02 \text{ N/mm}^2$

Minimum combined compressive stress $= +0.298 \text{ N/mm}^2$

Case (b) dead plus superimposed plus wind loading – at base level

Design axial stress due to G_k and Q_k
$$= +\frac{\gamma_f G_k + \gamma_f Q_k}{A} = +\frac{(1.2 \times 91.9) + (1.2 \times 38.48)}{0.46 \times 10^3} = +0.341 \text{ N/mm}^2$$

Design axial stress due to post-tension force, P_k
$$= +\frac{\gamma_f P_k}{A} = +\frac{1.2 \times 187 \times 10^3}{0.46 \times 10^6} = +0.488 \text{ N/mm}^2$$

Design flexural stresses due to post-tensioning force, P_k
$$= +\frac{\gamma_f P_k e}{Z_1} = +\frac{1.2 \times 187 \times 10^3 \times 124}{0.061 \times 10^9} = +0.456 \text{ N/mm}^2$$

or
$$= -\frac{\gamma_f P_k e}{Z_2} = -\frac{1.2 \times 187 \times 10^3 \times 124}{0.12 \times 10^9} = -0.232 \text{ N/mm}^2$$

Design flexural stresses due to applied moment
$$= +\frac{M_b}{Z_1} = +\frac{31.91 \times 10^6}{0.061 \times 10^9} = +0.523 \text{ N/mm}^2$$

or
$$= -\frac{M_b}{Z_2} = -\frac{31.91 \times 10^6}{0.12 \times 10^9} = -0.266 \text{ N/mm}^2$$

Hence:
Maximum combined compressive stress $= +1.808 \text{ N/mm}^2$
Minimum combined compressive stress $= +0.331 \text{ N/mm}^2$

Case (a) dead plus wind loading at $\frac{3}{8} h$ level

Design axial stress due to own weight, G_k
$$= +\frac{\gamma_f G_k}{A} = +\frac{1.4 \times 34.46 \times 10^3}{0.46 \times 10^6} = +0.105 \text{ N/mm}^2$$

Design axial stress due to post-tension force, P_k
$$= +\frac{\gamma_f P_k}{A} = +\frac{1.4 \times 187 \times 10^3}{0.46 \times 10^6} = +0.598 \text{ N/mm}^2$$

Design flexural stresses due to post-tensioning force, P_k
$$= +\frac{\gamma_f P_k e}{Z_1} = +\frac{1.4 \times 187 \times 10^3 \times 124}{0.061 \times 10^9} = +0.532 \text{ N/mm}^2$$

or
$$= -\frac{\gamma_f P_k e}{Z_2} = -\frac{1.4 \times 187 \times 10^3 \times 124}{0.12 \times 10^9} = -0.270 \text{ N/mm}^2$$

Design flexural stresses due to applied moment
$$= -\frac{M_w}{Z_1} = -\frac{20.95 \times 10^6}{0.061 \times 10^9} = -0.343 \text{ N/mm}^2$$

or
$$= +\frac{M_w}{Z_2} = +\frac{20.95 \times 10^6}{0.12 \times 10^9} = +0.175 \text{ N/mm}^2$$

Hence:
Maximum combined compressive stress $= +0.892 \text{ N/mm}^2$
Minimum combined compressive stress $= +0.608 \text{ N/mm}^2$

Case (b) dead plus superimposed plus wind loading at $\frac{3}{8} h$ level

Design axial stress due to G_k and Q_k
$$= +\frac{\gamma_f G_k + \gamma_f Q_k}{A} = +\frac{(1.2 \times 34.46) + (1.2 \times 38.48)}{0.46 \times 10^3} = +0.190 \text{ N/mm}^2$$

Design axial stress due to post-tension force, P_k
$$= +\frac{\gamma_f P_k}{A} = +\frac{1.2 \times 187 \times 10^3}{0.46 \times 10^6} = +0.488 \text{ N/mm}^2$$

Design flexural stresses due to post-tensioning force, P_k	$= +\dfrac{\gamma_f P_k e}{Z_1}$	$= +\dfrac{1.2 \times 187 \times 10^3 \times 124}{0.061 \times 10^9}$	$= +0.456 \text{ N/mm}^2$
or	$= -\dfrac{\gamma_f P_k e}{Z_2}$	$= -\dfrac{1.2 \times 187 \times 10^3 \times 124}{0.12 \times 10^9}$	$= -0.232 \text{ N/mm}^2$
Design flexural stresses due to applied moment	$= -\dfrac{M_w}{Z_1}$	$= -\dfrac{17.95 \times 10^6}{0.061 \times 10^9}$	$= -0.294 \text{ N/mm}^2$
or	$= +\dfrac{M_w}{Z_2}$	$= +\dfrac{17.95 \times 10^6}{0.12 \times 10^9}$	$= +0.150 \text{ N/mm}^2$

Hence:
Maximum combined compressive stress $= +0.840 \text{ N/mm}^2$
Minimum combined compressive stress $= +0.596 \text{ N/mm}^2$

Combined compressive stresses for case (ii) pressure

Case (a) dead plus wind loading – at base level

Design axial stress due to own weight G_k	$=$ as case (i)	$= +0.280 \text{ N/mm}^2$
Design axial stress due to post-tension force, P_k	$=$ as case (i)	$= +0.598 \text{ N/mm}^2$
Design flexural stresses due to post-tensioning force, P_k	$=$ as case (i)	$= +0.532 \text{ N/mm}^2$
or	$=$ as case (i)	$= -0.270 \text{ N/mm}^2$
Design flexural stresses due to applied moment	$= -\dfrac{M_b}{Z_1}$ $= -\dfrac{53.86 \times 10^6}{0.061 \times 10^9}$	$= -0.883 \text{ N/mm}^2$
or	$= +\dfrac{M_b}{Z_2}$ $= +\dfrac{53.86 \times 10^6}{0.12 \times 10^9}$	$= +0.449 \text{ N/mm}^2$

Hence:
Maximum combined compressive stress $= +1.057 \text{ N/mm}^2$
Minimum combined compressive stress $= +0.527 \text{ N/mm}^2$

Case (b) dead plus superimposed plus wind loading – at base level

Design axial stress due to G_k and Q_k	$=$ as case (i)	$= +0.341 \text{ N/mm}^2$
Design axial stress due to post-tension force, P_k	$=$ as case (i)	$= +0.488 \text{ N/mm}^2$
Design flexural stresses due to post-tensioning force, P_k	$=$ as case (i)	$= +0.456 \text{ N/mm}^2$
or	$=$ as case (i)	$= -0.232 \text{ N/mm}^2$
Design flexural stresses due to applied moment	$= -\dfrac{M_b}{Z_1}$ $= -\dfrac{46.16 \times 10^6}{0.061 \times 10^9}$	$= -0.757 \text{ N/mm}^2$
or	$= +\dfrac{M_b}{Z_2}$ $= +\dfrac{46.16 \times 10^6}{0.12 \times 10^9}$	$= +0.385 \text{ N/mm}^2$

Hence:
Maximum combined compressive stress $= +0.982 \text{ N/mm}^2$
Minimum combined compressive stress $= +0.528 \text{ N/mm}^2$

Case (a) dead plus wind loading at $\frac{3}{8}h$ *level*

Design axial stress due to own weight G_k	= as case (i)		$= +0.105 \text{ N/mm}^2$
Design axial stress due to post-tension force, P_k	= as case (i)		$= +0.598 \text{ N/mm}^2$
Design flexural stresses due to post-tensioning force, P_k	= as case (i)		$= +0.532 \text{ N/mm}^2$
or	= as case (i)		$= -0.270 \text{ N/mm}^2$

Design flexural stresses due to applied moment
$$= +\frac{M_w}{Z_1} = +\frac{30.30 \times 10^6}{0.061 \times 10^9} = +0.497 \text{ N/mm}^2$$

or
$$= -\frac{M_w}{Z_2} = -\frac{30.30 \times 10^6}{0.12 \times 10^9} = -0.253 \text{ N/mm}^2$$

Hence:
Maximum combined compressive stress $= +1.732 \text{ N/mm}^2$
Minimum combined compressive stress $= +0.180 \text{ N/mm}^2$

Case (b) dead plus superimposed plus wind loading at $\frac{3}{8}h$ *level*

Design axial stress due to G_k and Q_k	= as case (i)		$= +0.190 \text{ N/mm}^2$
Design axial stress due to post-tension force, P_k	= as case (i)		$= +0.488 \text{ N/mm}^2$
Design flexural stresses due to post-tensioning force, P_k	= as case (i)		$= +0.456 \text{ N/mm}^2$
or	= as case (i)		$= -0.232 \text{ N/mm}^2$

Design flexural stresses due to applied moment
$$= +\frac{M_w}{Z_1} = +\frac{25.97 \times 10^6}{0.061 \times 10^9} = +0.426 \text{ N/mm}^2$$

or
$$= -\frac{M_w}{Z_2} = -\frac{25.97 \times 10^6}{0.12 \times 10^9} = -0.216 \text{ N/mm}^2$$

Hence:
Maximum combined compressive stress $= +1.560 \text{ N/mm}^2$
Minimum combined compressive stress $= +0.230 \text{ N/mm}^2$

The two most critical design cases, for wind suction and wind pressure are shown in Figure 15.76 to demonstrate the method of stress calculations on the preceding pages. It would be more obvious to the experienced designer that a number of the preceding design cases did not require calculation to arrive at the two critical conditions stated.

15.12.12 Design flexural compressive strengths of wall: after losses
At base level, where $\beta = 1.0$ by inspection, design flexural compressive strength
$$= \frac{1.25f_k}{\gamma_m} = \frac{1.25 \times 7.6}{2.5} = 3.8 \text{ N/mm}^2$$

At $\frac{3}{8}h$ level, where $\beta = 0.58$ by calculation, design flexural compressive strength
$$= \frac{1.25\beta f_k}{\gamma_m} = \frac{1.25 \times 0.58 \times 7.6}{2.5} = 2.2 \text{ N/mm}^2$$

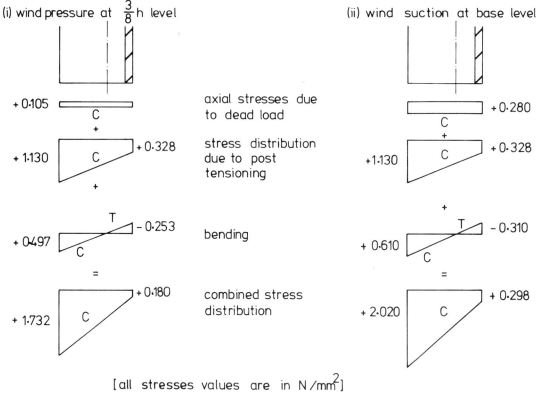

(i) wind pressure at $\frac{3}{8}h$ level

(ii) wind suction at base level

+0.105   axial stresses due to dead load   +0.280

+1.130 +0.328   stress distribution due to post tensioning   +1.130 +0.328

+0.497 T −0.253   bending   +0.610 T −0.310

+1.732 +0.180   combined stress distribution   +2.020 +0.298

[all stresses values are in N/mm^2]

Figure 15.76

Comparison of the design flexural compressive strengths with the previously calculated combined compressive stresses shows that the wall is acceptable for all loading cases considered this far.

15.12.13 Check overall stability of wall
The wall will be checked for overall stability, both before and after losses in the post-tensioning force, for the loading combination of dead plus superimposed plus post-tensioning force.

Consideration has already been given to local stability of the fin and flange in the design of flexural compressive strengths. In this design check consideration is given to buckling of the section as a whole.

As was discussed in Chapter 13, in the design of the plain fin wall, it is questionable whether the narrow flange of the T profile is able to offer full resistance to buckling about the axis perpendicular to the flange. The robustness of the fin section in comparison to the relatively flimsy flange section suggests that local instability of the flange would occur before it was able to develop its full apparent stiffness about that axis. Hence, it is proposed that the actual thickness of the fin, 327 mm for this example, should be used to give a safe design pending research into the buckling properties of such composite sections.

The effective height of the wall may be taken as $0.85h$ on the basis of full lateral restraint being provided at base level, partial lateral restraint at roof level and additional unquantified restraint from the contribution of the flange. The combination of $0.85h$ for the effective height and the fin thickness for the effective thickness about that axis is considered to provide a safe design basis and one which it would be expected to improve upon following a programme of suitable research work.

$$\text{Slenderness ratio, SR} = \frac{0.85 \times 10.0 \times 10^3}{327}$$

$$= 26$$

The eccentricity of the loading about this axis may be taken as zero.
Hence, from BS 5628, Table 7 (see Table 5.15) $\beta = 0.45$.

Design strength of wall – after losses:

$$= \frac{\beta f_k}{\gamma_m} \quad = \frac{0.45 \times 7.6}{2.5} \quad = 1.368 \text{ N/mm}^2$$

Design strength of wall – before losses:

$$= \frac{1.2\beta f_k}{\gamma_m} = \frac{1.2 \times 0.45 \times 7.6}{2.5} = 1.642 \text{ N/mm}^2$$

(where $f_{ki} = f_k$ for this example)

Design stress in wall – after losses:

$$= \frac{(1.4G_k) + (1.6Q_k) + (1.4P_k)}{A}$$

$$= \frac{(1.4 \times 122.68) + (1.6 \times 38.48) + (1.4 \times 187)}{0.46 \times 10^3}$$

$$= 1.076 \text{ N/mm}^2$$

where G_k = ow masonry   = 91.90

$$\text{roof dead load} = 0.6 \times 3.8 \times \frac{27}{2} = \underline{\quad 30.78 \quad}$$

$$= 122.68 \text{ kN}$$

The design strength after losses exceeds the design stress, hence, the wall is acceptable for this loading condition.

Design stress in wall – before losses:

$$= \frac{(1.4G_k) + (1.6Q_k) + (1.4P_k/0.8)}{A}$$

$$= \frac{(1.4 \times 122.68) + (1.6 \times 38.48) + (1.4 \times 187/0.8)}{0.46 \times 10^3} \quad = 1.22 \text{ N/mm}^2$$

where P_k is made initially 20% greater than the design value of 187 kN in anticipation of 20% losses.

The design strength before losses exceeds the design stress, hence, the wall is acceptable for this loading condition.

The stiffness of the fin about the other axis is considerably greater although the effect of the post-tensioning force is eccentric.

The net eccentricity of G_k, Q_k and P_k may be calculated by taking moments of these loads and forces about the flange face.

$$\frac{(122.68 \times 425) + (38.48 \times 425) + (187 \times 301)}{122.68 + 38.48 + 187} = \frac{124.78 \times 10^3}{348.16}$$

$$= 358 \text{ mm from flange face}$$

hence, net eccentricity of G_k, Q_k and P_k about neutral axis of section:

$$e = 301 - 358 = 57 \text{ mm}$$

Expressed in terms of the length of the fin the eccentricity

$$e = 0.064 \times 890$$

This small net eccentricity related to the considerable stiffness of the fin about this axis, compared to the design strengths calculated for the previous axis of buckling, indicates that the section will not be critical, from the point of view of overall stability, about this axis.

15.12.14 Design of post-tensioning rods

As for Example 7 the design stress in the high yield steel rods will be limited to

$$\frac{0.7f_y}{\gamma_m} = \frac{0.7 \times 410}{1.15} = 250 \, \text{N/mm}^2$$

The post-tensioning force required before losses:

$$= \frac{187}{0.8} = 233.75 \, \text{kN}$$

$$\text{steel area required} = \frac{233.75 \times 10^3}{250} = 935 \, \text{mm}^2$$

This will be provided by two high yield rods each of 25 mm diameter positioned as shown in Figure 15.77 to give an equivalent eccentricity of 124 mm.

The design of the torque required for the rods and the spreader plate should follow similar principles to those given for Example 7.

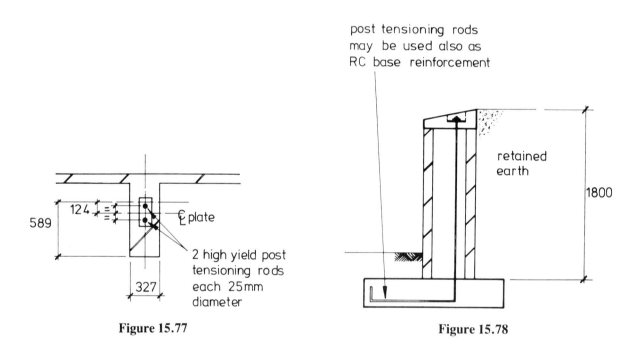

Figure 15.77   **Figure 15.78**

15.13 EXAMPLE 9: POST-TENSIONED, BRICK DIAPHRAGM, RETAINING WALL

The retaining wall designed in Example 3 (see 15.6) as a reinforced brick wall will now be designed as a post-tensioned diaphragm wall. The wall is 1.80 m high and is to be constructed with facing bricks with a crushing strength of 27.5 N/mm^2 and a water absorption of 8% set in a designation (i) mortar. A density of 19 kN/m^3 will be assumed. The same surcharge loading will be included.

For this retaining wall, bending from the retained earth occurs in one direction only and therefore an eccentric post-tensioning force will be applied to counteract this bending. A section through the wall is shown in Figure 15.78.

15.13.1 Design procedure

The design procedure is broadly similar to that used for the design of the post-tensioned fin wall in Example 8. The stress diagrams representing the design process are shown in Figure 15.79.

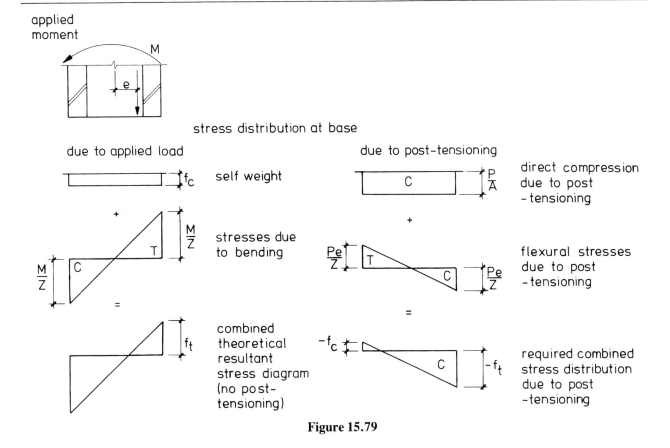

Figure 15.79

The design equations for the required post-tensioning force and its eccentricity may then be derived as follows:

$$\frac{P}{A} + \frac{Pe}{Z_1} = f_t$$

$$\frac{P}{A} - \frac{Pe}{Z_2} = -f_c$$

but $Z_1 = Z_2$ for a diaphragm wall, hence:

$$\frac{PZ}{A} + Pe = f_t Z$$

$$\frac{PZ}{A} - Pe = -f_c Z$$

Adding

$$\frac{2PZ}{A} = (f_t - f_c)Z$$

Rearranging

$$P = (f_t - f_c)\ \frac{A}{2}$$

This is the minimum value of P necessary to produce the required post-tensioning stress. The corresponding value of its eccentricity, e, may now be calculated by substituting P into one of the original equations thus:

$$\frac{P}{A} + \frac{Pe}{Z} = +f_t$$

$$\frac{Pe}{Z} = f_t - \frac{P}{A}$$

$$e = \left(f_t - \frac{P}{A}\right)\frac{Z}{P}$$

$$e = \left(\frac{f_t}{P} - \frac{1}{A}\right)Z$$

This equation provides the eccentricity corresponding to the minimum value of P already calculated. The eccentricity may be found to be larger than can be accommodated within the trial section selected. In such a case, the maximum value of e which can be accommodated should be inserted into the latter equation and a revised value of P obtained. This revised value of P will be larger than that originally calculated.

When the required post-tensioning force and its eccentricity have been calculated, the maximum combined compressive stresses (axial plus flexural) should be checked. The section must also be checked to ensure that no flexural tensile stresses are developed before losses of the post-tensioning force and also the overall stability of the wall when subject to the post-tensioning force alone.

Finally, the moments and stresses in the rc foundation should be considered. However, this aspect of the design is outside the scope of this book.

15.13.2 Design loads
The retained material will be assumed to comprise fine dry sand with an angle of internal friction of 30° and a density of 17 kN/m³. Adequate land drainage will be assumed to have been provided to ensure no build up of water pressure. A surcharge above the retained sand, equivalent to 1.0 m depth of retained material, will also be applied. Hence, from Rankines formula:

earth pressure at any level $= k_1 \times$ density $\times$ height

where k_1 $= \dfrac{1 - \sin\theta}{1 + \sin\theta}$

$= 0.33$

pressure at top of wall $= 0.33 \times 17 \times 1.0$
$= 5.6\ \text{kN/m}^2$

pressure at base of wall $= 0.33 \times 17 \times 2.8$
$= 15.7\ \text{kN/m}^2$

The loading from the retained sand will be treated as a superimposed load.

own weight of masonry $= \dfrac{19 \times 1.8}{10^3}$

$= 0.0342\ \text{N/mm}^2$ at base level

For the loading combination dead plus superimposed the partial safety factors for dead and superimposed loads are 1.4 and 1.6 respectively, when checking combined compressive stresses and 0.9 and 1.6 respectively when checking for the theoretical flexural tensile stresses.

Design superimposed loads:

at top of wall $= 1.6 \times 5.6 = 9.0\ \text{kN/m}^2$

at bottom of wall $= 1.6 \times 15.7 = 25.1\ \text{kN/m}^2$

The design loading diagram is shown in Figure 15.80

Hence, design bending moment at base level, maximum

$$= \left(9.0 \times 1.8 \times \frac{1.8}{2}\right) + \left(16.1 \times \frac{1.8}{2} \times \frac{1.8}{3}\right) = 23.3 \, \text{kN·m}$$

Minimum design dead load stress at base level

$f_c = 9.0 \times 0.0342 = 0.031 \, \text{N/mm}^2$

Design shear force at base level, maximum

$$= (9 \times 1.8) + \left((16.1 \times \frac{1.8}{2}\right) = 30.69 \, \text{kN/m}$$

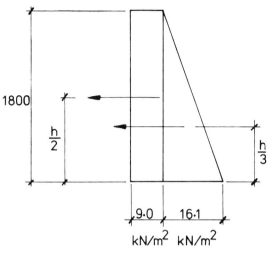

pressure diagram

Figure 15.80

Figure 15.81 Earth pressure diagram

1.0m height of flange designed to span horizontally supporting an average design pressure of $19.3 \, \text{kN/m}^2$

15.13.3 Trial section
There are three dimensional variables which require consideration in order to arrive at a trial section for fuller analysis, these being:

(a) the overall depth of the wall, D,
(b) the thickness of the wall flanges, T, and spacing of the cross-ribs, B_r,
(c) the thickness of the cross-ribs, t_r.

(a) Overall depth, D
There is limited guidance which can be given for a reasonable assessment of the overall depth. Experience and familiarity with the design processes will indicate to the designer the benefits of a deeper wall section which is required to be balanced against space requirements, quantity of walling materials and the effect on the size of the post-tensioning rods and magnitude of the post-tensioning force. Greater depth of section will also assist in resisting shear forces which can be critical and are considered in the third of the three dimensional variables. For this example, the overall depth, D, will be taken as 558 mm.

(b) Flange thickness, T and cross-rib spacing, B_r
The wall flange, on the earth face, is required to support the earth and transfer its pressure to the cross-ribs by spanning horizontally between the cross-ribs. It is considered unreasonable that the maximum pressure, at the base of the wall, should be taken as the load to be supported on the horizontal span and there is likely to be considerable resistance provided by the foundation and the flange will tend to cantilever for a certain height rather than span horizontally. Figure 15.81 shows the design pressure diagram with the loading, assessed by the authors, as being that appropriate to the horizontal span of the flange.

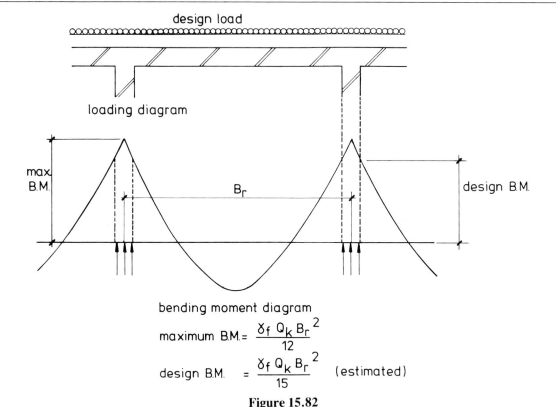

design load

loading diagram

max. B.M.

B_r

design B.M.

bending moment diagram

$$\text{maximum B.M.} = \frac{\gamma_f\, Q_k\, B_r{}^2}{12}$$

$$\text{design B.M.} = \frac{\gamma_f\, Q_k\, B_r{}^2}{15} \quad \text{(estimated)}$$

Figure 15.82

The bending moment diagram for the design of the flange is shown in Figure 15.82 in which the design bending moment intensity has been estimated as $\gamma_f Q_k B_r^2/15$ and occurs, as shown, at the intersection of the flange with the face of the cross-rib.

$$\text{Maximum } M_A = \frac{\gamma_f Q_k B_r^2}{12}$$

$$\text{Design } M_A = \frac{\gamma_f Q_k B_r^2}{15} \text{ (estimated)} = \frac{19.3 B_r^2}{15} = 1.29 B_r^2$$

$$\text{Design MR} = \frac{f_{kx} Z}{\gamma_m} = \frac{1.5 \times 1.0 \times 0.102^2 \times 10^3}{2.5 \times 6} = 1.04 \text{ kN·m}$$

Design MR $\geqslant$ design M_A, therefore $1.04 = 1.29 B_r^2$, hence

$$B_r = \sqrt{\left(\frac{1.04}{1.29}\right)}$$

$$= 0.898 \text{ m} \qquad = \text{maximum span of flange}$$

but the horizontal shear resistance is generally the most significant factor in assessing the size and centres of the cross-ribs in a retaining wall design.

Hence, to suit an acceptable bonding arrangement the cross-ribs will be spaced, for trial purposes, at 675 mm centres which is obviously within the capacity of the span of the flange as designed above.

(c) Cross-rib thickness, t_r
This may be assessed from a consideration of the horizontal shear force from the retained material which has been previously calculated as 30.69 kN/m at base level.

Shear force per cross-rib, $V = 30.69 \times 0.675 = 20.70$ kN

The maximum horizontal shear stress occurs on the centroid of the overall section and may be derived from the formula:

$$v_h = \frac{V A_1 \bar{y}}{I_{na} t_r}$$

where
   v_h = horizontal shear stress
   V = applied horizontal shear force
   A_1 = half the cross-sectional area (as shown hatched in Figure 15.83)
   $\bar{y}$ = distance from NA to centroid of area A_1 (see Figure 15.83)
   I_{na} = second moment of area about neutral axis
   t_r = thickness of cross-ribs.

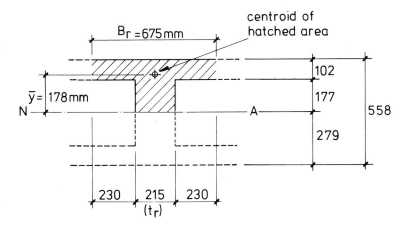

Figure 15.83 Section dimensions

For this example:
   $V = 20.70$ kN per cross-rib
   $A_1 = 0.107$ m^2
   $\bar{y} = 0.178$ m
   $I_{na} = 8.07 \times 10^{-3}$ m^4
   $t_r = 0.215$ m (assessed for trial purposes).

Hence

$$v_h = \frac{20.70 \times 0.107 \times 0.178}{8.07 \times 10^{-3} \times 0.215}$$

$$= 0.227 \text{ N/mm}^2$$

Shear resistance = f_v/γ_m where $f_v = 0.35 + 0.6g_A$ (for this example BS 5628, clause 25). The unknown factor in this equation, at this stage of the design process, is the value of g_A which is the summation of the own weight of the masonry plus the post-tensioning force. However, a minimum post-tensioning force required to provide the horizontal shear resistance may be calculated and later checked against the minimum post-tensioning force applied to eliminate the theoretical flexural tensile stress due to bending. The larger of the two forces calculated should then be used in the design.

$$v_h = \frac{0.35}{\gamma_m} + \frac{0.6g_A}{\gamma_m}$$

therefore

$$0.227 = \frac{0.35}{2.5} + \frac{0.6g_A}{2.5}$$

and

$$g_A = 0.3625 \text{ N/mm}^2$$

But

$$g_A = g_d + \text{design post-tensioning stress}$$

therefore

$$g_d = \frac{0.9G_k}{A} = \frac{6.59 \times 10^3}{0.214 \times 10^6} = 0.031 \text{ N/mm}^2$$

where $G_k = 0.214 \times 19 \times 1.8 = 7.32$ kN/cell = 10.8 kN/m

Hence

$$g_A = 0.031 + \text{design post-tensioning stress}$$
$$0.3625 - 0.031 = \text{design post-tensioning stress}$$
$$0.3315 = \text{design post-tensioning stress}$$

The design post-tensioning stress varies across the section owing to its eccentricity, however the maximum value of horizontal shear occurs where the design post-tensioning stress has its average value.

Hence:

$$\text{design post-tensioning force} = 0.3315 \times A$$
$$= 0.3315 \times 0.214 \times 10^3$$
$$= 71 \text{ kN per cross-rib}$$
$$= 105 \text{ kN/m}$$

At this stage, the design post-tensioning force calculated is that applicable only to the development of horizontal shear resistance. The trial section derived is shown in Figure 15.84 and this section will now be used to check masonry stresses and to design the post-tensioning rods.

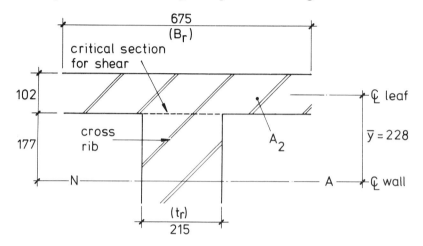

Figure 15.84 Trial section

15.13.4 Calculate theoretical flexural tensile stresses

$$f_t = \frac{0.9G_k}{A} - \frac{M_b}{Z}$$

$$= \frac{9.75 \times 10^3}{0.317 \times 10^6} - \frac{23.3 \times 10^6}{42.82 \times 10^6}$$

$$= 0.031 - 0.54$$

$$f_t = -0.509 \text{ N/mm}^2 \text{ and } f_c = +0.031 \text{ N/mm}^2$$

Hence an eccentric post-tensioning force to produce stresses of $f_c = -0.031$ N/mm^2 and $f_t = +0.509$ N/mm^2 is required.

15.13.5 Minimum required post-tensioning force based on bending stresses

$$P = (f_t - f_c)\frac{A}{2}$$

where
$f_t = 0.509$ N/mm^2
$f_c = -0.031$ N/mm^2
$A = 0.317 \times 10^6$ mm^2/m

hence

$$P = (0.509 - 0.031)\frac{0.317 \times 10^6}{2} = 76 \text{ kN/m}$$

The minimum post-tensioning force required to eliminate the theoretical flexural tensile stresses (76 kN/m) is less than that required to ensure adequate shear resistance (105 kN/m) as calculated in section 15.13.3. The higher value of the two is adopted for subsequent calculations. It is first necessary to determine the eccentricity at which the post-tensioning force should be placed. This is given by the equation:

$$e = \left(\frac{f_t}{P} - \frac{1}{A} \right) Z$$

where $Z = 42.82 \times 10^6 \text{ mm}^3$, hence

$$e = \left(\frac{0.509}{105 \times 10^3} - \frac{1}{0.317 \times 10^6} \right) \times 42.82 \times 10^6$$

$$= 73 \text{ mm}$$

The maximum practical eccentricity which can be provided in a wall section 558 mm deep is:

$$e_{max} = \frac{558}{2} - 102 - \frac{\text{rod diameter}}{2}$$

$$= 177 - \frac{\text{rod diameter}}{2}$$

Say maximum practical eccentricity = 150 mm.

The required eccentricty is within this limit, that is, by placing the calculated force, 105 kN/m, at an eccentricity of 73 mm, both the bending and shear stresses are limited to acceptable values.

If the calculation had indicated that an eccentricity greater than 150 mm were required, then it would have been necessary to use an increased post-tensioning force at a reduced eccentricity. The new force required would be determined by substituting the maximum practical value of e in the formula:

$$P = \frac{f_t}{(1/A) + (e/Z)}$$

15.13.6 Characteristic post-tensioning force, P_k
In order to achieve a minimum design post-tensioning force of 105 kN/m, a characteristic post-tensioning force, P_k of $105/\gamma_f$ should be provided in the rods where γ_f would be taken as 0.9 as discussed in 15.10.9.

Hence, characteristic post-tensioning force $P_k = 105/0.9 = 117$ kN/m. This force is now used to check the design compressive stresses in the wall and to establish the size of the post-tensioning rods.

15.13.7 Capacity reduction factors
The overall stability of the wall section and the local stability of the flanges (leaves) will be checked under combined axial and flexural loading.

(a) Overall stability
Effective height, h_{ef} $= 2h = 2 \times 1.8$ $= 3600$ mm
Effective thickness, t_{ef} = actual thickness $= 558$ mm
(Note: the effective thickness for diaphragm walls was discussed in chapter 13.)

Slenderness ratio, SR $= \dfrac{3600}{558} = 6.5$

Eccentricity of dead load plus characteristic post-tensioning force:

$$\text{effective eccentricity} = \frac{P_k e}{P_k + \text{design dead load}}$$

$$= \frac{117 \times 0.073}{117 + 9.75} = 0.067 \text{ m}$$

$$\frac{\text{effective eccentricity}}{t_{ef}} = \frac{67}{558}$$

$$= 0.120$$

Hence, for SR = 6.5 and $e_x = 0.120t$ from BS 5628, Table 7 (see Table 5.15), $\beta = 0.84$.

(b) Local stability
The flange (leaf) is restrained against buckling by the cross-ribs which may be taken to constitute enhanced resistance to lateral movement.

Hence

$$\text{slenderness ratio} = \frac{h_{ef}}{t_{ef}}$$

$$= \frac{0.75 \times 675}{102}$$

$$= 4.9$$

The combined axial and flexural stresses may be considered to be applied to the flange with zero eccentricity for the purpose of the calculation of β.

Hence, for SR = 4.9 and $e_x = 0$ from BS 5628, Table 7, $\beta = 1.0$ (see Table 5.15).

15.13.8 Check combined compressive stresses
Consider wall subject to dead loading plus post-tensioning force only – before losses.

In anticipation of 20% loss of post-tensioning force the characteristic post-tensioning force, P_k, should initially be increased by 20%, hence:

Post-tensioning force before losses $\qquad = \dfrac{117}{0.8} \qquad = 146 \text{ kN/m}$

Design post-tensioning force before losses $\quad = \gamma_f \times 146 \quad = 204 \text{ kN/m}$

Maximum design dead load $\qquad = 1.4G_k \quad = 1.4 \times 10.8 = 15.2 \text{ kN/m}$

Maximum flexural compressive stress due to post-tensioning force, $P_k = \dfrac{P}{A} + \dfrac{Pe}{Z}$

$$= \frac{204 \times 10^3}{0.317 \times 10^6} + \frac{204 \times 10^3 \times 73}{42.82 \times 10^6}$$

$$= 0.644 + 0.348$$

$$= 0.992 \text{ N/mm}^2$$

Minimum flexural compressive stress due to post-tensioning force $\quad = \dfrac{P}{A} - \dfrac{Pe}{Z}$

$$= 0.644 - 0.348 = 0.296 \text{ N/mm}^2$$

Axial stress due to maximum design dead load $\quad = \dfrac{15.2 \times 10^3}{0.317 \times 10^6} = 0.048 \text{ N/mm}^2$

Combined stresses due to G_k and post-tensioning force:

Maximum combined stress $\qquad = 0.048 + 0.992 = 1.04 \text{ N/mm}^2$

Minimum combined stress $\qquad = 0.048 + 0.296 = 0.344 \text{ N/mm}^2$

Design strength before losses $\qquad = 1.2 \left(\dfrac{1.25\beta f_{ki}}{\gamma_m} \right)$

$$= 1.2 \left(\frac{1.25 \times 1.0 \times 9.2}{2.5} \right)$$

$$= 5.52 \ \text{N/mm}^2$$

in which the wall is assumed to have achieved its full characteristic compressive strength at the time that the post-tensioning is carried out, hence, $f_{ki} = f_k$; also the capacity reduction factor relates to the local stability of the flange in this instance and the effect on the overall stability will be checked in due course.

By inspection the design strength far exceeds the combined compressive stress, hence, the wall is acceptable for this loading condition.

The reserve of strength available in this example before losses indicates that there is no need to check for the same loading condition after losses.

Consider stability of overall wall section under dead loading plus post-tensioning force after losses.

Design axial load
$$= (\gamma_f G_k) + (\gamma_f P_k)$$
$$= (1.4 \times 10.8) + (1.4 \times 117)$$
$$= 178.9 \ \text{kN/m}$$

Design axial stress
$$= \frac{178.9 \times 10^3}{0.317 \times 10^6}$$
$$= 0.564 \ \text{N/mm}^2$$

Design strength of wall
$$= \frac{\beta f_k}{\gamma_m}$$
$$= \frac{0.84 \times 9.2}{2.5}$$
$$= 3.09 \ \text{N/mm}^2$$

in which the effect of the eccentricity of the post-tensioning force has been taken into account in the calculation of β.

The design strength exceeds the design stress; hence, the wall is acceptable for this loading condition.

Consider wall subject to dead loading plus superimposed loading plus post-tensioning force.

Axial stress due to G_k $= \dfrac{\gamma_f G_k}{A} = \dfrac{1.4 \times 10.8 \times 10^3}{0.317 \times 10^6}$ $= +0.048 \ \text{N/mm}^2$

Maximum flexural stress due to post-tensioning force, P_k $= \dfrac{\gamma_f P_k}{A} + \dfrac{\gamma_f P_k e}{Z}$

$= \dfrac{1.4 \times 117 \times 10^3}{0.317 \times 10^6} + \dfrac{1.4 \times 117 \times 10^3 \times 73}{42.82 \times 10^6} = +0.795 \ \text{N/mm}^2$

Minimum flexural stress due to post-tensioning force, P_k $= \dfrac{\gamma_f P_k}{A} - \dfrac{\gamma_f P_k e}{Z}$

$= 0.516 - 0.279$ $= +0.237 \ \text{N/mm}^2$

Flexural stress due to applied moment $= \pm \dfrac{M_b}{Z} = \dfrac{23.3 \times 10^6}{42.82 \times 10^6}$ $= \pm 0.544 \ \text{N/mm}^2$

Hence:
Maximum combined compressive stress $= +0.829 \ \text{N/mm}^2$
Minimum combined compressive stress $= +0.299 \ \text{N/mm}^2$

The stress diagrams for this working condition are shown in Figure 15.85.

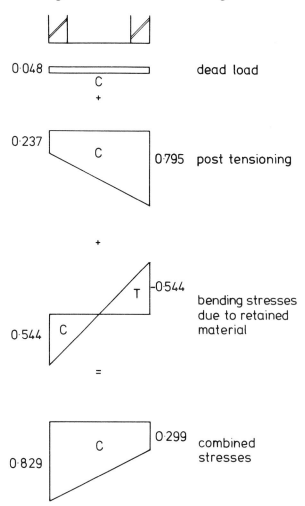

[all stress values in N/mm^2]

Figure 15.85

$$\text{Design strength of wall} = \frac{1.25\beta f_k}{\gamma_m}$$

$$= \frac{1.25 \times 1.0 \times 9.2}{2.5}$$

$$= 4.6 \, \text{N/mm}^2$$

The design strength exceeds the combined stress hence, the wall is acceptable for this loading condition.

Check minimum combined stress in which γ_f for dead and post-tensioning force $= 0.9$.

Axial stress due to G_k $= \dfrac{\gamma_f G_k}{A} = \dfrac{0.9 \times 10.8 \times 10^3}{0.317 \times 10^6}$ $= 0.031 \, \text{N/mm}^2$

Maximum flexural stress due to post-tensioning force, P_k $= \dfrac{\gamma_f P_k}{A} + \dfrac{\gamma_f P_k e}{Z}$

$$= \frac{0.9 \times 117 \times 10^3}{0.317 \times 10^6} + \frac{0.9 \times 117 \times 10^3 \times 73}{42.82 \times 10^6}$$

$$= 0.332 + 0.180 \qquad\qquad = 0.512 \, \text{N/mm}^2$$

| Minimum flexural stress due to post-tensioning force, P_k | $= 0.332 - 0.180$ | $= 0.152 \text{ N/mm}^2$ |

Flexural stresses due to applied moment $= \pm \dfrac{M_b}{Z}$ $= \pm 0.544 \text{ N/mm}^2$

Minimum combined compressive stress $= \text{zero N/mm}^2$
Hence, no tensile stresses are developed.

15.13.9 Check shear between leaf and cross-rib
Another critical section for checking shear stresses is in the vertical plane at the junction of the cross-ribs and the leaves as shown in Figure 15.86.

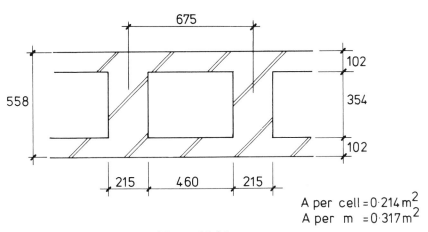

A per cell = 0·214 m²
A per m = 0·317 m²

Figure 15.86

$$\text{Shear stress, } v_h = \frac{VA_2\bar{y}}{I_{na}t_r}$$

where
- V = design shear force = 20.7 kN/cross-rib
- A_2 = area of leaf = $102 \times 675 = 0.069 \times 10^6 \text{ mm}^2$

$\bar{y} = 177 + \dfrac{102}{2} = 228 \text{ mm}$

- I_{na} = moment of inertia = $8.07 \times 10^9 \text{ mm}^4$
- t_r = thickness of cross-rib = 215 mm.

Hence

$$v_h = \frac{20.7 \times 10^3 \times 0.069 \times 10^6 \times 228}{8.07 \times 10^9 \times 215}$$

$$= 0.188 \text{ N/mm}^2$$

The cross-ribs are assumed to be half-bonded into the leaf at alternate courses as shown in the bonding diagram Figure 15.87. For shear failure to occur at the junction of the cross-rib with the leaf, the bonded bricks would need to snap. The Code of Practice does not give values for such a shear failure although research work is being carried out to establish such values.

By inspection of the design vertical shear stress calculated and in comparison with the compressive strength of the bricks being used (27.5 N/mm) the stress appears to be within the likely shear strength of the bricks. The horizontal shear stress has already been shown to be acceptable in 15.13.3.

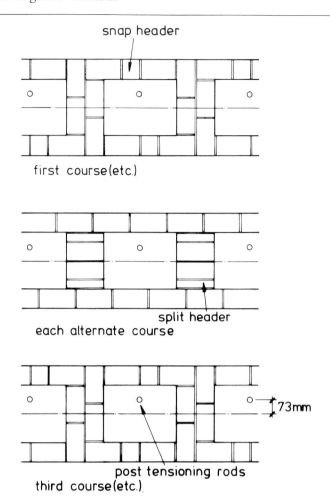

Figure 15.87 Bonding details

15.13.10 Design of post-tensioning rods

Post-tensioning force required = 146 kN/m before losses. One post-tensioning rod per cell of the diaphragm wall will be used, hence

$$\text{post-tensioning force per cell} = 146 \times 0.675 = 98.6 \, \text{kN}$$

$$\text{design strength of rods} = \frac{0.7f_y}{\gamma_{ms}} = 250 \, \text{N/mm}^2$$

$$\text{area of rod required per cell} = \frac{96.6 \times 10^3}{250} = 394 \, \text{mm}^2$$

One high yield post-tensioning rod of 25 mm diameter (area provided = 491 mm^2) will be used in each cell of the diaphragm wall.

The remainder of the rod design should follow similar principles to the previous two examples. A capping beam will be provided to spread the post-tensioning force throughout the wall section.

ARCHES

Brick arches were built in Egypt more than 5000 years ago. They are one of the oldest and most attractive structural forms.

Since arches are basically required to resist compressive forces they are well suited to masonry construction. There have been, in the past, exciting developments from the arch to the vault and the dome. They have been economical, durable and aesthetically pleasing – and almost forgotten by modern engineers. A revival in their use would be invaluable in developing countries with indigenous supplies of stone, brick and local masons. They would also be useful in developed countries in areas where the visual environment needs something more attractive than a plain steel or concrete beam bridge. To give some idea of their potential, a simple masonry arch can easily span 20 m or more. The common terms used in arch design are depicted in Figures 16.1–16.5 and may be useful to the engineer in discussions with architects and builders and in the production of working drawings.

16.1 GENERAL DESIGN

The following discussion deals, for simplicity, with an arch ring unconnected with a spandrel wall – which could be considered as a wall with an arch 'cut out'; or a narrow arch with spandrels on both faces which could be considered similar to a U-shaped section.

Most masonry arches are considered to be 'fixed' arches, i.e. there are no hinges and they are not considered to be capable of resisting tensile stresses. The downward load on the arch creates lateral and compression thrusts in the arch span (see Figure 16.6), which pushes the masonry units against each other and compresses them, and in turn the arch thrusts against the abutments.

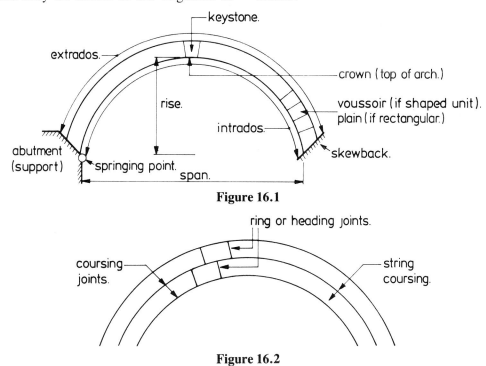

Figure 16.1

Figure 16.2

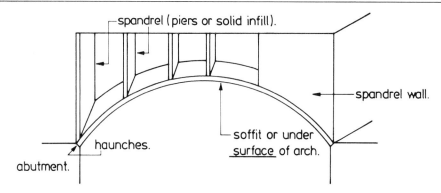

Figure 16.3

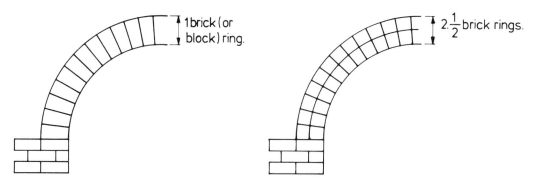

Figure 16.4

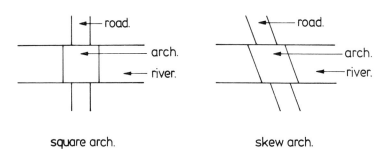

Figure 16.5

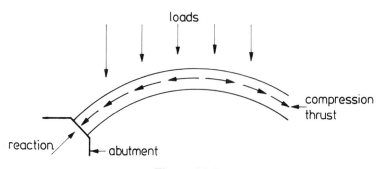

Figure 16.6

If the line of the thrust is on the centre of the arch, the arch ring is under uniform compressive stress (see Figure 16.7).

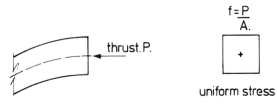

$$f = \frac{P}{A}.$$

uniform stress

Figure 16.7

As will be seen later, the line of the thrust does not always pass along the centre line of the arch, and the arch is not then in uniform compressive stress (see Figure 16.8).

The compressive stress on the arch due to P is P/bd, i.e. P/A, and the stress due to the moment Pe, is Pe/Z.

thrust. P
e
f = P/A + Pe/Z.
f = P/A - Pe/Z
b
d
cross
section.

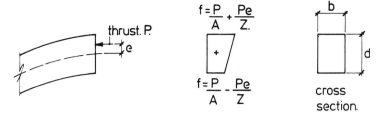

Figure 16.8

The total stress, then, is

$$f = \frac{P}{A} \pm \frac{Pe}{Z}$$

Z for a rectangular section = $bd^2/6$ and A = bd.

At the limit, for no tension (see Figure 16.9):

$$f = 0 = \frac{P}{A} - \frac{Pe}{Z}$$

therefore $\dfrac{P}{A} = \dfrac{Pe}{Z}$

therefore $e = \dfrac{Z}{A} = \dfrac{bd^2/6}{bd} = \dfrac{d}{6}$

P.
e = d/6
d
f = P/A - Pe/Z = 0
stress distribution.

Figure 16.9

Provided that the line of thrust does not pass outside a distance $d/6$ either side of the centre line of the arch, no tension stresses will develop. This, of course, is the well-known 'middle third' rule (see Figure 16.10).

d/6
d/3 middle third.
d/6

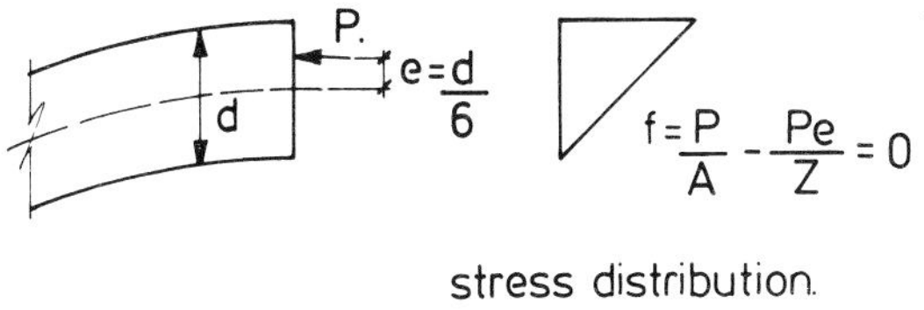

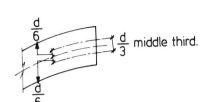

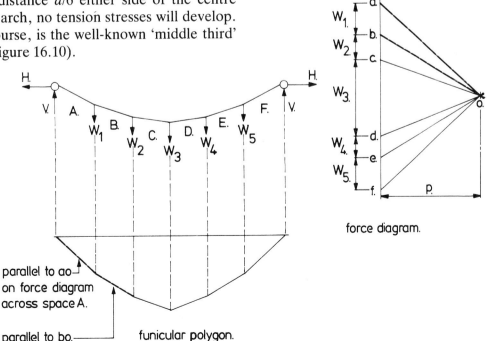

Figure 16.10

In fact, the line of thrust can lie outside the middle third, tensile stresses can develop and cracks can occur. The line of thrust can move to the edge of the arch ring and a 'hinge' will develop, but the arch will not necessarily collapse. This will be discussed later – for simplicity the design of the fixed arches with no tension will be discussed first.

16.1.1 Linear arch

The linear arch (see Figure 16.11) is analogous to a steel cable, and a consideration of the behaviour of the cable will clarify the behaviour of an arch.

H.
V.
A.
B.
C.
D.
E.
F.
W1
W2
W3
W4
W5
parallel to ao
on force diagram
across space A.
parallel to bo.
funicular polygon.

a.
b.
c.
d.
e.
f.
W1.
W2.
W3.
W4.
W5.
o.
p.
force diagram.

Figure 16.11

A force diagram is drawn, for the cable shown in Figure 16.11, from which the funicular polygon can be constructed which shows the magnitude of the forces in the cable. The value of P depends upon the magnitude of the horizontal reaction, H. As H increases, the sag in the cable lessens, but the tension in the cable increases. The force diagram (using the well-known Bow's notation) enables the value of the tension force in the cable to be determined.

If the point loads, W_1, W_2, etc., are replaced by the own weight of the cable, the cable takes up the shape of a catenary curve. When a cable is loaded with a uniformly distributed load, the deflected shape forms a parabola. (It should be noted that the own weight alone of a cable does not produce a uniformly distributed load.) If the steel cable, shown in Figure 16.11, is replaced by a series of short steel rods pin-jointed at their connections at load points W_1, W_2, W_3, etc., and then inverted, the rods would be in compression. The forces in the rods would be the same as in the cable, and have the same vertical and horizontal components. The inverted funicular polygon is known as a linear arch (see Figure 16.12).

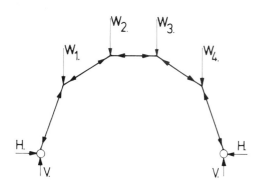

linear arch.

Figure 16.12

The horizontal thrust at the abutments of the arch, and the compression in the rods, depend not only on the magnitudes of W_1, W_2, etc., but also on the rise of the arch (in the same way as the sag in the cable depended on a horizontal pull at its supports) and on the magnitude of the loads (see Figure 16.13). Roman arches were nearly always semi-circular, so that the rise was half the span. Many experts were alarmed when Brunel built the Maidenhead Bridge with a span of 38 m and a rise of only 8 m. The bridge apparently now carries ten times the load Brunel envisaged.

Similarly, if the arch is uniformly loaded, its most efficient shape is parabolic (see Figure 16.14). In some cases, in practice, the line of thrust is near enough for design purposes to the arc of a circle.

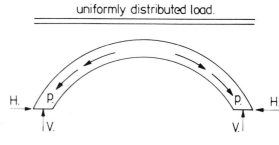

Figure 16.14

Changes in the magnitude of the thrust, H, affect the magnitude of the compression force, P. It will also alter its position within the arch ring.

In a uniformly loaded arch, P is assumed to act horizontally at the crown (top) of the arch and in the centre of the arch ring (see Figure 16.15(a)).

P is calculated (admittedly crudely) by cutting through the arch at the crown and taking moments about A, assuming A to act as a pin joint:

when $P \times r = W_1 \times X_1 + W_2 \times X_2 + W_3 \times X_3$ (see Figure 16.15(b)).

This assumes that there is no restraining moment at A, but this error gives the maximum possible value of P and is therefore on the safe side.

Since the theory of fixed arches is based on assumptions of doubtful validity, in practice (see

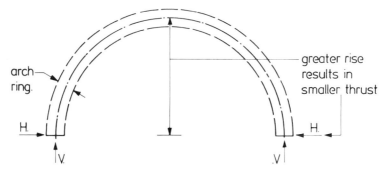

Figure 16.13

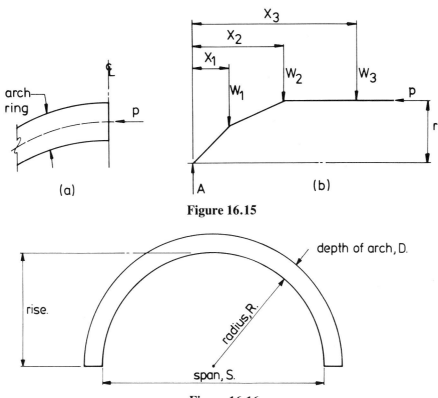

Figure 16.15

Figure 16.16

below) many engineers think it prudent to determine the linear arch which lies nearest to the line of the arch. This will be explained in later examples. From experience, it has been found that graphical analysis of trial sections is more satisfactory, certainly for spans up to 15 m, than the questionable and tedious mathematical analyses. It gives too, a better 'feel' of the structure and forces to the designer.

16.1.2 Trial sections

A number of Victorian engineers derived formulae for trial sections for brickwork based on experience, and some testing, and examples of these are given below (for S, R and D, see Figure 16.16).

Rankine:
$$D = \sqrt{(0.12R)}$$

Trautwine:
$$D = 0.27 + 0.33\sqrt{\left(\frac{S}{2} + R\right)}$$

Depuit:
$$D = \sqrt{(0.074S)} \text{ (for segmental arches)}$$
$$D = \sqrt{(0.13S)} \text{ (for semi-circular)}$$

Sejourne:
D (for ellipses)
$$= 0.15\left(\frac{4}{3 + \dfrac{2 \times \text{rise}}{\text{span}}}\right)(3.28 + \sqrt{(3.28S)})$$

D (for semi-circular)
$$= 0.15\,(3.25 + \sqrt{(3.28S)})$$

Hurst
$$D = 0.4\sqrt{R}$$

These formulae are based on Imperial measurements, and would need conversion to SI units.

If the foregoing 'rules' are applied to a semicircular arch of 50' span (therefore, $R = 25'$ and rise $= 25'$), the results are surprisingly consistent.

Rankine: $D = 2.45'$
Trautwine: $D = 2.58'$
Depuit: $D = 2.55'$ Average = 2.39 ft
Sejourne: $D = 2.4'$ (approximately 0.8 m)
Hurst: $D = 2.0'$

With better bricks, stricter control of workmanship, and a greater understanding of structural behaviour, the trial section could be reduced. The 'rules' were based on relatively thick arches supporting massive earth filling. The influence of the rolling imposed loads from horses and carts was negligible (see Figure 16.17).

Many Victorian masonry arches over railways and canals, built to carry horse-drawn carts and carriages, had to be checked before the D-Day invasion in the Second World War to determine whether they were capable of supporting the

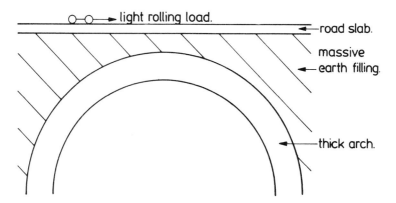

Figure 16.17

massive loadings from tanks and artillery. Practically all of them were more than adequate.

16.1.3 Mathematical analysis

Professor Pippard (I.C.E. Jan '39) stated that the 'middle third' rule was unduly pessimistic and suggested the use of the 'middle half'. In later work, he found that the thrust could pass not just outside the middle half, but outside the arch ring without the arch collapsing.

Professor Hardy Cross (of moment distribution fame) found on concrete arches (University of Illinois, Bulletin No. 203) that half the total stress in the arch was due to its own dead weight. He later went on to say: 'Some of the problems of arch analysis, however, cannot be solved by either rational or empirical methods alone. They are problems in probability in which the range of uncertainty of certain fundamental variables is a matter for observation, but the probable uncertainty in the results consequent upon accidental combination of these variables can scarcely be

determined by experiments'. This view may now be outdated, but good research, applied with appropriate partial factors of safety, would be valuable to practising engineers.

The mathematical analysis of masonry arches has, at times, attracted mathematicians and structural theorists. The resulting highly complex mathematics have not been acceptable to many practising designers, perhaps because the assumptions made by the analysis do not occur in practice. Typical assumptions and the authors' rejection of the analyses are given below:

The abutments do not move

In practice the abutments can be subject to the following movements:

(a) Spread of the abutments

The arch will thrust against the abutments causing them to spread apart (see Figure 16.18). The arch will crack at the crown and springings and form a statically determinate three-pinned arch

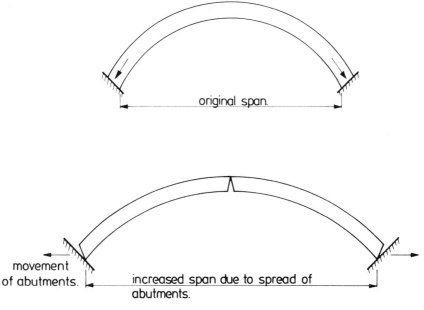

Figure 16.18

(as distinct from the assumed fixed arch). The hinges would show up as haircracks in the mortar joints, and the line of thrust would pass through the hinge points. The arch, acting as a three-pinned arch, would still be safe and stable, provided that the compressive resistance of the masonry is not exceeded.

(b) 'Squashing' together of the abutments

The pressure from retained earth and excessive surcharge loading can push the abutments together and reduce the span of the arch. Again, the arch could crack and form a three-pinned arch as shown in Figure 16.19.

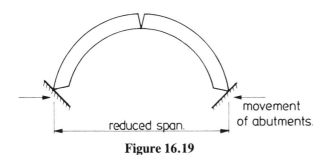

Figure 16.19

(c) Differential settlement of the abutments

It is rare in practice for both abutments to settle by exactly the same amount. Even minute differential settlements would distort many mathematical analyses. The most thorough soil surveys, investigations and analyses only give estimates of settlement.

Arches are very stable structures and are not over-sensitive to foundation movement. Clare Bridge at Cambridge, for example, is appreciably distorted due to movement of the abutment which took place a long time ago. It is still standing and is still safe.

The arch is elastic

Masonry, whilst at low stresses can act elastically, even though it is not an elastic material.

The arch does not change its profile

The elements of the arch can contract due to compression forces, creep action, shrinkage of the mortar (or concrete blocks), etc. These contractions will shrink the arch and cause it to sag. The sagging would in most cases, be minute, and can in practice be ignored.

The elements of the arch can expand due to moisture and thermal movement, expansion of bricks, chemical changes, etc. These expansions will 'lengthen' the arch and cause it to rise. Again the rise would generally be minute and of little practical consequence.

Arches, as any other structural member, are rarely built to an exact profile. There are the normal construction tolerances and small imperfections. Arch formwork will deform slightly during construction of the arch, mortar joints and masonry units have permissible tolerances, setting out even, if possible, to accuracies of 1 in 100000 is not 'accurate'.

The material is homogeneous and isotropic

Arches are made of masonry and mortar which have different properties. Mortar does not glue the masonry together, but helps to transfer the compressive load from one masonry unit to another uniformly, and not just at the high points. Masonry does not have an orthogonal ratio of unity.

The loading is uniform

When the dead loads and superimposed loadings are uniform, the arch has its highest compressive stresses. The worst case for tensile stresses is usually when only one half of its span is subject to superimposed load (see Figure 16.20).

If the superimposed load is sufficiently excessive, four hinges could form and a collapse mechanism develop.

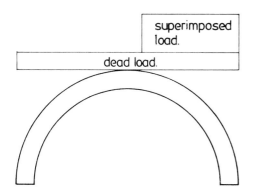

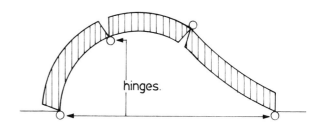

Figure 16.20

Changes of profile, span, length, etc., which will happen, have a very significant effect on complex mathematical analyses and very little effect on practical design or strength and stability of the arch. Masonry, like steel and reinforced concrete, has sufficient elasticity to withstand small structural movements without damage.

Mathematical analysis is, of course, invaluable to the engineer when it allows him to design safe and economical structures, and the above comments should not be regarded as denigrating analysis but as helping to give a sense of proportion. The simple graphical analysis should produce a safe arch. Some mathematical analysis may result in a thinner arch, which would make for extra economy. However, it should be appreciated that much of the cost of arch bridges is independent of the thickness of the arch. For example, the costs of the abutments, temporary centring formwork, road slab, filling and spandrels are hardly affected by whether the arch is three rings or four rings thick.

Most masonry designers would have more confidence in design tolerance based on extensive study of the structural behaviour of actual masonry arches, than those based on theoretical assumptions.

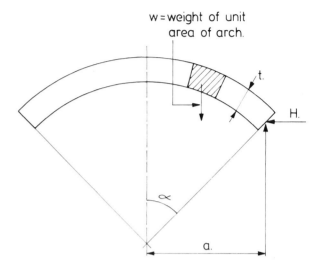

Figure 16.22

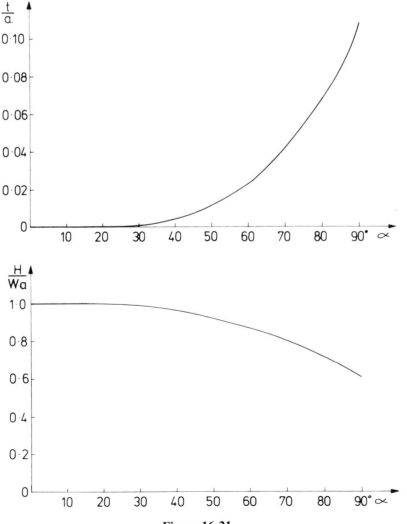

Figure 16.21

One of the simplest analyses (and in the opinion of the authors, one of the best) is probably that of Professor Heyman of Cambridge, in his excellent book, *Equilibrium of Shell Structures*. This deals with an arch of minimum thickness supporting its own weight only, and the depth of the calculated arch must be increased for trial section analyses to carry further dead load and any imposed loading. He has produced two graphs, (see Figure 16.21) to determine the thickness of the arch and horizontal component of the abutment thrust.

Notation in Figure 16.21 is shown in Figure 16.22.

16.2 DESIGN PROCEDURES

1. Choose rise (between ¼ to ½ span)
2. Choose shape (preferably parabolic or arc of a circle)
3. Choose trial section
4. Carry out graphical analysis

(a) Divide the arch and the filling above into a number of segments such as A, B, C, D, E (see Figures 16.23 and 16.24).
(b) Determine the dead load of the arch, filling and imposed load on each segment. Take both cases of imposed loading, i.e. whole span loaded and from abutment to crown only loaded.
(c) Treat the distributed load as a series of point loads, W_1, W_2, etc.
(d) Calculate thrust H_z at crown by taking moments about X:

$H_z \times r = W_1 \times X_1 + W_2 \times X_2 + W_3 \times X_3$ etc. (for this calculation, assume that H_z is horizontal and that the moment at the crown is zero, i.e. a three-pinned arch).

(e) Plot the force diagram, using Bow's notation (see Figure 16.25).
(f) Draw the line of thrust on the arch profile thus:
   (i)   start at crown of arch and plot the horizontal thrust line acting through

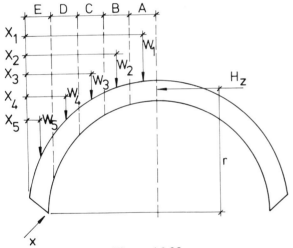

Figure 16.23

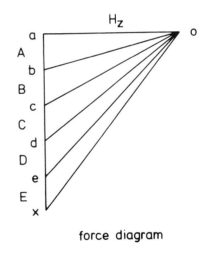

force diagram

Figure 16.25

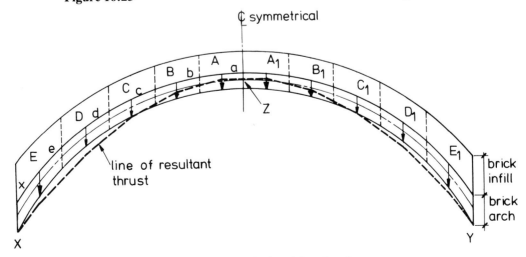

uniformly distributed dead load only

Figure 16.24

the centre of the depth of the arch ring.

(ii) through space A draw line parallel to ao (i.e. horizontal thrust line H_z) changing slope at 'a' and becoming parallel to bo.

(iii) continue line parallel to bo through space B and change slope at b to become parallel to co and so on.

(g) The resulting line, known as the line of resultant thrust, is checked to determine whether it passes outside the middle third of the depth of the arch ring.

5. Check stresses at the critical locations on the arch ring using the procedure given in the following examples.

6. Check 'cracked section' analysis if unacceptably high tensile stresses are indicated in operation 5.

7. Redesign as necessary; note that if the line of resultant thrust is shown, in operation 4(g), to be considerably outside the middle third of the depth of the arch ring, it is likely that the shape of the arch will require adjustment and operations 1 to 4 should be repeated for the new arch profile.

8. Having established a suitable arch profile, and checked the stress levels in operations 5 and 6, choose masonry unit and mortar strength.

16.3 DESIGN EXAMPLES

The following design examples will simply relate loads to stresses and will not be presented in 'limit state' terms. It is considered that this will provide the designer with a clearer picture of the mechanics of arch design and will highlight the need for experienced judgement and adjustment of the trial section.

16.3.1 Example 1: Footbridge arch

Segmental brick arch of 10 m span and 2 m rise will be assumed to be 330 mm deep and is subject to various loading conditions (see Figure 16.26).

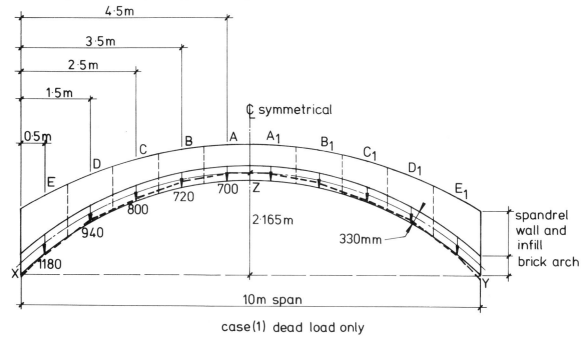

case (1) dead load only

Figure 16.26

The arch has been divided into ten segments each of equal length (equal length segments were chosen merely to simplify the calculations and this is not a pre-requisite of the process). The mass of each segment is then calculated and a density of 23.5 kN/m³ has been assumed for the masonry in each of the following examples.

Calculate mass of arch segments

segment A (A$_1$) 0.70×1.0 $= 0.70$ m³
segment B (B$_1$) 0.72×1.0 $= 0.72$ m³

segment C (C_1) 0.80×1.0 $= 0.80\,m^3$
segment D (D_1) 0.94×1.0 $= 0.94\,m^3$
segment E (E_1) 1.18×1.0 $= 1.18\,m^3$

Hence, using a density of 23.5 kN/m^3
mass   A (A_1) $= 0.70 \times 23.5$ $= 16.45\,kN$
mass   B (B_1) $= 0.72 \times 23.5$ $= 16.92\,kN$
mass   C (C_1) $= 0.80 \times 23.5$ $= 18.80\,kN$
mass   D (D_1) $= 0.94 \times 23.5$ $= 22.09\,kN$
mass   E (E_1) $= 1.18 \times 23.5$ $= 27.73\,kN$

Case (1): dead loading only
Having determined the mass of each segment, moments of the masses are taken about the point X at the base of the arch. The summation of these moments is equated to the moment of the thrust, acting at the point Z at the crown of the arch. In this example the thrust is thus shown to be 105 kN as follows.

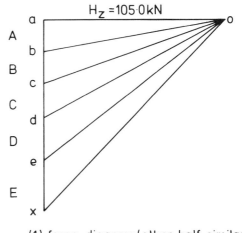

(1) force diagram (other half similar)

Figure 16.27

Take moments about X:
mass   A $= 16.45 \times 4.5$ $=$ 74.03
mass   B $= 16.92 \times 3.5$ $=$ 59.22
mass   C $= 18.80 \times 2.5$ $=$ 47.00
mass   D $= 22.09 \times 1.5$ $=$ 33.14
mass   E $= 27.73 \times 0.5$ $=$ 13.87
Total $= 227.26$ kN·m

$H_z \times 2.165 = 227.26$
$H_z = 105.0$ kN

The base line of the force diagram can now be plotted to a suitable scale with a horizontal line ao representing the thrust H_z in magnitude and direction (see Figure 16.27). From point 'a' the mass of each segment is plotted on a vertical line with a–b representing the mass of segment A, b–c representing the mass of segment B, etc., until point x is reached. The lines of thrust in each segment are now established by connecting point o to each of the points b, c, d, e and x. Line xo represents the resultant reaction at the base of the arch.

The forces from the force diagram are now transferred to the arch profile following the procedure given earlier in operation 4(f). The line of resultant thrust is symmetrical about the centre line of the arch for this symmetrical loading condition. By inspection of the line of resultant thrust there appears to be a potential problem near to the springing point where the thrust line is touching the edge of the arch ring (see Figure 16.26). Hence, the thrust has a large eccentricity and tensile stresses within the arch ring may be excessive. However, the thrust at the crown was assumed to be applied on the centre of the depth of the arch ring and it is likely that if this were moved upwards to the edge of the middle third of the depth of the arch ring, the stress levels at this location would still be acceptable. In doing so the line of resultant thrust at the springing point would move nearer to the centre line of the arch, thus reducing the effect of its eccentricity. The line of thrust will, of course, take is own course in practice and the object of the design process is to set out a theoretical line of thrust and adjust it until a situation is reached where all stresses are within acceptable limits. This may not necessarily be the most efficient line for the thrust to take, in that the stress levels in some zones of the arch ring may be disproportionately higher than in other zones. However, the design will have demonstrated that the section is adequate. In adjusting the point of application of the thrust at the crown, the assumptions made at the outset will be incorrect, i.e. the dimensions used to calculate the reactions will differ slightly and there will now be a bending moment at the crown where zero moment was assumed for the calculation of the reactions. However, it is considered that, provided stresses are checked throughout and shown to be acceptable, these errors may be ignored and the suggested design method should produce a safe and satisfactory solution.

Cases (2) and (3): dead plus superimposed loading.
Case (2) is based on a crowd superimposed loading of 4 kN/m in addition to the dead loads as calculated for case (1). The horizontal thrust H_z for this loading case is calculated to be 128.14 kN and the line of resultant thrust is shown in Figure 16.28 having been plotted from the force diagram shown in Figure 16.29.

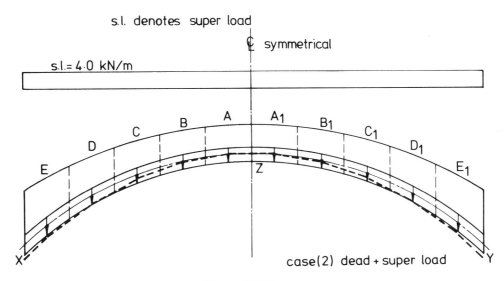

Figure 16.28

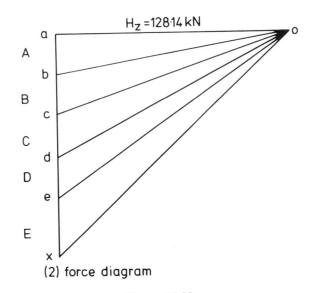

Figure 16.29

Similarly, for case (3) loading, where the superimposed load is increased to 8.0 kN/m, the relevant arch and force diagrams are shown in Figures 16.30 and 16.31 respectively.

The calculation of stresses within the masonry will be dealt with in Example 2.

16.3.2 Example 2: Segmental arch carrying traffic loading
A circular arch has been selected which, from experienced judgement, provides the most suitable profile to support the heavier loading involved. The span of the arch is 10.0 m and the rise will be made 4.0 m, almost semi-circular. The loading information and arch profile is shown in Figure 16.32 and the thickness of the masonry will be taken as 1.0 m.

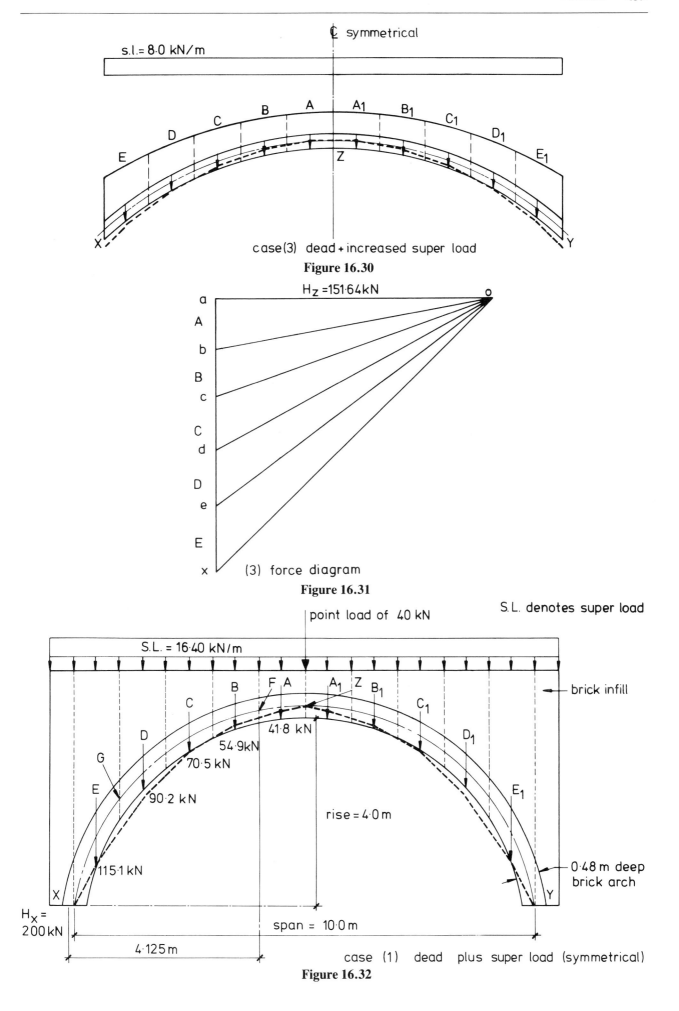

s.l.= 8.0 kN/m

ℓ symmetrical

case(3) dead+increased super load

Figure 16.30

$H_z = 151.64$ kN

(3) force diagram

Figure 16.31

S.L. denotes super load

point load of 40 kN

S.L. = 16.40 kN/m

brick infill

41.8 kN

54.9kN

70.5 kN

90.2 kN

rise = 4.0 m

115.1 kN

0.48 m deep
brick arch

$H_x =$
200 kN

span = 10.0 m

4.125 m

case (1) dead plus super load (symmetrical)

Figure 16.32

The design procedure is similar to that used for the previous example and for convenience the arch has been divided into ten equal segments.

Case 1: dead plus superimposed loading

Calculate thrust, H_z, at crown

	Dead load (kN)	Superimposed load (kN)	Total load (kN)
segment A	25.4	16.4	41.8
segment B	38.5	16.4	54.9
segment C	54.1	16.4	70.5
segment D	73.8	16.4	90.2
segment E	98.7	16.4	115.1

(excluding 40 kN point load)   Total = 372.5

total vertical reaction at X = 372.5 + 20.0   = 392.5

(where 20 kN = half centre point load)

Take moments about X:

mass A   = 41.8×4.5 = 188.10
mass B   = 54.9×3.5 = 192.15
mass C   = 70.5×2.5 = 176.25
mass D   = 90.2×1.5 = 135.30
mass E   = 115.1×0.5 = 57.55
half point
load   = 20.0×5.0 = 100.00

$$\text{Total} = 849.35 \text{ kN·m}$$
$$H_z \times 4.24 = 849.35$$
$$H_z = 200 \text{ kN}$$

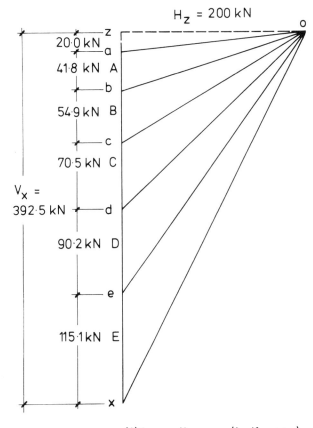

(1) force diagram (half span)

Figure 16.33

Plot force diagram
The force diagram, shown in Figure 16.33, may now be plotted. The one difference in the setting out of the force diagram is the need to provide an inclined thrust at the crown to generate a vertical component equal and opposite to the point load applied at the centre of the span. Half the point load has been assumed to influence each half of the symmetrical profile. Hence, line ao is inclined by the introduction, on the force diagram of the 20 kN external force, z–a.

The forces are now transferred back to the arch profile shown in Figure 16.32 and once again the initial commencement point is at the centre of the depth of the arch at the crown. By inspection, a logical adjustment of this line of resultant thrust would be to raise the point of application of the thrust at the crown which would very likely result in the thrust line being contained within the arch depth throughout its length. However the calculation of stresses in the masonry will be carried out using the line of thrust shown in Figure 16.32 based upon the thrust magnitudes given in the force diagram (Figure 16.33).

Calculate masonry stresses
For convenience, as numerous repetitive calculations are required, the masonry stresses may be computed in tabular form as shown in Table 16.1

Table 16.1 Masonry stresses (Two typical locations, F and G, are considered)

	Point F	Point G
Thrust (P)	213.0 kN	342 kN
Bending moment (M)	*anticlockwise* $54.9 \times 0.5 \quad = 27.45$ $70.5 \times 1.5 \quad = 105.75$ $90.2 \times 2.5 \quad = 225.50$ $115.1 \times 3.5 \quad = 402.85$ $200 \times 4.13 \quad = \underline{826.00}$ $\quad$ total $= 1587.55$ kN·m *clockwise* $392.5 \times 4.0 = 1570.0$ kN·m *imbalance* $1587.55 - 1570 = 17.55$ kN·m	*anticlockwise* $115.1 \times 0.5 \quad = 57.55$ $\underline{200 \quad \times 2.445} = 489.00$ $\quad$ total $= 546.55$ kN·m *clockwise* $392.5 \times 1.0 = 392.5$ kN·m *imbalance* $546.55 - 392.5 = 154.05$ kN·m
Calculated eccentricity (e) $e = M/P$	$e = \dfrac{17.55 \times 10^6}{213.0 \times 10^3}$	$e = \dfrac{154.05 \times 10^6}{342 \times 10^3}$
Eccentricity scaled from drg.	120 mm	410 mm
P/A	$= \dfrac{213 \times 10^3}{480 \times 10^3} = 0.444$ N/mm²	$= \dfrac{342 \times 10^3}{480 \times 10^3} = 0.713$ N/mm²
M/Z	$= \dfrac{17.55 \times 10^6}{38.4 \times 10^3} = 0.457$ N/mm²	$= \dfrac{154.05 \times 10^6}{38.4 \times 10^3} = 4.012$ N/mm²
Maximum stress $\dfrac{P}{A} + \dfrac{M}{Z}$	$0.444 + 0.457$ $= +0.901$ N/mm² (compressive)	$0.713 + 4.012$ $= +4.725$ N/mm² (compressive)
Minimum stress $\dfrac{P}{A} - \dfrac{M}{Z}$	$0.444 - 0.457$ $= -0.013$ N/mm² (tensile)	$0.713 - 4.012$ $= -3.299$ N/mm² (tensile)

The stresses shown in Table 16.1 should now be compared with those allowable for the selected bricks and mortar using the basic principles given in Chapter 11. It is already evident that high tensile stresses exist at location G and a 'cracked section' analysis will almost certainly have to be carried out at this location.

Case 2: dead plus partial superimposed loading
Figure 16.34 shows the same arch with superimposed loading on one half only of the arch span.

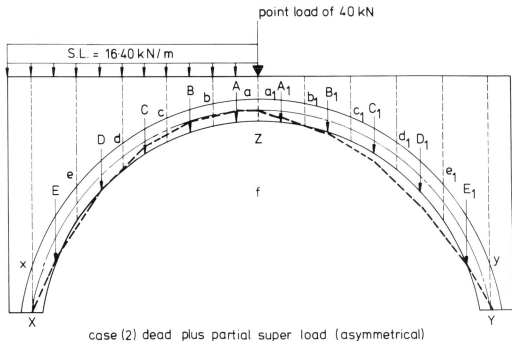

case (2) dead plus partial super load (asymmetrical)

Figure 16.34

The design procedure is the same and the force diagram, for the full span since the loading is asymmetrical, is shown in Figure 16.35.

Calculate thrust, H_z, at crown

	Dead load (kN)	Superimposed load (kN)	Total load (kN)
segment A	25.4	16.4	41.8
segment B	38.5	16.4	54.9
segment C	54.1	16.4	70.5
segment D	73.8	16.4	90.2
segment E	98.7	16.4	115.1
segment A_1	25.4	0	25.4
segment B_1	38.5	0	38.5
segment C_1	54.1	0	54.1
segment D_1	73.8	0	73.8
segment E_1	98.7	0	98.7

(excluding 40 kN point load)   Total = 663 kN
total vertical reaction at X = 372.0 kN
total vertical reaction at Y = 331.0 kN
$\Sigma V = 0$ $V_x + V_y = 703$ (1)
$\Sigma H = 0$ $H_x - H_y = 0$ (2)
Hence
$$0 = 4.24H_x + (41.8 \times 0.5) + (54.9 \times 1.5) + (70.5 \times 2.5) + (90.2 \times 3.5) + (115.1 \times 4.5) - (5.0V_x)$$

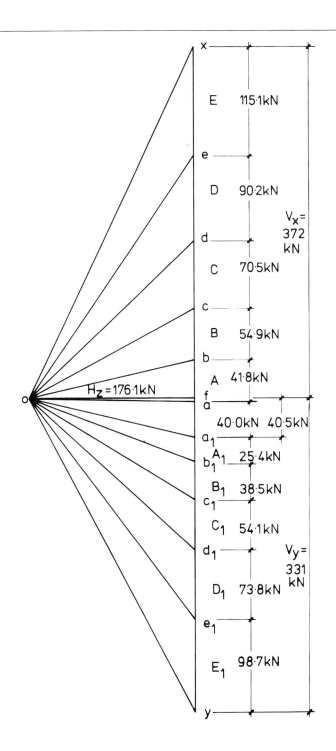

(2) force diagram (full span)

Figure 16.35

$$0 = (4.24H_x) + 1113.3 - (5.0V_x) \qquad (3)$$

and

$$0 = 4.24H_y + (25.4 \times 0.5) + (38.5 \times 1.5) + (54.1 \times 2.5) + (73.8 \times 3.5) + (98.7 \times 4.5) - (5.0V_y)$$

$$0 = (4.24H_y) + 908.2 - (5.0V_y) \qquad (4)$$

Now adding (3) + (4)

$$0 = 4.24(H_x + H_y) + 2021.5 - 5(V_x + V_y) \qquad (5)$$

and substituting (1) into (5)

$$0 = 4.24(H_x + H_y) + 2021.5 - 3515$$

therefore: $\qquad\qquad\qquad\qquad H_x + H_y = 352.2 \text{ kN} \qquad (6)$

and $\qquad\qquad\qquad\qquad\qquad\quad H_x - H_y = 0 \qquad (2)$

hence $\qquad\qquad\qquad\qquad\quad H_x = H_y = 176.1 \text{ kN}$

Vertical component of thrust at crown, V_z

to left of Z:
$$V_x - 372.5 - 40 - V_z = 0$$
$$V_z = 40.5 \text{ kN}$$

check to right of Z:
$$V_y - 290.5 + V_z = 0$$
$$V_z = 40.5 \text{ kN}$$

Horizontal component of thrust at crown, H_z

Taking moments at X

$(40 \times 5.0) + (115.1 \times 0.5) + (90.2 \times 1.5) + (70.5 \times 2.5) + (54.9 \times 3.5) + (41.8 \times 4.5) - (V_z \times 5.0) - (4.24 H_z) = 0$

$4.24 H_z = 949.35 - 202.5$

$\quad H_z = 176.1 \text{ kN}$

16.3.3 Example 3: Repeat Example 2 using a pointed arch

For the same loading and dimensional criterior as Example 2, a pointed arch will now be analysed to demonstrate its unsuitability.

Case 1: dead plus superimposed loading

Figure 16.36 shows the arch profile and the line of resultant thrust for case 1 loading (dead plus superimposed) which, again, has been plotted from the force diagram shown in Figure 16.37.

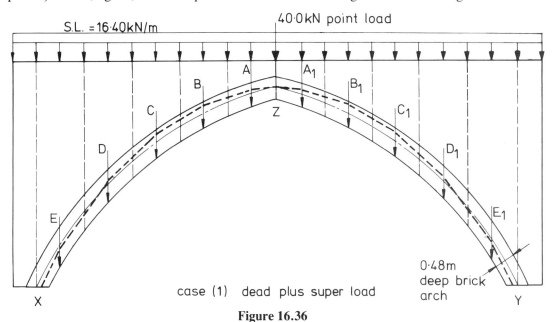

Figure 16.36

Calculate thrust, H_z, at crown

Take moments about X:

segment A $= (25.4 + 16.4) \times 4.5 = 188.10$
segment B $= (38.5 + 16.4) \times 3.5 = 192.15$
segment C $= (54.1 + 16.4) \times 2.5 = 176.25$
segment D $= (73.8 + 16.4) \times 1.5 = 135.30$
segment E $= (98.7 + 16.4) \times 0.5 = 57.55$
point load $= 40.0/2 \times 5.0 = \underline{100.00}$

$$\text{Total} = 849.35 \text{ kN·m}$$
$$H_z \times 4.24 = 849.35$$
$$H_z = 200.0 \text{ kN}$$

The force diagram, shown in Figure 16.37, may now be plotted in the same manner as for Example 2. By inspection of the line of thrust, a considerable eccentricity exists at approximately quarter span indicating that the profile of the arch is not as suitable as the circular arch analysed in the previous example.

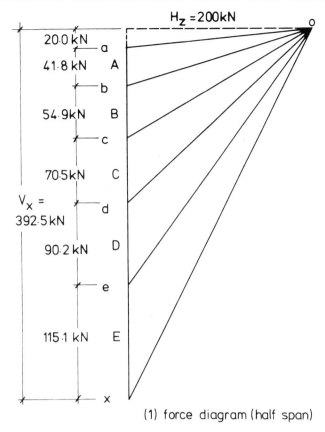

(1) force diagram (half span)

Figure 16.37

Case 2: dead plus partial superimposed loading

Figure 16.38 shows the same arch profile with the superimposed loading on only half of the span and the revised line of resultant thrust which, clearly, is an even more critical design condition than that produced from the case 1 loading previously calculated

The force diagram, shown in Figure 16.39, once again covers the full span owing to the asymmetrical loading.

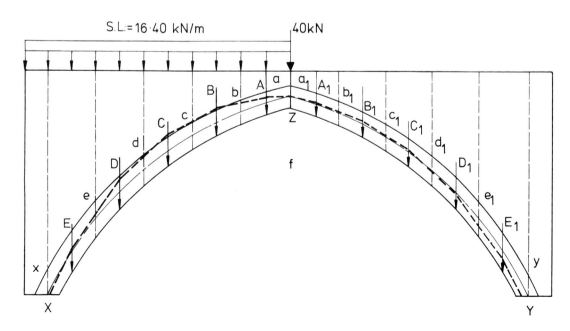

(2) dead plus partial super load

Figure 16.38

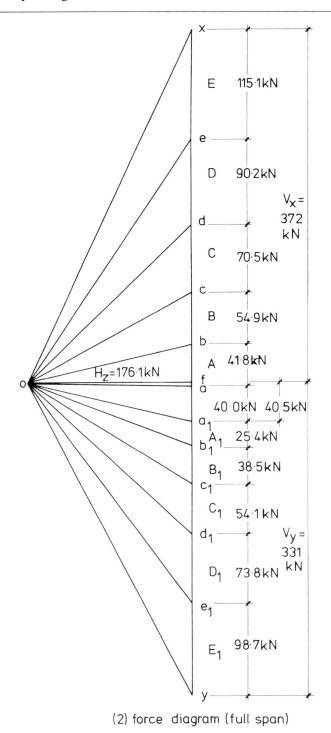

(2) force diagram (full span)

Figure 16.39

The calculation of V_x, V_y, V_z, H_x, H_y and H_z follows the same procedure as for Example 2 and, in fact, gives identical values i.e.

$V_x = 372 \text{ kN}$ $V_y = 331 \text{ kN}$ $V_z = 40.5 \text{ kN}$
$H_x = 176.1 \text{ kN}$ $H_y = 176.1 \text{ kN}$ $H_z = 176.1 \text{ kN}$

MATERIALS

A1.1 CLAY BRICKS

BS 3921 describes a brick as a walling unit laid in mortar, and being not more than 337.5 mm long × 225 mm wide × 112.5 mm high.

A1.1.1 Size

Many of the advantages of building with bricks stem from their easily handled size and weight, which have changed little over the centuries, and the human scale they bring to the appearance of a wall, however large and massive.

According to the current BS 3921, there is only one standard clay brick size: 215 mm × 102.5 mm × 65 mm. This is the actual size of the

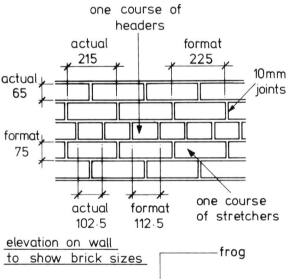

elevation on wall to show brick sizes

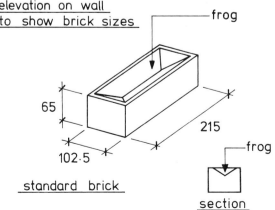

standard brick

Figure A1.1

brick. However, as bricks are laid in mortar, it is convenient to consider the size of the brick plus its share of the mortar joints. This measurement is termed the 'format size'. As a 10 mm joint is normal, the standard format size becomes 225 mm × 112.5 mm × 75 mm, as illustrated in Figure A1.1

Although there is only one British Standard size of brick, some manufacturers also produce modular bricks to meet the needs of dimensional coordination. Format sizes are as follows:

2M 200 × 100 × 100 mm
    200 × 100 × 75 mm

3M 300 × 100 × 100 mm
    300 × 100 × 75 mm

As bricks are made from clay, which is fired in kilns at very high temperatures, some variation in size is to be expected. Such differences are not generally significant in large areas of brickwork, but may be so in smaller elements. Problems can arise if two different suppliers are used for the opposite leaves of a cavity wall, in that there may be difficulties in tying the two leaves together.

A1.1.2 Classification

All clay bricks may be classified under the headings of variety, quality and type, as defined in BS 3921.

Variety

Three varieties are defined.

(i) Common: any brick for general building work, but not specifically chosen for attractive appearance.
(ii) Facing: any brick specially made or selected for its appearance.
(iii) Engineering: a brick having a dense and strong semi-vitreous body conforming to defined absorption and strength limits (unlike common or facing bricks which only

have a minimum strength requirement, and no particular absorption limits).

Quality
  (i) Internal: bricks for internal use only. Internal quality bricks may require protection during construction in winter.
 (ii) Ordinary: less durable than special quality bricks, but normally durable on the external face of a building above dpc level.
(iii) Special quality: durable in conditions of severe exposure where they may be liable to be wet and frozen, e.g. below dpc level, retaining walls, etc.

Type
Six types are defined (see also Figure A1.2).
  (i) Solid: having holes not exceeding 25% of the brick's volume, or frogs not exceeding 20% (a frog is illustrated in Figure A1.1).
 (ii) Perforated: having holes in excess of 25% of the brick's volume, provided the holes are less than 20 mm wide or 500 mm^2 in area with up to three hand holds within the 25% total.
(iii) Hollow: having holes in excess of 25% and larger than defined in (ii).
(iv) Cellular: having holes closed at one end which exceed 20% of the volume of the brick.
 (v) Special shapes.
(vi) Standard specials (two typical standard specials are shown in Figure A1.3).

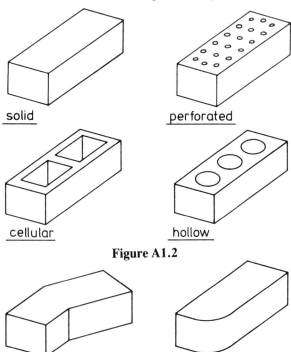

solid   perforated

cellular   hollow

Figure A1.2

'dogleg'   single bullnose

Figure A1.3 Standard specials

Note that the classifications of variety, quality and type are not related. For example, a common brick could be of internal quality or special quality, and could be of a solid type or cellular. Thus due to the permutations of variety, quality and type, plus the added variables of colour and texture, the range of bricks available is extremely wide. Designers should always take advantage of the wide choice available, and exercise care in selecting exactly the right brick for the job in hand. By far the majority of cases of unsatisfactory performance in use are attributable to an incorrect choice of brick.

Many other terms, either traditional and/or relating to manufacturing processes, are still sometimes used to describe particular kinds of bricks. These, however, do not provide a sufficiently accurate description for design engineering purposes.

A1.1.3 Strength and durability
From the structural viewpoint, the main classification of a brick is according to its compressive strength. The strength must, of course, be maintained for the required design life. So, the durability of a brick is just as important as its compressive strength, which latter is not necessarily an index of durability. Bricks having a compressive strength of over 48.5 N/mm^2 are usually durable, but there are bricks approaching this value which decay rapidly if exposed to frost in wet conditions. Conversely, there are many weaker bricks which are more durable. Durability should always be checked with the manufacturer.

It should also be remembered that walls are required to fulfil many other functions than that of providing resistance to direct compression loading and, for some of these purposes, the compressive strength of a particular brick is not necessarily the prime consideration.

Strength
The compressive strength of a brick relates to the characteristic strength value which may be used for design in accordance with BS 5628. The strength requirements for bricks, as set out in BS 3921, are given in Table A1.1.

Due to the effect of the mortar, there is no direct relationship between the compressive strength of a particular brick and the strength of a wall built with it. Obviously, though, a wall built with bricks of high compressive strength will have a

Table A1.1 Classification of bricks for use in load-bearing brickwork (BS 3921)

	Class	Minimum average compressive strength (N/mm²)
Engineering	A	69.0
	B	48.5
Loadbearing	15	103.5
brick for	10	69.0
brickwork	7	48.5
designed	5	34.5
to BS 5628	4	27.5
	3	20.5
	2	14.0
	1	7.0

greater loadbearing capacity than an identical wall built with bricks of lower compressive strength.

Durability

Having considered the strength of bricks, and in order to ensure that their design strength is maintained, the question of durability must be considered. The main factors which can cause problems with bricks and brickwork are sulphate attack, frost attack and crystallisation of soluble salts.

Clay bricks may contain sulphates derived from either the original clay or its reaction with the sulphur compounds from the fuel used in firing. Sulphates may also be present in mortars, soils, gypsum plasters and polluted atmospheres.

In persistently wet conditions, sulphates react slowly with tricalcium aluminate – a constituent of Portland cement and hydraulic lime – causing it to expand, and thus bring about cracking or spalling of the mortar joints and, possibly, spalling of the bricks themselves. Sulphate attack is liable to occur in brickwork which remains wet for long periods, e.g. below dpc level, and parapets above roof level.

Thus it is important to ensure the best possible protection by: (a) a correct choice of brick; (b) correct detailing practice. Regarding the first of these provisions, maximum acceptable levels of sulphate content for bricks to be used in exposed situations are recommended in BS 3921. In addition, resistance to attack can be further reduced by the use of fairly rich mortar mixes (see Tables A1.5 and A1.6). Brickwork which suffers sulphate attack will expand, and the provision of joints in the work can reduce the resulting prob-

lems. Further information on this aspect is provided in Appendix 3.

The resistance of a particular type of brick to frost attack is best measured by prolonged exposure to the conditions likely to be met in use. Special quality bricks, as defined in BS 3921, should have performed satisfactorily for three years under similar conditions to those occurring in use. A critical condition arises in situations where bricks may be frozen whilst saturated. As in the case of sulphate attack, sound detailing and a correct choice of brick are required. It is also particularly important that work in the course of construction should be protected, since sections not specifically designed for high resistance to frost may easily become saturated and then frozen.

The crystallisation of soluble salts in bricks often causes a white deposit, known as efflorescence, to appear on the surface of brickwork. Whilst not being particularly harmful in general, efflorescence is unsightly. Occasionally, though, it can lead to the decay of underfired bricks, if the salts crystallise out beneath the faces.

The liability to efflorescence depends on the soluble salt content of the bricks (see BS 3921, and refer to manufacturers data) and on the wetting and drying conditions. Again, the risks can be substantially reduced by sound detailing and the correct choice of brick to suit the exposure conditions.

A1.1.4 Testing

It has been found from test results that the allowance in BS 3921 of up to 25% of holes in a solid brick is acceptable. Bricks with not more than this percentage of holes may be treated in the same way as bricks without holes when relating crushing strength to characteristic strength for design purposes.

A method for testing the crushing strength of clay bricks is described in BS 3921. A representative sample of ten bricks, taken at the works or on site, is crushed to failure, and the compressive strength is defined as the mean of ten results. The compressive strength for each unit is the load at failure divided by the gross area – thus the effect of any perforations is automatically allowed for. However, bricks with frogs should have the frogs filled with mortar. Failure to do this leads to reduced values of compressive strength. In practice, frogged bricks should be laid with the frog uppermost (or the larger frog if frogged on two faces) to ensure that it is completely filled with mortar.

A1.2 CALCIUM SILICATE BRICKS

Calcium silicate (sand-lime and flint-lime) bricks should conform to the requirements of BS 187. Although, in the past, they were mainly used for non-loadbearing work, comparatively recent improvements in manufacturing techniques accompanied by an increase in compressive strength now make them suitable for loadbearing masonry.

Although possessing broadly similar functional properties to clay bricks, they have different movement characteristics, and the manufacturer's advice should be sought on the spacing of movement joints (see also Appendix 3).

A1.3 CONCRETE BRICKS

Concrete bricks, which are similar in size and shape to clay and calcium silicate bricks, should comply with the requirements of BS 1180. Note that these, too, have different movement characteristics from clay bricks (see Appendix 3).

A1.4 STONE MASONRY

Stone masonry should comply with the requirements of BS 5390. Because of its high initial cost and the expense of skilled labour, stonework is rarely used structurally in industrialised countries, and is mainly restricted to facing veneers on prestige buildings. However, it could still be an economical proposition in developing countries with adequate indigenous supplies of good stone and experienced masons.

A1.5 CONCRETE BLOCKS

Concrete blocks are produced in a great variety of sizes, and are generally described in BS 2028, 1364 as being a walling unit with any of its dimensions greater than those specified for a brick in BS 3921 (see A1.1.1), except that the height should not exceed the length or six times the thickness.

A1.5.1 Sizes
The sizes specified in BS 2028, 1364 are dimensionally coordinated – the coordinating sizes being the size of the basic space occupied by a unit plus its share of the mortar joints (as with the format size in brickwork). The work size is the size of the block itself. The sizes specified in BS 2028, 1364 are provided in Table A1.2.

Table A1.2 Range of concrete block sizes (BS 2028, 1364)

Block	Length and height (mm)		Thickness (mm)
	Coordinating size	Work size	Work size
Type A	400 × 100	390 × 90	75, 90, 100
	400 × 200	390 × 190	140 and 190
	450 × 225	440 × 215	75, 90, 100, 140 190 and 215
Type B	400 × 100	390 × 90	75, 90, 100
	400 × 200	390 × 190	140 and 190
	450 × 200	440 × 190	75, 90, 100, 140 190 and 215
	450 × 225	440 × 215	
	450 × 300	440 × 290	
	600 × 200	590 × 190	
	600 × 225	590 × 215	
Type C	400 × 200	390 × 190	60 and 75
	450 × 200	440 × 190	
	450 × 225	440 × 215	
	450 × 300	440 × 290	
	600 × 200	590 × 190	
	600 × 225	590 × 215	

It can be seen that there is a very wide and confusing variety of standard block sizes. Not altogether surprisingly, the current trend is for manufacturers to reduce their range of standard sizes – the blocks at the larger end of the scale tending to be discarded.

Ideally, the sizes should be such as to facilitate ease of handling. The larger blocks require lifting with both hands, and this tends both to slow down the rate of laying and to destroy the blocklayer's working rhythm. In addition, the thicker blocks tend to be difficult to grasp, unless provided with special hand grips. On the other hand, however, the economics of manufacturing require the largest possible units to be produced by the blockmaking machines. Thus the actual sizes of the blocks tend to be a compromise between these conflicting requirements.

Whilst it is slower to lay blocks than bricks, since they are larger than bricks the rate of walling production is not necessarily adversely affected.

A1.5.2 Classification
The current British Standard gives more of a

performance specification for blocks, rather than detailed descriptions of their manufacture. Thus virtually any suitable materials may be used, provided the blocks meet the specifications. Three different block types are given, each being defined according to specified properties and uses.

A1.5.3 Density
The three types are termed A, B and C, and differ essentially only in density. Type A blocks are 'dense', having a density not less than $1\,500$ kg/m^3. Types B and C are 'lightweight', with a density less than $1\,500$ kg/m^3. The required properties and uses of the three types are as follows:

Type A: for general use in building, including use below ground level.

Type B: for general use in building, including use below the ground dpc, in internal walls and the inner leaf of external cavity walls, or in the external leaves of cavity walls protected by tanking. In positions such as below the ground level dpc or the outer leaf of external cavity walls, blocks (whether solid, hollow or cellular, see A1.5.4) should be made with dense aggregate to BS 882 or BS 1047, or have an average compressive strength not less than 7.0 N/mm^2. Blocks not satisfying these requirements may be used if the manufacturer supplies authoritative evidence of their suitability.

Type C: these are primarily for internal non-loadbearing walls, such as partitions and panels in framed construction.

A1.5.4 Form
The three types are produced in three basic forms: solid, hollow or cellular (see Figure A1.4).

Solid blocks are basically voidless, although grooves or cavities may be included to facilitate handling or reduce weight provided these do not exceed 25% of the total volume. Hollow blocks may have large holes or cavities which pass through the block, but the volume of voids must not exceed 50% of the total volume. Cellular blocks are similar to hollow blocks, but with the cavities closed at one end.

A1.5.5 Strength
The strength of type A and B blocks is determined by compressive testing – the method being described in BS 2028, 1364. The load at failure during the test is divided by the gross cross-sectional area of the block. This means that the compressive load carrying capacity is equivalent for solid, hollow or cellular blocks. The specified compressive strengths for types A and B are provided in Table A1.3.

Table A1.3 Compressive strengths of concrete blocks type A and type B

Block types and designation	Minimum compressive strength	
	Average of ten blocks (N/mm^2)	Lowest individual block (N/mm^2)
A (3.5)	3.5	2.8
A (7)	7.0	5.6
A (10.5)	10.5	8.4
A (14)	14.0	11.2
A (21)	21.0	16.8
A (28)	28.0	22.4
A (35)	35.0	28.0
B (2.8)	2.8	2.25
B (7)	7.0	5.6

Type C blocks, primarily intended for non-loadbearing walls, are tested by a transverse breaking load test described in BS 2028, 1364. A sample of five blocks is used. Each block in the sample is tested in turn by placing it horizontally in the testing machine with a support at each end. A load is then applied in the middle of the block, the direction of loading being perpendicular to the side of the block. The breaking loads specified in BS 2028, 1364 are given in Table A1.4.

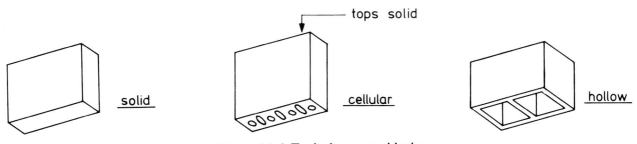

Figure A1.4 Typical concrete blocks

Table A1.4 Transverse breaking loads for concrete blocks type C

Work size		Minimum transverse breaking load	
Height (mm)	Thickness (mm)	Average of five blocks (N)	Lowest individual block (N)
190	60	776	621
190	75	1210	968
215	60	876	701
215	75	1368	1094
290	60	1184	947
290	75	1850	1480

A1.5.6 Durability

As with brickwork, one of the main requirements to ensure durability is correct construction detailing. Correctly detailed blockwork is generally durable, whatever the type of block – provided, of course, that the use is as specified in BS 2028, 1364.

The durability of concrete blocks is comparable to that of good concrete. If suitable joints are provided to cope with thermal and moisture movements, as described in Appendix 3, serious deterioration is unlikely.

The appearance of blockwork, particularly if open-textured blocks are used, can be marred by the effects of pollution. A problem of algae growth on the face of blockwork, during construction, has been encountered by the authors, but such effects are unlikely to affect the strength of blockwork.

A1.6 MORTARS

The role of the mortar between the bricks or blocks used in a structural element is very important and complex. There are requirements to be met by the mortar, both in the freshly made and hardened states. During construction, it must be easily workable – it must spread easily and remain plastic long enough to enable lining and levelling of the units. It must also retain water, so that it does not dry out and stiffen too quickly with absorbent units. It must then harden in a reasonable time to prevent squeezing out under the weight of the units laid above. On completion of the day's work, the mortar must have gained sufficient strength to resist frost.

When hardened, in the finished structure, the mortar must transfer the compressive, tensile and shear stresses between adjacent units, and it must be sufficiently durable to continue to do so. However, whilst adequate strength is essential, only the weakest mortar consistent with the strength and durability of the bricks or blocks should be used. When a suitably matched mortar is used, any cracking from thermal or other movements will occur at the joints. Cracks in the mortar tend to be smaller and easier to repair than cracks in the masonry units. In any case, the use of a stronger mortar does not necessarily produce a stronger structural element, because mortar strength is not directly related to the strength of the masonry built with that mortar (see Figure A1.5).

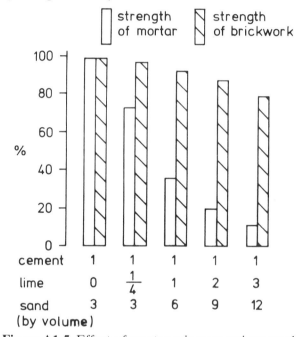

Figure A1.5 Effect of mortar mix proportions on the crushing strengths of mortar brickwork built with medium strength bricks. Strengths are shown relative to the strength of a 1:3 cement:sand mortar and the brickwork built with it

For any particular strength of unit, there is an optimum mortar strength, and a stronger mortar will not increase the strength of the brickwork or blockwork. Particular care is needed when choosing a mortar for use with the lower strength blocks to ensure that it is of sufficiently low strength to confine any cracking to the joints. On the other hand, richer mortars, which develop strength quickly enough to resist frost, are obviously to be preferred for winter working in that, if a lean mortar is specified, additional precautions will be required.

Masonry should be laid on a *full* bed of mortar and, if bed joints are raked out for pointing, allowance must be made in the design for the decreased width as well as the resulting loss of strength.

A1.6.1 Constituents
Mortars generally consist of sand and water in combination with one or more of the following:

(i) lime
(ii) Portland cement
(iii) sulphate-resisting Portland cement
(iv) masonry cement
(v) high alumina cement
(vi) plasticisers or other additives
(vii) pigments

Portland cement, in one or other of its several forms, is the principal binding agent in mortars. It is used because of its comparatively rapid strength gain and quick setting rate. Very high strengths are obtainable from cement:sand mortars (e.g. 1:3 cement:sand), but these are not generally required except for very exposed conditions, such as below dpc level or in retaining walls.

Lime is normally added to mortars to improve their workability and bonding properties, although this does result in some loss of strength. Probably the most commonly used mix is a 1:1:6 cement:lime:sand, which is suitable for most applications.

As an alternative to lime, plasticisers, which entrain air into the mortar, are often used to improve workability. Lime requires special handling, and the use of plasticisers can show economies in labour and material costs. Amongst other things, the entrainment of air is also claimed to improve frost resistance. However, the entrainment of air bubbles inevitably reduces the strength of the mortar. This can be a difficult problem to control on site, in that plasticisers are often added by the masons themselves, somewhat indiscriminately. For this reason, and others such as the effects on wall ties, reinforcement, and the long-term weakening of the mortar, care should be taken to obtain up-to-date and reliable information before considering the use of any plasticiser or similar additive.

Note that frost inhibitors based on calcium chloride, or calcium chloride itself, should never be used, since these cause long-term weakening of the mortar and excessive corrosion of wall ties and reinforcement.

Masonry cement consists of Portland cement

with the addition of a very fine mineral filler and an air-entraining agent. Masonry cement should be used with caution. The presence of the mineral filler reduces strength, and the comments above on the use of plasticisers also apply.

High alumina cement should not be used.

A1.6.2 Choice of mortar
Tables A1.5 and A1.6, reproduced from *BRE Digest* 160, give guidance on the selection of particular mortars for various applications. To avoid any confusion on site, the number of different mixes to be used on any one project should be kept to a minimum. It is also worth noting that the extra cost of making a good mortar is an insignificant proportion of the total cost of a wall. There is little point – indeed it is false economy – in trying to produce a cheap inferior mortar.

A1.6.3 Proportioning and mixing
Mortar is usually mixed on site in small batches, and strict control must be kept on the quality. Positive measures should be taken to ensure that only the specified materials are used and are mixed in the correct proportions. Where large areas of structural masonry are being constructed, weigh batching should be employed.

Allowance must be made for the increase in volume and weight of the sand when it is damp – whatever method is used for gauging the proportions of the mix. On large contracts, consideration should be given to the production of several trial mixes so as to ensure the quality of the mortar.

If weigh batching is not justified by the quantity of the work, gauge boxes should be used. These should be filled level to the top in order to provide the correct mix proportions.

Lime and sand, termed the 'coarse stuff', may be obtained ready-mixed for delivery to the site. However, as mixing is done off the site, some degree of control is lost. The use of coarse stuff is preferable to mixing the cement, lime and sand dry, because the bulking of the sand can be allowed for, and the lime becomes more plastic when soaked overnight. BS 4721: Specification for ready-mixed lime:sand mortar, gives mix proportions for use when batching by volume is employed. It should be noted that, under the conditions of BS 4721, it is permissible for the supplier of ready-mixed lime:sand to incorporate

Table A1.5 Mortar mixes (proportions by volume)

	Mortar group	Cement: lime: sand	Masonry-cement: sand	Cement: sand, with plasticiser
Increasing strength but decreasing ability to accommodate movements caused by settlement, shrinkage, etc.	i	1:0–¼:3	—	—
	ii	1:½:4–4½	1:2½–3½	1:3–4
	iii	1:1:5–6	1:4–5	1:5–6
	iv	1:2:8–9	1:5½–6½	1:7–8
	v	1:3:10–12	1:6½–7	1:8

	equivalent strengths ⟵————————⟶ within each group
Direction of changes in properties	increasing frost resistance ————————⟶
	improving bond and resistance ⟵———————— to rain penetration

Where a range of sand contents is given, the larger quantity should be used for sand that is well graded and the smaller for coarse or uniformly fine sand.

Because damp sands bulk, the volume of damp sand used may need to be increased. For cement:lime:sand mixes, the error due to bulking is reduced if the mortar is prepared from lime:sand coarse stuff and cement in appropriate proportions; in these mixes 'lime' refers to non-hydraulic or semi-hydraulic lime and the proportions given are for lime putty. If hydrated lime is batched dry, the volume may be increased by up to 50 per cent to get adequate workability.

an admixture without prior permission. This may not be particularly desirable.

A1.6.4 Testing

As noted earlier, the properties of freshly mixed and hardened mortar are both very important in ensuring that the design requirements are met. Various testing methods for both conditions are dealt with in BS 4551: Methods of testing mortar cubes. Samples of the mortar are taken on site and three cubes are prepared, similar to those taken for the testing of concrete except that they are $100 \times 100 \times 100$ mm. After curing the cubes are tested to ascertain their crushing strength, one cube being tested at 7 days, and two at 28 days. The results are then compared with the values specified for the work, in order to determine whether the mortar is acceptable.

Chemical analysis of mortar can be made, and is useful for the following purposes:

1. Assessment of the efficiency of mixing and the accuracy of batching on sites, in mixing plants and laboratories.
2. Analysis of recently placed mortar for assessment of compliance with specification requirements, and the investigation of failures.
3. Analysis of old mortars for the investigation of failures and, in the case of very old mortars, for assessing their type and chemical composition.

Chemical analysis may thus be used to provide a further check if compressive testing of cubes is producing low results. It should be remembered, however, that all such testing methods have limitations, especially when applied to a material of the nature of mortar. Test results should only be used as a guide, and not as a final judgement, depending, of course, on the size and nature of the work involved.

Table A1.6 Selection of mortar groups

Type of brick:	Clay		Concrete and calcium silicate	
Early frost hazard[a]	no	yes	no	yes
Internal walls	(v)	(iii) or (iv)[b]	(v)[c]	(iii) or plast (iv)[b]
Inner leaf of cavity walls	(v)	(iii) or (iv)[b]	(v)[c]	(iii) or plast (iv)[b]
Backing to external solid walls	(iv)	(iii) or (iv)[b]	(iv)	(iii) or plast (iv)[b]
External walls; outer leaf of cavity walls:				
– above damp proof course	(iv)[d]	(iii)[d]	(iv)	(iii)
– below damp proof course	(iii)[e]	(iii)[b, e]	(iii)[e]	(iii)[e]
Parapet walls; domestic chimneys:				
– rendered	(iii)[f, g]	(iii)[f, g]	(iv)	(iii)
– not rendered	(ii)[h] or (iii)	(i)	(iii)	(iii)
External free-standing walls	(iii)	(iii)[b]	(iii)	(iii)
Sills; copings	(i)	(i)	(ii)	(ii)
Earth-retaining walls (back-filled with free-draining material)	(i)	(i)	(ii)[e]	(ii)[e]

[a] During construction, before mortar has hardened (say seven days after laying) or before the wall is completed and protected against the entry of rain at the top.

[b] If the bricks are to be laid wet, a plasticiser may improve frost resistance (see also A1.6.1).

[c] If not plastered use group (iv).

[d] If to be rendered, use group (iii) mortar made with sulphate-resisting cement.

[e] If sulphates are present in the ground water, use sulphate-resisting cement.

[f] Parapet walls of clay units should not be rendered on both sides; if this is unavoidable, select mortar as though not rendered.

[g] Use sulphate-resisting cement.

[h] With special quality bricks, or with bricks that contain appreciable quantities of soluble sulphates.

COMPONENTS

In determining the suitability of any structural components, it is first essential to consider the purpose for which they are to be used, the practicality of construction, the control over workmanship, and the life expectancy of the material in relation to the life requirements of the structure. In loadbearing masonry, there are a number of components, and care should be exercised in their choice and specification.

A2.1 WALL TIES

Wall ties are mainly used to tie together un-bonded leaves of masonry, and there are a number of different types and qualities. In cases where the component is required to tie the leaves across the cavity, and to provide some interaction between them, a vertical twist tie is most suitable. In locations where differential movements are to be expected between the leaves, and little interaction is required, a more flexible type of tie is desirable.

In special circumstances, where high shear resistance is required across the tied joint, purpose-designed shear ties may be necessary.

In all cases, durability in relation to the severity of the corrosive environment is an important factor that merits close attention. As far as possible, the environment should be controlled, and thought should be given to the corrosive effects of building materials – particularly calcium chloride (see Appendix 1, A1.6.1) and certain colouring agents, the use of which should be avoided wherever possible.

In some locations, such as at junctions where restraint is required but an unbonded joint is desirable, tie bars or standard bed joint reinforcement can be used to provide the necessary tie action. In situations of severe exposure, or where required by building regulations, stainless steel or suitable non-ferrous ties should be used. Indeed, when considering durability in general,

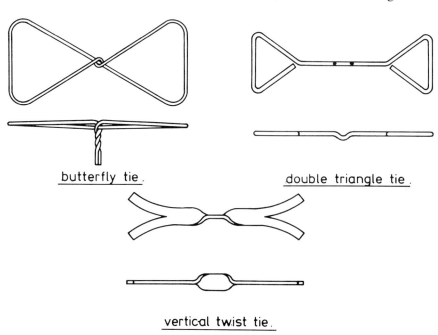

butterfly tie.

double triangle tie.

vertical twist tie.

Figure A2.1 British standard types of wall ties

there is a very sound argument for the use of such ties, even in normal external cavity walling.

Standard wall ties should conform to the requirements of BS 1243, and maximum centres for spacing should be in accordance with BS 5628, or a lesser figure to suit the design conditions. Minimum embedment should also comply with BS 5628. Examples of various ties are shown in Figure A2.1.

A2.2 DAMP PROOF COURSES

The purpose of damp proof courses is to form a barrier to cut off the movement of dampness from an external source to the building fabric. But, through necessity, they sometimes have to be located where structural forces must also be transferred, and care is needed to ensure that the chosen membrane can transfer these forces. Two common examples are:

(a) horizontal dpc to prevent vertical movement of moisture, located in a position where high compression and bending stresses are to be resisted;
(b) vertical dpc to prevent horizontal movement from outer to inner leaves of brickwork, located in a position where vertical shear forces are to be resisted.

Thus a dpc must not squeeze out under vertical loading, nor slide under the horizontal loading.

Damp proof courses can be made from a wide variety of materials such as: bitumen felt, metals, slate, plastic, brick, etc., and the choice must be based on the required performance related to the material's proven performance. Bitumen felt type dpcs, for example, are usually the least expensive but suffer from poor resistance to compressive forces and can, therefore, squeeze out under load. They can also be damaged by careless workmanship. On the other hand, the flexibility of felt or plastic dpcs can be of great importance where movements such as mining settlement, etc., are to be expected. A brick dpc can be very advantageous in providing resistance to tensile stresses at a critical cross-section, a property which few other membranes can provide. Brick dpcs are formed by specially selected engineering bricks (see Appendix 1, A1.1.2) built in a number of bonded courses to provide an impermeable barrier. Some typical horizontal dpcs are shown in Figure A2.2.

A2.3 FIXINGS

Components for providing fixings in brickwork are numerous, and the designer should be prudent when selecting these from catalogues. They can be very expensive, and the designer should carefully consider the type, sizes and costs of the various products available before specifying. Samples of fixings should be available in the design office for inspection, since, often, a component that appears in a catalogue to be a lightweight and economical fixing can be very disappointing in reality. In other cases, the designer may become aware of a weakness in the fixing, or see a practical construction problem related to the proposed use.

The forces to be taken on fixings must be considered since, if the masonry in the area around the fixing is weaker than that of the fixing itself, local failure of the masonry could occur. Particular care is needed for connections on the upper course of lightly loaded walls, and in locations where uplift forces are being resisted. Wherever possible, fixings should be located under similar conditions to that of the test specimen on which the manufacturer's performance data are based, otherwise additional testing or allowance for the differences will have to be made.

As with other components, the durability of

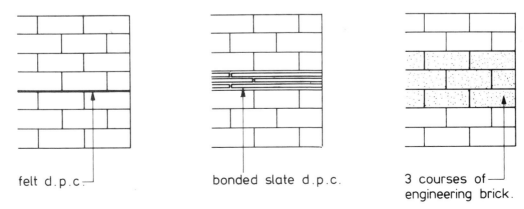

Figure A2.2 Some typical alternative damp proof courses

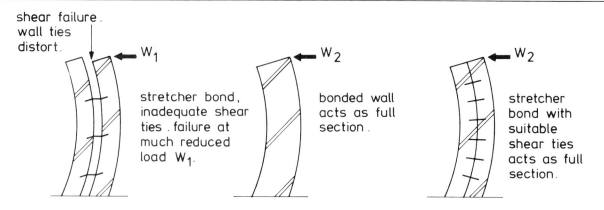

Figure A2.3 Effects of bonding masonry on lateral loading

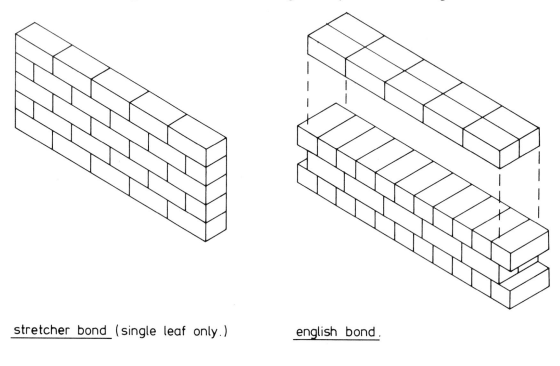

stretcher bond (single leaf only.)   english bond.

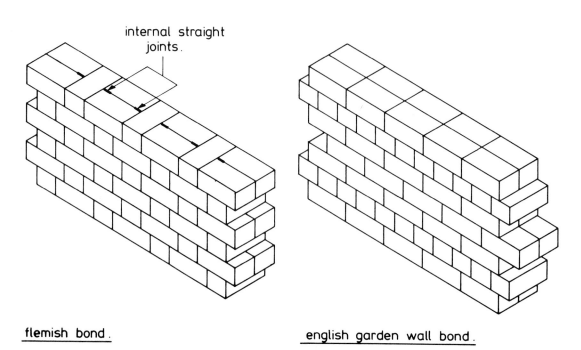

flemish bond.   english garden wall bond.

Figure A2.4 Some standard bonds

fixings must be considered in the light of the expected environment. The choice of fixing materials must be compatible with other materials in the locality if problems such as electrolytic action, etc., are to be avoided.

A2.4 BRICK BONDS

Although the bonding of masonry units is a technique rather than a component, it was nevertheless felt appropriate to discuss it in this Appendix.

For loadbearing walls, properly bonded masonry is essential. For single-leaf masonry, and ordinary cavity walling, stretcher bond is the only choice available. This can create engineering problems if not taken into account at an early stage in the design (see Figure A2.3).

With thicker, solid, walls, there appears to be little difference in the structural performance of the various standard bonds that have been used for many years. For walls of double-leaf thickness and over, English bond is, perhaps, ideal. On the other hand, Flemish and its derivative English Garden Wall bond appear to give a similar structural performance under normal loading conditions and are, therefore, usually acceptable. (See Figure A2.4.)

Stretcher bonds on walls of double-leaf thickness or over require special consideration, and may necessitate the use of special shear ties.

MOVEMENT JOINTS

All materials move, and the designer must take account of the movements and make due allowance for them. Like other materials, the movement of masonry is caused by variations in the environmental conditions such as: thermal changes, changes in moisture content, changes in loading conditions, chemical changes, foundation settlement, frost action, etc. (see Figure A3.1).

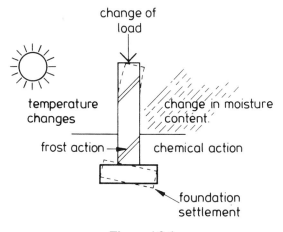

Figure A3.1

The causes of movement may operate singly, or in combination to supplement or oppose one another, and it is often very difficult to forecast precisely the movement that will occur in a particular situation. Nevertheless, the designer must try to anticipate the type and magnitude of movements, and the effect they are likely to have on the building.

If the movements act upon elements unable to contain the forces resulting from the movements, cracking is likely to occur. Where materials with different movement characteristics are bonded together, cracking is again likely to occur. Wherever possible, movements should be allowed to occur with minimum resistance, and carefully detailed movement joints should be provided to ensure that:

(a) the structural stability and performance of the jointed building is adequate;
(b) damage to the building and its finishes is kept within tolerable limits.

With all building materials, it is almost inevitable that some damage will occur, since the requirement to allow for movement frequently clashes with other needs – structural and non-structural – of the building. A compromise of carefully considered joints which control the damage within acceptable limits should be the designer's aim.

Where care is not taken, cracks and/or bulging may occur. In some situations, this can result in instability and become dangerous. In others it may be unsightly, and in some cases the condition deteriorates from the ingress of moisture

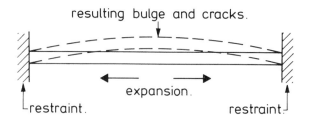

plan on wall restrained at ends.

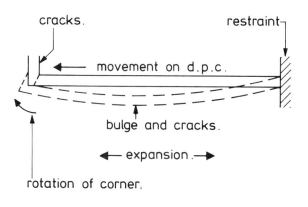

expansion failure.

Figure A3.2

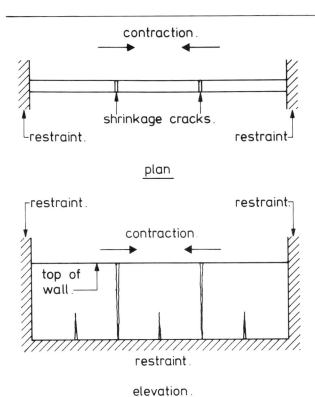

Figure A3.3

into cracks and/or from the loosening of wall ties. Some typical failure conditions are shown in Figures A3.2 and A3.3.

Restraints which aggravate cracking may arise from: the wall being fixed to, or built tightly

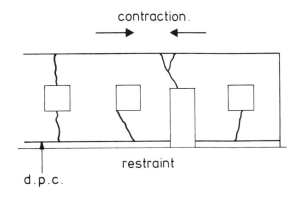

wall elevation cracking at openings.

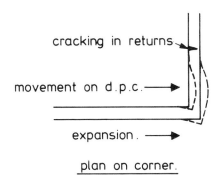

plan on corner.

Figure A3.4

around, some rigid unyielding feature at its ends; from some fixture to the wall; or where the wall incorporates materials of dissimilar properties. Cracking is most likely to occur at weakened sections, where the vertical or horizontal section of the wall changes abruptly (see Figure A3.4).

Correctly located and detailed control joints, taking into account restraints, the movement characteristics of the masonry, and any differential movements with other materials, will help to mitigate the detrimental effects of movement. Assessment of the likely locations of cracking needs particular care, in that a badly located joint might still allow critical cracking to occur at a more susceptible position.

A3.1 MOVEMENT DUE TO THERMAL EXPANSION AND CONTRACTION

The major problems caused by thermal expansion and construction of masonry occur on long walls which are subjected to large variations in temperature, such as external walls exposed to sun and frost, walls around boilerhouses or refrigeration plants, etc. In the case of internal walls subjected to a reasonably constant temperature, there are few problems, even on long walls, unless they are connected to other materials which are moving differentially, e.g. shrinking at the time the expansion of the masonry is occurring.

Particular care should be taken in the design of thin walls exposed directly to the sun, where surface wall temperatures may rise to 50°C (120°F). Such temperatures can give rise to bending as well as expansion of the wall, due to the thermal gradient through the wall thickness.

When considering expansion, it is important to obtain correct information on both the mortar and the brick or block being used. Some data on the coefficients of thermal expansion of various materials are given in CP 121: Part 1: 1973, Appendix C, and are reproduced below. Information should also be obtained from the manufacturers of the materials to be used.

Extract from CP 121: Part 1: 1973, Appendix C
   'The following data are provided to offer, in those cases where the normal recommendations of the Code are judged to be insufficient or inappropriate, further information on the movements of materials that constitute brickwork and blockwork. The natures

of the materials are such that ranges of properties are needed to convey useful information, although the ranges quoted are not intended to cover the extreme values which may sometimes be met.' (See Table A3.1.)

'A warning must be given that it is impracticable merely to sum the values given here in order to predict the likely movement of a given structure. It can be taken that most of the results quoted are for an unrestrained condition, and the imposition of restraint, whether from within the wall by friction or from without by restriction, will modify the calculated likely movements considerably.'

The unrestrained thermal movement of a wall may be estimated very approximately from the likely change in mean wall temperature and the coefficient of thermal expansion, which is often taken as 5×10^{-6} per ° C for fired-clay brickwork in a horizontal direction, and may be up to one and a half times this value for clay brickwork vertically, and for calcium silicate brickwork and concrete blockwork.

Thermal movements vertically in walls are generally reversible. However, horizontal movements are unlikely to be completely reversible since some form of restraint, particularly near the bottom of the wall, does tend to prevent the masonry from returning to its original length.

A3.2 MOVEMENT DUE TO MOISTURE

A3.2.1 Fired clay units
Clay bricks expand and contract with increases or decreases in moisture content, and these movements are normally negligible. However, superimposed over these changes, there is a permanent moisture expansion which depends on the type of clay and the degree of firing. The rate of

Table A3.1 Thermal expansion (CP 121, Table 10)

Material		Coefficient of linear thermal expansion $\times 10^{-6}$	
		per °C*	per °F*
Fired-clay bricks and blocks	: length	4–8	2–4
	: width and height	8–12	4–7
Concrete bricks and blocks† (depending on the aggregate and mix proportions)		7–14	4–8
Calcium silicate bricks	: length	11–15	6–8
	: width	14–22	8–12
Mortars†† (designations (iii) and (iv))		11–13	6–7

* The figures quoted have been rounded. The figures per °C and equivalents per °F are not exact. Repeated freezing and thawing before determination has been shown to increase these values.
† Increasing cement content increases the coefficient of linear thermal expansion.
†† Appreciably less if based on calcareous sand.

this permanent expansion decreases with time. It starts to occur during cooling in the kiln and, in many cases, up to 50% of the first two years expansion takes place during the first two days (see Table A3.2).

The solution is to avoid using kiln-fresh bricks, and bricks manufactured from clays with an unusually high moisture movement, in critical locations (see Table A3.3).

A3.2.2 Concrete and calcium silicate units
Whilst fired-clay units expand after manufacture due to the increase in moisture content, concrete and calcium silicate units dry out and shrink (see

Table A3.2 Moisture expansion of fired-clay brickwork (CP 121, Table 14)

	Expansion at constant temperature (per cent)			
Walls built of bricks made from:	Total after 15 days	Rate per 10 days after 15 days*	Total after 300 days	Rate per 10 days after 300 days*
Glacial clay	0.015	0.0026	0.039	0.0004
Coal measure shale	0.014	0.0027	0.050	0.0008

* Covering the 5 day period either side of the 15 and 300 day period.

Table A3.3 Expansions of fired-clay units resulting from changes in moisture content (CP 121, Table 11)

Clay from which units were made	Irreversible expansion* (per cent calculated on original dry length) for bricks fired to average works temperature		Wetting movement† (per cent)
	from kiln hot to 2 days	from 3 days to 128 days	
Lower Oxford	0.03	0.03	
London stock	0.05	0.02	
London clay	0.02	0.02	Generally less
Keuper marl	0.03	0.02	than 0.02
Weald clay	0.08	0.04	unless
Carboniferous shale	0.04	0.07	under-fired
Devonian shale	0.03	0.05	
Gault	0.02	0.01	

* The expansions quoted have been obtained from measurements made on unrestrained specimens. The bricks were removed from the furnace at 200°C, cooled in a desiccator and measured immediately they were cold.
† Measured by the method that was described in BS 1257 'Methods of testing clay building bricks' (now withdrawn).

Tables A3.4 and A3.5). If wetted, the units will expand again – but only part of the initial drying shrinkage is reversible. In addition, non-autoclaved units are subject to a slow non-reversible carbonation shrinkage, and should be stored for at least four weeks at normal temperature (longer in cold weather), and exposed to the wind but protected from rain prior to use. Auto-claved concrete and calcium silicate products need only a sufficient storage period to allow them to cool, but they should be kept dry prior to and during construction – which can, at times, be very difficult. For units in locations where large variations in temperature and humidity are to be expected, special precautions are necessary. For example, for short term variations, the use of plaster or render on both sides of the units can considerably reduce the effects. Reference to CP 211 and CP 221 is recommended for further information. It must be emphasised that the drying shrinkage of calcium silicate and concrete units can be a major problem, and requires particular care in design and construction.

A3.3 MOVEMENT DUE TO CHEMICAL INTERACTION OF MATERIALS (SULPHATE ATTACK)

The causes and effects of sulphate attack were outlined in Appendix 1, section A1.1.3. One of the most damaging effects is expansion of the mortar. This can cause deformation of the masonry – a common example being domestic chimneys (see Figure A3.5).

Table A3.4 Drying shrinkage of concrete and calcium silicate units (the upper limits are those set by the relevant British Standards) (CP 121, Table 12)

Material	Shrinkage (per cent)
Concrete bricks* or Type A concrete blocks†	0.02–0.06
Lightweight concrete blocks†	0.04–0.09
Calcium silicate †† (including sandlime)	0.01–0.035

* Measured according to BS 1180.
† Measured according to BS 2028, 1364.
†† Measured according to BS 187: Part 2.

Table A3.5 Shrinkage of mortar resulting from changes in moisture content (CP 121, Table 13)

Shrinkage of mortars (per cent)	
Reversible	Irreversible
0.03–0.06	0.04–0.10

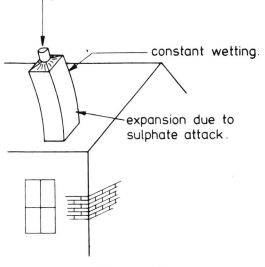

sulphur compounds from the flue gases dissolve and condense on the flue wall attacking the masonry.

constant wetting.

expansion due to sulphate attack.

Figure A3.5

In some cases, the condition can become dangerous. However, careful detailing and choice of materials in line with the recommendations of this Appendix can prevent that situation arising. For example, for the chimney shown, the use of bricks with a low sulphate content, and a sulphate-resisting cement:sand mortar, plus the addition of a flue liner in the chimney would probably have prevented the occurrence.

A3.4 DIFFERENTIAL MOVEMENT WITH DISSIMILAR MATERIALS AND MEMBERS

Where a wall has a concrete roof, floors or beams spanning onto it, consideration must be given to the potential effects of shrinkage and/or expansion of the concrete. For example, a large concrete roof on a loadbearing clay brick structure will be subjected to drying shrinkage for the first few years of its life, together with some expansion and contraction due to the varying temperature range to which it will be exposed. At the same time, the clay bricks below will be subjected to moisture expansion and thermal effects. At certain times, therefore, the materials could be attempting to move in opposite directions and, unless precautions are taken, unsightly cracking will result (see Figure A3.6). To reduce

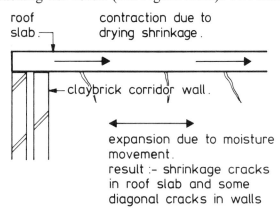

Figure A3.6 Movement of dissimilar materials

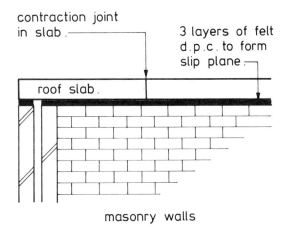

Figure A3.7 Slip plane provision

the effect of this movement, certain details can be incorporated in the construction (see Figure A3.7).

At the end of a long run of beams seated on piers, there is a danger of vertical cracking in the piers, due to shrinkage of the concrete beams (see Figure A3.8).

In those locations, a suitable padstone should be used, and a slip plane provided between the pad and the beam seating (see Figure A3.8).

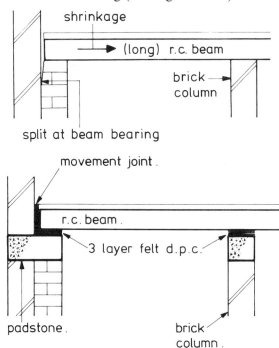

Figure A3.8 Shrinkage of concrete beam (upper diagram); provision of a slip plane (lower diagram)

With long runs of masonry built of concrete nibs, there is a danger of unsightly cracking, particularly at changes in direction and/or the end of runs – see the elevation in Figure A3.9. The detail should incorporate a slip plane between the nib seating and the first brick course, for example, a dpc membrane – see section in Figure A3.9. The masonry should also be jointed vertically, in accordance with the recommendations of this Appendix, to reduce the amount of differential movement to an acceptable level.

Where a wall is built on a floor which may deflect significantly under load, the wall should be separated from the floor, including any screed, by a separating layer, and should be strong enough to span between the points of least deflection. This applies in particular to concrete block walls supported on concrete floors and beams, where relatively small deflections of the supporting

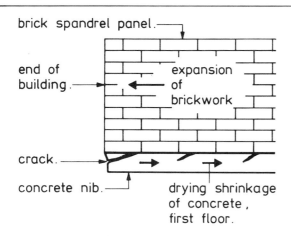

elevation on panel showing
typical cracking to concrete nib.

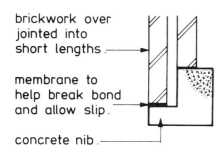

section through panel showing
preventative measures.

Figure A3.9

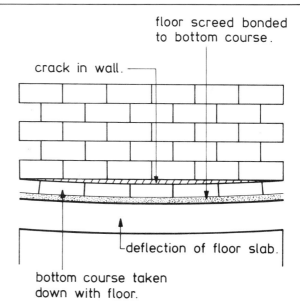

elevation on cracked wall.

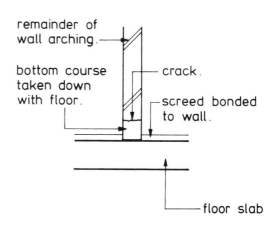

section through cracked wall.

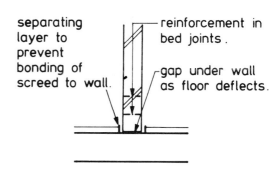

detail required to prevent failure.

Figure A3.10

members will result in the wall arching and cracking (see Figure A3.10).

It is not only when other members come into contact with masonry that differential movement occurs. For example, consider a cavity wall with concrete bricks for the internal leaf and clay bricks for the outer leaf (see Figure A3.11).

The outer leaf will expand due to moisture movement and temperature changes, while the inner leaf will contract due to drying shrinkage and load strain. The changes in length will need to be considered since, in multi-storey work, the build-up of vertical movement could cause buckling of the outer leaf and/or loosening of the wall ties. Joints should be incorporated to accommodate this movement. For example, by the introduction of a compression joint below a support slab at alternate or every third floor within the building (see Figure A3.11), the joint location and thickness being designed to absorb the anticipated movement without damage. If necessary, brick slip tiles can be adhered to the face of the concrete slab to achieve a continuous ceramic finish.

A3.5 FOUNDATION SETTLEMENT

For foundations where differential settlement is within reasonable limits, and where the correct mortar mixes are used (see A3.8), masonry structures are generally flexible enough to accommodate the movement without any de-

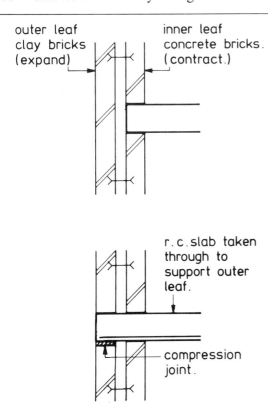

outer leaf
clay bricks
(expand)

inner leaf
concrete bricks.
(contract.)

r.c.slab taken
through to
support outer
leaf.

compression
joint.

Figure A3.11 Cavity wall detail

trimental effects. In locations where more severe settlements are likely, such as mining areas, large buildings should be jointed into smaller independent units where large strains and bending moments can be kept under control. For example, in areas of future mine workings, the smaller the unit the more economical the design of the foundations. On the other hand, the cost of providing the joints in the superstructure, and the problems of providing stability, will increase in proportion to the number of joints. It is important, therefore, to reach a reasonable compromise which, in mining areas, would generally be to limit the length of a unit to 20 m. Provided that the foundations and joints (see A3.6), are then designed for the particular site conditions, in accordance with good practice, problems should not occur.

A3.6 MOVEMENT JOINTS AND ACCOMMODATION OF MOVEMENT

In sections A3.1 to A3.5, movements due to changes in temperature, changes in moisture content, chemical interaction, dissimilar materials in contact with each other, and foundation settlements have been discussed. In some cases, the solutions to the problems have also been considered. But the location and type of joint which should be used have not. The reason for this seeming omission is that whilst the causes of

the various movements are different, there is, on most buildings, a need to accommodate a number of differing forms of movement. These movements may occur at different times, or at the same time; they may be additive, or they may cancel each other out. Joints designed for one form of movement may accommodate another. On the other hand, they may aggravate other problems in the building. The design of joints should, therefore, be based on assessment of all the movements likely in the particular building, and all the requirements that may be aggravated by the inclusion of joints.

This is probably best clarified by considering an example. A long block of four-storey flats is to be constructed in a mining area, mined by modern, deep, long wall techniques. The flats are to be built in loadbearing brickwork. Clay bricks are to be used for the outer leaf of the external cavity walls and the loadbearing crosswalls, but concrete bricks will be used for the inner leaf of the external walls. Floors are to be *in situ* concrete construction.

Considerations for this building would be:

(a) To joint the length of flats into units less than 20 m long, in order to bring the ground strains resulting from the mining within acceptable limits that can be accommodated within, say, a 75 mm wide joint between units.

(b) To check that expansion and contraction from temperature changes are controlled by suitable expansion joints. This condition could normally be expected to be controlled within the joint provided for mining strain and, to check this, a calculation for the total mining strain plus the effects of expansion would need to be compared with the amount of strain that can be accommodated by the joint and the jointing materials.

(c) To check the effects of moisture movement, a combination of conditions must be considered, i.e. moisture movement and the movement of dissimilar materials. For example the outer leaf of clay bricks will be expanding due to moisture expansion after firing, but the load strain on this leaf will partially cancel out this growth. At the same time, the inner leaf of concrete bricks will be contracting due to drying shrinkage and load strain (see Figure A3.12).

These movements will not only be occurring

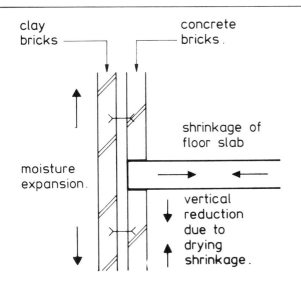

clay bricks

concrete bricks.

moisture expansion.

shrinkage of floor slab

vertical reduction due to drying shrinkage.

Figure A3.12 Differential movements

vertically, but will also have an effect horizontally where shrinkage of the concrete floor slab will be adding to the problem.

The typical floor plan and vertical sections shown in Figure A3.13 indicate how vertical control joints can be accommodated in the concrete brickwork, how horizontal control joints can be provided, and how the floor slab can be jointed to reduce the detrimental effects of shrinkage.

It should be noted that, since these joints affect the structural stability of the building, it is important that this is considered when locating them, and the effects of weak zones within the building, since movement will occur in these areas. Having located the joints, a stability check should be made taking all the joints into account.

The jointing shown in Figure A3.13 is only provided as a typical example. It must be emphasised that, for any particular building, the joints must be specifically designed to suit the materials being used and the conditions to which the structure will be subjected.

A3.7 JOINTING MATERIALS AND TYPICAL DETAILS

Whilst each joint must be designed for the movement, or movements, to be accommodated, there are a number of general points to consider. Joints which are required to accommodate expansion of the building or compressive ground strains, must be kept free from obstruction, and any compressible fillers must be capable of absorbing all of the strain.

Drying shrinkage, contraction and any other movement which could create tensile stresses in the masonry, need consideration if tensile cracks are to be avoided. A generous number of joints is always preferable to extending the spacing beyond that recommended.

It is, perhaps, surprising how many engineering designers keep the design stresses for the normal loading conditions well under control, but fail to keep the secondary stresses from movement within allowable limits. Thus quite often, a good basic design fails after construction, due to a lack of consideration of small details around the control joints, and/or lack of adequate site supervision in keeping debris out of the joints.

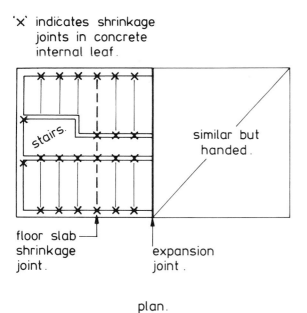

'x' indicates shrinkage joints in concrete internal leaf.

stairs.

similar but handed.

floor slab shrinkage joint.

expansion joint.

plan.

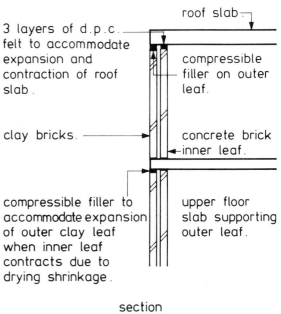

3 layers of d.p.c. felt to accommodate expansion and contraction of roof slab.

clay bricks.

compressible filler to accommodate expansion of outer clay leaf when inner leaf contracts due to drying shrinkage.

roof slab

compressible filler on outer leaf.

concrete brick inner leaf.

upper floor slab supporting outer leaf.

section

Figure A3.13 Typical example of movement joints

Some typical joint details are shown in Figure A3.14. Recommendations for materials for use in joints are provided in CP 121: Part 1.

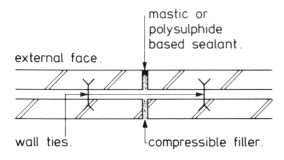

typical expansion joint in clay brickwork external cavity wall.

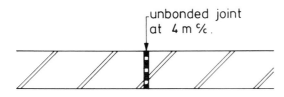

typical control joint in sand lime bricks for drying shrinkage internal wall.

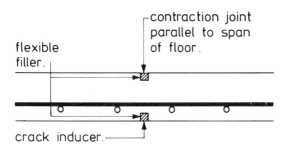

typical shrinkage joint in suspended reinforced concrete slab.

Figure A3.14

Vertical joints for thermal and moisture movement are formed by butting the masonry units against a flexible separator, such as a patent compressible strip, polythene sheet, bituminous felt or plastic strips. U-shaped copper strips have been used very successfully for years, but these have become relatively expensive and many new and more economical types of filler are now on the market. The problem of sealing the joint has been eased by the use of polysulphide-based sealants manufactured to BS 4254. The joint must go right through, not only the structural elements, but also the finishes such as plaster or screeds. Vertical joints should be about 12 mm thick, and spaced at 10–15 m for clay bricks, but should generally be reduced below 6 m for concrete blocks and bricks and for calcium silicate bricks (see Figure A3.15).

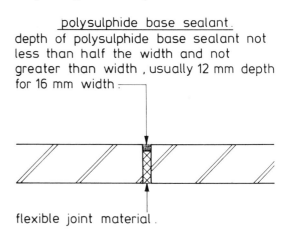

Figure A3.15 Vertical movement joint

A3.8 MORTARS IN ASSISTING MOVEMENT CONTROL

Whilst specially designed joints are important in the control of movement and prevention of excessive damage, the designer must bear in mind that each mortar joint between the masonry units can be just as important in preventing critical damage.

In the past, when all mortar mixes tended to be weak, the mortar joints were the main control for movement – and they usually performed very well. Since the general use of stronger cement mortars and weaker bricks and blocks, the problems resulting from movements have increased. The majority of unsightly cracks in brickwork are those which pass through the bricks as well as the mortar joints. Many of these cracks would have remained unnoticed, and would have been less harmful, if a weaker mortar had been used, since the movement would have tended to disperse between the numerous mortar joints – leaving the masonry units undamaged. There is still a tendency to use too strong a mortar mix relative to the brick or block strength, and it cannot be over-emphasised that the mortar strength should generally be much weaker than the bricks or blocks. See also Appendix 1, section A1.6.

PROVISION FOR SERVICES

Whilst the majority of service engineers would think twice about cutting a chase along a pre-stressed concrete beam, or cutting a hole through a steel column, they tend to think it much less serious when doing the same to structural masonry. This presented no problems in the past, when structural masonry was not of such slender construction, nor so highly stressed. However, modern masonry design calls for the same care and consideration as the prestressed beam and the steel column.

It is not difficult to make provision for services, provided that it is pre-planned. As far as possible, all service runs should be planned (see Figure A4.1), and coordinated by the design team before site work starts, and it should not be left to the services sub-contractors to cut runs indiscriminately in finished work.

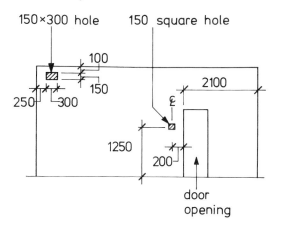

Figure A4.1 Typical elevation on wall showing builder's work details to be built in

The most common service provision is the cutting of chases for electrical conduits, etc. If these are cut horizontally, they obviously decrease the wall's effective thickness and cross-sectional area, and thus increase the stress in the wall and its tendency to buckle. If the stress would increase above the permissible limit, the chase should obviously not be allowed. However, even

if the chase does not overstress the wall, and is allowable, care must be taken in forming it. A labourer banging away with a hammer and chisel, or pneumatic hammer, will not only cut the chase but may also shatter the surrounding masonry units and mortar. The chase must be gently sawn with a power saw.

Vertical chases may not appear to be such a problem, but most research carried out on test walls, loaded to destruction, shows that the walls split vertically (see Figure A4.2).

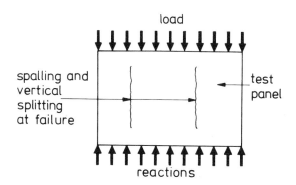

Figure A4.2 Typical failure for wall subjected to compressive loading

Where, for example, vertical chases are formed near doorways, they can produce isolated columns of masonry under lintols and beams, so that the bearing stress is concentrated rather than dispersed through the wall (see Figure A4.3).

A vertical chase could then easily become a pre-formed crack. As with horizontal chases, vertical chases should be gently sawn, and only cut in walls or parts of walls which are not highly stressed.

Holes through walls to allow the passage of heating pipes, etc., should be formed during construction by leaving out masonry units and, if necessary, the joints around the holes should be

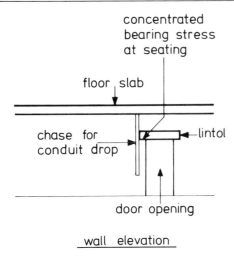

wall elevation

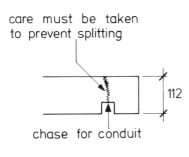

plan on chase

Figure A4.3 Vertical chases

reinforced to spread and reduce the stress concentration (see Figure A4.4). The pipes, etc. passing through the pre-formed holes can be sleeved.

Vertical services can be routed through voids in hollow block and diaphragm walls, and through the cavities of cavity walls. However, such methods make access to the services difficult, and pre-planned service ducts in walls and room lay-outs, with proper access, are more desirable for maintenance purposes (see Figure A4.5).

Openings through floor slabs for vertical ducts can be very useful for setting out on site, since they form check points for ensuring that the next lift of masonry on top of the slab lines up with the walls and columns already built below.

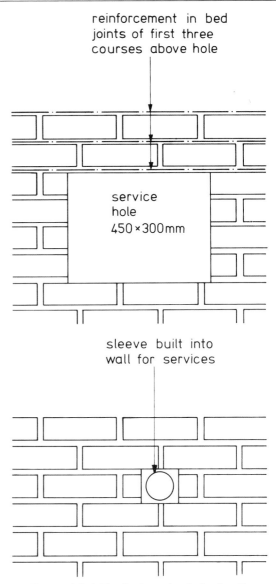

Figure A4.4 Typical service hole details

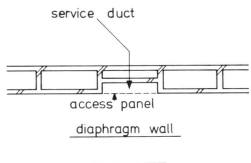

diaphragm wall

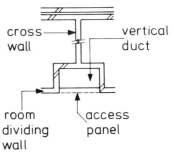

Figure A4.5 Sectional plans on typical vertical service ducts

TABLES OF DESIGN LOADS

The following tables are provided as an aid to designers giving the design vertical load-carrying capacities in kN/m run of 102.5 mm thick and 210 mm thick brick walls at various effective heights. Values are given for four basic brick strengths, laid in mortar designations (i) to (iv), for each value of partial factor of safety applicable to materials. Alternative values are given for differing values of eccentricity of loading, e. Values are for eccentricities up to $0.05t$, where t is the actual wall thickness and eccentricity equal to $0.2t$.

Tables A5.1 to A5.16 are for 102.5 mm thick solid walls and Tables A5.17 to A5.32 for 210 mm thick solid walls constructed in normally bonded masonry. Values will not necessarily apply for two separate leaves of masonry tied together.

Values are given for the following ranges of variables:

Partial factors of safety – BS 5628, Table 4 (see
on material γ_m Table 5.11)
Mortar designations   – BS 5628, Table 1 (see
(i) to (iv)   Table 5.3)
Brick strengths 20, – BS 5628, Table 2(a)
27.5, 35 and 50 N/mm^2 (see Table 5.4)

The capacity reduction factor has been calculated from equation (4) of BS 5628, Appendix B2, using equations (1) and (2) of Appendix B1. Values of β may hence differ slightly from those given in Table 7 of BS 5628 (see Table 5.15) since no rounding-up has been carried out. Clause 23.1.2 of BS 5628 has been applied to the 102.5 mm thick wall. The final effective height value in each table corresponds to a slenderness ratio of 27.

EXAMPLE 1

In a multi-storey structure an internal loadbearing wall is required to support a design load of 250 kN/m. The wall is 3 m high with enhanced lateral restraint top and bottom. There is no eccentricity of loading. The partial factor of safety on materials is common throughout the building with a value of $\gamma_m = 2.8$. The mortar designation is varied at differing floor levels and on the level in question is (iii). Determine a suitable wall thickness and unit strength:

Design load = 250 kN/m run.
$\gamma_m = 2.8$ with mortar designation (iii).
Actual height = 3000 mm
Effective height = 0.75×3000
                 = 2250 mm
From Table A5.7, for a 102.5 mm thick brick wall, the design load capacity for an effective height of 2.4 m, = 251 kN/m run with 50 N/mm^2 units.

$\therefore$ Adopt 102.5 mm thick wall built with 50 N/mm^2 units.

EXAMPLE 2

If a stronger mortar, e.g. designation (i), were to be used in Example 1 what strength of unit would be required?
For mortar designation (i), from Table A5.5, 35.0 N/mm^2 units would be required.

Table A5.1
Wall thickness = 102.5 mm

Partial safety factor for material strength = 2.5

Mortar designation (i)

Effective height (m)	Brick strength (N/mm²)							
	20.0		27.5		35.0		50.0	
	$e < 0.05t$	$e = 0.2t$	$e < 0.05t$	$e = 0.2t$	$e < 0.05t$	$e = 0.2t$	$e < 0.05t$	$e = 0.2t$
0.90	348.9	230.3	433.8	286.3	537.5	354.8	707.3	466.8
1.20	328.5	230.3	408.3	286.3	506.0	354.8	665.8	466.8
1.50	303.8	230.3	377.7	286.3	468.0	354.8	615.8	466.8
1.80	273.7	204.6	340.2	254.3	421.6	315.1	554.7	414.7
2.10	238.0	169.0	295.9	210.0	366.7	260.3	482.5	342.5
2.40	196.9	127.9	244.8	159.0	303.4	197.0	399.2	259.2
2.70	150.4	81.3	186.9	101.0	231.6	125.2	304.8	164.8
2.77	139.1	70.0	173.0	87.1	214.3	107.9	282.0	142.0

Table A5.2
Wall thickness = 102.5 mm

Partial safety factor for material strength = 2.5

Mortar designation (ii)

Effective height (m)	Brick strength (N/mm²)							
	20.0		27.5		35.0		50.0	
	$e < 0.05t$	$e = 0.2t$	$e < 0.05t$	$e = 0.2t$	$e < 0.05t$	$e = 0.2t$	$e < 0.05t$	$e = 0.2t$
0.90	301.8	199.2	372.5	245.8	443.2	292.5	575.2	379.7
1.20	284.1	199.2	350.6	245.8	417.2	292.5	541.5	379.7
1.50	262.7	199.2	324.3	245.8	385.9	292.5	500.8	379.7
1.80	236.7	176.9	292.1	218.4	347.6	259.9	451.2	337.3
2.10	205.9	146.1	254.1	180.4	302.4	214.6	392.4	278.5
2.40	170.3	110.6	210.2	136.5	250.2	162.4	324.7	210.8
2.70	130.0	70.3	160.5	86.8	191.0	103.2	247.9	134.0
2.77	120.3	60.6	148.5	74.8	176.7	89.0	229.4	115.5

Table A5.3
Wall thickness = 102.5 mm

Partial safety factor for material strength = 2.5

Mortar designation (iii)

Effective height (m)	Brick strength (N/mm²)							
	20.0		27.5		35.0		50.0	
	$e < 0.05t$	$e = 0.2t$	$e < 0.05t$	$e = 0.2t$	$e < 0.05t$	$e = 0.2t$	$e < 0.05t$	$e = 0.2t$
0.90	273.5	180.5	334.8	220.9	400.8	264.5	499.8	329.9
1.20	257.4	180.5	315.1	220.9	377.3	264.5	470.5	329.9
1.50	238.1	180.5	291.5	220.9	349.0	264.5	435.2	329.9
1.80	214.5	160.3	262.6	196.3	314.3	235.0	392.0	293.0
2.10	186.6	132.4	228.4	162.1	273.4	194.1	341.0	242.0
2.40	154.4	100.2	189.0	122.7	226.2	146.9	282.1	183.1
2.70	117.9	63.7	144.3	78.0	172.7	93.4	215.4	116.4
2.77	109.0	54.9	133.5	67.2	159.8	80.5	199.3	100.3

Table A5.4
Wall thickness = 102.5 mm

Partial safety factor for material strength = 2.5

Mortar designation (iv)

Effective height (m)	Brick strength (N/mm^2)							
	20.0		27.5		35.0		50.0	
	$e < 0.05t$	$e = 0.2t$	$e < 0.05t$	$e = 0.2t$	$e < 0.05t$	$e = 0.2t$	$e < 0.05t$	$e = 0.2t$
0.90	245.2	161.8	292.3	192.9	344.2	227.2	424.4	280.1
1.20	230.8	161.8	275.2	192.9	324.0	227.2	399.5	280.1
1.50	213.5	161.8	254.5	192.9	299.7	227.2	369.5	280.1
1.80	192.3	143.8	229.3	171.4	270.0	201.8	332.8	248.8
2.10	167.3	118.7	199.4	141.6	234.8	166.7	289.5	205.5
2.40	138.4	89.8	165.0	107.1	194.3	126.1	239.5	155.5
2.70	105.7	57.1	126.0	68.1	148.3	80.2	182.9	98.9
2.77	97.8	49.2	116.6	58.7	137.2	69.1	169.2	85.2

Table A5.5
Wall thickness = 102.5 mm

Partial safety factor for material strength = 2.8

Mortar designation (i)

Effective height (m)	Brick strength (N/mm^2)							
	20.0		27.5		35.0		50.0	
	$e < 0.05t$	$e = 0.2t$	$e < 0.05t$	$e = 0.2t$	$e < 0.05t$	$e = 0.2t$	$e < 0.05t$	$e = 0.2t$
0.90	311.5	205.6	387.3	255.6	479.9	316.7	631.5	416.8
1.20	293.3	205.6	364.6	255.6	451.8	316.7	594.4	416.8
1.50	271.2	205.6	337.2	255.6	417.9	316.7	549.8	416.8
1.80	244.3	182.7	303.8	227.1	376.4	281.4	495.3	370.2
2.10	212.5	150.9	264.2	187.5	327.4	232.4	430.8	305.8
2.40	175.8	114.2	218.6	141.9	270.9	175.9	356.4	231.4
2.70	134.3	72.6	166.9	90.2	206.8	111.8	272.1	147.1
2.77	124.2	62.5	154.4	77.8	191.4	96.3	251.8	126.8

Table A5.6
Wall thickness = 102.5 mm

Partial safety factor for material strength = 2.8

Mortar designation (ii)

Effective height (m)	Brick strength (N/mm^2)							
	20.0		27.5		35.0		50.0	
	$e < 0.05t$	$e = 0.2t$	$e < 0.05t$	$e = 0.2t$	$e < 0.05t$	$e = 0.2t$	$e < 0.05t$	$e = 0.2t$
0.90	269.4	177.8	332.6	219.5	395.7	261.2	513.6	339.0
1.20	253.6	177.8	313.1	219.5	372.5	261.2	483.5	339.0
1.50	234.6	177.8	289.6	219.5	344.6	261.2	447.2	339.0
1.80	211.3	158.0	260.8	195.0	310.4	232.0	402.8	301.1
2.10	183.8	130.5	226.9	161.0	270.0	191.6	350.4	248.7
2.40	152.1	98.7	187.7	121.9	223.4	145.0	289.9	188.2
2.70	116.1	62.8	143.3	77.5	170.5	92.2	221.3	119.6
2.77	107.4	54.1	132.6	66.8	157.8	79.4	204.8	103.1

Table A5.7
Wall thickness = 102.5 mm

Partial safety factor for material strength = 2.8

Mortar designation (iii)

Effective height (m)	Brick strength (N/mm²)							
	20.0		27.5		35.0		50.0	
	$e < 0.05t$	$e = 0.2t$	$e < 0.05t$	$e = 0.2t$	$e < 0.05t$	$e = 0.2t$	$e < 0.05t$	$e = 0.2t$
0.90	244.2	161.2	298.9	197.3	357.8	236.2	446.2	294.5
1.20	229.9	161.2	281.4	197.3	336.9	236.2	420.1	294.5
1.50	212.6	161.2	260.2	197.3	311.6	236.2	388.5	294.5
1.80	191.5	143.2	234.4	175.2	280.7	209.8	350.0	261.6
2.10	166.6	118.2	203.9	144.7	244.1	173.3	304.4	216.1
2.40	137.8	89.5	168.7	109.5	202.0	131.1	251.9	163.5
2.70	105.2	56.9	128.8	69.6	154.2	83.4	192.3	104.0
2.77	97.4	49.0	119.2	60.0	142.7	71.8	177.9	89.6

Table A5.8
Wall thickness = 102.5 mm

Partial safety factor for material strength = 2.8

Mortar designation (iv)

Effective height (m)	Brick strength (N/mm²)							
	20.0		27.5		35.0		50.0	
	$e < 0.05t$	$e = 0.2t$	$e < 0.05t$	$e = 0.2t$	$e < 0.05t$	$e = 0.2t$	$e < 0.05t$	$e = 0.2t$
0.90	218.9	144.5	261.0	172.3	307.3	202.8	378.9	250.1
1.20	206.1	144.5	245.7	172.3	289.3	202.8	356.7	250.1
1.50	190.6	144.5	227.3	172.3	267.6	202.8	329.9	250.1
1.80	171.7	128.3	204.7	153.0	241.0	180.2	297.2	222.1
2.10	149.3	106.0	178.1	126.4	209.7	148.8	258.5	183.5
2.40	123.6	80.2	147.3	95.6	173.5	112.6	213.9	138.8
2.70	94.3	51.0	112.5	60.8	132.4	71.6	163.3	88.3
2.77	87.3	43.9	104.1	52.4	122.5	61.7	151.1	76.1

Table A5.9
Wall thickness = 102.5 mm

Partial safety factor for material strength = 3.1

Mortar designation (i)

Effective height (m)	Brick strength (N/mm²)							
	20.0		27.5		35.0		50.0	
	$e < 0.05t$	$e = 0.2t$	$e < 0.05t$	$e = 0.2t$	$e < 0.05t$	$e = 0.2t$	$e < 0.05t$	$e = 0.2t$
0.90	281.4	185.7	349.8	230.9	433.5	286.1	570.4	376.4
1.20	264.9	185.7	329.3	230.9	408.1	286.1	536.9	376.4
1.50	245.0	185.7	304.6	230.9	377.4	286.1	496.6	376.4
1.80	220.7	165.0	274.4	205.1	340.0	254.2	447.3	334.4
2.10	192.0	136.3	238.7	169.4	295.7	209.9	389.1	276.2
2.40	158.8	103.1	197.5	128.2	244.7	158.8	321.9	209.0
2.70	121.3	65.5	150.8	81.5	186.8	101.0	245.8	132.9
2.77	112.2	56.5	139.5	70.2	172.8	87.0	227.4	114.5

Table A5.10
Wall thickness = 102.5 mm

Partial safety factor for material strength = 3.1

Mortar designation (ii)

Effective	Brick strength (N/mm^2)							
height	20.0		27.5		35.0		50.0	
(m)	$e < 0.05t$	$e = 0.2t$	$e < 0.05t$	$e = 0.2t$	$e < 0.05t$	$e = 0.2t$	$e < 0.05t$	$e = 0.2t$
0.90	243.4	160.6	300.4	198.3	357.4	235.9	463.9	306.2
1.20	229.1	160.6	282.8	198.3	336.5	235.9	436.7	306.2
1.50	211.9	160.6	261.5	198.3	311.2	235.9	403.9	306.2
1.80	190.9	142.7	235.6	176.1	280.3	209.6	363.8	272.0
2.10	166.0	117.8	204.9	145.5	243.8	173.1	316.5	224.6
2.40	137.4	89.2	169.6	110.1	201.7	131.0	261.8	170.0
2.70	104.9	56.7	129.5	70.0	154.0	83.3	199.9	108.1
2.77	97.0	48.9	119.8	60.3	142.5	71.8	185.0	93.1

Table A5.11
Wall thickness = 102.5 mm

Partial safety factor for material strength = 3.1

Mortar designation (iii)

Effective	Brick strength (N/mm^2)							
height	20.0		27.5		35.0		50.0	
(m)	$e < 0.05t$	$e = 0.2t$	$e < 0.05t$	$e = 0.2t$	$e < 0.05t$	$e = 0.2t$	$e < 0.05t$	$e = 0.2t$
0.90	220.5	145.6	270.0	178.2	323.2	213.3	403.1	266.0
1.20	207.6	145.6	254.1	178.2	304.3	213.3	379.4	266.0
1.50	192.0	145.6	235.1	178.2	281.4	213.3	350.9	266.0
1.80	173.0	129.3	211.7	158.3	253.5	189.5	316.1	236.3
2.10	150.5	106.8	184.2	130.7	220.5	156.5	275.0	195.2
2.40	124.5	80.8	152.4	98.9	182.4	118.4	227.5	147.7
2.70	95.0	51.4	116.3	62.9	139.3	75.3	173.7	93.9
2.77	87.9	44.3	107.7	54.2	128.9	64.9	160.7	80.9

Table A5.12
Wall thickness = 102.5 mm

Partial safety factor for material strength = 3.1

Mortar designation (iv)

Effective	Brick strength (N/mm^2)							
height	20.0		27.5		35.0		50.0	
(m)	$e < 0.05t$	$e = 0.2t$	$e < 0.05t$	$e = 0.2t$	$e < 0.05t$	$e = 0.2t$	$e < 0.05t$	$e = 0.2t$
0.90	197.7	130.5	235.8	155.6	277.6	183.2	342.2	225.9
1.20	186.1	130.5	221.9	155.6	261.3	183.2	322.2	225.9
1.50	172.2	130.5	205.3	155.6	241.7	183.2	298.0	225.9
1.80	155.1	115.9	184.9	138.2	217.7	162.7	268.4	200.6
2.10	134.9	95.7	160.8	114.2	189.4	134.4	233.5	165.7
2.40	111.6	72.5	133.1	86.4	156.7	101.7	193.2	125.4
2.70	85.2	46.1	101.6	54.9	119.6	64.7	147.5	79.7
2.77	78.8	39.7	94.0	47.3	110.7	55.7	136.5	68.7

Table A5.13
Wall thickness = 102.5 mm

Partial safety factor for material strength = 3.5

Mortar designation (i)

Effective height (m)	Brick strength (N/mm²)							
	20.0		27.5		35.0		50.0	
	$e < 0.05t$	$e = 0.2t$	$e < 0.05t$	$e = 0.2t$	$e < 0.05t$	$e = 0.2t$	$e < 0.05t$	$e = 0.2t$
0.90	249.2	164.5	309.8	204.5	383.9	253.4	505.2	333.4
1.20	234.6	164.5	291.7	204.5	361.4	253.4	475.6	333.4
1.50	217.0	164.5	269.8	204.5	334.3	253.4	439.9	333.4
1.80	195.5	146.1	243.0	181.7	301.1	225.1	396.2	296.2
2.10	170.0	120.7	211.4	150.0	261.9	185.9	344.6	244.6
2.40	140.7	91.3	174.9	113.5	216.7	140.7	285.1	185.1
2.70	107.4	58.1	133.5	72.2	165.5	89.4	217.7	117.7
2.77	99.4	50.0	123.5	62.2	153.1	77.1	201.4	101.4

Table A5.14
Wall thickness = 102.5 mm

Partial safety factor for material strength = 3.5

Mortar designation (ii)

Effective height (m)	Brick strength (N/mm²)							
	20.0		27.5		35.0		50.0	
	$e < 0.05t$	$e = 0.2t$	$e < 0.05t$	$e = 0.2t$	$e < 0.05t$	$e = 0.2t$	$e < 0.05t$	$e = 0.2t$
0.90	215.5	142.3	266.1	175.6	316.6	208.9	410.9	271.2
1.20	202.9	142.3	250.5	175.6	298.0	208.9	386.8	271.2
1.50	187.7	142.3	231.7	175.6	275.6	208.9	357.7	271.2
1.80	169.1	126.4	208.7	156.0	248.3	185.6	322.3	240.9
2.10	147.0	104.4	181.5	128.8	216.0	153.3	280.3	199.0
2.40	121.7	79.0	150.2	97.5	178.7	116.0	231.9	150.6
2.70	92.9	50.2	114.7	62.0	136.4	73.7	177.1	95.7
2.77	85.9	43.3	106.1	53.4	126.2	63.6	163.8	82.5

Table A5.15
Wall thickness = 102.5 mm

Partial safety factor for material strength = 3.5

Mortar designation (iii)

Effective height (m)	Brick strength (N/mm²)							
	20.0		27.5		35.0		50.0	
	$e < 0.05t$	$e = 0.2t$	$e < 0.05t$	$e = 0.2t$	$e < 0.05t$	$e = 0.2t$	$e < 0.05t$	$e = 0.2t$
0.90	195.3	128.9	239.1	157.8	286.3	188.9	357.0	235.6
1.20	183.9	128.9	225.1	157.8	269.5	188.9	336.1	235.6
1.50	170.1	128.9	208.2	157.8	249.3	188.9	310.8	235.6
1.80	153.2	114.5	187.5	140.2	224.5	167.8	280.0	209.3
2.10	133.3	94.6	163.1	115.8	195.3	138.6	243.6	172.9
2.40	110.3	71.6	135.0	87.6	161.6	104.9	201.5	130.8
2.70	84.2	45.5	103.0	55.7	123.4	66.7	153.8	83.2
2.77	77.9	39.2	95.3	48.0	114.1	57.5	142.4	71.7

Table A5.16
Wall thickness = 102.5 mm

Partial safety factor for material strength = 3.5

Mortar designation (iv)

Effective height (m)	Brick strength (N/mm^2)							
	20.0		27.5		35.0		50.0	
	$e < 0.05t$	$e = 0.2t$	$e < 0.05t$	$e = 0.2t$	$e < 0.05t$	$e = 0.2t$	$e < 0.05t$	$e = 0.2t$
0.90	175.1	115.6	208.8	137.8	245.9	162.3	303.1	200.1
1.20	164.9	115.6	196.6	137.8	231.4	162.3	285.3	200.1
1.50	152.5	115.6	181.8	137.8	214.1	162.3	263.9	200.1
1.80	137.4	102.7	163.8	122.4	192.8	144.1	237.7	177.7
2.10	119.5	84.8	142.5	101.1	167.7	119.0	206.8	146.8
2.40	98.8	64.2	117.9	76.5	138.8	90.1	171.1	111.1
2.70	75.5	40.8	90.0	48.6	106.0	57.3	130.6	70.6
2.77	69.8	35.2	83.3	41.9	98.0	49.4	120.9	60.8

Table A5.17
Wall thickness = 210.0 mm

Partial safety factor for material strength = 2.5

Mortar designation (i)

Effective height (m)	Brick strength (N/mm^2)							
	20.0		27.5		35.0		50.0	
	$e < 0.05t$	$e = 0.2t$	$e < 0.05t$	$e = 0.2t$	$e < 0.05t$	$e = 0.2t$	$e < 0.05t$	$e = 0.2t$
0.90	621.6	410.3	772.8	510.0	957.6	632.0	1260.0	831.6
1.20	621.6	410.3	772.8	510.0	957.6	632.0	1260.0	831.6
1.50	621.6	410.3	772.8	510.0	957.6	632.0	1260.0	831.6
1.80	621.6	410.3	772.8	510.0	957.6	632.0	1260.0	831.6
2.10	606.3	410.3	753.7	510.0	934.0	632.0	1228.9	831.6
2.40	588.8	410.3	732.1	510.0	907.1	632.0	1193.6	831.6
2.70	569.1	410.3	707.5	510.0	876.7	632.0	1153.5	831.6
3.00	547.0	410.3	680.0	510.0	842.6	632.0	1108.7	831.6
3.30	522.5	399.5	649.6	496.6	805.0	615.4	1059.2	809.7
3.60	495.8	372.7	616.4	463.4	763.8	574.2	1005.0	755.5
3.90	466.7	343.6	580.3	427.2	719.0	529.4	946.1	696.6
4.20	435.3	312.3	541.2	388.2	670.6	481.0	882.4	632.9
4.50	401.6	278.5	499.3	346.3	618.7	429.1	814.1	564.6
4.80	365.6	242.5	454.5	301.5	563.2	373.5	741.0	491.5
5.10	327.2	204.1	406.8	253.8	504.0	314.4	663.2	413.7
5.40	286.5	163.4	356.2	203.2	441.3	251.7	580.7	331.2
5.67	247.9	124.8	308.2	155.1	381.8	192.2	502.4	252.9

Table A5.18
Wall thickness = 210.0 mm

Partial safety factor for material strength = 2.5

Mortar designation (ii)

Effective height (m)	Brick strength (N/mm²) 20.0		27.5		35.0		50.0	
	$e < 0.05t$	$e = 0.2t$	$e < 0.05t$	$e = 0.2t$	$e < 0.05t$	$e = 0.2t$	$e < 0.05t$	$e = 0.2t$
0.90	537.6	354.8	663.6	438.0	789.6	521.1	1024.8	676.4
1.20	537.6	354.8	663.6	438.0	789.7	521.1	1024.8	676.4
1.50	537.6	354.8	663.6	438.0	789.6	521.1	1024.8	676.4
1.80	537.6	354.8	663.6	438.0	789.6	521.1	1024.8	676.4
2.10	524.3	354.8	647.2	438.0	770.1	521.1	999.5	676.4
2.40	509.3	354.8	628.6	438.0	748.0	521.1	970.8	676.4
2.70	492.2	354.8	607.5	438.0	722.9	521.1	938.2	676.4
3.00	473.0	354.8	583.9	438.0	694.8	521.1	901.7	676.4
3.30	451.9	345.5	557.8	426.5	663.8	507.4	861.5	658.6
3.60	428.8	322.4	529.3	397.9	629.8	473.5	817.4	614.5
3.90	403.7	297.2	498.3	366.9	592.9	436.5	769.5	566.6
4.20	376.5	270.1	464.7	333.3	553.0	396.6	717.7	514.8
4.50	347.3	240.9	428.7	297.3	510.1	353.8	662.1	459.2
4.80	316.2	209.7	390.3	258.9	464.4	308.0	602.7	399.8
5.10	283.0	176.5	349.3	217.9	415.6	259.3	539.4	336.5
5.40	247.8	141.3	305.8	174.4	363.9	207.6	472.3	269.4
5.67	214.4	107.9	264.6	133.2	314.9	158.5	408.6	205.7

Table A5.19
Wall thickness = 210.0 mm

Partial safety factor for material strength = 2.5

Mortar designation (iii)

Effective height (m)	Brick strength (N/mm²) 20.0		27.5		35.0		50.0	
	$e < 0.05t$	$e = 0.2t$	$e < 0.05t$	$e = 0.2t$	$e < 0.05t$	$e = 0.2t$	$e < 0.05t$	$e = 0.2t$
0.90	487.2	321.6	596.4	393.6	714.0	471.2	890.4	587.7
1.20	487.2	321.6	596.4	393.6	714.0	471.2	890.4	587.7
1.50	487.2	321.6	596.4	393.6	714.0	471.2	890.4	587.7
1.80	487.2	321.6	596.4	393.6	714.0	471.2	890.4	587.7
2.10	475.2	321.6	581.7	393.6	696.4	471.2	868.4	587.7
2.40	461.5	321.6	565.0	393.6	676.4	471.2	843.5	587.7
2.70	446.0	321.6	546.0	393.6	653.6	471.2	815.1	587.7
3.00	428.7	321.6	524.8	393.6	628.3	471.2	783.5	587.7
3.30	409.6	313.1	501.4	383.3	600.2	458.8	748.5	572.2
3.60	388.6	292.1	475.7	357.6	569.5	428.1	710.2	533.9
3.90	365.8	269.3	447.8	329.7	536.1	394.7	668.6	492.3
4.20	341.2	244.7	417.7	299.6	500.0	358.7	623.6	447.3
4.50	314.8	218.3	385.3	267.2	461.3	319.9	575.3	399.0
4.80	286.5	190.1	350.7	232.6	419.9	278.5	523.6	347.3
5.10	256.4	160.0	313.9	195.8	375.8	234.4	468.7	292.4
5.40	224.5	128.1	274.9	156.8	329.1	187.7	410.4	234.1
5.67	194.3	97.8	237.8	119.7	284.7	143.3	355.0	178.7

Table A5.20
Wall thickness = 210.0 mm

Partial safety factor for material strength = 2.5

Mortar designation (iv)

Effective height (m)	Brick strength (N/mm²)							
	20.0		27.5		35.0		50.0	
	$e < 0.05t$	$e = 0.2t$	$e < 0.05t$	$e = 0.2t$	$e < 0.05t$	$e = 0.2t$	$e < 0.05t$	$e = 0.2t$
0.90	436.8	288.3	520.8	343.7	613.2	404.7	756.0	499.0
1.20	436.8	288.3	520.8	343.7	613.2	404.7	756.0	499.0
1.50	436.8	288.3	520.8	343.7	613.2	404.7	756.0	499.0
1.80	436.8	288.3	520.8	343.7	613.2	404.7	756.0	499.0
2.10	426.0	288.3	508.0	343.7	598.1	404.7	737.4	499.0
2.40	413.8	288.3	493.3	343.7	580.9	404.7	716.1	499.0
2.70	399.9	288.3	476.8	343.7	561.4	404.7	692.1	499.0
3.00	384.4	288.3	458.3	343.7	539.6	404.7	665.2	499.0
3.30	367.2	280.7	437.8	334.7	515.5	394.1	635.5	485.8
3.60	348.4	261.9	415.4	312.3	489.1	367.7	603.0	453.3
3.90	328.0	241.5	391.0	287.9	460.4	339.0	567.6	417.9
4.20	305.9	219.4	364.7	261.6	429.4	308.0	529.5	379.8
4.50	282.2	195.7	336.5	233.4	396.2	274.8	488.4	338.7
4.80	256.9	170.4	306.3	203.2	360.6	239.2	444.6	294.9
5.10	229.9	143.4	274.1	171.0	322.8	201.3	397.9	248.2
5.40	201.3	114.8	240.0	136.9	282.6	161.2	348.4	198.7
5.67	174.2	87.7	207.7	104.6	244.5	123.1	301.5	151.8

Table A5.21
Wall thickness = 210.0 mm

Partial safety factor for material strength = 2.8

Mortar designation (i)

Effective height (m)	Brick strength (N/mm²)							
	20.0		27.5		35.0		50.0	
	$e < 0.05t$	$e = 0.2t$	$e < 0.05t$	$e = 0.2t$	$e < 0.05t$	$e = 0.2t$	$e < 0.05t$	$e = 0.2t$
0.90	555.0	366.3	690.0	455.4	855.0	564.3	1125.0	742.5
1.20	555.0	366.3	690.0	455.4	855.0	564.3	1125.0	742.5
1.50	555.0	366.3	690.0	455.4	855.0	564.3	1125.0	742.5
1.80	555.0	366.3	690.0	455.4	855.0	564.3	1125.0	742.5
2.10	541.3	366.3	673.0	455.4	833.9	564.3	1097.3	742.5
2.40	525.7	366.3	653.6	455.4	809.9	564.3	1065.7	742.5
2.70	508.1	366.3	631.7	455.4	782.7	564.3	1029.9	742.5
3.00	488.4	366.3	607.1	455.4	752.3	564.3	989.9	742.5
3.30	466.6	356.7	580.0	443.4	718.7	549.5	945.7	723.0
3.60	442.7	332.8	550.4	413.7	682.0	512.7	897.3	674.6
3.90	416.7	306.8	518.1	381.5	642.0	472.7	844.7	621.9
4.20	388.7	278.8	483.2	346.6	598.8	429.5	787.9	565.1
4.50	358.6	248.7	445.8	309.2	552.4	383.1	726.8	504.1
4.80	326.4	216.5	405.8	269.2	502.8	333.5	661.6	438.8
5.10	292.1	182.2	363.2	226.6	450.0	280.7	592.1	369.4
5.40	255.8	145.9	318.0	181.4	394.1	224.8	518.5	295.7
5.67	221.3	111.4	275.1	138.5	340.9	171.6	448.6	225.8

Table A5.22
Wall thickness = 210.0 mm

Partial safety factor for material strength = 2.8

Mortar designation (ii)

Effective height (m)	Brick strength (N/mm²)							
	20.0		27.5		35.0		50.0	
	$e < 0.05t$	$e = 0.2t$	$e < 0.05t$	$e = 0.2t$	$e < 0.05t$	$e = 0.2t$	$e < 0.05t$	$e = 0.2t$
0.90	480.0	316.8	592.5	391.1	705.0	465.3	915.0	603.9
1.20	480.0	316.8	592.5	391.1	705.0	465.3	915.0	603.9
1.50	480.0	316.8	592.5	391.1	705.0	465.3	915.0	603.9
1.80	480.0	316.8	592.5	391.1	705.0	465.3	915.0	603.9
2.10	468.2	316.8	577.9	391.1	687.6	465.3	892.4	603.9
2.40	454.7	316.8	561.3	391.1	667.8	465.3	866.8	603.9
2.70	439.4	316.8	542.4	391.1	645.4	465.3	837.7	603.9
3.00	422.4	316.8	521.4	391.1	620.3	465.3	805.1	603.9
3.30	403.5	308.5	498.1	380.8	592.7	453.1	769.2	588.0
3.60	382.9	287.8	472.6	355.3	562.3	422.7	729.8	548.6
3.90	360.4	265.4	444.9	327.6	529.3	389.8	687.0	505.9
4.20	336.2	241.1	414.9	297.6	493.7	354.1	640.8	459.6
4.50	310.1	215.1	382.8	265.5	455.5	315.9	591.2	410.0
4.80	282.3	187.2	348.4	231.1	414.6	275.0	538.1	356.9
5.10	252.6	157.6	311.9	194.5	371.1	231.5	481.6	300.4
5.40	221.2	126.2	273.1	155.8	324.9	185.3	421.7	240.5
5.67	191.4	96.4	236.3	118.9	281.1	141.5	364.9	183.7

Table A5.23
Wall thickness = 210.0 mm

Partial safety factor for material strength = 2.8

Mortar designation (iii)

Effective height (m)	Brick strength (N/mm²)							
	20.0		27.5		35.0		50.0	
	$e < 0.05t$	$e = 0.2t$	$e < 0.05t$	$e = 0.2t$	$e < 0.05t$	$e = 0.2t$	$e < 0.05t$	$e = 0.2t$
0.90	435.0	287.1	532.5	351.5	637.5	420.8	795.0	524.7
1.20	435.0	287.1	532.5	351.5	637.5	420.8	795.0	524.7
1.50	435.0	287.1	532.5	351.5	637.5	420.8	795.0	524.7
1.80	435.0	287.1	532.5	351.5	637.5	420.8	795.0	524.7
2.10	424.3	287.1	519.4	351.5	621.8	420.8	775.4	524.7
2.40	412.1	287.1	504.4	351.5	603.9	420.8	753.1	524.7
2.70	398.2	287.1	487.5	351.5	583.6	420.8	727.8	524.7
3.00	382.8	287.1	468.6	351.5	561.0	420.8	699.5	524.7
3.30	365.7	279.5	447.6	342.2	535.9	409.7	668.3	510.9
3.60	347.0	260.8	424.7	319.3	508.5	382.3	634.1	476.7
3.90	326.6	240.5	399.8	294.4	478.7	352.4	596.9	439.5
4.20	304.6	218.5	372.9	267.5	446.5	320.2	556.8	399.4
4.50	281.0	194.9	344.0	238.6	411.9	285.7	513.6	356.2
4.80	255.8	169.7	313.2	207.7	374.9	248.7	467.5	310.1
5.10	229.0	142.8	280.3	174.8	335.6	209.3	418.5	261.0
5.40	200.5	114.4	245.4	140.0	293.8	167.6	366.4	209.0
5.67	173.5	87.3	212.3	106.9	254.2	128.0	317.0	159.6

Table A5.24
Wall thickness = 210.0 mm

Partial safety factor for material strength = 2.8

Mortar designation (iv)

Effective height (m)	Brick strength (N/mm²)							
	20.0		27.5		35.0		50.0	
	$e < 0.05t$	$e = 0.2t$	$e < 0.05t$	$e = 0.2t$	$e < 0.05t$	$e = 0.2t$	$e < 0.05t$	$e = 0.2t$
0.90	390.0	257.4	465.0	306.9	547.5	361.4	675.0	445.5
1.20	390.0	257.4	465.0	306.9	547.5	361.4	675.0	445.5
1.50	390.0	257.4	465.0	306.9	547.5	361.4	675.0	445.5
1.80	390.0	257.4	465.0	306.9	547.5	361.4	675.0	445.5
2.10	380.4	257.4	453.5	306.9	534.0	361.4	658.4	445.5
2.40	369.4	257.4	440.5	306.9	518.6	361.4	639.4	445.5
2.70	357.0	257.4	425.7	306.9	501.2	361.4	617.9	445.5
3.00	343.2	257.4	409.2	306.9	481.8	361.4	593.9	445.5
3.30	327.8	250.6	390.9	298.8	460.3	351.8	567.4	433.8
3.60	311.1	233.8	370.9	278.8	436.7	328.3	538.4	404.7
3.90	292.8	215.6	349.1	257.1	411.1	302.7	506.8	373.2
4.20	273.1	195.9	325.7	233.6	383.4	275.0	472.7	339.1
4.50	252.0	174.8	300.4	208.4	353.7	245.3	436.1	302.5
4.80	229.4	152.1	273.5	181.4	322.0	213.6	397.0	263.3
5.10	205.3	128.1	244.8	152.7	288.2	179.8	355.3	221.6
5.40	179.7	102.5	214.3	122.2	252.3	143.9	311.1	177.4
5.67	155.5	78.3	185.4	93.3	218.3	109.9	269.2	135.5

Table A5.25
Wall thickness = 210.0 mm

Partial safety factor for material strength = 3.1

Mortar designation (i)

Effective height (m)	Brick strength (N/mm²)							
	20.0		27.5		35.0		50.0	
	$e < 0.05t$	$e = 0.2t$	$e < 0.05t$	$e = 0.2t$	$e < 0.05t$	$e = 0.2t$	$e < 0.05t$	$e = 0.2t$
0.90	501.3	330.9	623.2	411.3	772.3	509.7	1016.1	670.6
1.20	501.3	330.9	623.2	411.3	772.3	509.7	1016.1	670.6
1.50	501.3	330.9	623.2	411.3	772.3	509.7	1016.1	670.6
1.80	501.3	330.9	623.2	411.3	772.3	509.7	1016.1	670.6
2.10	488.9	330.9	607.9	411.3	753.2	509.7	991.1	670.6
2.40	474.9	330.9	590.4	411.3	731.5	509.7	962.6	670.6
2.70	458.9	330.9	570.5	411.3	707.0	509.7	930.2	670.6
3.00	441.1	330.9	548.4	411.3	679.5	509.7	894.1	670.6
3.30	421.4	322.1	523.9	400.5	649.2	496.3	854.2	653.0
3.60	399.8	300.6	497.1	373.7	616.0	463.1	810.5	609.3
3.90	376.4	277.1	467.9	344.5	579.8	426.9	763.0	561.8
4.20	351.1	251.8	436.5	313.1	540.8	387.9	711.6	510.4
4.50	323.9	224.6	402.7	279.3	498.9	346.0	656.5	455.3
4.80	294.8	195.5	366.5	243.1	454.2	301.2	597.6	396.4
5.10	263.9	164.6	328.0	204.6	406.5	253.6	534.8	333.6
5.40	231.0	131.8	287.2	163.8	355.9	203.0	468.3	267.1
5.67	199.9	100.6	248.5	125.1	307.9	155.0	405.2	204.0

Table A5.26

Wall thickness = 210.0 mm

Partial safety factor for material strength = 3.1

Mortar designation (ii)

Effective height (m)	Brick strength (N/mm^2)							
	20.0		27.5		35.0		50.0	
	$e < 0.05t$	$e = 0.2t$	$e < 0.05t$	$e = 0.2t$	$e < 0.05t$	$e = 0.2t$	$e < 0.05t$	$e = 0.2t$
0.90	433.5	286.1	533.2	353.2	636.8	420.3	826.5	545.5
1.20	433.5	286.1	535.2	353.2	636.8	420.3	826.5	545.5
1.50	433.5	286.1	535.2	353.2	636.8	420.3	826.5	545.5
1.80	433.5	286.1	535.2	353.2	636.8	420.3	826.5	545.5
2.10	422.9	286.1	522.0	353.2	621.1	420.3	806.1	545.5
2.40	410.7	286.1	506.9	353.2	603.2	420.3	782.9	545.5
2.70	396.9	286.1	489.9	353.2	582.9	420.3	756.6	545.5
3.00	381.5	286.1	470.9	353.2	560.3	420.3	727.2	545.5
3.30	364.5	278.6	449.9	343.9	535.3	409.2	694.7	531.1
3.60	345.8	260.0	426.9	320.9	507.9	381.8	659.2	495.6
3.90	325.5	239.7	401.8	295.9	478.1	352.0	620.5	456.9
4.20	303.6	217.8	374.8	268.8	446.0	319.9	578.8	415.2
4.50	280.1	194.3	345.8	239.8	411.4	285.3	534.0	370.3
4.80	255.0	169.1	314.7	208.8	374.5	248.4	486.0	322.4
5.10	228.2	142.4	281.7	175.7	335.2	209.1	435.0	271.4
5.40	199.8	114.0	246.6	140.7	293.5	167.4	380.9	217.3
5.67	172.9	87.0	213.4	107.4	253.9	127.8	329.5	165.9

Table A5.27

Wall thickness = 210.0 mm

Partial safety factor for material strength = 3.1

Mortar designation (iii)

Effective height (m)	Brick strength (N/mm^2)							
	20.0		27.5		35.0		50.0	
	$e < 0.05t$	$e = 0.2t$	$e < 0.05t$	$e = 0.2t$	$e < 0.05t$	$e = 0.2t$	$e < 0.05t$	$e = 0.2t$
0.90	392.9	259.3	481.0	317.4	575.8	380.0	718.1	473.9
1.20	392.9	259.3	481.0	317.4	575.8	380.0	718.1	473.9
1.50	392.9	259.3	481.0	317.4	575.8	380.0	718.1	473.9
1.80	392.9	259.3	481.0	317.4	575.8	380.0	718.1	473.9
2.10	383.2	259.3	469.1	317.4	561.6	380.0	700.4	473.9
2.40	372.2	259.3	455.6	317.4	545.4	380.0	680.2	473.9
2.70	359.7	259.3	440.3	317.4	527.1	380.0	657.4	473.9
3.00	345.7	259.3	423.2	317.4	506.7	380.0	631.8	473.9
3.30	330.3	252.5	404.3	309.1	484.0	370.0	603.6	461.5
3.60	313.4	235.6	383.6	288.4	459.3	345.3	572.7	430.6
3.90	295.0	217.2	361.1	265.9	432.3	318.3	539.2	397.0
4.20	275.2	197.4	336.8	241.6	403.3	289.2	502.9	360.7
4.50	253.8	176.1	310.7	215.5	372.0	258.0	463.9	321.8
4.80	231.1	153.3	282.9	187.6	338.6	224.6	422.3	280.1
5.10	206.8	129.0	253.2	157.9	303.1	189.1	378.0	235.8
5.40	181.1	103.3	221.7	126.4	265.4	151.4	330.9	188.8
5.67	156.7	78.9	191.8	96.6	229.6	115.6	286.3	144.2

Table A5.28
Wall thickness = 210.0 mm

Partial safety factor for material strength = 3.1

Mortar designation (iv)

Effective height (m)	Brick strength (N/mm^2)							
	20.0		27.5		35.0		50.0	
	$e < 0.05t$	$e = 0.2t$	$e < 0.05t$	$e = 0.2t$	$e < 0.05t$	$e = 0.2t$	$e < 0.05t$	$e = 0.2t$
0.90	352.3	232.5	420.0	277.2	494.5	326.4	609.7	402.4
1.20	352.3	232.5	420.0	277.2	494.5	326.4	609.7	402.4
1.50	352.3	232.5	420.0	277.2	494.5	326.4	609.7	402.4
1.80	352.3	232.5	420.0	277.2	494.5	326.4	609.7	402.4
2.10	343.6	232.5	409.6	277.2	482.3	326.4	594.6	402.4
2.40	333.7	232.5	397.9	277.2	468.4	326.4	577.5	402.4
2.70	322.5	232.5	384.5	277.2	452.7	326.4	558.1	402.4
3.00	310.0	232.5	369.6	277.2	435.1	326.4	536.5	402.4
3.30	296.1	226.4	353.1	269.9	415.7	317.8	512.5	391.8
3.60	281.0	211.2	335.0	251.8	394.4	296.5	486.3	365.6
3.90	264.5	194.7	315.4	232.2	371.3	273.4	457.8	337.1
4.20	246.7	177.0	294.1	211.0	346.3	248.4	427.0	306.3
4.50	227.6	157.8	271.4	188.2	319.5	221.6	393.9	273.2
4.80	207.2	137.4	247.0	163.8	290.8	192.9	358.5	237.8
5.10	185.4	115.7	221.1	137.9	260.3	162.4	320.9	200.2
5.40	162.3	92.6	193.6	110.4	227.9	130.0	281.0	160.3
5.67	140.5	70.7	167.5	84.3	197.2	99.3	243.1	122.4

Table A5.29
Wall thickness = 210.0 mm

Partial safety factor for material strength = 3.5

Mortar designation (i)

Effective height (m)	Brick strength (N/mm^2)							
	20.0		27.5		35.0		50.0	
	$e < 0.05t$	$e = 0.2t$	$e < 0.05t$	$e = 0.2t$	$e < 0.05t$	$e = 0.2t$	$e < 0.05t$	$e = 0.2t$
0.90	444.0	293.0	552.0	364.3	684.0	451.4	900.0	594.0
1.20	444.0	293.0	552.0	364.3	684.0	451.4	900.0	594.0
1.50	444.0	293.0	552.0	364.3	684.0	451.4	900.0	594.0
1.80	444.0	293.0	552.0	364.3	684.0	451.4	900.0	594.0
2.10	433.0	293.0	538.4	364.3	667.1	451.4	877.8	594.0
2.40	420.6	293.0	522.9	364.3	647.9	451.4	852.5	594.0
2.70	406.5	293.0	505.3	364.3	626.2	451.4	823.9	594.0
3.00	390.7	293.0	485.7	364.3	601.9	451.4	791.9	594.0
3.30	373.2	285.3	464.0	354.7	575.0	439.6	756.6	578.4
3.60	354.1	266.2	440.3	331.0	545.6	410.1	717.9	539.7
3.90	333.4	245.5	414.5	305.2	513.6	378.1	675.8	497.6
4.20	310.9	223.0	386.6	277.3	479.0	343.6	630.3	452.1
4.50	286.9	198.9	356.6	247.3	441.9	306.5	581.5	403.3
4.80	261.1	173.2	324.6	215.3	402.3	266.8	529.3	351.1
5.10	233.7	145.8	290.5	181.3	360.0	224.6	473.7	295.5
5.40	204.6	116.7	254.4	145.1	315.2	179.8	414.8	236.6
5.67	177.0	89.1	220.1	110.8	272.7	137.3	358.9	180.7

Table A5.30
Wall thickness = 210.0 mm

Partial safety factor for material strength = 3.5

Mortar designation (ii)

Effective height (m)	Brick strength (N/mm^2)							
	20.0		27.5		35.0		50.0	
	$e < 0.05t$	$e = 0.2t$	$e < 0.05t$	$e = 0.2t$	$e < 0.05t$	$e = 0.2t$	$e < 0.05t$	$e = 0.2t$
0.90	384.0	253.4	474.0	312.8	564.0	372.2	732.0	483.1
1.20	384.0	253.4	474.0	312.8	564.0	372.2	732.0	483.1
1.50	384.0	253.4	474.0	312.8	564.0	372.2	732.0	483.1
1.80	384.0	253.4	474.0	312.8	564.0	372.2	732.0	483.1
2.10	374.5	253.4	462.3	312.8	550.1	372.2	713.9	483.1
2.40	363.8	253.4	449.0	312.8	534.3	372.2	693.4	483.1
2.70	351.5	253.4	433.9	312.8	516.3	372.2	670.1	483.1
3.00	337.9	253.4	417.1	312.8	496.3	372.2	644.1	483.1
3.30	322.8	246.8	398.5	304.6	474.1	362.4	615.3	470.4
3.60	306.3	230.3	378.1	284.2	449.9	338.2	583.9	438.9
3.90	288.3	212.3	355.9	262.0	423.5	311.8	549.6	404.7
4.20	268.9	192.9	332.0	238.1	395.0	283.3	512.6	367.7
4.50	248.1	172.1	306.2	212.4	364.4	252.7	472.9	328.0
4.80	225.8	149.8	278.8	184.9	331.7	220.0	430.5	285.5
5.10	202.1	126.1	249.5	155.6	296.9	185.2	385.3	240.4
5.40	177.0	100.9	218.5	124.6	259.9	148.3	337.4	192.4
5.67	153.1	77.1	189.0	95.2	224.9	113.2	291.9	146.9

Table A5.31
Wall thickness = 210.0 mm

Partial safety factor for material strength = 3.5

Mortar designation (iii)

Effective height (m)	Brick strength (N/mm^2)							
	20.0		27.5		35.0		50.0	
	$e < 0.05t$	$e = 0.2t$	$e < 0.05t$	$e = 0.2t$	$e < 0.05t$	$e = 0.2t$	$e < 0.05t$	$e = 0.2t$
0.90	348.0	229.7	426.0	281.2	510.0	336.6	636.0	419.8
1.20	348.0	229.7	426.0	281.2	510.0	336.6	636.0	419.8
1.50	348.0	229.7	426.0	281.2	510.0	336.6	636.0	419.8
1.80	348.0	229.7	426.0	281.2	510.0	336.6	636.0	419.8
2.10	339.4	229.7	415.5	281.2	497.4	336.6	620.3	419.8
2.40	329.7	229.7	403.5	281.2	483.1	336.6	602.5	419.8
2.70	318.6	229.7	390.0	281.2	466.9	336.6	582.2	419.8
3.00	306.2	229.7	374.8	281.2	448.8	336.6	559.6	419.8
3.30	292.5	223.6	358.1	273.8	428.7	327.7	534.6	408.7
3.60	277.6	208.7	339.8	255.4	406.8	305.8	507.3	381.4
3.90	261.3	192.4	319.9	235.5	382.9	282.0	477.5	351.6
4.20	243.7	174.8	298.3	214.0	357.2	256.2	445.4	319.5
4.50	224.8	155.9	275.2	190.9	329.5	228.5	410.9	285.0
4.80	204.7	135.8	250.5	166.2	299.9	198.9	374.0	248.1
5.10	183.2	114.3	224.2	139.9	268.4	167.5	334.8	208.8
5.40	160.4	91.5	196.3	112.0	235.0	134.1	293.1	167.2
5.67	138.8	69.9	169.9	85.5	203.4	102.4	253.6	127.7

Table A5.32
Wall thickness = 210.0 mm

Partial safety factor for material strength = 3.5

Mortar designation (iv)

Effective height (m)	Brick strength (N/mm²)							
	20.0		27.5		35.0		50.0	
	$e < 0.05t$	$e = 0.2t$	$e < 0.05t$	$e = 0.2t$	$e < 0.05t$	$e = 0.2t$	$e < 0.05t$	$e = 0.2t$
0.90	312.0	205.9	372.0	245.5	438.0	289.1	540.0	356.4
1.20	312.0	205.9	372.0	245.5	438.0	289.1	540.0	356.4
1.50	312.0	205.9	372.0	245.5	438.0	289.1	540.0	356.4
1.80	312.0	205.9	372.0	245.5	438.0	289.1	540.0	356.4
2.10	304.3	205.9	362.8	245.5	427.2	289.1	526.7	356.4
2.40	295.5	205.9	352.4	245.5	414.9	289.1	511.5	356.4
2.70	285.6	205.9	340.6	245.5	401.0	289.1	494.4	356.4
3.00	274.5	205.9	327.3	245.5	385.4	289.1	475.2	356.4
3.30	262.3	200.5	312.7	239.1	368.2	281.5	453.9	347.0
3.60	248.9	187.1	296.7	223.1	349.4	262.6	430.7	323.8
3.90	234.3	172.5	279.3	205.7	328.9	242.1	405.5	298.5
4.20	218.5	156.7	260.5	186.9	306.7	220.0	378.2	271.3
4.50	201.6	139.8	240.3	166.7	283.0	196.3	348.9	242.0
4.80	183.5	121.7	218.8	145.1	257.6	170.9	317.6	210.6
5.10	164.2	102.4	195.8	122.1	230.5	143.8	284.2	177.3
5.40	143.8	82.0	171.4	97.8	201.9	115.1	248.9	142.0
5.67	124.4	62.6	148.3	74.7	174.7	87.9	215.3	108.4

DATA

PROPERTIES OF SECTIONS.

1. Half Parabola.

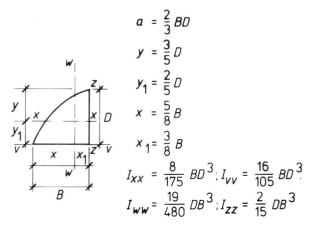

$$a = \frac{2}{3}BD$$
$$y = \frac{3}{5}D$$
$$y_1 = \frac{2}{5}D$$
$$x = \frac{5}{8}B$$
$$x_1 = \frac{3}{8}B$$
$$I_{xx} = \frac{8}{175}BD^3 \; ; \; I_{vv} = \frac{16}{105}BD^3$$
$$I_{ww} = \frac{19}{480}DB^3 \; ; \; I_{zz} = \frac{2}{15}DB^3$$

2. Complement of Half Parabola.

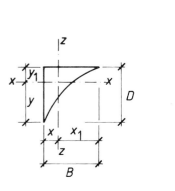

$$a = \frac{1}{3}BD$$
$$y = \frac{7}{10}D$$
$$y_1 = \frac{3}{10}D$$
$$x = \frac{1}{4}B$$
$$x_1 = \frac{3}{4}B$$
$$I_{xx} = \frac{37}{2100}BD^3$$
$$I_{zz} = \frac{DB^3}{80}$$

3. Hollow Rectangle.

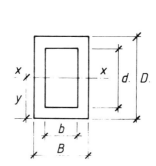

$$a = BD - bd$$
$$y = \frac{D}{2}$$
$$I_{xx} = \frac{BD^3 - bd^3}{12}$$
$$Z_{xx} = \frac{BD^3 - bd^3}{6D}$$
$$k_{xx} = \sqrt{\frac{BD^3 - bd^3}{12a}}$$

4. Triangle with Axis on Base.

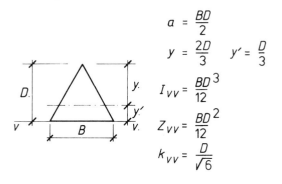

$$a = \frac{BD}{2}$$
$$y = \frac{2D}{3} \qquad y' = \frac{D}{3}$$
$$I_{vv} = \frac{BD^3}{12}$$
$$Z_{vv} = \frac{BD^2}{12}$$
$$k_{vv} = \frac{D}{\sqrt{6}}$$

5. Circle with Axis through Centre.

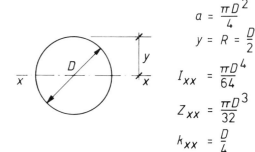

$$a = \frac{\pi D^2}{4}$$
$$y = R = \frac{D}{2}$$
$$I_{xx} = \frac{\pi D^4}{64}$$
$$Z_{xx} = \frac{\pi D^3}{32}$$
$$k_{xx} = \frac{D}{4}$$

6. Semi-circle with Axis through C. of G.

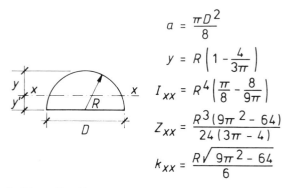

$$a = \frac{\pi D^2}{8}$$
$$y = R\left(1 - \frac{4}{3\pi}\right)$$
$$I_{xx} = R^4\left(\frac{\pi}{8} - \frac{8}{9\pi}\right)$$
$$Z_{xx} = \frac{R^3(9\pi^2 - 64)}{24(3\pi - 4)}$$
$$k_{xx} = \frac{R\sqrt{9\pi^2 - 64}}{6}$$

7. Parabola.

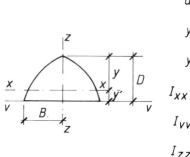

$$a = \frac{4}{3}BD$$
$$y = \frac{3}{5}D$$
$$y' = \frac{2}{5}D$$
$$I_{xx} = \frac{16}{175}BD^3$$
$$I_{vv} = \frac{32}{105}BD^3$$
$$I_{zz} = \frac{4}{15}DB^3$$

8. Ring Section.

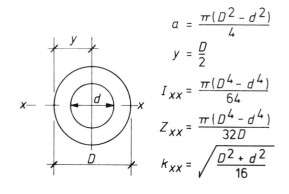

$$a = \frac{\pi(D^2 - d^2)}{4}$$
$$y = \frac{D}{2}$$
$$I_{xx} = \frac{\pi(D^4 - d^4)}{64}$$
$$Z_{xx} = \frac{\pi(D^4 - d^4)}{32D}$$
$$k_{xx} = \sqrt{\frac{D^2 + d^2}{16}}$$

PROPERTIES OF SECTIONS - (contd.)

9. Hollow Square

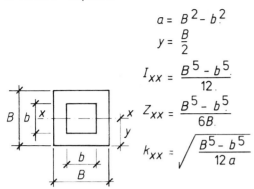

$$a = B^2 - b^2$$

$$y = \frac{B}{2}$$

$$I_{xx} = \frac{B^5 - b^5}{12}$$

$$Z_{xx} = \frac{B^5 - b^5}{6B}$$

$$k_{xx} = \sqrt{\frac{B^5 - b^5}{12\,a}}$$

10. H Section.

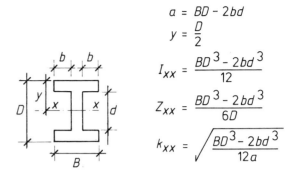

$$a = BD - 2bd$$

$$y = \frac{D}{2}$$

$$I_{xx} = \frac{BD^3 - 2bd^3}{12}$$

$$Z_{xx} = \frac{BD^3 - 2bd^3}{6D}$$

$$k_{xx} = \sqrt{\frac{BD^3 - 2bd^3}{12a}}$$

11. T Section.

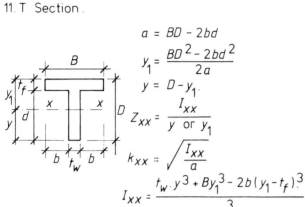

$$a = BD - 2bd$$

$$y_1 = \frac{BD^2 - 2bd^2}{2a}$$

$$y = D - y_1$$

$$Z_{xx} = \frac{I_{xx}}{y \text{ or } y_1}$$

$$k_{xx} = \sqrt{\frac{I_{xx}}{a}}$$

$$I_{xx} = \frac{t_w \cdot y^3 + B y_1^3 - 2b(y_1 - t_f)^3}{3}$$

12. L Section.

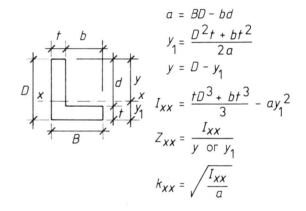

$$a = BD - bd$$

$$y_1 = \frac{D^2 t + bt^2}{2a}$$

$$y = D - y_1$$

$$I_{xx} = \frac{tD^3 + bt^3}{3} - ay_1^2$$

$$Z_{xx} = \frac{I_{xx}}{y \text{ or } y_1}$$

$$k_{xx} = \sqrt{\frac{I_{xx}}{a}}$$

13. U Section.

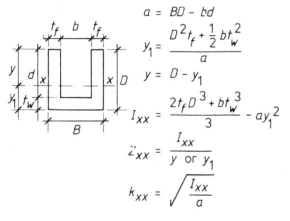

$$a = BD - bd$$

$$y_1 = \frac{D^2 t_f + \frac{1}{2} bt_w^2}{a}$$

$$y = D - y_1$$

$$I_{xx} = \frac{2t_f D^3 + bt_w^3}{3} - ay_1^2$$

$$Z_{xx} = \frac{I_{xx}}{y \text{ or } y_1}$$

$$k_{xx} = \sqrt{\frac{I_{xx}}{a}}$$

> Where k = radius of gyration.

THE GREEK ALPHABET.

Letters from the Greek alphabet are frequently used as symbols in engineering problems.

Alpha	A	α	Iota	I	ι	Rho	P	ρ
Beta	B	β	Kappa	K	κ	Sigma	Σ	σ
Gamma	Γ	γ	Lambda	Λ	λ	Tau	T	τ
Delta	Δ	δ	Mu	M	μ	Upsilon	Y	υ
Epsilon	E	ε	Nu	N	ν	Phi	Φ	φ
Zeta	Z	ζ	Xi	Ξ	ξ	Chi	X	χ
Eta	H	η	Omicron	O	o	Psi	Ψ	ψ
Theta	Θ	θ	Pi	Π	π	Omega	Ω	ω

NEUTRAL AXIS OF T SECTIONS.

The depth of the neutral axis y is given by —

$$Y = \frac{r_d}{2}\left[\frac{R_1^2 + R_2 + 2R_2R_1}{R_1 + R_2}\right] = \frac{r_d}{2} \times K_4$$

where $R_1 = \dfrac{f_t}{r_d}$ (Fig 1.) and $R_2 = \dfrac{r_t}{f_w}$ (Fig 1.)

Fig 2. gives values of the factor K_4 for various values of R_2 and R_1.

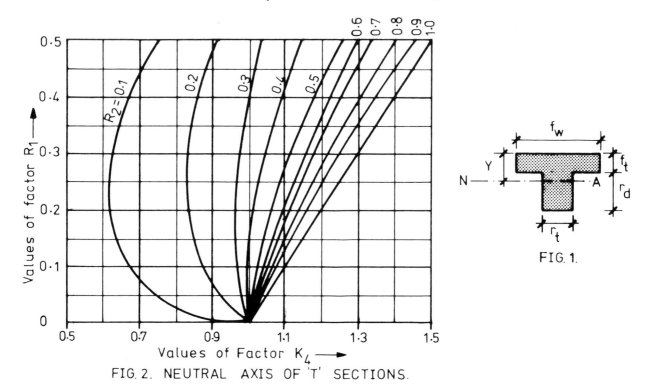

FIG. 2. NEUTRAL AXIS OF 'T' SECTIONS.

FIG. 1.

PROPERTIES OF THE CIRCLE.

For the circle, radius R, a segment of which is shown in FIG. 3

$$\text{Area} = \pi.R^2$$

$$\text{Circumference} = 2\pi.R.$$

Length of arc

$$= \text{number of degrees subtended by arc} \times \text{radius} \times \frac{\pi}{180} = \frac{\theta.R.\pi}{180}$$

$$V = \text{versed sine} = R - \sqrt{R^2 - C^2} \quad \text{where } C = \frac{1}{2} \text{ chord length}.$$

$$R = \text{radius} \qquad = \frac{V^2 + C^2}{2V}$$

If O is any ordinate distance X from centre —

$$O = \sqrt{R^2 - X^2} - (R - V) \text{ or } X = \sqrt{R^2 - (O + R - V)^2}$$

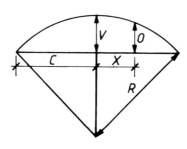

Fig 3.

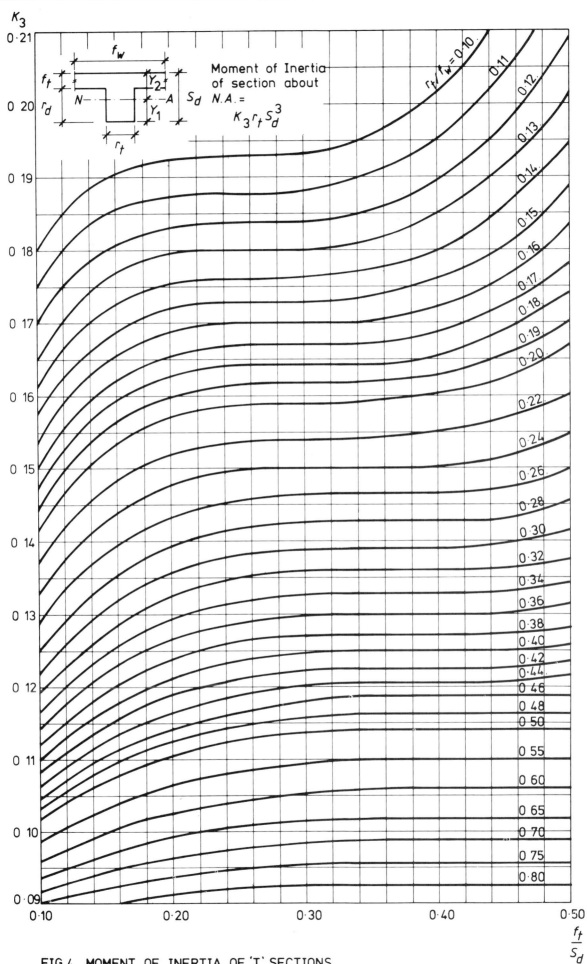

FIG 4. MOMENT OF INERTIA OF 'T' SECTIONS.

CONSTRUCTION OF THE PARABOLA.

Method 1.

It is required to draw a parabola on base *AB* and of height *DC* (Fig.5.)

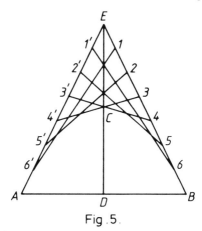

Fig.5.

Produce *DC* to *E* making *CE = CD*. Join *AE* and *BE* ; divide *AE* and *BE* into an equal number of divisions ,*1,2,3,4,*etc. and ,*1',2',3'*,etc. Join points *1-6', 2-5', 3-4',* etc. The intersections so formed are points on the parabola , and the curve is completed by drawing a smooth curve through the intersections.

Method 2.

It is required to draw a parabola on base *AB* and of height *DC* (Fig.6.)

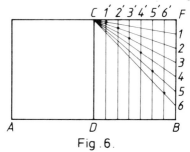

Fig.6.

Divide the height *BF* and the half span *CF* into an equal number of divisions *1,2,3,4* and *1',2',3',4',*etc. as shown. Vertical lines are dropped from points *1',2',3',4',* etc. The apex *C* is joined to each of the points *1,2,3,4,*etc. The intersections of these radial lines with the vertical lines are points on the parabola.It is completed by drawing a smooth curve through the intersections.

Construction of the other half is completed in a similar manner.

PROPERTIES OF THE PARABOLA

For the parabola, base AC, height DB, in Fig. 7.

$$\text{Area} = \tfrac{2}{3}\,\text{base} \times \text{height} = \tfrac{2}{3}\,AC.DB.$$

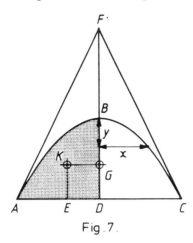

Fig. 7.

Centres of gravity.

$$DG = \tfrac{2}{5}.DB.$$

For half parabola — shown shaded —

$$KG = \tfrac{3}{8}.AD.$$

Tangents.

To draw a tangent, produce DB to F, making $BF = BD$, and join FA and FC. FA and FC are tangents, at A and C.

$$x = \text{any ordinate} = \frac{AD\sqrt{y}}{\sqrt{DB}} \qquad\qquad y = \text{any abscissa} = \frac{DB.x^2}{AD^2}$$

General Equation.

$$y = ax^2 + bx + C.$$

where a, b and C are constants.

STATIC FRICTION

The second law of friction states —

1. The limiting friction is a constant fraction of the normal pressure.

2. This constant fraction, called the coefficient of friction (μ), depends only on the nature of the two substances in contact, and does *not* depend on the extent of the surfaces in contact; it is also independent of the velocities of the bodies within reasonable limits.

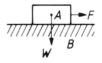

Fig. 8.

$F = \mu W$ and is the force required to make the body A, of weight W, move on the surface B, where μ is the coefficient of friction between the material A and the material B.

	μ
Timber on timber, fibres parallel to the motion.	0.4
Timber on timber, fibres at 90° to the motion.	0.5
Metal on timber	0.2
Metal on metal	0.15—0.2
Timber on stone	0.4
Metal on masonry	0.3—0.5
Masonry on masonry (hard)	0.2—0.3
Masonry on masonry (soft)	0.4—0.6
Masonry on dry clay } Masonry on wet clay } depends on shear strength of clay.	
Well lubricated hard smooth surfaces (bearings)	0.05

ALGEBRAICAL FORMULAE

Products and factors

$$(x + a)(x + b) = x^2 + (a + b)x + ab$$
$$(x + a)(x - a) = x^2 - a^2$$
$$(x \pm a)(x^2 \pm ax + a^2) = x^3 \pm a^3$$
$$(x \pm a)^2 = x^2 \pm 2ax + a^2$$
$$(x \pm a)^3 = x^3 \pm 3x^2a + 3xa^2 \pm a^3$$

General Solution to a Quadratic Equation.

If

$$ax^2 + bx + c = 0$$
$$x = \frac{-b \pm \sqrt{b^2 - 4ac}}{2a}$$

Indices.

$$a^m \times a^n = a^{m+n}$$
$$a^m \div a^n = a^{m-n}$$
$$(a^m)^n = a^{mn}$$
$$\log(ab) = \log a + \log b$$
$$\log(a/b) = \log a - \log b$$
$$\log(a^n) = n \log a$$
$$\log \sqrt[n]{a} = \log a \times \frac{1}{n}$$

EXPANSION OF MATERIALS.

Change in length = $l \times t \times e$.   where l = original length.
t = change in temperature.
e = coefficient of expansion.

Remember that the coefficient (e) must be in the same temperature scale as the temperature change (t)

$$1°F = \frac{5}{9}°C$$

Material.	$e/1°C$	Material.	$e/1°C$
Glass	0·0000026	Tin	0·0000067
Steel	0·0000037	Fire brick	0·0000015
Copper	0·0000053	Facing brick	0·0000017
Lead	0·0000088	Water	0·000049
Brass	0·0000058	Concrete	0·0000030
Zinc	0·0000090	Granite	0·0000024

For buildings in U.K. the temperature range is usually taken as from -2°C to 33°C i.e. 35°C

For slabs in contact with the ground and exposed to the sun, it is suggested that a temperature range of 45°C should be allowed for.

Shrinkage of Concrete.

The shrinkage of concrete due to ageing is approximately —

Time.	Shrinkage (% of original length.)
After 28 days	0·00025
After 3 months	0·00035
After 12 months.	0·00050

INDEX

|||||||||||||||||||